Digital Signal Pr
System Analysis and Design

Digital signal processing lies at the heart of the communications revolution and is an essential element of key technologie.
book covers all the major topics in digital
supported by MATLAB® examples and ... ., sity of Ireland,
explain clearly and concisely why and how .   ... gnal p
how to approximate a desired transfer funct... .haracteristic v
ratios of polynomials; why an appropriate mapping of a tran... function
suitable structure is important for practical applications; and how to analyze, repre...
and explore the trade-off between time and frequency representation of signals. An
ideal textbook for students, it will also be a useful reference for engineers working on
the development of signal processing systems.

Paulo S. R. Diniz teaches at the undergraduate Department of Electronics and Computer Engineering at the Federal University of Rio de Janeiro (UFRJ), and at the graduate Program of Electrical Engineering at COPPE/UFRJ, where he is a professor. He is also a visiting professor at the Helsinki University of Technology. He is an active Fellow Member of the IEEE and an associate editor of two journals: *IEEE Transactions on Signal Processing* and *Circuits, Systems and Digital Signal Processing*. He is also serving as distinguished lecturer of the IEEE.

Eduardo A. B. da Silva is an Associate Professor at the undergraduate Department of Electronics and Computer Engineering at the Federal University of Rio de Janeiro (UFRJ) and at the graduate Program of Electrical Engineering at COPPE/UFRJ.

Sergio L. Netto is an Associate Professor at the undergraduate Department of Electronics and Computer Engineering at the Federal University of Rio de Janeiro (UFRJ), and at the graduate Program of Electrical Engineering at COPPE/UFRJ.

Digital Signal Processing

System Analysis and Design

Paulo S. R. Diniz
Eduardo A. B. da Silva
and Sergio L. Netto

Federal University of Rio de Janeiro

CAMBRIDGE
UNIVERSITY PRESS

CAMBRIDGE UNIVERSITY PRESS
Cambridge, New York, Melbourne, Madrid, Cape Town, Singapore, São Paulo

Cambridge University Press
The Edinburgh Building, Cambridge CB2 2RU, UK

Published in the United States of America by Cambridge University Press, New York

www.cambridge.org
Information on this title: www.cambridge.org/9780521781756

First published 2002
This digitally printed first paperback version 2006

A catalogue record for this publication is available from the British Library

Library of Congress Cataloguing in Publication data

Diniz, Paulo Sergio Ramirez, 1956–
Digital signal processing : system analysis and design / by Paulo S. R. Diniz,
Eduardo A. B. da Silva, and Sergio L. Netto.
    p.   cm.
Includes bibliographical references and index.
ISBN 0 521 78175 2
1. Signal processing–Digital techniques.   I. Silva, Eduardo A. B. da (Eduardo Antônio Barros da, 1963–
II. Netto, Sergio L. (Sergio Lima), 1967–   III. Title.
TK5102.9.D63 2001
621.382′2–dc21   2001025447

ISBN-13 978-0-521-78175-6 hardback
ISBN-10 0-521-78175-2 hardback

ISBN-13 978-0-521-02513-3 paperback
ISBN-10 0-521-02513-3 paperback

MATLAB is a registered trademark of The MathWorks Inc.
Verilog is a registered trademark of Cadence Design Automation Co.
SHARC is a registered trademark of Analog Devices Inc.
All ADSP-21XX, ADSP-219X, and TigerSHARC DSPs are trademarks of Analog Devices Inc.
All DSP5600X, DSP5630X, DSP566XX, and MSC81XX, DSPs, including the MSC8101, are trademarks of
Motorola Inc.
All TMS320C10, TMS320C20X (including the TMS320C25), TMS320C50, TMS320C3X (including the
TMS320C30 and TMS320VC33-150), TMS320C40, TMS320C8X, TMS320C54X, TMS320C55X,
TMS320C62X, TMS320C64X, and TMS320C67X DSPs are trademarks of Texas Instruments Inc.

For MATLAB product information, please contact:
The MathWorks, Inc.
3 Apple Hill Drive
Natick, MA, 01760-2098 USA
Tel: 508-647-7000
Fax: 508-647-7101
e-mail: info@mathworks.com
web: www.mathworks.com

To our families, our parents, and our students

Contents

4   Digital filters   148

10   Efficient FIR structures

11   Efficient IIR structures

Preface

This book originated from a training course for engineers at the research and development center of TELEBRAS, the former Brazilian telecommunications holding. That course was taught by the first author back in 1987, and its main goal was to present efficient digital filter design methods suitable for solving some of their engineering problems. Later on, this original text was used by the first author as the basic reference for the digital filters and digital signal processing courses of the Electrical Engineering Program at COPPE/Federal University of Rio de Janeiro.

For many years, former students asked why the original text was not transformed into a book, as it presented a very distinct view that they considered worth publishing. Among the numerous reasons not to attempt such task, we could mention that there were already a good number of well-written texts on the subject; also, after many years of teaching and researching on this topic, it seemed more interesting to follow other paths than the painful one of writing a book; finally, the original text was written in Portuguese and a mere translation of it into English would be a very tedious task.

In recent years, the second and third authors, who had attended the signal processing courses using the original material, were continuously giving new ideas on how to proceed. That was when we decided to go through the task of completing and updating the original text, turning it into a modern textbook. The book then took on its present form, updating the original text, and including a large amount of new material written for other courses taught by the three authors during the past few years.

This book is mainly written for use as a textbook on a digital signal processing course for undergraduate students who have had previous exposure to basic discrete-time linear systems, or to serve as textbook on a graduate-level course where the most advanced topics of some chapters are covered. This reflects the structure we have at the Federal University of Rio de Janeiro, as well as at a number of other universities we have contact with. The book includes, at the end of most chapters, a brief section aimed at giving a start to the reader on how to use MATLAB® as a tool for the analysis and design of digital signal processing systems. After many discussions, we decided that having explanations about MATLAB inserted in the main text would in some cases distract the reader, making him or her lose focus on the subject.

The distinctive feature of this book is to present a wide range of topics in digital signal processing design and analysis in a concise and complete form, while allowing the reader to fully develop practical systems. Although this book is primarily intended as an undergraduate and graduate textbook, its origins on training courses for industry warrant its potential usefulness to engineers working in the development of signal processing systems. In fact, our objective is to equip the readers with the tools that enable them to understand why and how to use digital signal processing systems; to show them how to approximate a desired transfer function characteristic using polynomials and ratios of polynomials; to teach them why an appropriate mapping of a transfer function into a suitable structure is important for practical applications; to show how to analyze, represent, and explore the tradeoff between the time and frequency representations of signals. For all that, each chapter includes a number of examples and end-of-chapter problems to be solved, aimed at assimilating the concepts as well as complementing the text.

Chapters 1 and 2 review the basic concepts of discrete-time signal processing and z transforms. Although many readers may be familiar with these subjects, they could benefit from reading these chapters, getting used to the notation and the authors' way of presenting the subject. In Chapter 1, we review the concepts of discrete-time systems, including the representation of discrete-time signals and systems, as well as their time- and frequency-domain responses. Most importantly, we present the sampling theorem, which sets the conditions for the discrete-time systems to solve practical problems related to our real continuous-time world. Chapter 2 is concerned with the z and Fourier transforms which are useful mathematical tools for representation of discrete-time signals and systems. The basic properties of the z and Fourier transforms are discussed, including a stability test in the z transform domain.

Chapter 3 discusses discrete transforms, with special emphasis given to the discrete Fourier transform (DFT), which is an invaluable tool in the frequency analysis of discrete-time signals. The DFT allows a discrete representation of discrete-time signals in the frequency domain. Since the sequence representation is natural for digital computers, the DFT is a very powerful tool, because it enables us to manipulate frequency-domain information in the same way as we can manipulate the original sequences. The importance of the DFT is further increased by the fact that computationally efficient algorithms, the so-called fast Fourier transforms (FFTs), are available to compute the DFT. This chapter also presents real coefficient transforms, such as cosine and sine transforms, which are widely used in modern audio and video coding, as well as in a number of other applications. A section includes a discussion on the several forms of representing the signals, in order to aid the reader with the available choices.

Chapter 4 addresses the basic structures for mapping a transfer function into a digital filter. It is also devoted to some basic analysis methods and properties of digital filter structures.

Chapter 5 introduces several approximation methods for filters with finite-duration impulse response (FIR), starting with the simpler frequency sampling method and the widely used windows method. This method also provides insight to the windowing strategy used in several signal processing applications. Other approximation methods included are the maximally flat filters and those based on the weighted-least-squares (WLS) method. This chapter also presents the Chebyshev approximation based on a multivariable optimization algorithm called the Remez exchange method. This approach leads to linear-phase transfer functions with minimum order given a pre-scribed set of frequency response specifications. This chapter also discusses the WLS–Chebyshev method which leads to transfer functions where the maximum and the total energy of the approximation error are prescribed. This approximation method is not widely discussed in the open literature but appears to be very useful for a number of applications.

Chapter 6 discusses the approximation procedures for filters with infinite-duration impulse response (IIR). We start with the classical continuous-time transfer-function approximations, namely the Butterworth, Chebyshev, and elliptic approximations, that can generate discrete-time transfer functions by using appropriate transformations. Two transformation methods are then presented which are the impulse-invariance and the bilinear transformation methods. The chapter also includes a section on frequency transformations in the discrete-time domain. The simultaneous magnitude and phase approximation of IIR digital filters using optimization techniques is also included, providing a tool to design transfer functions satisfying more general specifications. The chapter closes by addressing the issue of time-domain approximations.

Chapter 7 includes the models that account for quantization effects in digital filters. We discuss several approaches to analyze and deal with the effects of representing signals and filter coefficients with finite wordlength. In particular, we study the effects of quantization noise in products, signal scaling that limits the internal signal dynamic range, coefficient quantization in the designed transfer function, and the nonlinear oscillations which may occur in recursive realizations. These analyses are used to indicate the filter realizations that lead to practical finite-precision implementations of digital filters.

Chapter 8 deals with basic principles of discrete-time systems with multiple sampling rates. In this chapter, we emphasize the basic properties of multirate systems, thoroughly addressing the decimation and interpolation operations, giving examples of their use for efficient digital filter design.

Chapter 9 presents several design techniques for multirate filter banks, including several forms of 2-band filter banks, cosine-modulated filter banks, and lapped transforms. It also introduces the concept of multiresolution representation of signals through wavelet transforms, and discusses the design of wavelet transforms using filter banks. In addition, some design techniques to generate orthogonal, as well as biorthogonal bases for signal representation, are presented.

In Chapter 10, we present some techniques to reduce the computational complexity of FIR filters with demanding specifications. In particular, we introduce the prefilter and interpolation methods which are mainly useful in designing narrowband lowpass and highpass filters. In addition, we present the frequency response masking approach, for designing filters with narrow transition bands satisfying more general specifications, and the quadrature method, for narrow bandpass and bandstop filters.

Chapter 11 presents a number of efficient realizations for IIR filters. For these filters, a number of realizations considered efficient from the finite-precision effects point of view are presented and their salient features are discussed in detail. These realizations will equip the reader with a number of choices for the design of good IIR filters. Several families of structures are considered in this chapter, namely: parallel and cascade designs using direct-form second-order sections; parallel and cascade designs using section-optimal and limit-cycle-free state-space sections; lattice filters; and several forms of wave digital filters.

In Chapter 12, the most widely used implementation techniques for digital signal processing are briefly introduced. This subject is too large to fit in a chapter of a book; in addition, it is changing so fast that it is not possible for a textbook on implementation to remain up to date for long. To cope with that, we have chosen to analyze the most widely used implementation techniques which have been being employed for digital signal processing in the last decade and to present the current trends, without going into the details of any particular implementation strategy. Nevertheless, the chapter should be enough to assist any system designer in choosing the most appropriate form of implementing a particular digital signal processing algorithm.

This book contains enough material for an undergraduate course on digital signal processing and a first-year graduate course. There are many alternative ways to compose these courses; however, we recommend that an undergraduate course should include most parts of Chapters 1, 2, 3, and 4. It could also include the non-iterative approximation methods of Chapters 5 and 6, namely, the frequency sampling and window methods described in Chapter 5, the analog-based approximation methods, and also the continuous-time to discrete-time transformation methods for IIR filtering of Chapter 6. At the instructor's discretion the course could also include selected parts of Chapters 7, 8, 10, and 11.

As a graduate course textbook, Chapters 1 to 4 could be seen as review material, and the other chapters should be covered in depth.

This book would never be written if people with a wide vision of how an academic environment should be were not around. In fact, we were fortunate to have Professors L. P. Calôba and E. H. Watanabe as colleagues and advisors. The staff of COPPE, in particular Mr M. A. Guimarães and Ms F. J. Ribeiro, supported us in all possible ways to make this book a reality. Also, the first author's early students, J. C. Cabezas, R. G. Lins, and J. A. B. Pereira (in memoriam) wrote, with him, a computer package that generated several of the examples in this book. The engineers of CPqD helped us to correct the early version of this text. In particular, we would like to thank the engineer J. Sampaio for his complete trust in this work. We benefited from working in an environment with a large signal processing group where our colleagues always helped us in various ways. Among them, we should mention Profs. L. W. P. Biscainho, M. L. R. de Campos, G. V. Mendonça, A. C. M. de Queiroz, F. G. V. de Resende Jr, and J. M. de Seixas, and the entire staff of the Signal Processing Lab. (www.lps.ufrj.br). We would like to thank our colleagues at the Federal University of Rio de Janeiro, in particular at the Department of Electronics of the School of Engineering, the undergraduate studies department, and at the Electrical Engineering Program of COPPE, the graduate studies department, for their constant support during the preparation of this book.

The authors would like to thank many friends from other institutions whose influence helped in shaping this book. In particular, we may mention Prof. A. S. de la Vega of Fluminense Federal University; Prof. M. Sarcinelli Fo. of Federal University of Espírito Santo; Profs. P. Agathoklis, A. Antoniou, and W.-S. Lu of University of Victoria; Profs. I. Hartimo, T. I. Laakso, and Dr V. Välimäki of Helsinki University of Technology; Prof. T. Saramäki of Tampere University of Technology; Prof. Y. C. Lim of National University of Singapore; Dr R. L. de Queiroz of Xerox Corporation; Dr H. S. Malvar of Microsoft Corporation; Prof. Y.-F. Huang of University of Notre Dame; Prof. J. E. Cousseau of Univerdad Nacional del Sur; Prof. B. Nowrouzian of University of Alberta; Dr M. G. de Siqueira of Cisco Systems; Profs. R. Miscow Fo. and E. Viegas of Military Institute of Engineering in Rio de Janeiro; Dr E. Cabral of University of São Paulo.

This acknowledgement list would be incomplete without mentioning the staff of Cambridge University Press, in particular our editors, Dr Philip Meyler and Mr Eric Willner, and our copy editor, Dr Jo Clegg.

The authors would like to thank their families for their endless patience and support. In particular, Paulo would like to express his deepest gratitude to Mariza, Paula, and Luiza, and to his mother Hirlene. Eduardo would like to mention that the sweetness of Luis Eduardo and Isabella, the continuing love and friendship from his wife Cláudia, and the strong and loving background provided by his parents, Zélia and Bismarck, were in all respects essential to the completion of this task. Sergio would like to express his deepest gratitude to his parents, Sergio and Maria Christina, and his sincere love to Luciana and Bruno. All the authors would also like to thank their families for bearing with them working together.

Paulo S. R. Diniz
Eduardo A. B. da Silva
Sergio L. Netto

Introduction

When we hear the word signal we may first think of a phenomenon that occurs continuously over time that carries some information. Most phenomena observed in nature are continuous in time, as for instance, our speech or heart beating, the temperature of the city where we live, the car speed during a given trip, the altitude of the airplane we are traveling in – these are typical continuous-time signals. Engineers are always devising ways to design systems, which are in principle continuous time, for measuring and interfering with these and other phenomena.

One should note that, although continuous-time signals pervade our daily life, there are also several signals which are originally discrete time, as, for example, the stock-market weekly financial indicators, the maximum and minimum daily temperatures in our cities, and the average lap speed of a racing car.

If an electrical or computer engineer has the task of designing systems to interact with natural phenomena, his/her first impulse is to convert some quantities from nature into electric signals through a transducer. Electric signals, which are represented by voltages or currents, have a continuous-time representation. Since digital technology constitutes an extremely powerful tool for information processing, it is natural to think of processing the electric signals generated using it. However, continuous-time signals cannot be processed using computer technology (digital machines), which are especially suited to deal with sequential computation involving numbers. Fortunately, this fact does not prevent the use of digital integrated circuits (which is the technology behind the computer technology revolution we witness today) in signal processing systems designs. This is because many signals taken from nature can be fully represented by their sampled versions, where the sampled signals coincide with the original continuous-time signals at some instants in time. If we know how fast the important information changes, we can always sample and convert continuous-time information into discrete-time information which, in turn can be converted into a sequence of numbers and transferred to a digital machine.

The main advantages of digital systems relative to analog systems are high reliability, suitability for modifying the system's characteristics, and low cost. These advantages motivated the digital implementation of many signal processing systems which used to be implemented with analog circuit technology. In addition, a number of new applications became viable once the very large scale integration (VLSI) technology was available. Usually in the VLSI implementation of a digital signal processing system the concern is to reduce power consumption and/or area, or to increase the circuit's speed in order to meet the demands of high-throughput applications.

Currently, a single digital integrated circuit may contain millions of logic gates operating at very high speeds, allowing very fast digital processors to be built at a reasonable cost. This technological development opened avenues to the introduction of powerful general-purpose computers which can be used to design and implement digital signal processing systems. In addition, it allowed the design of microprocessors with special features for signal processing, namely the digital signal processors (DSPs). As a consequence, there are several tools available to implement very complex digital signal processing systems. In practice, a digital signal processing system is implemented either by software on a general-purpose digital computer or DSP, or by using application-specific hardware, usually in the form of an integrated circuit.

For the reasons explained above, the field of digital signal processing has developed so fast in the last decades that it has been incorporated into the graduate and undergraduate programs of virtually all universities. This is confirmed by the number of good textbooks available in this area (Oppenheim & Schafer, 1975; Rabiner & Gold, 1975; Peled & Liu, 1985; Roberts & Mullis, 1987; Oppenheim & Schafer, 1989; Ifeachor & Jervis, 1993; Antoniou, 1993; Jackson, 1996; Proakis & Manolakis, 1996; Mitra, 1998, to mention just a few). The present book is aimed at equipping readers with tools that will enable them to design and analyze most digital signal processing systems. The building blocks for digital signal processing systems considered here are used to process signals which are discrete in time and in amplitude. The main tools emphasized in this text are:

- Discrete-time signal representations.
- Discrete transforms and their fast algorithms.
- Design and implementation of digital filters.
- Multirate systems, filter banks, and wavelets.
- Implementation of digital signal processing systems.

Transforms and filters are the main parts of linear signal processing systems. Although the techniques we deal with are directly applicable to processing deterministic signals, many statistical signal processing methods employ similar building blocks in some way.

Digital signal processing is extremely useful in many areas. In the following, we enumerate a few of the disciplines where the topics covered by this book have found application.

(a) Image processing: An image is essentially a space-domain signal, that is, it represents a variation of light intensity and color in space. Therefore, in order to process an image using an analog system, it has to be converted into a time-domain signal, using some form of scanning process. However, to process an image digitally, there is no need to perform this conversion, for it can be processed directly in the spatial domain, as a matrix of numbers. This lends the digital image processing techniques enormous power. In fact, in image processing, two-dimensional signal representation, filtering, and transforms play a central role (Jain, 1989).

(b) Multimedia systems: Such systems deal with different kinds of information sources such as image, video, audio, and data. In such systems, the information is essentially represented in digital form. Therefore, it is crucial to remove redundancy from the information, allowing compression, and thus efficient transmission and storage (Jayant & Noll, 1984; Gersho & Gray, 1992; Bhaskaran & Konstantinides, 1997). The Internet is a good application example where information files are transferred in a compressed manner. Most of the compression standards for image use transforms and quantizers.

The transforms, filter banks, and wavelets are very popular in compression applications because they are able to exploit the high correlation of signal sources such as audio, still images and video (Malvar, 1992; Fliege, 1994; Vetterli & Kovačević, 1995; Strang & Nguyen, 1996; Mallat, 1998).

(c) Communication systems: In communication systems, the compression and coding of the information sources are also crucial, since by reducing the amount of data to be transmitted services can be provided at higher speed or to more users. In addition, channel coding, which consists of inserting redundancy in the signal to compensate for possible channel distortions, may also use special types of digital filtering (Stüber, 1996).

Communication systems usually include fixed filters, as well as some self-designing filters for equalization and channel modeling which fall in the class of adaptive filtering systems (Diniz, 1997). Although these filters employ a statistical signal processing framework (Manolakis et al., 2000) to determine how their parameters should change, they also use some of the filter structures and in some cases the transforms introduced in this book.

Many filtering concepts take part on modern multiuser communication systems employing coded division multiple access (CDMA) (Verdu, 1998).

Wavelets, transforms, and filter banks also play a crucial role in the conception of orthogonal frequency division multiplexing (OFDM) (Akansu & Medley, 1999), which is used in digital audio and TV broadcasting.

(d) Audio signal processing: In statistical signal processing the filters are designed based on observed signals, which might imply that we are estimating the parameters of the model governing these signals (Kailath et al., 2000). Such estimation techniques can be employed in digital audio restoration (Godsill & Rayner, 1998), where the resulting models can be used to restore lost information. However, these estimation models can be simplified and made more effective if we use some kind of sub-band processing with filter banks and transforms (Kahrs & Brandenburg,

1998). In the same field digital filters were found to be suitable for reverberation algorithms and as models for musical instruments (Kahrs & Brandenburg, 1998).

In addition to the above applications, digital signal processing is at the heart of modern developments in speech analysis and synthesis, mobile radio, sonar, radar, biomedical engineering, seismology, home appliances, and instrumentation, among others. These developments occurred in parallel with the advances in the technology of transmission, processing, recording, reproduction, and general treatment of signals through analog and digital electronics, as well as other means such as acoustics, mechanics, and optics.

We expect that with the digital signal processing tools described in this book, the reader will be able to proceed further, not only exploring and understanding some of the applications described above, but developing new ones as well.

1 Discrete-time systems

1.1 Introduction

Digital signal processing is the discipline that studies the rules governing the behavior of discrete signals, as well as the systems used to process them. It also deals with the issues involved in processing continuous signals using digital techniques. Digital signal processing pervades modern life. It has applications in compact disc players, computer tomography, geological processing, mobile phones, electronic toys, and many others.

In analog signal processing, we take a continuous signal, representing a continuously varying physical quantity, and pass it through a system that modifies this signal for a certain purpose. This modification is, in general, continuously variable by nature, that is, it can be described by differential equations.

Alternatively, in digital signal processing, we process sequences of numbers using some sort of digital hardware. We usually call these sequences of numbers digital or discrete-time signals.[1] The power of digital signal processing comes from the fact that, once a sequence of numbers is available to appropriate digital hardware we can carry out any form of numerical processing on them. For example, suppose we need to perform the following operation on a continuous-time signal:

$$y(t) = \frac{\cosh\left[\ln(|x(t)|) + x^3(t) + \cos^3\left(\sqrt{|x(t)|}\right)\right]}{5x^5(t) + e^{x(t)} + \tan(x(t))} \tag{1.1}$$

This would be clearly very difficult to implement using analog hardware. However, if we sample the analog signal $x(t)$ and convert it into a sequence of numbers $x(n)$, it can be input to a digital computer, which can perform the above operation easily and reliably, generating a sequence of numbers $y(n)$. If the continuous-time signal $y(t)$ can be recovered from $y(n)$, then the desired processing has been successfully performed.

This simple example highlights two important points. One is how powerful is digital signal processing. The other is that, if we want to process an analog signal using this sort of resource, we must have a way of converting a continuous-time signal into a

---

[1] Throughout this text we refer to discrete-time and continuous-time signals. However, the signals do not necessarily vary with time. For example, if the signal represents a temperature variation along a metal rod, it can be a continuous-space or discrete-space signal. Therefore, one should bear in mind that whenever we refer to the independent variable as time, we do this without loss of generality, as it could in fact be any other variable.

discrete-time one, such that the continuous-time signal can be recovered from the discrete-time signal. However, it is important to note that very often discrete-time signals do not come from continuous-time signals, that is, they are originally discrete-time, and the results of their processing are only needed in digital form.

In this chapter, we study the basic concepts of the theory of discrete-time signals and systems. We emphasize the treatment of discrete-time systems as separate entities from continuous-time systems. We first define discrete-time signals and, based on this, we define discrete-time systems, highlighting the properties of an important subset of these systems, namely, linearity and time invariance, as well as their description by discrete-time convolutions. We then study their time-domain response by characterizing them using difference equations. We close the chapter with Nyquist's sampling theorem, which tells us how to generate, from a continuous-time signal, a discrete-time signal from which the continuous-time signal can be completely recovered. Nyquist's sampling theorem forms the basis of the digital processing of continuous-time signals.

1.2   Discrete-time signals

A discrete-time signal is one that can be represented by a sequence of numbers. For example, the sequence

$$\{x(n), \ n \in \mathbb{Z}\} \tag{1.2}$$

where $\mathbb{Z}$ is the set of integer numbers, can represent a discrete-time signal where each number $x(n)$ corresponds to the amplitude of the signal at an instant nT. Since n is an integer, T represents the interval between two consecutive points at which the signal is defined. It is important to note that T is not necessarily a time unit. For example, if $x(n)$ represents the temperature at sensors placed uniformly along a metal rod, then T is a length unit.

In this text, we usually represent a discrete-time signal using the notation in equation (1.2), where $x(n)$ is referred to as the nth sample of the signal. An alternative notation, used in many texts, is to represent the signal as

$$\{x(nT), \ n \in \mathbb{Z}\} \tag{1.3}$$

where the time interval between samples is explicitly shown, that is, $x(nT)$ is the sample at time nT. Note that, using the notation in equation (1.2), a discrete-time signal whose adjacent samples are 0.03 seconds apart would be represented as

$$\dots \ x(0), x(1), x(2), x(3), x(4), \ \dots \tag{1.4}$$

whereas, using equation (1.3), it would be represented as

$$\dots \ x(0), x(0.03), x(0.06), x(0.09), x(0.12), \ \dots \tag{1.5}$$

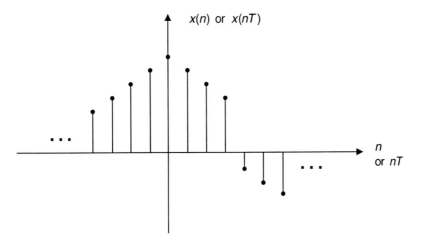

Figure 1.1 General representation of a discrete-time signal.

The graphical representation of a general discrete-time signal is shown in Figure 1.1. In what follows, we describe some of the most important discrete-time signals.

Unit impulse: See Figure 1.2a

$$\delta(n) = \begin{cases} 1, & n = 0 \\ 0, & n \neq 0 \end{cases} \tag{1.6}$$

Delayed unit impulse: See Figure 1.2b

$$\delta(n - m) = \begin{cases} 1, & n = m \\ 0, & n \neq m \end{cases} \tag{1.7}$$

Unit step: See Figure 1.2c

$$u(n) = \begin{cases} 1, & n \geq 0 \\ 0, & n < 0 \end{cases} \tag{1.8}$$

Cosine function:

$$x(n) = \cos(\omega n) \tag{1.9}$$

The angular frequency of this sinusoid is ω rad/sample and its frequency is $\frac{\omega}{2\pi}$ cycles/sample. For example, in Figure 1.2d, the cosine function has angular frequency $\omega = \frac{2\pi}{16}$ rad/sample. This means that it completes one cycle, that equals 2π radians, in 16 samples. If the sample separation represents time, ω can be given in rad/(time unit).

Real exponential function: See Figure 1.2e

$$x(n) = e^{an} \tag{1.10}$$

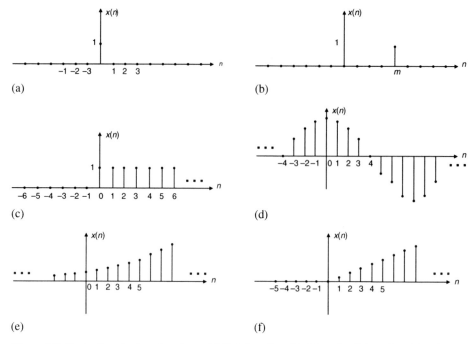

Figure 1.2 Basic discrete-time functions. (a) Unit impulse; (b) delayed unit impulse; (c) unit step; (d) cosine function with $\omega = \frac{2\pi}{16}$ rad/sample; (e) real exponential function with $a = 0.2$; (f) unit ramp.

Unit ramp: See Figure 1.2f

$$r(n) = \begin{cases} n, & n \geq 0 \\ 0, & n < 0 \end{cases} \tag{1.11}$$

By examining Figure 1.2b–f, we notice that any discrete-time signal is equivalent to a sum of shifted impulses multiplied by a constant, that is, the impulse shifted by k samples is multiplied by $x(k)$. This can also be deduced from the definition of a shifted impulse in equation (1.7). For example, the unit step $u(n)$ in equation (1.8) can also be expressed as

$$u(n) = \sum_{k=0}^{\infty} \delta(n - k) \tag{1.12}$$

Likewise, any discrete-time signal $x(n)$ can be expressed as

$$x(n) = \sum_{k=-\infty}^{\infty} x(k)\delta(n - k) \tag{1.13}$$

An important class of discrete-time signals or sequences is that of periodic sequences. A periodic sequence $x(n)$ is periodic if and only if there is an integer $N \neq 0$

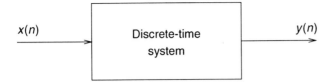

Figure 1.3 Representation of a discrete-time system.

such that $x(n) = x(n + N)$ for all n. In such a case, N is called the period of the sequence. Note that, using this definition and referring to equation (1.9), the period of the cosine function is an integer N such that

$$\cos(\omega n) = \cos(\omega(n + N)), \quad \forall n \in \mathbb{Z} \tag{1.14}$$

This happens only if there is $k \in \mathbb{Z}$ such that $\omega N = 2\pi k$, or equivalently

$$N = \frac{2\pi}{\omega}k \tag{1.15}$$

Therefore, we notice that not all discrete cosine sequences are periodic. For example, if $\omega = \sqrt{2}$ the cosine sequence is not periodic. An example of a periodic cosine sequence with period equal to 16 samples is given in Figure 1.2d.

1.3   Discrete-time systems

A discrete-time system maps an input sequence $x(n)$ to an output sequence $y(n)$, such that

$$y(n) = \mathcal{H}\{x(n)\} \tag{1.16}$$

where the operator $\mathcal{H}\{\cdot\}$ represents a discrete-time system, as shown in Figure 1.3. Depending on the properties of $\mathcal{H}\{\cdot\}$, the discrete-time system can be classified in several ways, the most basic ones being either linear or nonlinear, time invariant or time variant, causal or noncausal. These classifications will be discussed in what follows.

1.3.1   Linearity

A discrete-time system is linear if and only if

$$\mathcal{H}\{ax(n)\} = a\mathcal{H}\{x(n)\} \tag{1.17}$$

and

$$\mathcal{H}\{x_1(n) + x_2(n)\} = \mathcal{H}\{x_1(n)\} + \mathcal{H}\{x_2(n)\} \tag{1.18}$$

for any constant a, and any sequences $x(n)$, $x_1(n)$, and $x_2(n)$.

1.3.2 Time invariance

A discrete-time system is time invariant if and only if, for any input sequence $x(n)$ and integer n_0, then

$$\mathcal{H}\{x(n - n_0)\} = y(n - n_0) \tag{1.19}$$

with $y(n) = \mathcal{H}\{x(n)\}$

Some texts refer to the time-invariance property as the shift-invariance property, since a discrete system can represent samples of a function not necessarily in time.

1.3.3 Causality

A discrete-time system is causal if and only if, when $x_1(n) = x_2(n)$ for $n < n_0$, then

$$\mathcal{H}\{x_1(n)\} = \mathcal{H}\{x_2(n)\}, \quad \text{for } n < n_0 \tag{1.20}$$

In other words, causality means that the output of a system at instant n does not depend on any input occurring after n.

It is important to note that, usually, in the case of a discrete-time signal, a noncausal system is not implementable in real time. This is because we would need input samples at instants of time greater than n in order to compute the output at time n. This would be allowed only if the time samples were pre-stored, as in offline or batch implementations. However, in some cases a discrete signal does not consist of time samples. For example, in Section 1.1, we mentioned a discrete signal that corresponds to the temperature at sensors uniformly spaced along a metal rod. For this discrete signal, a processor can have access to all its samples simultaneously. Therefore, in this case, even a noncausal system can be easily implemented.

EXAMPLE 1.1

Characterize the following systems as being either linear or nonlinear, time invariant or time varying, causal or noncausal:

(a) $y(n) = (nT + bT)x(nT - 4T)$
(b) $y(n) = x^2(n + 1)$

SOLUTION

(a) • Linearity:

$$\begin{aligned}
\mathcal{H}\{ax(n)\} &= (nT + bT)ax(nT - 4T) \\
&= a(nT + bT)x(nT - 4T) \\
&= a\mathcal{H}\{x(n)\}
\end{aligned} \tag{1.21}$$

$$\mathcal{H}\{x_1(n) + x_2(n)\} = (nT + bT)[x_1(nT - 4T) + x_2(nT - 4T)]$$
$$= (nT + bT)x_1(nT - 4T) + (nT + bT)x_2(nT - 4T)$$
$$= \mathcal{H}\{x_1(n)\} + \mathcal{H}\{x_2(n)\} \qquad (1.22)$$

and therefore the system is linear.

- Time invariance:

$$y(n - n_0) = [(n - n_0)T + bT]x[(n - n_0)T - 4T] \qquad (1.23)$$
$$\mathcal{H}\{x(n - n_0)\} = (nT + bT)x[(n - n_0)T - 4T] \qquad (1.24)$$

then $y(n - n_0) \neq \mathcal{H}\{x(n - n_0)\}$ and the system is time varying.

- Causality:

$$\mathcal{H}\{x_1(n)\} = (nT + bT)x_1(nT - 4T) \qquad (1.25)$$
$$\mathcal{H}\{x_2(n)\} = (nT + bT)x_2(nT - 4T) \qquad (1.26)$$

Therefore, if $x_1(nT) = x_2(nT)$, for $n < n_0$, then $\mathcal{H}\{x_1(n)\} = \mathcal{H}\{x_2(n)\}$, for $n < n_0$, and, consequently, the system is causal.

(b) • Linearity:

$$\mathcal{H}\{ax(n)\} = a^2 x^2(n + 1) \neq a\mathcal{H}\{x(n)\} \qquad (1.27)$$

and therefore the system is nonlinear.

- Time invariance:

$$\mathcal{H}\{x(n - n_0)\} = x^2[(n - n_0) + 1] = y(n - n_0) \qquad (1.28)$$

so the system is time invariant.

- Causality:

$$\mathcal{H}\{x_1(n)\} = x_1^2(n + 1) \qquad (1.29)$$
$$\mathcal{H}\{x_2(n)\} = x_2^2(n + 1) \qquad (1.30)$$

Therefore, if $x_1(n) = x_2(n)$, for $n < n_0$, and $x_1(n_0) \neq x_2(n_0)$, then, for $n = n_0 - 1 < n_0$,

$$\mathcal{H}\{x_1(n_0 - 1)\} = x_1^2(n_0) \qquad (1.31)$$
$$\mathcal{H}\{x_2(n_0 - 1)\} = x_2^2(n_0) \qquad (1.32)$$

and we have that $\mathcal{H}\{x_1(n)\} \neq \mathcal{H}\{x_2(n)\}$, and the system is noncausal.

$\triangle$

1.3.4   Impulse response and convolution sums

Suppose that $\mathcal{H}\{\cdot\}$ is a linear system, and we apply an excitation $x(n)$ to the system. Since, from equation (1.13), $x(n)$ can be expressed as a sum of shifted impulses

$$x(n) = \sum_{k=-\infty}^{\infty} x(k)\delta(n - k) \tag{1.33}$$

we can express its output as

$$y(n) = \mathcal{H}\left\{ \sum_{k=-\infty}^{\infty} x(k)\delta(n - k) \right\} = \sum_{k=-\infty}^{\infty} \mathcal{H}\{x(k)\delta(n - k)\} \tag{1.34}$$

Since $x(k)$ in the above equation is just a constant, the linearity of $\mathcal{H}\{\cdot\}$ also implies that

$$y(n) = \sum_{k=-\infty}^{\infty} x(k)\mathcal{H}\{\delta(n - k)\} = \sum_{k=-\infty}^{\infty} x(k)h_n(k) \tag{1.35}$$

where $h_n(k) = \mathcal{H}\{\delta(n - k)\}$ is the response of the system to an impulse at $n = k$.

If the system is also time invariant, and we define

$$\mathcal{H}\{\delta(n)\} = h_0(n) = h(n) \tag{1.36}$$

then $\mathcal{H}\{\delta(n - k)\} = h(n - k)$, and the expression in equation (1.35) becomes

$$y(n) = \sum_{k=-\infty}^{\infty} x(k)h(n - k) \tag{1.37}$$

indicating that a linear time-invariant system is completely characterized by its unit impulse response $h(n)$. This is a very powerful and convenient result, and will be explored further, which lends great importance and usefulness to the class of linear time-invariant discrete-time systems. One should note that, when the system is linear and time varying, we would need, in order to compute $y(n)$, the values of $h_n(k)$, which depend on both n and k. This makes the computation of the summation in equation (1.35) quite complex.

Equation (1.37) is called a convolution sum or a discrete-time convolution.[2] If we make the change of variables $l = n - k$, equation (1.37) can be written as

$$y(n) = \sum_{l=-\infty}^{\infty} x(n - l)h(l) \tag{1.38}$$

---

[2] This operation is often also referred to as a discrete-time linear convolution, in order to differentiate it from the discrete-time circular convolution, which will be defined in Chapter 3.

that is, we can interpret $y(n)$ as the result of the convolution of the excitation $x(n)$ with the system impulse response $h(n)$. A shorthand notation for the convolution operation, as described in equations (1.37) and (1.38), is

$$y(n) = x(n) * h(n) = h(n) * x(n) \tag{1.39}$$

Suppose now that the output $y(n)$ of a system with impulse response $h(n)$ is the excitation for a system with impulse response $h'(n)$. In this case, we have

$$y(n) = \sum_{k=-\infty}^{\infty} x(k)h(n-k) \tag{1.40}$$

$$y'(n) = \sum_{l=-\infty}^{\infty} y(l)h'(n-l) \tag{1.41}$$

Substituting equation (1.40) in equation (1.41), we have that

$$\begin{aligned}
y'(n) &= \sum_{l=-\infty}^{\infty} \left(\sum_{k=-\infty}^{\infty} x(k)h(l-k) \right) h'(n-l) \\
&= \sum_{k=-\infty}^{\infty} x(k) \left(\sum_{l=-\infty}^{\infty} h(l-k)h'(n-l) \right)
\end{aligned} \tag{1.42}$$

By performing the change of variables $l = n - r$, the above equation becomes

$$\begin{aligned}
y'(n) &= \sum_{k=-\infty}^{\infty} x(k) \left(\sum_{r=-\infty}^{\infty} h(n-r-k)h'(r) \right) \\
&= \sum_{k=-\infty}^{\infty} x(k) \left(h(n-k) * h'(n-k) \right) \\
&= \sum_{k=-\infty}^{\infty} x(n-k) \left(h(k) * h'(k) \right)
\end{aligned} \tag{1.43}$$

showing that the impulse response of a linear time-invariant system formed by the series (cascade) connection of two linear time-invariant subsystems is the convolution of the impulse responses of the two subsystems.

EXAMPLE 1.2

Compute $y(n)$ for the system depicted in Figure 1.4, as a function of the input signal and of the impulse responses of the subsystems.

SOLUTION

From the previous results, it is easy to conclude that

$$y(n) = (h_2(n) + h_3(n)) * h_1(n) * x(n) \tag{1.44}$$

$\triangle$

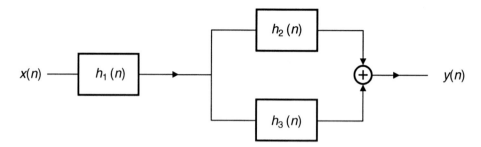

Figure 1.4 Linear time-invariant system composed of the connection of three subsystems.

1.3.5 Stability

A system is referred to as bounded-input bounded-output (BIBO) stable if, for every input limited in amplitude, the output signal is also limited in amplitude. For a linear time-invariant system, equation (1.38) implies that

$$|y(n)| \leq \sum_{k=-\infty}^{\infty} |x(n-k)||h(k)| \tag{1.45}$$

The input being limited in amplitude is equivalent to

$$|x(n)| \leq x_{max} < \infty, \quad \forall n \tag{1.46}$$

Therefore

$$|y(n)| \leq x_{max} \sum_{k=-\infty}^{\infty} |h(k)| \tag{1.47}$$

Hence, we can conclude that a sufficient condition for a system to be BIBO stable is

$$\sum_{k=-\infty}^{\infty} |h(k)| < \infty \tag{1.48}$$

since this condition forces $y(n)$ to be limited. To prove that this condition is also necessary, suppose that it does not hold, that is, the summation in equation (1.48) is infinite. If we choose an input such that

$$x(n_0 - k) = \begin{cases} +1, & \text{for } h(k) \geq 0 \\ -1, & \text{for } h(k) < 0 \end{cases} \tag{1.49}$$

we then have that

$$y(n_0) = |y(n_0)| = \sum_{k=-\infty}^{\infty} |h(k)| \tag{1.50}$$

that is, the output $y(n)$ is unbounded, thus completing the necessity proof.

We can then conclude that the necessary and sufficient condition for a linear time-invariant system to be BIBO stable is given in equation (1.48).

1.4   Difference equations and time-domain response

In most applications, discrete-time systems can be described by difference equations, which are the equivalent, for the discrete-time domain, to differential equations for the continuous-time domain. In fact, the systems that can be specified by difference equations are powerful enough to cover most practical applications. The input and output of a system described by a linear difference equation are generally related by (Gabel & Roberts, 1980)

$$\sum_{i=0}^{N} a_i y(n-i) - \sum_{l=0}^{M} b_l x(n-l) = 0 \tag{1.51}$$

This difference equation has an infinite number of solutions $y(n)$, like the solutions of differential equations in the continuous case. For example, suppose that a particular $y_p(n)$ satisfies equation (1.51), that is

$$\sum_{i=0}^{N} a_i y_p(n-i) - \sum_{l=0}^{M} b_l x(n-l) = 0 \tag{1.52}$$

and that $y_h(n)$ is a solution to the homogeneous equation, that is

$$\sum_{i=0}^{N} a_i y_h(n-i) = 0 \tag{1.53}$$

Then, from equations (1.51)–(1.53), we can easily infer that $y(n) = y_p(n) + y_h(n)$ is also a solution to the same difference equation.

The homogeneous solution $y_h(n)$ of a difference equation of order N, as in equation (1.51), has N degrees of freedom (depends on N arbitrary constants). Therefore, one can only determine a solution for a difference equation if one supplies N auxiliary conditions. One example of a set of auxiliary conditions is given by the values of $y(-1), y(-2), \ldots, y(-N)$. It is important to note that any N independent auxiliary conditions would be enough to solve a difference equation. However, in this text, we generally use N consecutive samples of $y(n)$ as auxiliary conditions.

EXAMPLE 1.3
Solve the following difference equation:

$$y(n) = e^{-\beta} y(n-1) + \delta(n) \tag{1.54}$$

SOLUTION
Any function of the form $y_h(n) = K e^{-\beta n}$ satisfies

$$y(n) = e^{-\beta} y(n-1) \tag{1.55}$$

and is therefore a solution of the homogeneous difference equation. Also, one can verify by substitution that $y_p(n) = e^{-\beta n}u(n)$ is a particular solution of equation (1.54).

Therefore, the general solution of the difference equation is given by

$$y(n) = y_p(n) + y_h(n) = e^{-\beta n}u(n) + Ke^{-\beta n} \tag{1.56}$$

where the value of K is determined by the auxiliary conditions. Since this difference equation is of first order, we need specify only one condition. For example, if we have that $y(-1) = \alpha$, the solution to equation (1.54) becomes

$$y(n) = e^{-\beta n}u(n) + \alpha e^{-\beta(n+1)} \tag{1.57}$$

$\triangle$

Since a linear system must satisfy equation (1.17), it is clear that for a linear system, $\mathcal{H}\{0\} = 0$, that is, the output for a zero input must be zero. Therefore, in order for a system described by a difference equation to be linear, it is necessary that its homogeneous solution be null. This is obtained only when the N auxiliary conditions are zero, since a nonzero auxiliary condition would necessarily lead to a nonzero homogeneous solution.

In order to guarantee the causality of a system described by equation (1.51), an additional relaxation restriction must be imposed on the system. A system is said to be initially relaxed if, when $x(n) = 0$, for $n \leq n_0$, then $y(n) = 0$, for $n \leq n_0$. This will be made clear in the following example.

EXAMPLE 1.4

For the linear system described by

$$y(n) = e^{-\beta}y(n-1) + u(n) \tag{1.58}$$

determine its output for the auxiliary conditions:

(a) $y(1) = 0$
(b) $y(-1) = 0$

and discuss the causality in both situations.

SOLUTION

The homogeneous solution of equation (1.58) is the same as in Example 1.3, that is

$$y_h(n) = Ke^{-\beta n} \tag{1.59}$$

The particular solution is of the form (Gabel & Roberts, 1980)

$$y_p(n) = (a + be^{-\beta n})u(n) \tag{1.60}$$

By direct substitution of $y_p(n)$ above in equation (1.58), we find that

$$a = \frac{1}{1 - e^{-\beta}}; \quad b = \frac{-e^{-\beta}}{1 - e^{-\beta}} \tag{1.61}$$

Thus, the general solution of the difference equation is given by

$$y(n) = \left(\frac{1 - e^{-\beta(n+1)}}{1 - e^{-\beta}} \right) u(n) + K e^{-\beta n} \tag{1.62}$$

(a) For the auxiliary condition $y(1) = 0$, we have that

$$y(1) = \left(\frac{1 - e^{-2\beta}}{1 - e^{-\beta}} \right) + K e^{-\beta} = 0 \tag{1.63}$$

yielding $K = 1 + e^{\beta}$, and the general solution becomes

$$y(n) = \left(\frac{1 - e^{-\beta(n+1)}}{1 - e^{-\beta}} \right) u(n) - \left(e^{-\beta n} + e^{-\beta(n-1)} \right) \tag{1.64}$$

Since for $n < 0$, $u(n) = 0$, $y(n)$ simplifies to

$$y(n) = - \left(e^{-\beta n} + e^{-\beta(n-1)} \right) \tag{1.65}$$

Clearly, in this case, $y(n) \neq 0$, for $n < 0$, whereas the input $u(n) = 0$ for $n < 0$. Thus, the system is not initially relaxed and therefore is noncausal.

Another way to verify that the system is noncausal is by noting that, if the input is doubled, becoming $x(n) = 2u(n)$, instead of $u(n)$, then the particular solution is also doubled. Hence, the general solution of the difference equation becomes

$$y(n) = \left(\frac{2 - 2e^{-\beta(n+1)}}{1 - e^{-\beta}} \right) u(n) + K e^{-\beta n} \tag{1.66}$$

If we require that $y(1) = 0$, then $K = 2 + 2e^{\beta}$, and, for $n < 0$, this yields

$$y(n) = -2 \left(e^{-\beta n} + e^{-\beta(n-1)} \right) \tag{1.67}$$

Since this is different from the value of $y(n)$ for $u(n)$ as input, we see that the output for $n < 0$ depends on the input for $n > 0$, and therefore the system is noncausal.

(b) For the auxiliary condition $y(-1) = 0$, we have that $K = 0$, yielding the solution

$$y(n) = \left(\frac{1 - e^{-\beta(n+1)}}{1 - e^{-\beta}} \right) u(n) \tag{1.68}$$

In this case, $y(n) < 0$, for $n < 0$, and the system is initially relaxed, and therefore causal.

$\triangle$

The above example shows that the system described by the difference equation is noncausal because it has nonzero auxiliary conditions prior to the application of the input to the system. To guarantee both causality and linearity for the solution of a difference equation, we have to impose zero auxiliary conditions for the samples preceding the application of the excitation to the system. This is the same as assuming that the system is initially relaxed. Therefore, an initially relaxed system described by a difference equation of the form in (1.51) has the highly desirable linearity, time invariance, and causality properties. In this case, time invariance can be easily inferred if we consider that, for an initially relaxed system, the history of the system up to the application of the excitation is the same irrespective of the time sample position at which the excitation is applied. This happens because the outputs are all zero up to, but not including, the time of the application of the excitation. Therefore, if time is measured having as a reference the time sample $n = n_0$, at which the input is applied, then the output will not depend on the reference n_0, because the history of the system prior to n_0 is the same irrespective of n_0. This is equivalent to saying that if the input is shifted by k samples, then the output is just shifted by k samples, the rest remaining unchanged, thus characterizing a time-invariant system.

Equation (1.51) can be rewritten, without loss of generality, considering that $a_0 = 1$, yielding

$$y(n) = -\sum_{i=1}^{N} a_i y(n - i) + \sum_{l=0}^{M} b_l x(n - l) \tag{1.69}$$

This equation can be interpreted as the output signal $y(n)$ being dependent both on samples of the input, $x(n), x(n - 1), \ldots, x(n - M)$, and on previous samples of the output $y(n - 1), y(n - 2), \ldots, y(n - N)$. Then, in this general case, we say that the system is recursive, since, in order to compute the output, we need past samples of the output itself. When $a_1 = a_2 = \cdots = a_N = 0$, then the output at sample n depends only on values of the input signal. In such case, the system is called nonrecursive, being distinctively characterized by a difference equation of the form

$$y(n) = \sum_{l=0}^{M} b_l x(n - l) \tag{1.70}$$

If we compare the above equation with the expression for the convolution sum given in equation (1.38), we see that equation (1.70) corresponds to a discrete system with impulse response $h(l) = b_l$. Since b_l is defined only for l between 0 and M, we can say that $h(l)$ is nonzero only for $0 \leq l \leq M$. This implies that the system in equation (1.70) has a finite-duration impulse response. Such discrete-time systems are often referred to as finite-duration impulse-response (FIR) filters.

In contrast, when $y(n)$ depends on its past values, as in equation (1.69), we have that the impulse response of the discrete system, in general, might not be zero when

$n \to \infty$. Therefore, recursive digital systems are often referred to as infinite-duration impulse-response (IIR) filters.

EXAMPLE 1.5
Find the impulse response of the system

$$y(n) - \frac{1}{\alpha}y(n - 1) = x(n) \tag{1.71}$$

supposing that it is initially relaxed.

SOLUTION
Since the system is initially relaxed, then $y(n) = 0$, for $n \leq -1$. Hence, for $n = 0$, we have that

$$y(0) = \frac{1}{\alpha}y(-1) + \delta(0) = \delta(0) = 1 \tag{1.72}$$

For $n > 0$, we have that

$$y(n) = \frac{1}{\alpha}y(n - 1) \tag{1.73}$$

and, therefore, $y(n)$ can be expressed as

$$y(n) = \left(\frac{1}{\alpha}\right)^n u(n) \tag{1.74}$$

Note that $y(n) \neq 0$, $\forall n \geq 0$, that is, the impulse response has infinite length.

$\triangle$

One should note that, in general, recursive systems have infinite-duration impulse responses, although there are some cases when recursive systems have finite-duration impulse responses. Illustrations can be found in Exercise 1.14 and in Section 10.5.

1.5   Sampling of continuous-time signals

In many cases, a discrete-time signal $x(n)$ consists of samples of a continuous-time signal $x_a(t)$, that is

$$x(n) = x_a(nT) \tag{1.75}$$

If we want to process the continuous-time signal $x_a(t)$ using a discrete-time system, then we first need to convert it using equation (1.75), process the discrete-time input digitally, and then convert the discrete-time output back to the continuous-time domain. Therefore, in order for this operation to be effective, it is essential that we have capability of restoring a continuous-time signal from its samples. In this section, we derive conditions under which a continuous-time signal can be recovered from its samples. We also devise ways of performing this recovery. To do so, we first introduce some basic concepts of analog signal processing that can be found in most standard books on linear systems (Gabel & Roberts, 1980; Oppenheim et al., 1983). Following that, we derive the sampling theorem, which gives the basis for digital signal processing of continuous-time signals.

1.5.1   Basic principles

The Fourier transform of a continuous-time signal $f(t)$ is given by

$$F(j\Omega) = \int_{-\infty}^{\infty} f(t)e^{-j\Omega t}\,dt \tag{1.76}$$

where Ω is referred to as the frequency and is measured in radians per second (rad/s). The corresponding inverse relationship is expressed as

$$f(t) = \frac{1}{2\pi}\int_{-\infty}^{\infty} F(j\Omega)e^{j\Omega t}\,d\Omega \tag{1.77}$$

An important property associated with the Fourier transform is that the Fourier transform of the product of two functions equals the convolution of their Fourier transforms. That is, if $x(t) = a(t)b(t)$, then

$$X(j\Omega) = \frac{1}{2\pi}A(j\Omega) * B(j\Omega) = \frac{1}{2\pi}\int_{-\infty}^{\infty} A(j\Omega - j\Omega')B(j\Omega')\,d\Omega' \tag{1.78}$$

where $X(j\Omega)$, $A(j\Omega)$, and $B(j\Omega)$ are the Fourier transforms of $x(t)$, $a(t)$, and $b(t)$, respectively.

In addition, if a signal $x(t)$ is periodic with period T, then we can express it by its Fourier series defined by

$$x(t) = \sum_{k=-\infty}^{\infty} a_k e^{j\frac{2\pi}{T}kt} \tag{1.79}$$

where the a_ks are called the series coefficients which are determined as

$$a_k = \frac{1}{T}\int_{-\frac{T}{2}}^{\frac{T}{2}} x(t)e^{-jk\frac{2\pi}{T}t}\,dt \tag{1.80}$$

1.5.2   Sampling theorem

Given a discrete-time signal $x(n)$ derived from a continuous-time signal $x_a(t)$ using equation (1.75), we define a continuous-time signal $x_i(t)$ consisting of a train of

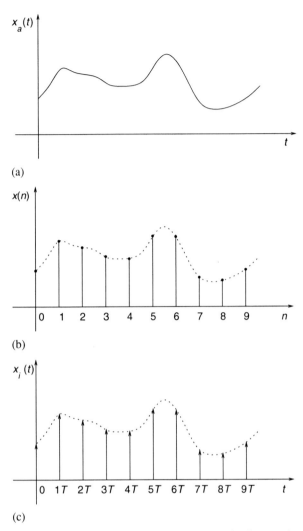

Figure 1.5 (a) Continuous-time signal $x_a(t)$; (b) discrete-time signal $x(n)$; (c) auxiliary continuous-time signal $x_i(t)$.

impulses at $t = nT$, each of energy equal to $x(n) = x_a(nT)$. Examples of signals $x_a(t)$, $x(n)$, and $x_i(t)$ are depicted in Figure 1.5, where the direct relationships between these three signals can be seen.

The signal $x_i(t)$ can be expressed as

$$x_i(t) = \sum_{n=-\infty}^{\infty} x(n)\delta(t - nT) \tag{1.81}$$

Since, from equation (1.75), $x(n) = x_a(nT)$, then equation (1.81) becomes

$$x_i(t) = \sum_{n=-\infty}^{\infty} x_a(nT)\delta(t - nT) = x_a(t) \sum_{n=-\infty}^{\infty} \delta(t - nT) = x_a(t)p(t) \tag{1.82}$$

indicating that $x_i(t)$ can also be obtained by multiplying the continuous-time signal $x_a(t)$ by a train of impulses $p(t)$ defined as

$$p(t) = \sum_{n=-\infty}^{\infty} \delta(t - nT) \tag{1.83}$$

In the equations above, we have defined a continuous-time signal $x_i(t)$ that can be obtained from the discrete-time signal $x(n)$ in a straightforward manner. In what follows, we relate the Fourier transforms of $x_a(t)$ and $x_i(t)$, and study the conditions under which $x_a(t)$ can be obtained from $x_i(t)$.

From equations (1.78) and (1.82), the Fourier transform of $x_i(t)$ is such that

$$X_i(j\Omega) = \frac{1}{2\pi}X_a(j\Omega) * P(j\Omega) = \frac{1}{2\pi}\int_{-\infty}^{\infty} X_a(j\Omega - j\Omega')P(j\Omega')d\Omega' \tag{1.84}$$

Therefore, in order to arrive at an expression for the Fourier transform of $x_i(t)$, we must first determine the Fourier transform of $p(t)$, $P(j\Omega)$. From equation (1.83), we see that $p(t)$ is a periodic function with period T and that we can decompose it in a Fourier series, as described in equations (1.79) and (1.80). Since, from equation (1.83), $p(t)$ in the interval $[-\frac{T}{2}, \frac{T}{2}]$ is equal to just an impulse $\delta(t)$, the coefficients a_k in the Fourier series of $p(t)$ are given by

$$a_k = \frac{1}{T}\int_{-\frac{T}{2}}^{\frac{T}{2}} \delta(t)e^{-jk\frac{2\pi}{T}t}dt = \frac{1}{T} \tag{1.85}$$

and the Fourier series for $p(t)$ becomes

$$p(t) = \frac{1}{T}\sum_{k=-\infty}^{\infty} e^{j\frac{2\pi}{T}kt} \tag{1.86}$$

As the Fourier transform of $f(t) = e^{j\Omega_0 t}$ is equal to $F(j\Omega) = 2\pi\delta(\Omega - \Omega_0)$, then, from equation (1.86), the Fourier transform of $p(t)$ becomes

$$P(j\Omega) = \frac{2\pi}{T}\sum_{k=-\infty}^{\infty} \delta\left(\Omega - \frac{2\pi}{T}k\right) \tag{1.87}$$

Substituting this expression for $P(j\Omega)$ in equation (1.84), we have that

$$\begin{aligned} X_i(j\Omega) &= \frac{1}{2\pi}X_a(j\Omega) * P(j\Omega) \\ &= \frac{1}{T}X_a(j\Omega) * \sum_{k=-\infty}^{\infty} \delta\left(\Omega - \frac{2\pi}{T}k\right) \\ &= \frac{1}{T}\sum_{k=-\infty}^{\infty} X_a\left(j\Omega - j\frac{2\pi}{T}k\right) \end{aligned} \tag{1.88}$$

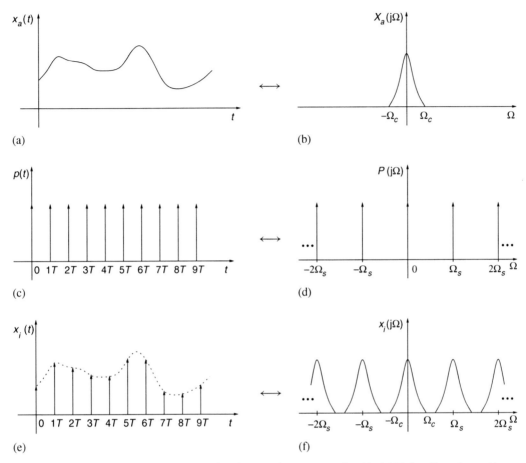

Figure 1.6 (a) Continuous-time signal $x_a(t)$; (b) spectrum of $x_a(t)$. (c) Train of impulses $p(t)$; (d) spectrum of $p(t)$. (e) Auxiliary continuous-time signal $x_i(t)$; (f) spectrum of $x_i(t)$.

where, in the last step, we used the fact that the convolution of a function $F(j\Omega)$ with a shifted impulse $\delta(\Omega - \Omega_0)$ is the shifted function $F(j\Omega - \Omega_0)$. Equation (1.88) shows that the spectrum of $x_i(t)$ is composed of infinite shifted copies of the spectrum of $x_a(t)$, with the shifts in frequency being multiples of the sampling frequency, $\Omega_s = \frac{2\pi}{T}$. Figure 1.6 shows examples of signals $x_a(t)$, $p(t)$, and $x_i(t)$, and their respective Fourier transforms.

From equation (1.88) and Figure 1.6f, we see that, in order to avoid the repeated copies of the spectrum of $x_a(t)$ interfering with one another, the signal should be band-limited. In addition, its bandwidth Ω_c should be such that the upper edge of the spectrum centered at zero is smaller than the lower edge of the spectrum centered at Ω_s, that is $\Omega_c \leq \Omega_s - \Omega_c$, implying that

$$\Omega_s \geq 2\Omega_c \qquad (1.89)$$

that is, the sampling frequency must be larger than double the one-sided bandwidth of the continuous-time signal. The frequency $\Omega = 2\Omega_c$ is called the Nyquist frequency of the continuous-time signal $x_a(t)$.

In addition, if the condition in equation (1.89) is satisfied, the original continuous signal $x_a(t)$ can be recovered by isolating the part of the spectrum of $x_i(t)$ that corresponds to the spectrum of $x_a(t)$.[3] This can be achieved by filtering the signal $x_i(t)$ with an ideal lowpass filter having bandwidth $\frac{\Omega_s}{2}$.

On the other hand, if the condition in equation (1.89) is not satisfied, the repetitions of the spectrum interfere with one another, and the continuous-time signal cannot be recovered from its samples. This superposition of the repetitions of the spectrum of $x_a(t)$ in $x_i(t)$, when the sampling frequency is smaller than $2\Omega_c$, is commonly referred to as aliasing. Figure 1.7b–d shows the spectra of $x_i(t)$ for Ω_s equal to, smaller than, and larger than $2\Omega_c$, respectively. The aliasing phenomenon is clearly identified in Figure 1.7c.

We are now ready to enunciate a very important result:

THEOREM 1.1 (SAMPLING THEOREM)
If a continuous-time signal $x_a(t)$ is band-limited, that is, its Fourier transform is such that $X_a(j\Omega) = 0$, for $|\Omega| > |\Omega_c|$, then $x_a(t)$ can be completely recovered from the discrete-time signal $x(n) = x_a(nT)$, if and only if the sampling frequency Ω_s satisfies $\Omega_s > 2\Omega_c$.
$\diamond$

As mentioned above, the original continuous-time signal $x_a(t)$ can be recovered from the signal $x_i(t)$ by filtering $x_i(t)$ with an ideal lowpass filter having cutoff frequency $\frac{\Omega_s}{2}$. More specifically, if the signal has bandwidth Ω_c, then it suffices that the cutoff frequency of the ideal lowpass filter is Ω_{LP}, such that $\Omega_c \leq \Omega_{LP} \leq \frac{\Omega_s}{2}$. Therefore, the Fourier transform of the impulse response of such a filter should be

$$H(j\Omega) = \begin{cases} T, & \text{for } |\Omega| \leq \Omega_{LP} \\ 0, & \text{for } |\Omega| > \Omega_{LP} \end{cases} \tag{1.90}$$

where the passband gain T compensates for the factor $\frac{1}{T}$ in equation (1.88). This ideal frequency response is illustrated in Figure 1.8a.

Computing the inverse Fourier transform of $H(j\Omega)$, using equation (1.77), we see that the impulse response $h(t)$ of the filter is

$$h(t) = \frac{T \sin(\Omega_{LP}t)}{\pi t} \tag{1.91}$$

which is depicted in Figure 1.8b.

---

[3] In fact, any of the spectrum repetitions have the full information about $x_a(t)$. However, if we isolate a repetition of the spectrum not centered at $\Omega = 0$, we get a modulated version of $x_a(t)$, which should be demodulated. Since its demodulation is equivalent to shifting the spectrum back to the origin, it is usually better to take the repetition of the spectrum centered at the origin in the first place.

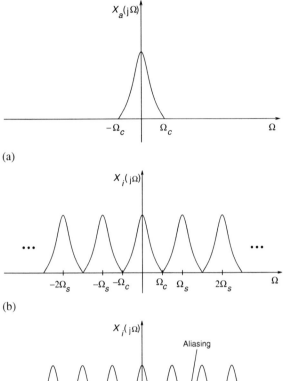

(a)

(b)

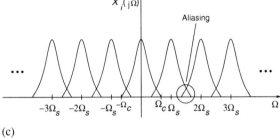

(c)

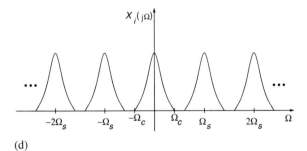

(d)

Figure 1.7 (a) Spectrum of the continuous-time signal. Spectra of $x_i(t)$ for: (b) $\Omega_s = 2\Omega_c$; (c) $\Omega_s < 2\Omega_c$; (d) $\Omega_s > 2\Omega_c$.

Then, given $h(t)$, the signal $x_a(t)$ can be recovered from $x_i(t)$ by the following convolution integral (Oppenheim et al., 1983)

$$x_a(t) = \int_{-\infty}^{\infty} x_i(\tau)h(t - \tau)d\tau \tag{1.92}$$

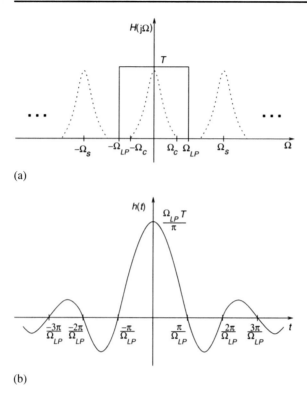

(a)

(b)

Figure 1.8 Ideal lowpass filter: (a) frequency response; (b) impulse response.

Replacing $x_i(t)$ by its definition in equation (1.81), we have that

$$x_a(t) = \int_{-\infty}^{\infty} \sum_{n=-\infty}^{\infty} x(n)\delta(\tau - nT)h(t - \tau)d\tau$$

$$= \sum_{n=-\infty}^{\infty} \int_{-\infty}^{\infty} x(n)\delta(\tau - nT)h(t - \tau)d\tau$$

$$= \sum_{n=-\infty}^{\infty} x(n)h(t - nT) \tag{1.93}$$

and using $h(t)$ from equation (1.91),

$$x_a(t) = \sum_{n=-\infty}^{\infty} x(n)\frac{T\sin(\Omega_{LP}(t - nT))}{\pi(t - nT)} \tag{1.94}$$

If we make Ω_{LP} equal to half the sampling frequency, that is, $\Omega_{LP} = \frac{\Omega_s}{2} = \frac{\pi}{T}$, then equation (1.94) becomes

$$x_a(t) = \sum_{n=-\infty}^{\infty} x(n) \frac{\sin\left(\frac{\Omega_s}{2}t - n\pi\right)}{\frac{\Omega_s}{2}t - n\pi} = \sum_{n=-\infty}^{\infty} x(n) \frac{\sin\left[\pi\left(\frac{t}{T} - n\right)\right]}{\pi\left(\frac{t}{T} - n\right)} \qquad (1.95)$$

Equations (1.94) and (1.95) represent interpolation formulas to recover the continuous-time signal $x_a(t)$ from its samples $x(n) = x_a(nT)$. However, since, in order to compute $x_a(t)$ at any time t_0, all the samples of $x(n)$ have to be known, these interpolation formulas are not practical. This is the same as saying that the lowpass filter with impulse response $h(t)$ is not realizable. This is because it is noncausal and cannot be implemented via any differential equation of finite order. Clearly, the closer the lowpass filter is to the ideal, the smaller is the error in the computation of $x(t)$ using equation (1.94). In Chapters 5 and 6 methods to approximate such ideal filters will be extensively studied.

From what we have studied in this section, we can develop a block diagram of the several phases constituting the processing of an analog signal using a digital system. Figure 1.9 depicts each step of the procedure.

The first block in the diagram of Figure 1.9 is a lowpass filter, that guarantees the analog signal is band-limited, with bandwidth $\Omega_c \leq \frac{\Omega_s}{2}$.

The second block is a sample-and-hold system, which samples the signal $x_a(t)$ at times $t = nT$ and holds the obtained value for T seconds, that is, until the value for the next time interval is sampled.

The third block, the encoder, converts each sample value $x_a^*(t)$, for $nT < t < (n+1)T$, at its input to a number $x(n)$. Since this number is input to digital hardware, it must be represented with a finite number of bits. This operation introduces an error in the signal, which gets smaller as the number of bits used in the representation increases. The second and third blocks constitute what we usually call the analog-to-digital (A/D) conversion.

The fourth block carries out the digital signal processing, transforming the discrete-time signal $x(n)$ into a discrete-time signal $y(n)$.

The fifth block transforms the numbers representing the samples $y(n)$ back into the analog domain, in the form of a train of impulses $y_i(n)$, constituting the process known as digital-to-analog (D/A) conversion.

The sixth block is a lowpass filter necessary to eliminate the repetitions of the spectrum contained in $y_i(t)$, in order to recover the analog signal $y_a(t)$ corresponding to $y(n)$. In practice, sometimes the fifth and sixth blocks are implemented in one operation. For example, we can transform the samples $y(n)$ into an analog signal $y_a(n)$ using an D/A converter plus a sample-and-hold operation, similar to the one

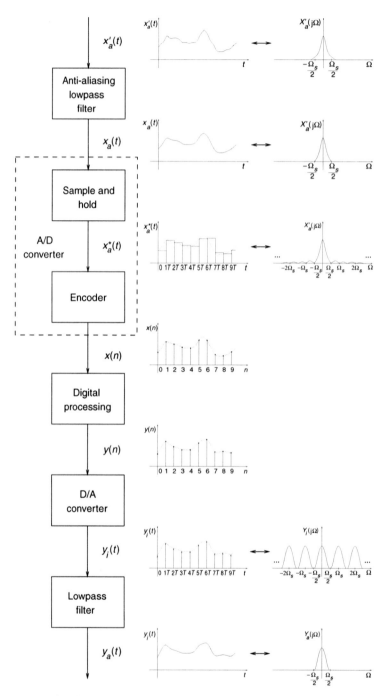

Figure 1.9 Digital processing of analog signals (all graphs on the left represent the signals in the time domain, and all graphs on the right, their corresponding Fourier transform).

of the second block. It is easy to show (see Exercise 1.21) that the sample-and-hold operation is equivalent to filtering the train of impulses

$$y_i(t) = \sum_{n-\infty}^{\infty} y(n)\delta(t - nT) \tag{1.96}$$

with a filter having impulse response

$$h(t) = \begin{cases} 1, & \text{for } 0 \le t \le T \\ 0, & \text{otherwise} \end{cases} \tag{1.97}$$

In this case, the recovery of the analog signal is not perfect, but is a good enough approximation in some practical cases.

1.6   Discrete-time signals and systems with MATLAB

The Signal Processing toolbox of MATLAB has several functions that aid in the processing of discrete signals. In this section, we give a brief overview of the ones that are most closely related to the material in this chapter.

- stem: Plots a data sequence as stems from the x axis terminated in circles, as in the plots of discrete-time signals shown in this chapter.
  Input parameters:

  - The ordinate values t at which the data is to be plotted;
  - The data values x to be plotted.

  Example:
  ```
  t=[0:0.1:2]; x=cos(pi*t+0.6); stem(t,x);
  ```

- conv: Performs the discrete convolution of two sequences.
  Input parameters: The vectors a and b holding the two sequences.

  Output parameter: The vector y holding the convolution of a and b.
  Example:
  ```
  a=[1 1 1 1 1]; b=[1 2 3 4 5 6 7 8 9];
  c=conv(a,b); stem(c);
  ```

- impz: Computes the impulse response of an initially relaxed system described by a difference equation.
  Input parameters (refer to equation (1.51)):

  - A vector b containing the values of $b_l, l = 0, \ldots, M$;
  - A vector a containing the values of $a_i, i = 0, \ldots, N$;
  - The desired number of samples n of the impulse response.

Output parameter: A vector h holding the impulse response of the system;
Example:

```
a=[1 -0.22 -0.21 0.017 0.01]; b=[1 2 1];
h=impz(b,a,20); stem(h);
```

1.7   Summary

In this chapter, we have seen the basic concepts referring to discrete-time systems
and have introduced the properties of linearity, time invariance, and causality. Systems
presenting these properties are of particular interest due to the numerous tools that are
available to analyze and to design them. We also discussed the concept of stability.
Linear time-invariant systems have been characterized using convolution sums. We
have also introduced difference equations as another way to characterize discrete-
time systems. The conditions necessary to carry out the discrete-time processing of
continuous-time signals without losing the information they carry have been studied.
Finally, we presented some MATLAB functions that are useful in the processing of
discrete-time signals seen in this chapter.

1.8   Exercises

1.1   Characterize the systems below as linear/nonlinear, causal/noncausal and time invari-
ant/time varying.

(a) $y(n) = (n + a)^2 x(n + 4)$
(b) $y(n) = ax(nT + T)$
(c) $y(n) = x(n + 1) + x^3(n - 1)$
(d) $y(n) = x(nT) \sin(\omega nT)$
(e) $y(n) = x(n) + \sin(\omega n)$
(f) $y(n) = \frac{x(n)}{x(n+3)}$
(g) $y(n) = y(n - 1) + 8x(n - 3)$
(h) $y(n) = 2ny(n - 1) + 3x(n - 5)$
(i) $y(n) = n^2 y(n + 1) + 5x(n - 2) + x(n - 4)$
(j) $y(n) = y(n - 1) + x(n + 5) + x(n - 5)$
(k) $y(n) = (2u(n - 3) - 1)y(n - 1) + x(n) + x(n - 1)$

1.2   For each of the discrete signals below, determine whether they are periodic or not.
Calculate the periods of those that are periodic.

(a) $x(n) = \cos^2\left(\frac{2\pi}{15}n\right)$

(b) $x(n) = \cos\left(\frac{4\pi}{5}n + \frac{\pi}{4}\right)$

(c) $x(n) = \cos\left(\frac{\pi}{27}n + 31\right)$

(d) $x(n) = \sin(100n)$

(e) $x(n) = \cos\left(\frac{11\pi}{12}n\right)$

(f) $x(n) = \sin[(5\pi + 1)n]$

1.3   Consider the system whose output $y(m)$ is described as a function of the input $x(m)$ by the following difference equations:

(a) $y(m) = \sum_{n=-\infty}^{\infty} x(n)\delta(m - nN)$

(b) $y(m) = x(m) \sum_{n=-\infty}^{\infty} \delta(m - nN)$

Determine whether the systems are linear and/or time invariant.

1.4   Compute the convolution sum of the following pairs of sequences:

(a) $x(n) = \begin{cases} 1, & 0 \le n \le 4 \\ 0, & \text{otherwise} \end{cases}$ and $h(n) = \begin{cases} a^n, & 0 \le n \le 7 \\ 0, & \text{otherwise} \end{cases}$

(b) $x(n) = \begin{cases} 1, & 0 \le n \le 2 \\ 0, & 3 \le n \le 6 \\ 1, & 7 \le n \le 8 \\ 0, & \text{otherwise} \end{cases}$ and $h(n) = \begin{cases} n, & 1 \le n \le 4 \\ 0, & \text{otherwise} \end{cases}$

(c) $x(n) = \begin{cases} a(n), & n \text{ even} \\ 0, & n \text{ odd} \end{cases}$ and $h(n) = \begin{cases} \frac{1}{2}, & n = -1 \\ 1, & n = 0 \\ \frac{1}{2}, & n = 1 \\ 0, & \text{otherwise} \end{cases}$

Check the results of item (b) using the MATLAB function conv.

1.5   For the sequence

$$x(n) = \begin{cases} 1, & 0 \le n \le 1 \\ 0, & \text{otherwise} \end{cases}$$

compute $y(n) = x(n) * x(n) * x(n) * x(n)$. Check your results using the MATLAB function conv.

1.6   Show that $x(n) = a^n$ is an eigenfunction of a linear time-invariant system, by computing the convolution summation of $x(n)$ and the impulse response of the system, $h(n)$. Determine the corresponding eigenvalue.

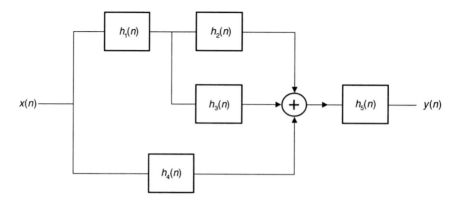

Figure 1.10 Linear time-invariant system.

1.7   Supposing that all systems in Figure 1.10 are linear and time invariant, compute $y(n)$ as a function of the input and the impulse responses of each system.

1.8   We define the even and odd parts of a sequence $x(n)$, $\mathcal{E}\{x(n)\}$, and $\mathcal{O}\{x(n)\}$, respectively, as

$$\mathcal{E}\{x(n)\} = \frac{x(n) + x(-n)}{2}$$

$$\mathcal{O}\{x(n)\} = \frac{x(n) - x(-n)}{2}$$

Show that

$$\sum_{n=-\infty}^{\infty} x^2(n) = \sum_{n=-\infty}^{\infty} \mathcal{E}\{x(n)\}^2 + \sum_{n=-\infty}^{\infty} \mathcal{O}\{x(n)\}^2$$

1.9   Find one solution for each of the difference equations below:

(a) $y(n) + 2y(n-1) + y(n-2) = 0$, $y(0) = 1$ and $y(1) = 0$
(b) $y(n) + y(n-1) + 2y(n-2) = 0$, $y(-1) = 1$ and $y(0) = 1$

1.10   Write a MATLAB program to plot the samples of the solutions of the difference equations in Exercise 1.9 from $n = 0$ to $n = 20$.

1.11   Show that a system described by equation (1.51) is linear if and only if the auxiliary conditions are zero. Show also that the system is time invariant if the auxiliary conditions represent the values of $y(-1), y(-2), \ldots, y(-N)$.

1.12   Compute the impulse responses of the systems below:

(a) $y(n) = 5x(n) + 3x(n-1) + 8x(n-2) + 3x(n-4)$
(b) $y(n) + \frac{1}{3}y(n-1) = x(n) + \frac{1}{2}x(n-1)$
(c) $y(n) - 3y(n-1) = x(n)$

(d) $y(n) + 2y(n-1) + y(n-2) = x(n)$

1.13 Write a MATLAB program to compute the impulse responses of the systems described by following difference equations:

(a) $y(n) + y(n-1) + y(n-2) = x(n)$
(b) $4y(n) + y(n-1) + 3y(n-2) = x(n) + x(n-4)$

1.14 Determine the impulse response of the following recursive system:

$$y(n) - y(n-1) = x(n) - x(n-5)$$

1.15 Determine the steady-state response for the input $x(n) = \sin(\omega n)u(n)$ of the filters described by

(a) $y(n) = x(n-2) + x(n-1) + x(n)$
(b) $y(n) - \frac{1}{2}y(n-1) = x(n)$
(c) $y(n) = x(n-2) + 2x(n-1) + x(n)$

What are the amplitude and phase of $y(n)$ as a function of ω in each case?

1.16 Write a MATLAB program to plot the solution to the three difference equations in Exercise 1.15 for $\omega = \frac{\pi}{3}$ and $\omega = \pi$.

1.17 Discuss the stability of the systems described by the impulse responses below:

(a) $h(n) = 2^{-n}u(n)$
(b) $h(n) = 1.5^n u(n)$
(c) $h(n) = 0.1^n$
(d) $h(n) = 2^{-n}u(-n)$
(e) $h(n) = 10^n u(n) - 10^n u(n-10)$
(f) $h(n) = 0.5^n u(n) - 0.5^n u(4-n)$

1.18 Show that $X_i(j\Omega)$ in equation (1.88) is a periodic function of Ω with period $\frac{2\pi}{T}$.

1.19 Suppose we want to process the continuous-time signal

$$x_a(t) = 3\cos(2\pi 1000t) + 7\sin(2\pi 1100t)$$

using a discrete-time system. The sampling frequency used is 4000 samples/s. The discrete-time processing carried out on the signal samples $x(n)$ is described by the following difference equation

$$y(n) = x(n) + x(n-2)$$

After the processing, the samples of the output $y(n)$ are converted back to continuous-time form using equation (1.95). Give a closed-form expression for the processed

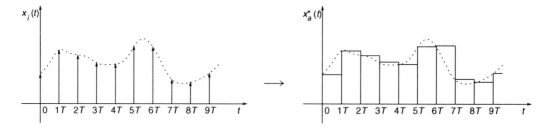

Figure 1.11 Sample-and-hold operation.

continuous-time signal $y_a(t)$. Interpret the effect this processing has on the input signal.

1.20 Write a MATLAB program to perform a simulation of the solution of Exercise 1.19.

Hints:

(i) Simulate the continuous-time signals $x_a(t)$ in MATLAB using sequences obtained by sampling them at 100 times the sampling frequency, that is, as $f_s(m) = f_a(m\frac{T}{100})$. The sampling process to generate the discrete-time signal is then equivalent to keeping one out of 100 samples of the original sequence, forming the signal $f(n) = f_s(100n) = f_a(nT)$.

(ii) The interpolation in equation (1.95) can be approximated by truncating each function $\frac{\sin[\pi(\frac{t}{T}-n)]}{\pi(\frac{t}{T}-n)}$, considering them to be zero outside the time interval $nT - 10T \leq t \leq nT + 10T$. Note that, since we are truncating the interpolation functions, the summation in equation (1.95) needs to be carried out only for a few terms.

1.21 In the conversion of a discrete-time to a continuous-time signal, the practical D/A converters, instead of generating impulses at the output, generate a series of pulses $g(t)$ described by

$$x_p(t) = \sum_{n=-\infty}^{\infty} x(n)g(t - nT) \tag{1.98}$$

where $x(n) = x_a(nT)$. For example, in the sample-and-hold operation, which is shown in Figure 1.11, the train of impulses has been replaced by a function $g(t)$ as

$$g(t) = \begin{cases} 1, & 0 \leq t \leq T \\ 0, & \text{otherwise} \end{cases}$$

Determine, for the above choice of pulse $g(t)$:

(a) An expression for the Fourier transform of $x_p(t)$ in equation (1.98) as a function of the Fourier transform of $x_a(t)$.

(b) The frequency response of an ideal lowpass filter that outputs $x_a(t)$ when $x_p(t)$ is applied to its input. Such filter would compensate for the artifacts introduced by the sample-and-hold operation in a D/A conversion.

1.22 Given a sinusoid $y(t) = A\cos(\Omega_c t)$, using equation (1.88), show that if $y(t)$ is sampled with a sampling frequency slightly above the Nyquist frequency, that is, $\Omega_s = 2\Omega_c + \epsilon$, where $\epsilon \ll \Omega_c$, then the envelope of the sampled signal will vary slowly, with a frequency of $\frac{\pi\epsilon}{2\Omega_c+\epsilon}$ rad/sample. This is often referred to as the Moiré effect. Write a MATLAB program to plot 100 samples of $y(n)$ for ϵ equal to $\frac{\Omega_s}{100}$ and confirm the above result.

2 The z and Fourier transforms

2.1 Introduction

In Chapter 1, we studied linear time-invariant systems, using both impulse responses and difference equations to characterize them. In this chapter, we study another very useful way to characterize discrete-time systems. It is linked with the fact that, when an exponential function is input to a linear time-invariant system, its output is an exponential function of the same type, but with a different amplitude. This can be deduced by considering that, from equation (1.38), a linear time-invariant discrete-time system with impulse response $h(n)$, when excited by an exponential $x(n) = z^n$, produces at its output a signal $y(n)$ such that

$$y(n) = \sum_{k=-\infty}^{\infty} x(n-k)h(k) = \sum_{k=-\infty}^{\infty} z^{n-k}h(k) = z^n \sum_{k=-\infty}^{\infty} h(k)z^{-k} \qquad (2.1)$$

that is, the signal at the output is also an exponential z^n, but with an amplitude multiplied by the complex function

$$H(z) = \sum_{k=-\infty}^{\infty} h(k)z^{-k} \qquad (2.2)$$

In this chapter, we characterize linear time-invariant systems using the quantity $H(z)$ in the above equation, commonly known as the z transform of the discrete-time sequence $h(n)$. As we will see later in this chapter, with the help of the z transform, linear convolutions can be transformed into simple algebraic equations. The importance of this for discrete-time systems parallels that of the Laplace transform for continuous-time systems.

The case when z^n is a complex sinusoid with frequency ω, that is, $z = e^{j\omega}$, is of particular importance. In this case, equation (2.2) becomes

$$H(e^{j\omega}) = \sum_{k=-\infty}^{\infty} h(k)e^{-j\omega k} \qquad (2.3)$$

which can be represented in polar form as $H(e^{j\omega}) = |H(e^{j\omega})|e^{j\Theta(\omega)}$, yielding, from equation (2.1), an output signal $y(n)$ such that

$$y(n) = H(e^{j\omega})e^{j\omega n} = |H(e^{j\omega})|e^{j\Theta(\omega)}e^{j\omega n} = |H(e^{j\omega})|e^{j\omega n + j\Theta(\omega)} \qquad (2.4)$$

This relationship implies that the effect of a linear system characterized by $H(e^{j\omega})$ on a complex sinusoid is to multiply its amplitude by $|H(e^{j\omega})|$ and to add $\Theta(\omega)$ to its phase. For this reason, the descriptions of $|H(e^{j\omega})|$ and $\Theta(\omega)$ as functions of ω are widely used to characterize linear time-invariant systems, and are known as their magnitude and phase responses, respectively. The complex function $H(e^{j\omega})$ in equation (2.4) is also known as the Fourier transform of the discrete-time sequence $h(n)$. The Fourier transform is as important for discrete-time systems as it is for continuous-time systems.

In this chapter, we will study the z and Fourier transforms for discrete-time signals. We begin by defining the z transform, discussing issues related to its convergence and its relation to the stability of discrete-time systems. Then we present the inverse z transform, as well as several z-transform properties. Next, we show how to transform discrete-time convolutions into algebraic equations, and introduce the concept of a transfer function. We then present an algorithm to determine, given the transfer function of a discrete-time system, whether the system is stable or not and go on to discuss how the frequency response of a system is related to its transfer function. At this point, we give a formal definition of the Fourier transform of discrete-time signals, highlighting its relations to the Fourier transform of continuous-time signals. An expression for the inverse Fourier transform is also presented. Its main properties are then shown as particular cases of those of the z transform. We close the chapter by presenting some MATLAB functions which are related to z and Fourier transforms, and which aid in the analysis of transfer functions of discrete-time systems.

2.2   Definition of the z transform

The z transform of a sequence $x(n)$ is defined as

$$X(z) = \mathcal{Z}\{x(n)\} = \sum_{n=-\infty}^{\infty} x(n)z^{-n} \tag{2.5}$$

where z is a complex variable. Note that $X(z)$ is only defined for the regions of the complex plane in which the summation on the right converges.

Very often, the signals we work with start only at $n = 0$, that is, they are nonzero only for $n \geq 0$. Because of that, some textbooks define the z transform as

$$X_U(z) = \sum_{n=0}^{\infty} x(n)z^{-n} \tag{2.6}$$

which is commonly known as the one-sided z transform, while equation (2.5) is referred to as the two-sided z transform. Clearly, if the signal $x(n)$ is nonzero for $n < 0$, then the one-sided and two-sided z transforms are different. In this text, we work only with the two-sided z transform, which is referred to, without any risk of ambiguity, just as the z transform.

As mentioned above, the z transform of a sequence exists only for those regions of the complex plane in which the summation in equation (2.5) converges. The example below clarifies this point.

EXAMPLE 2.1
Compute the z transform of the sequence $x(n) = Ku(n)$.

SOLUTION
By definition, the z transform of $Ku(n)$ is

$$X(z) = K \sum_{n=0}^{\infty} z^{-n} = K \sum_{n=0}^{\infty} (z^{-1})^n \tag{2.7}$$

Thus, $X(z)$ is the sum of a power series which converges only if $|z^{-1}| < 1$. In such a case, $X(z)$ can be expressed as

$$X(z) = \frac{K}{1 - z^{-1}} = \frac{Kz}{z - 1}, \quad |z| > 1 \tag{2.8}$$

Note that for $|z| < 1$, the nth term of the summation, z^{-n}, tends to infinity as $n \rightarrow \infty$, and therefore $X(z)$ is not defined. For $z = 1$, the summation is also infinite. For $z = -1$, the summation oscillates between 1 and 0. In none of these cases does the z transform converge.

$\triangle$

It is important to note that the z transform of a sequence is a Laurent series in the complex variable z (Churchill, 1975). Therefore, the properties of Laurent series apply directly to the z transform. As a general rule, we can apply a result from series theory stating that, given a series of the complex variable z

$$S(z) = \sum_{i=0}^{\infty} f_i(z) \tag{2.9}$$

such that $|f_i(z)| < \infty$, $i = 0, 1, \ldots$, and given the quantity

$$\alpha(z) = \lim_{n \to \infty} \left| \frac{f_{n+1}(z)}{f_n(z)} \right| \tag{2.10}$$

then the series converges absolutely if $\alpha(z) < 1$, and diverges if $\alpha(z) > 1$ (Kreyszig, 1979). Note that, for $\alpha(z) = 1$, the above procedure tells us nothing about the convergence of the series, which must be investigated by other means. One can justify this by noting that, if $\alpha(z) < 1$, the terms of the series are under an exponential a^n for some $a < 1$, and therefore their sum converges as $n \rightarrow \infty$. One should clearly note that, if $|f_i(z)| = \infty$, for some i, then the series is not convergent.

The above result can be extended for the case of two-sided series as in the equation below

$$S(z) = \sum_{i=-\infty}^{\infty} f_i(z) \tag{2.11}$$

if we express $S(z)$ above as the sum of two series $S_1(z)$ and $S_2(z)$ such that

$$S_1(z) = \sum_{i=0}^{\infty} f_i(z) \quad \text{and} \quad S_2(z) = \sum_{i=-\infty}^{-1} f_i(z) \tag{2.12}$$

then $S(z)$ converges if the two series $S_1(z)$ and $S_2(z)$ converge. Therefore, in this case, we have to compute the two quantities

$$\alpha_1(z) = \lim_{n \to \infty} \left| \frac{f_{n+1}(z)}{f_n(z)} \right| \quad \text{and} \quad \alpha_2(z) = \lim_{n \to -\infty} \left| \frac{f_{n+1}(z)}{f_n(z)} \right| \tag{2.13}$$

Naturally, $S(z)$ converges absolutely if $\alpha_1(z) < 1$ and $\alpha_2(z) > 1$. The condition $\alpha_1(z) < 1$ is equivalent to saying that, for $n \to \infty$, the terms of the series are under a^n for some $a < 1$. The condition $\alpha_2(z) > 1$ is equivalent to saying that, for $n \to -\infty$, the terms of the series are under b^n for some $b > 1$. One should note that, for convergence, we must also have $|f_i(z)| < \infty, \forall i$.

Applying these convergence results to the z-transform definition given in equation (2.5), we conclude that the z transform converges if

$$\alpha_1 = \lim_{n \to \infty} \left| \frac{x(n+1)z^{-n-1}}{x(n)z^{-n}} \right| = |z^{-1}| \lim_{n \to \infty} \left| \frac{x(n+1)}{x(n)} \right| < 1 \tag{2.14}$$

$$\alpha_2 = \lim_{n \to -\infty} \left| \frac{x(n+1)z^{-n-1}}{x(n)z^{-n}} \right| = |z^{-1}| \lim_{n \to -\infty} \left| \frac{x(n+1)}{x(n)} \right| > 1 \tag{2.15}$$

Defining

$$r_1 = \lim_{n \to \infty} \left| \frac{x(n+1)}{x(n)} \right| \tag{2.16}$$

$$r_2 = \lim_{n \to -\infty} \left| \frac{x(n+1)}{x(n)} \right| \tag{2.17}$$

then equations (2.14) and (2.15) are equivalent to

$$r_1 < |z| < r_2 \tag{2.18}$$

That is, the z transform of a sequence exists in an annular region of the complex plane defined by equation (2.18) and illustrated in Figure 2.1. It is important to note that, for some sequences, $r_1 = 0$ or $r_2 \to \infty$. In these cases, the region of convergence may or may not include $z = 0$ or $|z| = \infty$, respectively.

We now take a closer look at the convergence of z transforms for four important classes of sequences.

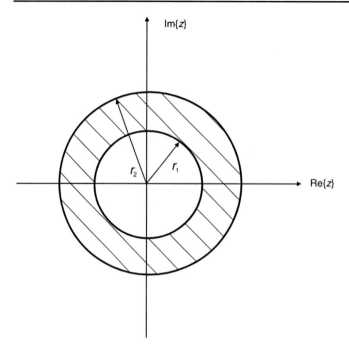

Figure 2.1 General region of convergence of the z transform.

- Right-handed, one-sided sequences: These are sequences such that $x(n) = 0$, for $n < n_0$, that is

$$X(z) = \sum_{n=n_0}^{\infty} x(n)z^{-n} \tag{2.19}$$

In this case, the z transform converges for $|z| > r_1$, where r_1 is given by equation (2.16). Since $|x(n)z^{-n}|$ must be finite, then, if $n_0 < 0$, the convergence region excludes $|z| = \infty$.

- Left-handed, one-sided sequences: These are sequences such that $x(n) = 0$, for $n > n_0$, that is

$$X(z) = \sum_{n=-\infty}^{n_0} x(n)z^{-n} \tag{2.20}$$

In this case, the z transform converges for $|z| < r_2$, where r_2 is given by equation (2.17). Since $|x(n)z^{-n}|$ must be finite, then, if $n_0 > 0$, the convergence region excludes $|z| = 0$.

- Two-sided sequences: In this case,

$$X(z) = \sum_{n=-\infty}^{\infty} x(n)z^{-n} \tag{2.21}$$

and the z transform converges for $r_1 < |z| < r_2$, where r_1 and r_2 are given by equations (2.16) and (2.17). Clearly, if $r_1 > r_2$, then the z transform does not exist.

- Finite-length sequences: These are sequences such that $x(n) = 0$, for $n < n_0$ and $n > n_1$, that is

$$X(z) = \sum_{n=n_0}^{n_1} x(n)z^{-n} \tag{2.22}$$

In such cases, the z transform converges everywhere except at the points such that $|x(n)z^{-n}| = \infty$. This implies that the convergence region excludes the point $z = 0$ if $n_1 > 0$ and $|z| = \infty$ if $n_0 < 0$.

EXAMPLE 2.2

Compute the z transforms of the following sequences, specifying their region of convergence:

(a) $x(n) = k2^n u(n)$
(b) $x(n) = u(-n + 1)$
(c) $x(n) = -k2^n u(-n - 1)$
(d) $x(n) = 0.5^n u(n) + 3^n u(-n)$
(e) $x(n) = 4^{-n} u(n) + 5^{-n} u(n + 1)$

SOLUTION

(a) $X(z) = \displaystyle\sum_{n=0}^{\infty} k2^n z^{-n}$

This series converges if $|2z^{-1}| < 1$, that is, for $|z| > 2$. In this case, $X(z)$ is the sum of a geometric series, and therefore

$$X(z) = \frac{k}{1 - 2z^{-1}} = \frac{kz}{z - 2}, \quad \text{for } 2 < |z| \le \infty \tag{2.23}$$

(b) $X(z) = \displaystyle\sum_{n=-\infty}^{1} z^{-n}$

This series converges if $|z^{-1}| > 1$, that is, for $|z| < 1$. Also, in order for the term z^{-1} to be finite, $|z| \ne 0$. In this case, $X(z)$ is the sum of a geometric series, such that

$$X(z) = \frac{z^{-1}}{1 - z} = \frac{1}{z - z^2}, \quad \text{for } 0 < |z| < 1 \tag{2.24}$$

(c) $X(z) = \displaystyle\sum_{n=-\infty}^{-1} -k2^n z^{-n}$

This series converges if $|\frac{z}{2}| < 1$, that is, for $|z| < 2$. In this case, $X(z)$ is the sum of a geometric series, such that

$$X(z) = \frac{-k\frac{z}{2}}{1 - \frac{z}{2}} = \frac{kz}{z - 2}, \quad \text{for } 0 \le |z| < 2 \tag{2.25}$$

(d) $X(z) = \sum_{n=0}^{\infty} 0.5^n z^{-n} + \sum_{n=-\infty}^{0} 3^n z^{-n}$

This series converges if $|0.5z^{-1}| < 1$ and $|3z^{-1}| > 1$, that is, for $0.5 < |z| < 3$. In this case, $X(z)$ is the sum of two geometric series, and therefore

$$X(z) = \frac{1}{1 - 0.5z^{-1}} + \frac{1}{1 - \frac{1}{3}z} = \frac{z}{z - 0.5} + \frac{3}{3 - z}, \quad \text{for } 0.5 < |z| < 3 \tag{2.26}$$

(e) $X(z) = \sum_{n=0}^{\infty} 4^{-n} z^{-n} + \sum_{n=-1}^{\infty} 5^{-n} z^{-n}$

This series converges if $|\frac{1}{4}z^{-1}| < 1$ and $|\frac{1}{5}z^{-1}| < 1$, that is, for $|z| > \frac{1}{4}$. Also, the term for $n = -1$, $\left(\frac{1}{5}z^{-1}\right)^{-1} = 5z$, is finite only for $|z| < \infty$. In this case, $X(z)$ is the sum of two geometric series, resulting in

$$X(z) = \frac{1}{1 - \frac{1}{4}z^{-1}} + \frac{5z}{1 - \frac{1}{5}z^{-1}} = \frac{4z}{4z - 1} + \frac{25z^2}{5z - 1}, \quad \text{for } \frac{1}{4} < |z| < \infty \tag{2.27}$$

In this example, although the sequences in items (a) and (c) are distinct, the expressions for their z transforms are the same, the difference being only in their regions of convergence. This highlights the important fact that, in order to completely specify a z transform, its region of convergence must be supplied. In Section 2.3, when we study the inverse z transform, this issue is dealt with in more detail.

$\triangle$

In several cases we deal with causal and stable systems. Since for a causal system its impulse response $h(n)$ is zero for $n < n_0$, then, from equation (1.48), we have that a causal system is also BIBO stable if

$$\sum_{n=n_0}^{\infty} |h(n)| < \infty \tag{2.28}$$

Applying the series convergence criterion seen above, we have that the system is stable only if

$$\lim_{n \to \infty} \left| \frac{h(n + 1)}{h(n)} \right| = r < 1 \tag{2.29}$$

This is equivalent to saying that $H(z)$, the z transform of $h(n)$, converges for $|z| > r$. Since, for stability, $r < 1$, then we conclude that the convergence region of the z

transform of the impulse response of a stable causal system includes the region outside the unit circle and the unit circle itself (in fact, if $n_0 < 0$, then this region excludes $|z| = \infty$).

A very important case is when $X(z)$ can be expressed as a ratio of polynomials, in the form

$$X(z) = \frac{N(z)}{D(z)} \tag{2.30}$$

We refer to the roots of $N(z)$ as the zeros of $X(z)$ and to the roots of $D(z)$ as the poles of $X(z)$. More specifically, in this case $X(z)$ can be expressed as

$$X(z) = \frac{N(z)}{\prod_{k=1}^{K}(z - p_k)^{m_k}} \tag{2.31}$$

where p_k is a pole of multiplicity m_k, and K is the total number of distinct poles. Since $X(z)$ is not defined at its poles, its convergence region must not include them. Therefore, given $X(z)$ as in equation (2.31), there is an easy way of determining its convergence region, depending on the type of sequence $x(n)$:

- Right-handed, one-sided sequences: The convergence region of $X(z)$ is $|z| > r_1$. Since $X(z)$ is not convergent at its poles, then its poles must be inside the circle $|z| = r_1$ (except for poles at $|z| = \infty$), and $r_1 = \max_{1 \le k \le K} \{|p_k|\}$. This is illustrated in Figure 2.2a.
- Left-handed, one-sided sequences: The convergence region of $X(z)$ is $|z| < r_2$. Therefore, its poles must be outside the circle $|z| = r_2$ (except for poles at $|z| = 0$), and $r_2 = \min_{1 \le k \le K} \{|p_k|\}$. This is illustrated in Figure 2.2b.
- Two-sided sequences: The convergence region of $X(z)$ is $r_1 < |z| < r_2$, and therefore some of its poles are inside the circle $|z| = r_1$ and some outside the circle $|z| = r_2$. In this case, the convergence region needs to be further specified. This is illustrated in Figure 2.2c.

2.3   Inverse z transform

Very often one needs to determine which sequence corresponds to a given z transform. A formula for the inverse z transform can be obtained from the residue theorem, which we state next.

THEOREM 2.1 (RESIDUE THEOREM)
Let $X(z)$ be a complex function that is analytic inside a closed contour C, including

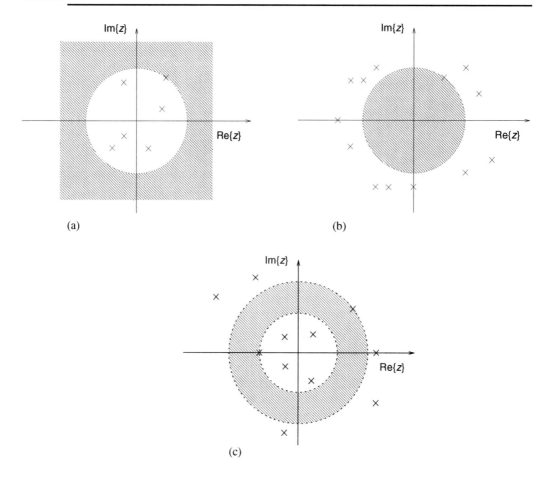

Figure 2.2 Regions of convergence of a z transform in relation to its poles: (a) right-handed, one-sided sequences; (b) left-handed, one-sided sequences; (c) two-sided sequences.

the contour itself, except in a finite number of singular points p_n inside C. In this case, the following equality holds:

$$\oint_C X(z)\mathrm{d}z = 2\pi\,\mathrm{j} \sum_{k=1}^{K} \mathop{\mathrm{res}}_{z=p_k} \{X(z)\}$$
(2.32)

with the integral evaluated counterclockwise around C.

If p_k is a pole of multiplicity m_k of $X(z)$, that is, if $X(z)$ can be written as

$$X(z) = \frac{P_k(z)}{(z - p_k)^{m_k}}$$
(2.33)

where $P_k(z)$ is analytic at $z = p_k$, then the residue of $X(z)$ with respect to p_k is given by

$$\operatorname*{res}_{z=p_k}\{X(z)\} = \frac{1}{(m_k-1)!} \frac{d^{(m_k-1)}[(z-p_k)^{m_k} X(z)]}{dz^{(m_k-1)}}\Bigg|_{z=p_k} \tag{2.34}$$

$\diamond$

Using the above theorem, one can show that, if C is a counterclockwise closed contour, containing the origin of the z plane, then

$$\frac{1}{2\pi j}\oint_C z^{n-1}dz = \begin{cases} 0, & \text{for } n \neq 0 \\ 1, & \text{for } n = 0 \end{cases} \tag{2.35}$$

and then we can derive that the inverse z transform of $X(z)$ is given by

$$x(n) = \frac{1}{2\pi j}\oint_C X(z)z^{n-1}dz \tag{2.36}$$

where C is a closed counterclockwise contour in the convergence region of $X(z)$.

PROOF

Since

$$X(z) = \sum_{n=-\infty}^{\infty} x(n)z^{-n} \tag{2.37}$$

by expressing $x(n)$ using the inverse z transform as in equation (2.36), then changing the order of the integration and summation, we have that

$$\frac{1}{2\pi j}\oint_C X(z)z^{m-1}dz = \frac{1}{2\pi j}\oint_C \sum_{n=-\infty}^{\infty} x(n)z^{-n+m-1}dz$$

$$= \frac{1}{2\pi j}\sum_{n=-\infty}^{\infty} x(n)\oint_C z^{-n+m-1}dz$$

$$= x(m) \tag{2.38}$$

$\square$

In the remainder of this section, we describe techniques for the computation of the inverse z transform in several practical cases.

2.3.1   Computation based on residue theorem

Whenever $X(z)$ is a ratio of polynomials, the residue theorem can be used very efficiently to compute the inverse z transform. In this case, equation (2.36) becomes

$$x(n) = \frac{1}{2\pi j}\oint_C X(z)z^{n-1}dz = \sum_{k=1}^{K} \operatorname*{res}_{z=p_k}\left\{X(z)z^{n-1}\right\} \tag{2.39}$$

where

$$X(z)z^{n-1} = \frac{N(z)}{\displaystyle\prod_{k=1}^{K}(z - p_k)^{m_k}} \tag{2.40}$$

Note that p_k, with multiplicity m_k, for $k = 1, \ldots, K$, are just the poles of $X(z)z^{n-1}$ that are inside the contour C, which must be contained in the convergence region of $X(z)$. It is important to observe that, in order to compute $x(n)$, for $n \leq 0$, we have to also consider the residues of the poles of $X(z)z^{n-1}$ at the origin.

EXAMPLE 2.3
Determine the inverse z transform of

$$X(z) = \frac{z^2}{(z - 0.2)(z + 0.8)} \tag{2.41}$$

considering that it represents the z transform of the impulse response of a causal system.

SOLUTION
One should note that a z transform is only completely specified if its region of convergence is given. In this example, since the system is causal, we have that its impulse response is right handed and one sided. Therefore, as seen earlier in this section, the convergence region of the z transform is characterized by $|z| > r_1$, its poles are inside the circle $|z| = r_1$ and therefore $r_1 = \max\limits_{1 \leq k \leq K}\{|p_k|\} = 0.8$.

We then need to compute

$$x(n) = \frac{1}{2\pi j}\oint_C H(z)z^{n-1}\,dz = \frac{1}{2\pi j}\oint_C \frac{z^{n+1}}{(z - 0.2)(z + 0.8)}\,dz \tag{2.42}$$

where C is any closed contour in the convergence region of $X(z)$, that is, involving the poles $z = 0.2$ and $z = 0.8$, as well as the poles at $z = 0$, for $n \leq -2$.

Since we want to use the residue theorem, there are two distinct cases. For $n \geq -1$, there are two poles inside C, $z = -0.2$ and $z = 0.8$; whereas for $n \leq -2$, there are three poles inside C, $z = -0.2$, $z = 0.8$, and $z = 0$. Therefore, we have that:

- For $n \geq -1$, equation (2.39) leads to

$$
\begin{aligned}
x(n) &= \operatorname*{res}_{z=0.2}\left\{\frac{z^{n+1}}{(z - 0.2)(z + 0.8)}\right\} + \operatorname*{res}_{z=-0.8}\left\{\frac{z^{n+1}}{(z - 0.2)(z + 0.8)}\right\} \\
&= \operatorname*{res}_{z=0.2}\left\{\frac{P_1(z)}{z - 0.2}\right\} + \operatorname*{res}_{z=-0.8}\left\{\frac{P_2(z)}{z + 0.8}\right\}
\end{aligned} \tag{2.43}
$$

where

$$P_1(z) = \frac{z^{n+1}}{z + 0.8} \quad \text{and} \quad P_2(z) = \frac{z^{n+1}}{z - 0.2} \tag{2.44}$$

From equation (2.34),

$$\operatorname*{res}_{z=0.2} \left\{ \frac{z^{n+1}}{(z - 0.2)(z + 0.8)} \right\} = P_1(0.2) = (0.2)^{n+1} \tag{2.45}$$

$$\operatorname*{res}_{z=-0.8} \left\{ \frac{z^{n+1}}{(z - 0.2)(z + 0.8)} \right\} = P_2(-0.8) = -(-0.8)^{n+1} \tag{2.46}$$

and then

$$x(n) = (0.2)^{n+1} - (-0.8)^{n+1}, \quad \text{for } n \geq -1 \tag{2.47}$$

- For $n \leq -2$, we also have a pole of multiplicity $(-n - 1)$ at $z = 0$, Therefore, we have to add the residue at $z = 0$ to the two residues in equation (2.47), such that

$$x(n) = (0.2)^{n+1} - (-0.8)^{n+1} + \operatorname*{res}_{z=0} \left\{ \frac{z^{n+1}}{(z - 0.2)(z + 0.8)} \right\}$$

$$= (0.2)^{n+1} - (-0.8)^{n+1} + \operatorname*{res}_{z=0} \left\{ P_3(z) z^{n+1} \right\} \tag{2.48}$$

where

$$P_3(z) = \frac{1}{(z - 0.2)(z + 0.8)} \tag{2.49}$$

From equation (2.34), since the pole $z = 0$ has multiplicity $m_k = (-n - 1)$, we have that

$$\operatorname*{res}_{z=0} \left\{ P_3(z) z^{n+1} \right\} = \frac{1}{(-n - 2)!} \left. \frac{d^{(-n-2)} P_3(z)}{dz^{(-n-2)}} \right|_{z=0}$$

$$= \frac{1}{(-n - 2)!} \left. \frac{d^{(-n-2)}}{dz^{(-n-2)}} \left\{ \frac{1}{(z - 0.2)(z + 0.8)} \right\} \right|_{z=0}$$

$$= \left\{ \frac{(-1)^{-n-2}}{(z - 0.2)^{-n-1}} - \frac{(-1)^{-n-2}}{(z + 0.8)^{-n-1}} \right\} \Bigg|_{z=0}$$

$$= (-1)^{-n-2} \left[(-0.2)^{n+1} - (0.8)^{n+1} \right]$$

$$= -(0.2)^{n+1} + (-0.8)^{n+1} \tag{2.50}$$

Substituting the above result into equation (2.48), we have that

$$x(n) = (0.2)^{n+1} - (-0.8)^{n+1} - (0.2)^{n+1} + (-0.8)^{n+1} = 0, \quad \text{for } n \leq -2 \tag{2.51}$$

From equations (2.47) and (2.51), we then have that

$$x(n) = \left[(0.2)^{n+1} - (-0.8)^{n+1}\right]u(n+1) \tag{2.52}$$

$\triangle$

From what we have seen in the above example, the computation of residues for the case of multiple poles at $z = 0$ involves computation of nth-order derivatives, which can very often become quite involved. Fortunately, these cases can be easily solved by means of a simple trick, which we describe next.

When the integral in

$$X(z) = \frac{1}{2\pi j} \oint_C X(z)z^{n-1}dz \tag{2.53}$$

involves computation of residues of multiple poles at $z = 0$, we make the change of variable $z = \frac{1}{v}$. If the poles of $X(z)$ are located at $z = p_i$, then the poles of $X\left(\frac{1}{v}\right)$ are located at $v = \frac{1}{p_i}$. Also, if $X(z)$ converges for $r_1 < |z| < r_2$, then $X\left(\frac{1}{v}\right)$ converges for $\frac{1}{r_2} < |v| < \frac{1}{r_1}$. The integral in equation (2.36) then becomes

$$x(n) = \frac{1}{2\pi j} \oint_C X(z)z^{n-1}dz = -\frac{1}{2\pi j} \oint_{C'} X\left(\frac{1}{v}\right)v^{-n-1}dv \tag{2.54}$$

Note that, if the contour C is traversed in the counterclockwise direction by z then the contour C' is traversed in the clockwise direction by v. Substituting the contour C' by an equal contour C'' that is traversed in the counterclockwise direction, the sign of the integral is reversed, and the above equation becomes

$$x(n) = \frac{1}{2\pi j} \oint_C X(z)z^{n-1}dz = \frac{1}{2\pi j} \oint_{C''} X\left(\frac{1}{v}\right)v^{-n-1}dv \tag{2.55}$$

If $X(z)z^{n-1}$ has multiple poles at the origin, then $X\left(\frac{1}{v}\right)v^{-n-1}$ has multiple poles at $|z| = \infty$, which are now outside the closed contour C''. Therefore, the computation of the integral on the right-hand side of equation (2.55) avoids the computation of nth-order derivatives. This fact is illustrated by Example 2.4 below, which revisits the computation of the z transform in Example 2.3.

EXAMPLE 2.4

Compute the inverse z transform of $X(z)$ in Example 2.3, for $n \leq -2$, using the residue theorem by employing the change of variables in equation (2.55).

SOLUTION

If we make the change of variables $z = \frac{1}{v}$, equation (2.42) becomes

$$x(n) = \frac{1}{2\pi j} \oint_C \frac{z^{n+1}}{(z-0.2)(z+0.8)} dz = \frac{1}{2\pi j} \oint_{C''} \frac{v^{-n-1}}{(1-0.2v)(1+0.8v)} dv \tag{2.56}$$

The convergence region of the integrand on the right is $|v| < \frac{1}{0.8}$, and therefore, for $n \leq -2$, no poles are inside the closed contour C''. Then, from equation (2.39), we conclude that

$$x(n) = 0, \quad \text{for } n \leq -2 \tag{2.57}$$

which is the result of Example 2.3 obtained in a straightforward manner. $\triangle$

2.3.2   Computation based on partial-fraction expansions

When the z transform of a sequence can be expressed as a ratio of polynomials, then we can carry out a partial-fraction expansion of $X(z)$ before computing the inverse z transform.

If $X(z) = \frac{N(z)}{D(z)}$ has K distinct poles $p_k, k = 1, \ldots, K$, each of multiplicity m_k, then the partial-fraction expansion of $X(z)$ is as follows (Kreyszig, 1979):

$$X(z) = \sum_{l=0}^{M-L} g_l z^l + \sum_{k=1}^{K} \sum_{i=1}^{m_k} \frac{c_{ki}}{(z - p_k)^i} \tag{2.58}$$

where M and L are the degrees of the numerator and denominator of $X(z)$, respectively.

The coefficients $g_l, l = 0, \ldots, M - L$, can be obtained from the quotient of the polynomials $N(z)$ and $D(z)$ as follows

$$X(z) = \frac{N(z)}{D(z)} = \sum_{l=0}^{M-L} g_l z^l + \frac{C(z)}{D(z)} \tag{2.59}$$

where the degree of $C(z)$ is smaller than the degree of $D(z)$. Clearly, if $M < N$, then $g_l = 0, \forall l$.

The coefficients c_{ki} are

$$c_{ki} = \frac{1}{(m_k - i)!} \frac{d^{(m_k-i)}[(z - p_k)^{m_k} X(z)]}{dz^{(m_k-i)}} \Bigg|_{z=p_k} \tag{2.60}$$

In the case of a simple pole, c_{k1} is given by

$$c_{k1} = (z - p_k)X(z)|_{z=p_k} \tag{2.61}$$

Since the z transform is linear and the inverse z transform of each of the terms $\frac{c_{ki}}{(z-p_k)^i}$ is easy to compute (being also listed in z-transform tables), then the inverse z transform follows directly from equation (2.58).

EXAMPLE 2.5
Solve Example 2.3 using the partial-fraction expansion of $X(z)$.

SOLUTION

We form

$$X(z) = \frac{z^2}{(z-0.2)(z+0.8)} = g_0 + \frac{c_1}{z-0.2} + \frac{c_2}{z+0.8} \tag{2.62}$$

where

$$g_0 = \lim_{|z|\to\infty} X(z) = 1 \tag{2.63}$$

and, using equation (2.34), we find that

$$c_1 = \left.\frac{z^2}{z+0.8}\right|_{z=0.2} = (0.2)^2 \tag{2.64}$$

$$c_2 = \left.\frac{z^2}{z-0.2}\right|_{z=-0.8} = -(0.8)^2 \tag{2.65}$$

such that

$$X(z) = 1 + \frac{(0.2)^2}{z-0.2} - \frac{(0.8)^2}{z+0.8} \tag{2.66}$$

Since $X(z)$ is the z transform of the impulse response of a causal system, then we have that the terms in the above equation correspond to a right-handed, one-sided power series. Thus, the inverse z transforms of each term are:

$$\mathcal{Z}^{-1}\{1\} = \delta(n) \tag{2.67}$$

$$\mathcal{Z}^{-1}\left\{\frac{(0.2)^2}{z-0.2}\right\} = \mathcal{Z}^{-1}\left\{\frac{(0.2)^2 z^{-1}}{1-0.2z^{-1}}\right\}$$

$$= \mathcal{Z}^{-1}\left\{0.2\sum_{n=1}^{\infty}(0.2z^{-1})^n\right\}$$

$$= 0.2(0.2)^n u(n-1)$$

$$= (0.2)^{n+1} u(n-1) \tag{2.68}$$

$$\mathcal{Z}^{-1}\left\{\frac{-(0.8)^2}{z+0.8}\right\} = \mathcal{Z}^{-1}\left\{\frac{-(0.8)^2 z^{-1}}{1+0.8z^{-1}}\right\}$$

$$= \mathcal{Z}^{-1}\left\{0.8\sum_{n=1}^{\infty}(-0.8z^{-1})^n\right\}$$

$$= 0.8(-0.8)^n u(n-1)$$

$$= -(-0.8)^{n+1} u(n-1) \tag{2.69}$$

Summing the three terms above (equations (2.67)–(2.69)), we have that the inverse z transform of $X(z)$ is

$$x(n) = \delta(n) + (0.2)^{n+1}u(n-1) - (-0.8)^{n+1}u(n-1)$$

$$= (0.2)^{n+1}u(n) - (-0.8)^{n+1}u(n) \tag{2.70}$$

$\triangle$

2.3.3 Computation based on polynomial division

Given $X(z) = \frac{N(z)}{D(z)}$, we can perform long division on the polynomials $N(z)$ and $D(z)$, and obtain that the coefficient of z^k corresponds to the value of $x(n)$ at $n = k$. One should note that this is possible only in the case of one-sided sequences. If the sequence is right handed, then the polynomials should be a function of z. If the sequence is left handed, the polynomials should be a function of z^{-1}. This is made clear in Examples 2.6 and 2.7.

EXAMPLE 2.6
Solve Example 2.3 using polynomial division.

SOLUTION
Since $X(z)$ is the z transform of a right-handed, one-sided (causal) sequence, we can express it as the ratio of polynomials in z, that is

$$X(z) = \frac{z^2}{(z-0.2)(z+0.8)} = \frac{z^2}{z^2 + 0.6z - 0.16} \tag{2.71}$$

Then the division becomes

$$
\begin{array}{r|l}
z^2 & z^2 + 0.6z - 0.16 \\
-z^2 - 0.6z + 0.16 & \overline{1 - 0.6z^{-1} + 0.52z^{-2} - 0.408z^{-3} + \cdots} \\
\hline
-0.6z + 0.16 & \\
0.6z + 0.36 - 0.096z^{-1} & \\
\hline
0.52 - 0.096z^{-1} & \\
-0.52 - 0.312z^{-1} + 0.0832z^{-2} & \\
\hline
-0.408z^{-1} + 0.0832z^{-2} & \\
\vdots &
\end{array}
$$

and therefore

$$X(z) = 1 + (-0.6)z^{-1} + (0.52)z^{-2} + (-0.408)z^{-3} + \cdots \tag{2.72}$$

This is the same as saying that

$$x(n) = \begin{cases} 0, & \text{for } n < 0 \\ 1, -0.6, 0.52, -0.408, \ldots & \text{for } n = 0, 1, 2, \ldots \end{cases} \tag{2.73}$$

The main difficulty with this method is to find a closed-form expression for $x(n)$. In the above case, we can check that indeed the above sequence corresponds to equation (2.52). △

EXAMPLE 2.7
Find the inverse z transform of $X(z)$ in Example 2.3 using polynomial division and supposing that the sequence $x(n)$ is left handed and one sided.

SOLUTION
Since $X(z)$ is the z transform of a left-handed, one-sided sequence, we should express it as

$$X(z) = \frac{z^2}{(z - 0.2)(z + 0.8)} = \frac{1}{-0.16z^{-2} + 0.6z^{-1} + 1} \tag{2.74}$$

Then the division becomes

$$
\begin{array}{r|l}
1 & \underline{-0.16z^{-2} + 0.6z^{-1} + 1} \\
\underline{-1 + 3.75z + 6.25z^2} & -6.25z^2 - 23.4375z^3 - 126.953\,125z^4 - \cdots \\
3.75z + 6.25z^2 & \\
\underline{-3.75z + 14.0625z^2 + 23.4375z^3} & \\
20.3125z^2 + 23.4375z^3 & \\
\vdots &
\end{array}
$$

yielding

$$X(z) = -6.25z^2 - 23.4375z^3 - 126.953\,125z^4 - \cdots \tag{2.75}$$

implying that

$$x(n) = \begin{cases} \ldots, -126.953\,125, -23.4375, -6.25, & \text{for } n = \ldots, -4, -3, -2 \\ 0, & \text{for } n > -2 \end{cases} \tag{2.76}$$

△

2.3.4   Computation based on series expansion

When a z transform is not expressed as a ratio of polynomials, we can try to perform its inversion using a Taylor series expansion around either $z^{-1} = 0$ or $z = 0$, depending on whether the convergence region includes $|z| = \infty$ or $z = 0$. For right-handed, one-sided sequences, we use the expansion of $X(z)$ using the variable z^{-1} around

$z^{-1} = 0$. The Taylor series expansion of $F(x)$ around $x = 0$ is given by

$$F(x) = F(0) + x \left. \frac{\mathrm{d}F}{\mathrm{d}x} \right|_{x=0} + \frac{x^2}{2!} \left. \frac{\mathrm{d}^2 F}{\mathrm{d}x^2} \right|_{x=0} + \frac{x^3}{3!} \left. \frac{\mathrm{d}^3 F}{\mathrm{d}x^3} \right|_{x=0} + \cdots$$

$$= \sum_{n=0}^{\infty} \frac{x^n}{n!} \left. \frac{\mathrm{d}^n F}{\mathrm{d}x^n} \right|_{x=0} \tag{2.77}$$

If we make $x = z^{-1}$, then the expansion above has the form of a z transform of a right-handed, one-sided sequence.

EXAMPLE 2.8
Find the inverse z transform of

$$X(z) = \ln \left(\frac{1}{1 - z^{-1}} \right) \tag{2.78}$$

Consider the sequence as right handed and one sided.

SOLUTION
Expanding $X(z)$ as in equation (2.77), using z^{-1} as the variable, we have that

$$X(z) = \sum_{n=1}^{\infty} \frac{z^{-n}}{n} \tag{2.79}$$

Therefore the inverse z transform of $X(z)$ is, by inspection,

$$x(n) = \frac{1}{n} u(n - 1) \tag{2.80}$$

$\triangle$

2.4   Properties of the z transform

In this section we state some of the more important z-transform properties.

2.4.1   Linearity

Given two sequences $x_1(n)$ and $x_2(n)$ and two arbitrary constants k_1 and k_2 such that $x(n) = k_1 x_1(n) + k_2 x_2(n)$, then

$$X(z) = k_1 X_1(z) + k_2 X_2(z) \tag{2.81}$$

where the region of convergence of $X(z)$ is the intersection of the regions of convergence of $X_1(z)$ and $X_2(z)$.

PROOF

$$X(z) = \sum_{n=-\infty}^{\infty} (k_1 x_1(n) + k_2 x_2(n)) z^{-n}$$

$$= k_1 \sum_{n=-\infty}^{\infty} x_1(n) z^{-n} + k_2 \sum_{n=-\infty}^{\infty} x_2(n) z^{-n}$$

$$= k_1 X_1(z) + k_2 X_2(z) \tag{2.82}$$

$\square$

2.4.2 Time-reversal

$$x(-n) \longleftrightarrow X(z^{-1}) \tag{2.83}$$

where, if the region of convergence of $X(z)$ is $r_1 < |z| < r_2$, then the region of convergence of $\mathcal{Z}\{x(-n)\}$ is $\frac{1}{r_2} < |z| < \frac{1}{r_1}$.

PROOF

$$\mathcal{Z}\{x(-n)\} = \sum_{n=-\infty}^{\infty} x(-n) z^{-n} = \sum_{m=-\infty}^{\infty} x(m) z^{m} = \sum_{m=-\infty}^{\infty} x(m)(z^{-1})^{-m} = X(z^{-1})$$

$$\tag{2.84}$$

implying that the region of convergence of $\mathcal{Z}\{x(-n)\}$ is $r_1 < |z^{-1}| < r_2$, which is equivalent to $\frac{1}{r_2} < |z| < \frac{1}{r_1}$. $\square$

2.4.3 Time-shift theorem

$$x(n + l) \longleftrightarrow z^{l} X(z) \tag{2.85}$$

where l is an integer. The region of convergence of $\mathcal{Z}\{x(n + l)\}$ is the same as the region of convergence of $X(z)$, except for the possible inclusion or exclusion of the regions $z = 0$ and $|z| = \infty$.

PROOF
By definition

$$\mathcal{Z}\{x(n + l)\} = \sum_{n=-\infty}^{\infty} x(n + l) z^{-n} \tag{2.86}$$

Making the change of variables $m = n + l$, we have that

$$\mathcal{Z}\{x(n + l)\} = \sum_{m=-\infty}^{\infty} x(m) z^{-(m-l)} = z^{l} \sum_{m=-\infty}^{\infty} x(m) z^{-m} = z^{l} X(z) \tag{2.87}$$

noting that the multiplication by z^l can either introduce or exclude poles at $z = 0$ and $|z| = \infty$. □

2.4.4 Multiplication by an exponential

$$\alpha^{-n}x(n) \longleftrightarrow X(\alpha z) \tag{2.88}$$

where, if the region of convergence of $X(z)$ is $r_1 < |z| < r_2$, then the region of convergence of $\mathcal{Z}\{\alpha^{-n}x(n)\}$ is $\frac{r_1}{|\alpha|} < |z| < \frac{r_2}{|\alpha|}$.

PROOF

$$\mathcal{Z}\{\alpha^{-n}x(n)\} = \sum_{n=-\infty}^{\infty} \alpha^{-n}x(n)z^{-n} = \sum_{n=-\infty}^{\infty} x(n)(\alpha z)^{-n} = X(\alpha z) \tag{2.89}$$

where the summation converges for $r_1 < |\alpha z| < r_2$, which is equivalent to $\frac{r_1}{|\alpha|} < |z| < \frac{r_2}{|\alpha|}$. □

2.4.5 Complex differentiation

$$nx(n) \longleftrightarrow -z\frac{\mathrm{d}X(z)}{\mathrm{d}z} \tag{2.90}$$

where the region of convergence of $\mathcal{Z}\{nx(n)\}$ is the same as the one of $X(z)$, that is $r_1 < |z| < r_2$.

PROOF

$$\mathcal{Z}\{nx(n)\} = \sum_{n=-\infty}^{\infty} nx(n)z^{-n}$$

$$= z \sum_{n=-\infty}^{\infty} nx(n)z^{-n-1}$$

$$= -z \sum_{n=-\infty}^{\infty} x(n)\left(-nz^{-n-1}\right)$$

$$= -z \sum_{n=-\infty}^{\infty} x(n)\frac{\mathrm{d}}{\mathrm{d}z}\{z^{-n}\}$$

$$= -z\frac{\mathrm{d}X(z)}{\mathrm{d}z} \tag{2.91}$$

From equations (2.16) and (2.17), we have that, if the region of convergence of $X(z)$

is $r_1 < |z| < r_2$, then

$$r_1 = \lim_{n \to \infty} \left| \frac{x(n+1)}{x(n)} \right| \qquad (2.92)$$

$$r_2 = \lim_{n \to -\infty} \left| \frac{x(n+1)}{x(n)} \right| \qquad (2.93)$$

Therefore, if the region of convergence of $\mathcal{Z}\{nx(n)\}$ is given by $r_1' < |z| < r_2'$, then

$$r_1' = \lim_{n \to \infty} \left| \frac{nx(n)}{(n-1)x(n-1)} \right| = \lim_{n \to \infty} \left| \frac{x(n)}{x(n-1)} \right| = r_1 \qquad (2.94)$$

$$r_2' = \lim_{n \to -\infty} \left| \frac{nx(n)}{(n-1)x(n-1)} \right| = \lim_{n \to -\infty} \left| \frac{x(n)}{x(n-1)} \right| = r_2 \qquad (2.95)$$

implying that the region of convergence of $\mathcal{Z}\{nx(n)\}$ is the same as the one of $X(z)$.
□

2.4.6 Complex conjugation

$$x^*(n) \longleftrightarrow X^*(z^*) \qquad (2.96)$$

The regions of convergence of $X(z)$ and $\mathcal{Z}\{x^*(n)\}$ being the same.

PROOF

$$\mathcal{Z}\{x^*(n)\} = \sum_{n=-\infty}^{\infty} x^*(n) z^{-n}$$

$$= \sum_{n=-\infty}^{\infty} \left[x(n) (z^*)^{-n} \right]^*$$

$$= \left[\sum_{n=-\infty}^{\infty} x(n) (z^*)^{-n} \right]^*$$

$$= X^*(z^*) \qquad (2.97)$$

from which it follows trivially that the region of convergence of $\mathcal{Z}\{x^*(n)\}$ is the same as the one of $X(z)$.
□

2.4.7 Real and imaginary sequences

$$\mathrm{Re}\{x(n)\} \longleftrightarrow \frac{1}{2} \left(X(z) + X^*(z^*) \right) \qquad (2.98)$$

$$\mathrm{Im}\{x(n)\} \longleftrightarrow \frac{1}{2j} \left(X(z) - X^*(z^*) \right) \qquad (2.99)$$

where $\mathrm{Re}\{x(n)\}$ and $\mathrm{Im}\{x(n)\}$ are the real and imaginary parts of the sequence $x(n)$, respectively. The regions of convergence of $\mathcal{Z}\{\mathrm{Re}\{x(n)\}\}$ and $\mathcal{Z}\{\mathrm{Im}\{x(n)\}\}$ are the same as the ones of $X(z)$.

PROOF

$$\mathcal{Z}\{\mathrm{Re}\{x(n)\}\} = \mathcal{Z}\left\{\frac{1}{2}\left(x(n)+x^*(n)\right)\right\} = \frac{1}{2}\left(X(z)+X^*\left(z^*\right)\right) \tag{2.100}$$

$$\mathcal{Z}\{\mathrm{Im}\{x(n)\}\} = \mathcal{Z}\left\{\frac{1}{2j}\left(x(n)-x^*(n)\right)\right\} = \frac{1}{2j}\left(X(z)-X^*\left(z^*\right)\right) \tag{2.101}$$

with the respective regions of convergence following trivially from the above expressions. $\qquad\square$

2.4.8   Initial value theorem

If $x(n) = 0$, for $n < 0$, then

$$x(0) = \lim_{z\to\infty} X(z) \tag{2.102}$$

PROOF
If $x(n) = 0$, for $n < 0$, then

$$\lim_{z\to\infty} X(z) = \lim_{z\to\infty} \sum_{n=0}^{\infty} x(n)z^{-n} = \sum_{n=0}^{\infty} \lim_{z\to\infty} x(n)z^{-n} = x(0) \tag{2.103}$$

$\qquad\square$

2.4.9   Convolution theorem

$$x_1(n) * x_2(n) \longleftrightarrow X_1(z)X_2(z) \tag{2.104}$$

The region of convergence of $\mathcal{Z}\{x_1(n) * x_2(n)\}$ is the intersection of the regions of convergence of $X_1(z)$ and $X_2(z)$. If a pole of $X_1(z)$ is canceled by a zero of $X_2(z)$, or vice versa, then the region of convergence of $\mathcal{Z}\{x_1(n) * x_2(n)\}$ can be larger than the ones of both $X_1(z)$ and $X_2(z)$.

PROOF

$$\mathcal{Z}\{x_1(n) * x_2(n)\} = \mathcal{Z}\left\{\sum_{l=-\infty}^{\infty} x_1(l)x_2(n-l)\right\}$$

$$= \sum_{n=-\infty}^{\infty}\left[\sum_{l=-\infty}^{\infty} x_1(l)x_2(n-l)\right]z^{-n}$$

$$= \sum_{l=-\infty}^{\infty} x_1(l) \sum_{n=-\infty}^{\infty} x_2(n-l)z^{-n}$$

$$= \left[\sum_{l=-\infty}^{\infty} x_1(l)z^{-l} \right] \left[\sum_{n=-\infty}^{\infty} x_2(n)z^{-n} \right]$$

$$= X_1(z)X_2(z) \tag{2.105}$$

$\square$

2.4.10 Product of two sequences

$$x_1(n)x_2(n) \longleftrightarrow \frac{1}{2\pi j} \oint_{C_1} X_1(v)X_2\left(\frac{z}{v}\right)v^{-1}dv = \frac{1}{2\pi j} \oint_{C_2} X_1\left(\frac{z}{v}\right)X_2(v)v^{-1}dv \tag{2.106}$$

where C_1 is a contour contained in the intersection of the regions of convergence of $X_1(v)$ and $X_2\left(\frac{z}{v}\right)$, and C_2 is a contour contained in the intersection of the regions of convergence of $X_1\left(\frac{z}{v}\right)$ and $X_2(v)$. Both C_1 and C_2 are assumed to be counterclockwise oriented.

If the region of convergence of $X_1(z)$ is $r_1 < |z| < r_2$ and the region of convergence of $X_2(z)$ is $r_1' < |z| < r_2'$, then the region of convergence of $\mathcal{Z}\{x_1(n)x_2(n)\}$ is

$$r_1 r_1' < |z| < r_2 r_2' \tag{2.107}$$

PROOF

By expressing $x_2(n)$ as a function of its z transform, $X_2(z)$ (equation (2.36)), then changing the order of the integration and summation, and using the definition of the z transform, we have that

$$\mathcal{Z}\{x_1(n)x_2(n)\} = \sum_{n=-\infty}^{\infty} x_1(n)x_2(n)z^{-n}$$

$$= \sum_{n=-\infty}^{\infty} x_1(n) \left[\frac{1}{2\pi j} \oint_{C_2} X_2(v)v^{(n-1)}dv \right] z^{-n}$$

$$= \frac{1}{2\pi j} \oint_{C_2} \sum_{n=-\infty}^{\infty} x_1(n)z^{-n}v^{(n-1)} X_2(v)dv$$

$$= \frac{1}{2\pi j} \oint_{C_2} \left[\sum_{n=-\infty}^{\infty} x_1(n) \left(\frac{v}{z}\right)^n \right] X_2(v)v^{-1}dv$$

$$= \frac{1}{2\pi j} \oint_{C_2} X_1\left(\frac{z}{v}\right) X_2(v)v^{-1}dv \tag{2.108}$$

If the region of convergence of $X_1(z)$ is $r_1 < |z| < r_2$, then the region of convergence of $X_1\left(\frac{z}{v}\right)$ is

$$r_1 < \frac{|z|}{|v|} < r_2 \qquad (2.109)$$

which is equivalent to

$$\frac{|z|}{r_2} < |v| < \frac{|z|}{r_1} \qquad (2.110)$$

In addition, if the region of convergence of $X_2(v)$ is $r_1' < |v| < r_2'$, then the contour C_2 must lie in the intersection of the two regions of convergence, that is, C_2 must be contained in the region

$$\max\left\{\frac{|z|}{r_2}, r_1'\right\} < |v| < \min\left\{\frac{|z|}{r_1}, r_2'\right\} \qquad (2.111)$$

Therefore, we must have

$$\min\left\{\frac{|z|}{r_1}, r_2'\right\} > \max\left\{\frac{|z|}{r_2}, r_1'\right\} \qquad (2.112)$$

which holds if $r_1 r_1' < |z| < r_2 r_2'$.

$\square$

Equation (2.106) is also known as the complex convolution theorem (Antoniou, 1993; Oppenheim & Schafer, 1975). Although at first it does not have the form of a convolution, if we express $z = \rho_1 e^{j\theta_1}$ and $v = \rho_2 e^{j\theta_2}$ in polar form, then it can be rewritten as

$$\mathcal{Z}\{x_1(n)x_2(n)\}|_{z=\rho_1 e^{j\theta_1}} = \frac{1}{2\pi} \int_{-\pi}^{\pi} X_1\left(\frac{\rho_1}{\rho_2} e^{j(\theta_1-\theta_2)}\right) X_2\left(\rho_2 e^{j\theta_2}\right) d\theta_2 \qquad (2.113)$$

which has the form of a convolution in θ_1 and θ_2.

2.4.11 Parseval's theorem

$$\sum_{n=-\infty}^{\infty} x_1(n)x_2^*(n) = \frac{1}{2\pi j} \oint_C X_1(v)X_2^*\left(\frac{1}{v^*}\right) v^{-1} dv \qquad (2.114)$$

where x^* denotes the complex conjugate of x and C is a contour contained in the intersection of the convergence regions of $X_1(v)$ and $X_2^*\left(\frac{1}{v^*}\right)$.

PROOF

We begin by noting that

$$\sum_{n=-\infty}^{\infty} x(n) = X(z)|_{z=1} \tag{2.115}$$

Therefore

$$\sum_{n=-\infty}^{\infty} x_1(n)x_2^*(n) = \mathcal{Z}\{x_1(n)x_2^*(n)\}\big|_{z=1} \tag{2.116}$$

By using equation (2.106) and the complex conjugation property in equation (2.96), we have that the above equation implies that

$$\sum_{n=-\infty}^{\infty} x_1(n)x_2^*(n) = \frac{1}{2\pi j} \oint_C X_1(v) X_2^* \left(\frac{1}{v^*}\right) v^{-1} dv \tag{2.117}$$

$\square$

2.4.12 Table of basic z transforms

Table 2.1 contains some commonly used sequences and their corresponding z transforms, along with their regions of convergence.

2.5 Transfer functions

As we have seen in Chapter 1, a discrete-time linear system can be characterized by a difference equation. In this section, we show how the z transform can be used to solve difference equations, and therefore characterize linear systems.

The general form of a difference equation associated with a linear system is given by equation (1.51), which we rewrite here for convenience

$$\sum_{i=0}^{N} a_i y(n-i) - \sum_{l=0}^{M} b_l x(n-l) = 0 \tag{2.118}$$

Applying the z transform on both sides and using the linearity property, we find that

$$\sum_{i=0}^{N} a_i \mathcal{Z}\{y(n-i)\} - \sum_{l=0}^{M} b_l \mathcal{Z}\{x(n-l)\} = 0 \tag{2.119}$$

Applying the time-shift theorem, we obtain

$$\sum_{i=0}^{N} a_i z^{-i} Y(z) - \sum_{l=0}^{M} b_l z^{-l} X(z) = 0 \tag{2.120}$$

Table 2.1. *The z transforms of commonly used sequences.*

$x(n)$	$X(z)$	Convergence region
$\delta(n)$	1	$z \in \mathbb{C}$
$u(n)$	$\dfrac{z}{(z-1)}$	$\lvert z \rvert > 1$
$(-a)^n u(n)$	$\dfrac{z}{(z+a)}$	$\lvert z \rvert > a$
$nu(n)$	$\dfrac{z}{(z-1)^2}$	$\lvert z \rvert > 1$
$n^2 u(n)$	$\dfrac{z(z+1)}{(z-1)^3}$	$\lvert z \rvert > 1$
$e^{an} u(n)$	$\dfrac{z}{(z-e^a)}$	$\lvert z \rvert > e^a$
$\cos(\omega n)u(n)$	$\dfrac{z[z-\cos(\omega)]}{z^2 - 2z\cos(\omega)+1}$	$\lvert z \rvert > 1$
$\sin(\omega n)u(n)$	$\dfrac{z\sin(\omega)}{z^2 - 2z\cos(\omega)+1}$	$\lvert z \rvert > 1$
$\dfrac{1}{n}u(n-1)$	$\ln\left(\dfrac{z}{z-1}\right)$	$\lvert z \rvert > 1$
$\sin(\omega n + \theta)u(n)$	$\dfrac{z^2 \sin(\theta) + z\sin(\omega - \theta)}{z^2 - 2z\cos(\omega)+1}$	$\lvert z \rvert > 1$
$e^{an}\cos(\omega n)u(n)$	$\dfrac{z^2 - ze^a \cos(\omega)}{z^2 - 2ze^a \cos(\omega) + e^{2a}}$	$\lvert z \rvert > e^a$
$e^{an}\sin(\omega n)u(n)$	$\dfrac{ze^a \sin(\omega)}{z^2 - 2ze^a \cos(\omega) + e^{2a}}$	$\lvert z \rvert > e^a$

Therefore, for a linear system, given $X(z)$, the z-transform representation of the input, and the coefficients of its difference equation, we can use equation (2.120) to find $Y(z)$, the z transform of the output. Applying the inverse z-transform relation in equation (2.36), the output $y(n)$ can be computed for all n.[1]

---

[1] One should note that, since equation (2.120) uses z transforms, which consist of summations for $-\infty < n < \infty$, then the system has to be describable by a difference equation for $-\infty < n < \infty$. This is the case only for initially relaxed systems, that is, systems that produce no output if the input is zero for $-\infty < n < \infty$. In our case, this does not restrict the applicability of equation (2.120), because we are only interested in linear systems, which, as seen in Chapter 1, must be initially relaxed.

Making $a_0 = 1$, without loss of generality, we then define

$$H(z) = \frac{Y(z)}{X(z)} = \frac{\displaystyle\sum_{l=0}^{M} b_l z^{-l}}{1 + \displaystyle\sum_{i=1}^{N} a_i z^{-i}} \tag{2.121}$$

as the transfer function of the system relating the output $Y(z)$ to the input $X(z)$.

Applying the convolution theorem to equation (2.121), we have that

$$Y(z) = H(z)X(z) \longleftrightarrow y(n) = h(n) * x(n) \tag{2.122}$$

that is, the transfer function of the system is the z transform of its impulse response. Indeed, equations (2.120) and (2.121) are the equivalent expressions of the convolution sum in the z-transform domain when the system is described by a difference equation.

Equation (2.121) gives the transfer function for the general case of recursive (IIR) filters. For nonrecursive (FIR) filters, all $a_i = 0$, for $i = 1, \ldots, N$, and the transfer function simplifies to

$$H(z) = \sum_{l=0}^{M} b_l z^{-l} \tag{2.123}$$

Transfer functions are widely used to characterize discrete-time linear systems. We can describe a transfer function through its poles p_i and zeros z_l, yielding

$$H(z) = H_0 \frac{\displaystyle\prod_{l=1}^{M}(1 - z^{-1} z_l)}{\displaystyle\prod_{i=1}^{N}(1 - z^{-1} p_i)} = H_0 z^{N-M} \frac{\displaystyle\prod_{l=1}^{M}(z - z_l)}{\displaystyle\prod_{i=1}^{N}(z - p_i)} \tag{2.124}$$

As discussed in Section 2.2, for a causal stable system, the convergence region of the z transform of its impulse response must include the unit circle. Indeed, this result is more general as for *any* stable system, the convergence region of its transfer function must necessarily include the unit circle. We can show this by noting that, for z_0 on the unit circle ($|z_0| = 1$), we have

$$|H(z_0)| = \left| \sum_{n=-\infty}^{\infty} z_0^{-n} h(n) \right| \leq \sum_{n=-\infty}^{\infty} |z_0^{-n} h(n)| = \sum_{n=-\infty}^{\infty} |h(n)| < \infty \tag{2.125}$$

which implies that $H(z)$ converges on the unit circle. Since in the case of a causal system the convergence region of the transfer function is defined by $|z| > r_1$, then all the poles of a causal stable system must be inside the unit circle. For a noncausal stable system, since the convergence region is defined for $|z| < r_2$, then all its poles must be outside the unit circle, with the possible exception of a pole at $z = 0$.

In the following section, we present a numerical method for assessing the stability of a linear system without explicitly determining the positions of its poles.

2.6 Stability in the z domain

In this section, we present a method for determining whether the roots of a polynomial are inside the unit circle of the complex plane. This method can be used to assess the BIBO stability of a causal discrete-time system.[2]

Given a polynomial in z

$$D(z) = a_n + a_{n-1}z + \cdots + a_0 z^n \tag{2.126}$$

with $a_0 > 0$, we have that a necessary and sufficient condition for its zeros (the poles of the given transfer function) to be inside the unit circle of the z plane is given by the following algorithm:

(i) Make $D_0(z) = D(z)$.
(ii) For $k = 0, 1, \ldots, n - 2$:

    (a) Form the polynomial $D_k^i(z)$ such that

$$D_k^i(z) = z^{n+k} D_k(z^{-1}) \tag{2.127}$$

    (b) Compute α_k and $D_{k+1}(z)$, such that

$$D_k(z) = \alpha_k D_k^i(z) + D_{k+1}(z) \tag{2.128}$$

    where the terms in z^j, $j = 0, \ldots, k$, of $D_{k+1}(z)$ are zero. In other words, $D_{k+1}(z)$ is the remainder of the division of $D_k(z)$ by $D_k^i(z)$, when it is performed upon the terms of smallest degree.

(iii) All the roots of $D(z)$ are inside the unit circle if the following conditions are satisfied:

- $D(1) > 0$
- $D(-1) > 0$ for n even and $D(-1) < 0$ for n odd
- $|\alpha_k| < 1$, for $k = 0, 1, \ldots, n - 2$.

EXAMPLE 2.9
Test the stability of the causal systems whose transfer functions possess the following denominator polynomials:

(a) $D(z) = 8z^4 + 8z^3 + 2z^2 - z - 1$
(b) $D(z) = 8z^4 + 4z^3 + 2z^2 - z - 1$

---

[2] There are several such methods described in the literature (Jury, 1973). We have chosen to present this method in particular because it is based on polynomial division, which we consider a very important tool in the analysis and design of discrete-time systems.

SOLUTION

(a) $D(z) = 8z^4 + 8z^3 + 2z^2 - z - 1$

- $D(1) = 16 > 0$.
- $n = 4$ is even and $D(-1) = 2 > 0$.
- Computation of α_0, α_1, and α_2:

$$D_0(z) = D(z) = 8z^4 + 8z^3 + 2z^2 - z - 1 \tag{2.129}$$

$$D_0^i(z) = z^4(8z^{-4} + 8z^{-3} + 2z^{-2} - z^{-1} - 1)$$

$$= 8 + 8z + 2z^2 - z^3 - z^4 \tag{2.130}$$

Since $D_0(z) = \alpha_0 D_0^i(z) + D_1(z)$,

$$
\begin{array}{c|c}
-1 - z + 2z^2 + 8z^3 + 8z^4 & 8 + 8z + 2z^2 - z^3 - z^4 \\
+1 + z + \frac{1}{4}z^2 - \frac{1}{8}z^3 - \frac{1}{8}z^4 & \overline{\quad -\frac{1}{8} \quad} \\
\hline
\frac{9}{4}z^2 + \frac{63}{8}z^3 + \frac{63}{8}z^4 &
\end{array}
$$

then $\alpha_0 = -\frac{1}{8}$ and

$$D_1(z) = \frac{9}{4}z^2 + \frac{63}{8}z^3 + \frac{63}{8}z^4 \tag{2.131}$$

$$D_1^i(z) = z^{4+1}\left(\frac{9}{4}z^{-2} + \frac{63}{8}z^{-3} + \frac{63}{8}z^{-4}\right) = \frac{63}{8}z + \frac{63}{8}z^2 + \frac{9}{4}z^3 \tag{2.132}$$

Since $D_1(z) = \alpha_1 D_1^i(z) + D_2(z)$,

$$
\begin{array}{c|c}
+\frac{9}{4}z^2 + \frac{63}{8}z^3 + \frac{63}{8}z^4 & \frac{63}{8}z + \frac{63}{8}z^2 + \frac{9}{4}z^3 \\
0 & 0 \\
\hline
+\frac{9}{4}z^2 + \frac{63}{8}z^3 + \frac{63}{8}z^4 &
\end{array}
$$

then $\alpha_1 = 0$ and

$$D_2(z) = \frac{9}{4}z^2 + \frac{63}{8}z^3 + \frac{63}{8}z^4 \tag{2.133}$$

$$D_2^i(z) = z^{4+2}\left(\frac{9}{4}z^{-2} + \frac{63}{8}z^{-3} + \frac{63}{8}z^{-4}\right) = \frac{63}{8}z^2 + \frac{63}{8}z^3 + \frac{9}{4}z^4 \tag{2.134}$$

Since $D_2(z) = \alpha_2 D_2^i(z) + D_3(z)$, we have that $\alpha_2 = \frac{9}{4}/\frac{63}{8} = \frac{2}{7}$.
Thus,

$$|\alpha_0| = \frac{1}{8} < 1, \quad |\alpha_1| = 0 < 1, \quad |\alpha_2| = \frac{2}{7} < 1 \tag{2.135}$$

and, consequently, the system is stable.

(b) $D(z) = 8z^4 + 4z^3 + 2z^2 - z - 1$

- $D(1) = 12 > 0$
- $n = 4$ is even and $D(-1) = 6 > 0$
- Computation of α_0, α_1, and α_2:

$$D_0(z) = D(z) = 8z^4 + 4z^3 + 2z^2 - z - 1 \tag{2.136}$$
$$D_0^i(z) = z^4(8z^{-4} + 4z^{-3} + 2z^{-2} - z^{-1} - 1)$$
$$= 8 + 4z + 2z^2 - z^3 - z^4 \tag{2.137}$$

Since $D_0(z) = \alpha_0 D_0^i(z) + D_1(z)$,

$$
\begin{array}{c|c}
-1 - z + 2z^2 + 4z^3 + 8z^4 & 8 + 4z + 2z^2 - z^3 - z^4 \\
+1 + \frac{1}{2}z + \frac{1}{4}z^2 - \frac{1}{8}z^3 - \frac{1}{8}z^4 & -\frac{1}{8} \\
\hline
-\frac{1}{2}z + \frac{9}{4}z^2 + \frac{31}{8}z^3 + \frac{63}{8}z^4 &
\end{array}
$$

then $\alpha_0 = -\frac{1}{8}$ and

$$D_1(z) = -\frac{1}{2}z + \frac{9}{4}z^2 + \frac{31}{8}z^3 + \frac{63}{8}z^4 \tag{2.138}$$
$$D_1^i(z) = z^{4+1}\left(-\frac{1}{2}z^{-1} + \frac{9}{4}z^{-2} + \frac{31}{8}z^{-3} + \frac{63}{8}z^{-4}\right)$$
$$= -\frac{1}{2}z^4 + \frac{9}{4}z^3 + \frac{31}{8}z^2 + \frac{63}{8}z \tag{2.139}$$

Since $D_1(z) = \alpha_1 D_1^i(z) + D_2(z)$,

$$
\begin{array}{c|c}
-\frac{1}{2}z + \frac{9}{4}z^2 + \frac{31}{8}z^3 + \frac{63}{8}z^4 & \frac{63}{8}z + \frac{31}{8}z^2 + \frac{9}{4}z^3 - \frac{1}{2}z^4 \\
+\frac{1}{2}z + \frac{31}{126}z^2 + \frac{1}{7}z^3 - \frac{2}{63}z^4 & -\frac{4}{63} \\
\hline
2.496z^2 + 4.018z^3 + 7.844z^4 &
\end{array}
$$

then $\alpha_1 = -\frac{4}{63}$ and

$$D_2(z) = 2.496z^2 + 4.018z^3 + 7.844z^4 \tag{2.140}$$
$$D_2^i(z) = z^{4+2}(2.496z^{-2} + 4.018z^{-3} + 7.844z^{-4})$$
$$= 2.496z^4 + 4.018z^3 + 7.844z^2 \tag{2.141}$$

Since $D_2(z) = \alpha_2 D_2^i(z) + D_3(z)$, we have that $\alpha_2 = \frac{2.496}{7.844} = 0.3182$. Thus,

$$|\alpha_0| = \frac{1}{8} < 1, \quad |\alpha_1| = \frac{4}{63} < 1, \quad |\alpha_2| = 0.3182 < 1 \tag{2.142}$$

and, consequently, the system is stable.

$\triangle$

The complete derivation of the above algorithm, as well as a method to determine the number of roots a polynomial $D(z)$ has inside the unit circle, can be found in Jury (1973).

2.7 Frequency response

As mentioned in Section 2.1, when an exponential z^n is input to a linear system with impulse response $h(n)$, then its output is an exponential $H(z)z^n$. Since, as seen above, stable systems are guaranteed to have the z transform on the unit circle, it is natural to try to characterize these systems on the unit circle. Complex numbers on the unit circle are of the form $z = e^{j\omega}$, for $0 \leq \omega \leq 2\pi$. This implies that the corresponding exponential sequence is a sinusoid $x(n) = e^{j\omega n}$. Therefore, we can state that if we input a sinusoid $x(n) = e^{j\omega n}$ to a linear system, then its output is also a sinusoid of the same frequency, that is

$$y(n) = H\left(e^{j\omega}\right)e^{j\omega n} \tag{2.143}$$

If $H(e^{j\omega})$ is a complex number with magnitude $|H(e^{j\omega})|$ and phase $\Theta(\omega)$, then $y(n)$ can be expressed as

$$y(n) = H(e^{j\omega})e^{j\omega n} = |H(e^{j\omega})|e^{j\Theta(\omega)}e^{j\omega n} = |H(e^{j\omega})|e^{j\omega n + j\Theta(\omega)} \tag{2.144}$$

indicating that the output of a linear system for a sinusoidal input is a sinusoid of the same frequency, but with its amplitude multiplied by $|H(e^{j\omega})|$ and phase increased by $\Theta(\omega)$. Thus, when we characterize a linear system in terms of $H(e^{j\omega})$, we are in fact specifying, for every frequency ω, the effect that the linear system has on the input signal's amplitude and phase. Therefore, $H(e^{j\omega})$ is commonly known as the frequency response of the system.

It is important to emphasize that $H(e^{j\omega})$ is the value of the z transform $H(z)$ on the unit circle. This implies that we need to specify it only for one turn around the unit circle, that is, for $0 \leq \omega \leq 2\pi$. Indeed, since, for $k \in \mathbb{Z}$

$$H(e^{j(\omega+2\pi k)}) = H(e^{j2\pi k}e^{j\omega}) = H(e^{j\omega}) \tag{2.145}$$

then $H(e^{j\omega})$ is periodic with period 2π.

Another important characteristic of a linear discrete-time system is its group delay. This is defined as the derivative of the phase of its frequency response,

$$\tau(\omega) = -\frac{d\Theta(\omega)}{d\omega} \tag{2.146}$$

The group delay $\tau(\omega)$ gives the delay, in samples, introduced by the system to a sinusoid of frequency ω.

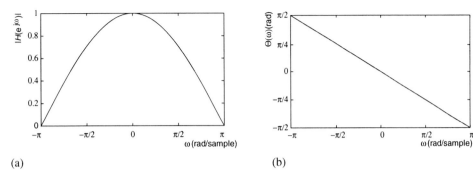

(a)   (b)

Figure 2.3 Frequency response of the moving average filter: (a) magnitude response; (b) phase response.

EXAMPLE 2.10
Find the frequency response and the group delay of the FIR filter characterized by the following difference equation

$$y(n) = \frac{x(n) + x(n-1)}{2} \tag{2.147}$$

SOLUTION
Taking the z transform of $y(n)$, we find

$$Y(z) = \frac{X(z) + z^{-1}X(z)}{2} = \frac{1}{2}(1 + z^{-1})X(z) \tag{2.148}$$

and then the transfer function of the system is

$$H(z) = \frac{1}{2}(1 + z^{-1}) \tag{2.149}$$

Making $z = e^{j\omega}$, the frequency response of the system becomes

$$H(e^{j\omega}) = \frac{1}{2}(1 + e^{-j\omega}) = \frac{1}{2}e^{-j\frac{\omega}{2}}\left(e^{j\frac{\omega}{2}} + e^{-j\frac{\omega}{2}}\right) = e^{-j\frac{\omega}{2}}\cos\left(\frac{\omega}{2}\right) \tag{2.150}$$

Since $\Theta(\omega) = -\frac{\omega}{2}$, then, from equation (2.146), the group delay, $\tau(\omega) = \frac{1}{2}$. This implies that the system delays all sinusoids equally by half a sample.

The magnitude and phase responses of $H(e^{j\omega})$ are depicted in Figure 2.3. Note that the frequency response is plotted for $-\pi \le \omega \le \pi$, rather than for $0 \le \omega \le 2\pi$. In practice, those two ranges are equivalent, since both comprise one period of $H(e^{j\omega})$.

$\triangle$

EXAMPLE 2.11
A discrete-time system with impulse response $h(n) = \left(\frac{1}{2}\right)^n u(n)$ is excited with $x(n) = \sin(\omega_0 n + \phi)$. Find the output $y(n)$ using the frequency response of the system.

SOLUTION

Since

$$x(n) = \sin(\omega_0 n + \phi) = \frac{e^{j(\omega_0 n + \phi)} - e^{-j(\omega_0 n + \phi)}}{2j} \qquad (2.151)$$

then, the output $y(n) = \mathcal{H}\{x(n)\}$ is

$$y(n) = \mathcal{H}\left\{\frac{e^{j(\omega_0 n + \phi)} - e^{-j(\omega_0 n + \phi)}}{2j}\right\}$$

$$= \frac{1}{2j}\left(\mathcal{H}\{e^{j(\omega_0 n + \phi)}\} - \mathcal{H}\{e^{-j(\omega_0 n + \phi)}\}\right)$$

$$= \frac{1}{2j}\left(H(e^{j\omega_0})e^{j(\omega_0 n + \phi)} - H(e^{-j\omega_0})e^{-j(\omega_0 n + \phi)}\right)$$

$$= \frac{1}{2j}\left(|H(e^{j\omega_0})|e^{j\Theta(\omega_0)}e^{j(\omega_0 n + \phi)} - |H(e^{-j\omega_0})|e^{-j\Theta(\omega_0)}e^{-j(\omega_0 n + \phi)}\right)$$

$$= |H(e^{j\omega_0})|\left[\frac{e^{j(\omega_0 n + \phi + \Theta(\omega_0))} - e^{-j(\omega_0 n + \phi + \Theta(\omega_0))}}{2j}\right]$$

$$= |H(e^{j\omega_0})|\sin(\omega_0 n + \phi + \Theta(\omega_0)) \qquad (2.152)$$

Since the system transfer function is

$$H(z) = \sum_{n=0}^{\infty}\left(\frac{1}{2}\right)^n z^{-n} = \frac{1}{1 - \frac{1}{2}z^{-1}} \qquad (2.153)$$

we have that

$$H(e^{j\omega}) = \frac{1}{1 - \frac{1}{2}e^{-j\omega}} = \frac{1}{\sqrt{\frac{5}{4} - \cos\omega}}e^{-j\arctan\left(\frac{\sin\omega}{2 - \cos\omega}\right)} \qquad (2.154)$$

and then

$$|H(e^{j\omega})| = \frac{1}{\sqrt{\frac{5}{4} - \cos\omega}} \qquad (2.155)$$

$$\Theta(\omega) = -\arctan\left(\frac{\sin\omega}{2 - \cos\omega}\right) \qquad (2.156)$$

Substituting these values of $|H(e^{j\omega})|$ and $\Theta(\omega)$ in equation (2.152), the output $y(n)$ becomes

$$y(n) = \frac{1}{\sqrt{\frac{5}{4} - \cos\omega_0}}\sin\left[\omega_0 n + \phi - \arctan\left(\frac{\sin\omega_0}{2 - \cos\omega_0}\right)\right] \qquad (2.157)$$

$\triangle$

In general, when we design a discrete-time system, we have to satisfy pre-determined magnitude, $|H(e^{j\omega})|$, and phase, $\Theta(\omega)$, characteristics. One should note that, when processing a continuous-time signal using a discrete-time system, we should translate the analog frequency Ω to the discrete-time frequency ω that is restricted to the interval $[-\pi, \pi]$. This can be done by noting that if an analog sinusoid $x_a(t) = e^{j\Omega t}$ is sampled as in equation (1.75) to generate a sinusoid $e^{j\omega n}$, that is, if $T = \frac{2\pi}{\Omega_s}$ is the sampling frequency, then

$$e^{j\omega n} = x(n) = x_a(nT) = e^{j\Omega nT} \tag{2.158}$$

Therefore, one can deduce that the relation between the digital frequency ω and the analog frequency Ω is

$$\omega = \Omega T = 2\pi \frac{\Omega}{\Omega_s} \tag{2.159}$$

indicating that the frequency interval $[-\pi, \pi]$ for the discrete-time frequency response corresponds to the frequency interval $[-\frac{\Omega_s}{2}, \frac{\Omega_s}{2}]$ in the analog domain.

EXAMPLE 2.12

The sixth-order discrete-time lowpass elliptic filter, whose frequency response is shown in Figure 2.4, is used to process an analog signal in a similar scheme to the one depicted in Figure 1.9. If the sampling frequency used in the analog-to-digital conversion is 8000 Hz, determine the passband of the equivalent analog filter. Consider the passband as the frequency range when the magnitude response of the filter is within 0.1 dB of its maximum value.

SOLUTION

From Figure 2.4c, we see that the digital bandwidth in which the magnitude response of the system is within 0.1 dB of its maximum value is approximately from $\omega_{p_1} = 0.755\pi$ rad/sample to $\omega_{p_2} = 0.785\pi$ rad/sample. As the sampling frequency is

$$f_s = \frac{\Omega_s}{2\pi} = 8000 \text{ Hz} \tag{2.160}$$

then the analog passband is such that

$$\Omega_{p_1} = 0.755\pi \frac{\Omega_s}{2\pi} = 0.755\pi \times 8000 = 6040\pi \text{ rad/s} \Rightarrow f_{p_1} = \frac{\Omega_{p_1}}{2\pi} = 3020 \text{ Hz} \tag{2.161}$$

$$\Omega_{p_2} = 0.785\pi \frac{\Omega_s}{2\pi} = 0.785\pi \times 8000 = 6280\pi \text{ rad/s} \Rightarrow f_{p_2} = \frac{\Omega_{p_2}}{2\pi} = 3140 \text{ Hz} \tag{2.162}$$

$\triangle$

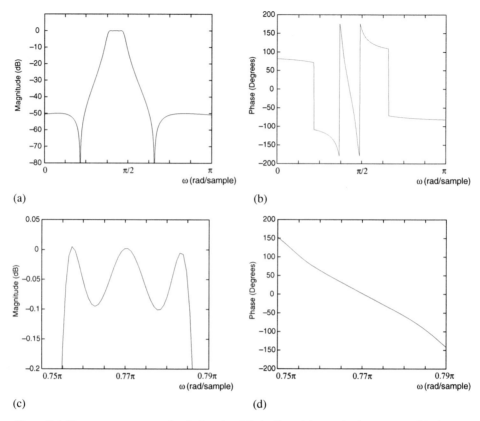

Figure 2.4 Frequency response of a sixth-order elliptic filter: (a) magnitude response; (b) phase response; (c) magnitude response in the passband; (d) phase response in the passband.

The positions of the poles and zeros of a transfer function are very useful in the determination of its characteristics. For example, one can determine the frequency response $H(e^{j\omega})$ using a geometric method. Expressing $H(z)$ as a function of its poles and zeros as in equation (2.124), we have that $H(e^{j\omega})$ becomes

$$H(e^{j\omega}) = H_0 e^{j\omega(N-M)} \frac{\displaystyle\prod_{l=1}^{M}(e^{j\omega} - z_l)}{\displaystyle\prod_{i=1}^{N}(e^{j\omega} - p_i)} \tag{2.163}$$

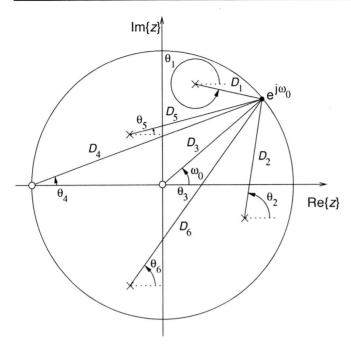

Figure 2.5 Determination of the frequency response of $H(z)$ from the position of its poles ($\times$) and zeros ($\circ$).

The magnitude and phase responses of $H(\mathrm{e}^{\mathrm{j}\omega})$ are then

$$|H(\mathrm{e}^{\mathrm{j}\omega})| = |H_0| \frac{\prod\limits_{l=1}^{M} |\mathrm{e}^{\mathrm{j}\omega} - z_l|}{\prod\limits_{i=1}^{N} |\mathrm{e}^{\mathrm{j}\omega} - p_i|} \tag{2.164}$$

$$\Theta(\omega) = \sum_{l=1}^{M} \angle(\mathrm{e}^{\mathrm{j}\omega} - z_l) - \sum_{i=1}^{N} \angle(\mathrm{e}^{\mathrm{j}\omega} - p_i) \tag{2.165}$$

where $\angle z$ denotes the angle of the complex number z. The terms of the form $|\mathrm{e}^{\mathrm{j}\omega} - c|$ represent the distance between the point $\mathrm{e}^{\mathrm{j}\omega}$ on the unit circle and the complex number c. The terms of the form $\angle(\mathrm{e}^{\mathrm{j}\omega} - c)$ represent the angle of the line segment joining $\mathrm{e}^{\mathrm{j}\omega}$ and c and the real axis, measured in the counterclockwise direction.

For example, for $H(z)$ having its poles and zeros as in Figure 2.5, we have that

$$|H(\mathrm{e}^{\mathrm{j}\omega_0})| = \frac{D_3 D_4}{D_1 D_2 D_5 D_6} \tag{2.166}$$

$$\Theta(\omega_0) = \theta_3 + \theta_4 - \theta_1 - \theta_2 - \theta_5 - \theta_6 \tag{2.167}$$

2.8   Fourier transform

In the previous section, we characterized linear discrete-time systems using the frequency response, which describes the behavior of a system when the input is a complex sinusoid. In this section, we present the Fourier transform of discrete-time signals, which is a generalization of the concept of frequency response. It is equivalent to the decomposition of a discrete-time signal into an infinite sum of complex discrete-time sinusoids.

In Chapter 1, when deducing the sampling theorem, we formed, from the discrete-time signal $x(n)$, a continuous-time signal $x_i(t)$ consisting of a train of impulses with amplitude $x(n)$ at $t = nT$ (see Figure 1.5c). Its expression is given by equation (1.81), which is repeated here for convenience

$$x_i(t) = \sum_{n=-\infty}^{\infty} x(n)\delta(t - nT) \tag{2.168}$$

Since the Fourier transform of $\delta(t - Tn)$ is $e^{-j\Omega Tn}$, the Fourier transform of the continuous-time signal $x_i(t)$ becomes

$$X_i(j\Omega) = \sum_{n=-\infty}^{\infty} x(n)e^{-j\Omega Tn} \tag{2.169}$$

where Ω represents the analog frequency. Comparing this equation with the z-transform definition given in equation (2.5), we conclude that the Fourier transform of $x_i(t)$ is such that

$$X_i(j\Omega) = X(e^{j\Omega T}) \tag{2.170}$$

This equation means that the Fourier transform at a frequency Ω of the continuous-time signal $x_i(t)$, that is generated by replacing each sample of the discrete-time signal $x(n)$ by an impulse of the same amplitude located at $t = nT$, is equal to the z transform of the signal $x(n)$ at $z = e^{j\Omega T}$. This fact implies that $X(e^{j\omega})$ holds the information about the frequency content of the signal $x_i(t)$, and therefore represents the frequency content of the discrete-time signal $x(n)$.

We can show that $X(e^{j\omega})$ does indeed represent the frequency content of $x(n)$ by applying the inverse z-transform formula in equation (2.36) with C being the closed contour $z = e^{j\omega}$, for $-\pi \leq \omega \leq \pi$, resulting in

$$
\begin{aligned}
x(n) &= \frac{1}{2\pi j} \oint_C X(z) z^{n-1} dz \\
&= \frac{1}{2\pi j} \oint_{z=e^{j\omega}} X(z) z^{n-1} dz \\
&= \frac{1}{2\pi j} \int_{-\pi}^{\pi} X(e^{j\omega}) e^{j\omega(n-1)} j e^{j\omega} d\omega \\
&= \frac{1}{2\pi} \int_{-\pi}^{\pi} X(e^{j\omega}) e^{j\omega n} d\omega
\end{aligned}
\tag{2.171}
$$

indicating that the discrete-time signal $x(n)$ can be expressed as an infinite summation of sinusoids, where the sinusoid of frequency ω, $e^{j\omega n}$, has a complex amplitude proportional to $X(e^{j\omega})$. Thus computing $X(e^{j\omega})$ is equivalent to decomposing the discrete-time signal $x(n)$ into a sum of complex discrete-time sinusoids.

This interpretation of $X(e^{j\omega})$ parallels the one of the Fourier transform of a continuous-time signal $x_a(t)$. The direct and inverse Fourier transforms of a continuous-time signal are given by

$$
\left.
\begin{aligned}
X_a(j\Omega) &= \int_{-\infty}^{\infty} x_a(t) e^{-j\Omega t} dt \\
x_a(t) &= \frac{1}{2\pi} \int_{-\infty}^{\infty} X_a(j\Omega) e^{j\Omega t} d\Omega
\end{aligned}
\right\}
\tag{2.172}
$$

This pair of equations indicates that a continuous-time signal $x_a(t)$ can be expressed as an infinite summation of continuous-time sinusoids, where the sinusoid of frequency Ω, $e^{j\Omega t}$, has a complex amplitude proportional to $X_a(j\Omega)$. Then computing $X_a(j\Omega)$ is equivalent to decomposing the continuous-time signal into a sum of complex continuous-time sinusoids.

From the above discussion, we see that $X(e^{j\omega})$ defines a Fourier transform of the discrete-time signal $x(n)$, with its inverse given by equation (2.171). Indeed, the direct and inverse Fourier transforms of a sequence $x(n)$ are formally defined as

$$
\left.
\begin{aligned}
X(e^{j\omega}) &= \sum_{n=-\infty}^{\infty} x(n) e^{-j\omega n} \\
x(n) &= \frac{1}{2\pi} \int_{-\pi}^{\pi} X(e^{j\omega}) e^{j\omega n} d\omega
\end{aligned}
\right\}
\tag{2.173}
$$

Naturally, the Fourier transform $X(e^{j\omega})$ of a discrete-time signal $x(n)$ is periodic with period 2π, since

$$
X(e^{j\omega}) = X(e^{j(\omega + 2\pi k)}), \quad \forall k \in \mathbb{Z}
\tag{2.174}
$$

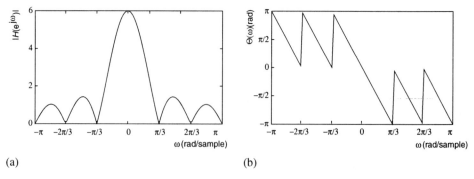

Figure 2.6 Fourier transform of the sequence in Example 2.13: (a) magnitude response; (b) phase response.

Therefore, the Fourier transform of a discrete-time signal requires specification only for a range of 2π, as, for example, $\omega \in [-\pi, \pi]$ or $\omega \in [0, 2\pi]$.

EXAMPLE 2.13

Compute the Fourier transform of the sequence

$$
x(n) = \begin{cases} 1, & 0 \le n \le 5 \\ 0, & \text{otherwise} \end{cases} \tag{2.175}
$$

SOLUTION

$$
X(e^{j\omega}) = \sum_{k=0}^{5} e^{-j\omega k} = \frac{1 - e^{-6j\omega}}{1 - e^{-j\omega}} = \frac{e^{-3j\omega}(e^{3j\omega} - e^{-3j\omega})}{e^{-j\frac{\omega}{2}}(e^{j\frac{\omega}{2}} - e^{-j\frac{\omega}{2}})} = e^{-j\frac{5\omega}{2}} \frac{\sin(3\omega)}{\sin\left(\frac{\omega}{2}\right)} \tag{2.176}
$$

The magnitude and phase responses of $X(e^{j\omega})$ are depicted in Figures 2.6a and 2.6b, respectively. Note that in Figure 2.6b the phase has been wrapped to fit in the interval $[-\pi, \pi]$.

$\triangle$

One should note that, in order for the Fourier transform of a discrete-time signal to exist, its z transform must converge for $|z| = 1$. From the discussion in Section 2.5, we have seen that whenever

$$
\sum_{n=-\infty}^{\infty} |x(n)| < \infty \tag{2.177}
$$

then the z transform converges on the unit circle, and therefore the Fourier transform exists. One example when equation (2.177) does not hold and the Fourier transform does not exist is given by the sequence $x(n) = u(n)$. The proof of this statement is left to the reader in Exercise 2.13. Also, in Exercise 2.14 there is an example in which equation (2.177) does not hold but the Fourier transform does exist. Therefore, the condition in equation (2.177) is sufficient, but not necessary for the Fourier transform to exist. In addition, we have that not all sequences which have a Fourier transform have a z transform. For example, the z transform of any sequence is continuous in its convergence region (Churchill, 1975), and hence the sequences whose Fourier transforms are discontinuous functions of ω do not have a z transform, as also seen in Exercise 2.14.

In Chapter 1, equation (1.88), we gave a relation between the Fourier transform of the signal $x_i(t)$ in equation (1.81) and the one of the original analog signal $x_a(t)$, which we rewrite here for convenience

$$X_i(j\Omega) = \frac{1}{T} \sum_{k=-\infty}^{\infty} X_a\left(j\Omega - j\frac{2\pi}{T}k\right) \tag{2.178}$$

If the discrete-time signal $x(n)$ is such that $x(n) = x_a(nT)$, then we can use this equation to derive the relation between the Fourier transforms of the discrete-time and continuous-time signals, $X(e^{j\omega})$ and $X_a(j\Omega)$. In fact, from equation (2.170)

$$X_i(j\Omega) = X(e^{j\Omega T}) \tag{2.179}$$

and making the change of variables $\Omega = \frac{\omega}{T}$ in equation (2.178), yields

$$X(e^{j\omega}) = X_i\left(j\frac{\omega}{T}\right) = \frac{1}{T} \sum_{k=-\infty}^{\infty} X_a\left(j\frac{\omega - 2\pi k}{T}\right) \tag{2.180}$$

that is, $X(e^{j\omega})$ is composed of copies of $X_a(j\frac{\omega}{T})$ repeated in intervals of 2π. Also, as seen in Chapter 1, one can recover the analog signal $x_a(t)$ from $x(n)$ provided that $X_a(j\Omega) = 0$ for $|\Omega| > \frac{\Omega_s}{2} = \frac{\pi}{T}$.

2.9   Properties of the Fourier transform

As seen previously, the Fourier transform $X(e^{j\omega})$ of a sequence $x(n)$ is equal to its z transform $X(z)$ at $z = e^{j\omega}$. Therefore, most properties of the Fourier transform derive from the ones of the z transform by simple substitution of z by $e^{j\omega}$. In what follows, we state them without proof, except in the cases where the properties do not have a z-transform correspondent.

2.9.1   Linearity

$$k_1 x_1(n) + k_2 x_2(n) \longleftrightarrow k_1 X_1(e^{j\omega}) + k_2 X_2(e^{j\omega}) \tag{2.181}$$

2.9.2   Time-reversal

$$x(-n) \longleftrightarrow X(e^{-j\omega}) \tag{2.182}$$

2.9.3 Time-shift theorem

$$x(n + l) \longleftrightarrow e^{j\omega l} X(e^{j\omega}) \qquad (2.183)$$

where l is an integer.

2.9.4 Multiplication by an exponential

$$e^{j\omega_0 n} x(n) \longleftrightarrow X(e^{j(\omega - \omega_0)}) \qquad (2.184)$$

2.9.5 Complex differentiation

$$nx(n) \longleftrightarrow -j\frac{dX(e^{j\omega})}{d\omega} \qquad (2.185)$$

2.9.6 Complex conjugation

$$x^*(n) \longleftrightarrow X^*(e^{-j\omega}) \qquad (2.186)$$

2.9.7 Real and imaginary sequences

$$\text{Re}\{x(n)\} \longleftrightarrow \frac{1}{2}\left(X(e^{j\omega}) + X^*(e^{-j\omega})\right) \qquad (2.187)$$

$$\text{Im}\{x(n)\} \longleftrightarrow \frac{1}{2j}\left(X(e^{j\omega}) - X^*(e^{-j\omega})\right) \qquad (2.188)$$

If $x(n)$ is real, then $\text{Im}\{x(n)\} = 0$. Hence, from equation (2.188),

$$X(e^{j\omega}) = X^*(e^{-j\omega}) \qquad (2.189)$$

that is, the Fourier transform of a real sequence is conjugate symmetric. The following properties for real $x(n)$ follow directly from equation (2.189):

- The real part of the Fourier transform of a real sequence is even:

$$\text{Re}\{X(e^{j\omega})\} = \text{Re}\{X(e^{-j\omega})\} \qquad (2.190)$$

- The imaginary part of the Fourier transform of a real sequence is odd:

$$\text{Im}\{X(e^{j\omega})\} = -\text{Im}\{X(e^{-j\omega})\} \qquad (2.191)$$

- The magnitude of the Fourier transform of a real sequence is even:

$$|X(e^{j\omega})| = |X(e^{-j\omega})| \tag{2.192}$$

- The phase of the Fourier transform of a real sequence is odd:

$$\angle[X(e^{j\omega})] = -\angle[X(e^{-j\omega})] \tag{2.193}$$

Similarly, if $x(n)$ is imaginary, then $\text{Re}\{x(n)\} = 0$. Hence, from equation (2.187),

$$X(e^{j\omega}) = -X^*(e^{-j\omega}) \tag{2.194}$$

Properties similar to the ones in equations (2.190)–(2.193) can be deduced for imaginary sequences, which is left as an exercise to the reader.

2.9.8 Symmetric and antisymmetric sequences

Before presenting the properties of the Fourier transforms of symmetric and antisymmetric sequences, we give precise definitions for them:

- A symmetric (even) sequence is such that $x(n) = x(-n)$.
- An antisymmetric (odd) sequence is such that $x(n) = -x(-n)$.
- A conjugate symmetric sequence is such that $x(n) = x^*(-n)$.
- A conjugate antisymmetric sequence is such that $x(n) = -x^*(-n)$.

Therefore, the following properties hold:

- If $x(n)$ is real and symmetric, $X(e^{j\omega})$ is also real and symmetric.

PROOF

$$
\begin{aligned}
X(e^{j\omega}) &= \sum_{n=-\infty}^{\infty} x(n)e^{-j\omega n} \\
&= \sum_{n=-\infty}^{\infty} x(-n)e^{-j\omega n} \\
&= \sum_{m=-\infty}^{\infty} x(m)e^{j\omega m} \\
&= X(e^{-j\omega}) \\
&= \sum_{m=-\infty}^{\infty} (x(m)e^{j\omega m})^* \\
&= X^*(e^{j\omega}) \tag{2.195}
\end{aligned}
$$

Hence, if $X(e^{j\omega}) = X(e^{-j\omega})$, then $X(e^{j\omega})$ is even, and if $X(e^{j\omega}) = X^*(e^{j\omega})$, then $X(e^{j\omega})$ is real.

□

- If $x(n)$ is imaginary and even, then $X(e^{j\omega})$ is imaginary and even.
- If $x(n)$ is real and odd, then $X(e^{j\omega})$ is imaginary and odd.
- If $x(n)$ is imaginary and odd, then $X(e^{j\omega})$ is real and odd.
- If $x(n)$ is conjugate symmetric, then $X(e^{j\omega})$ is real.

PROOF

$$
\begin{aligned}
X(e^{j\omega}) &= \sum_{n=-\infty}^{\infty} x(n)e^{-j\omega n} \\
&= \sum_{n=-\infty}^{\infty} x^*(-n)e^{-j\omega n} \\
&= \sum_{m=-\infty}^{\infty} x^*(m)e^{j\omega m} \\
&= \sum_{m=-\infty}^{\infty} (x(m)e^{-j\omega m})^* \\
&= X^*(e^{j\omega})
\end{aligned}
$$

(2.196)

and then $X(e^{j\omega})$ is real.

□

- If $x(n)$ is conjugate antisymmetric, then $X(e^{j\omega})$ is imaginary.

2.9.9 Convolution theorem

$$
x_1(n) * x_2(n) \longleftrightarrow X_1(e^{j\omega})X_2(e^{j\omega})
$$

(2.197)

2.9.10 Product of two sequences

$$
\begin{aligned}
x_1(n)x_2(n) \longleftrightarrow &\frac{1}{2\pi}\int_{-\pi}^{\pi} X_1(e^{j\Omega})X_2(e^{j(\omega-\Omega)})d\Omega = \frac{1}{2\pi}\int_{-\pi}^{\pi} X_1(e^{j(\omega-\Omega)})X_2(e^{j\Omega})d\Omega \\
= &\quad X_1(e^{j\omega}) \circledast X_2(e^{j\omega})
\end{aligned}
$$

(2.198)

2.9.11 Parseval's theorem

$$
\sum_{n=-\infty}^{\infty} x_1(n)x_2^*(n) = \frac{1}{2\pi}\int_{-\pi}^{\pi} X_1(e^{j\omega})X_2^*(e^{j\omega})d\omega
$$

(2.199)

If we make $x_1(n) = x_2(n) = x(n)$, then Parseval's theorem becomes

$$
\sum_{n=-\infty}^{\infty} |x(n)|^2 = \frac{1}{2\pi}\int_{-\pi}^{\pi} |X(e^{j\omega})|^2 d\omega
$$

(2.200)

The left-hand side of this equation corresponds to the energy of the sequence $x(n)$ and its right-hand side corresponds to the energy of $X(e^{j\omega})$ divided by 2π. Hence, equation (2.200) means that the energy of a sequence is the same as the energy of its Fourier transform divided by 2π.

2.10   Transfer functions with MATLAB

The Signal Processing toolbox of MATLAB has several functions that are related to transfer functions. In this section, we briefly describe some of them.

- `abs`: Finds the absolute value of the elements in a matrix of complex numbers.
  Input parameter: The matrix of complex numbers **x**.
  Output parameter: The matrix **y** with the magnitudes of the elements of **x**.
  Example:
  ```
  x=[0.3+0.4i -0.5+i 0.33+0.47i]; abs(x);
  ```

- `angle`: Computes the phase of each element of a matrix of complex numbers.
  Input parameter: The matrix of complex numbers **x**.
  Output parameter: The matrix **y** with the phases of the elements of **x** in radians.
  Example:
  ```
  x=[0.3+0.4i -0.5+i 0.33+0.47i]; angle(x);
  ```

- `freqz`: Gives the frequency response of a linear discrete-time system specified by its transfer function.
  Input parameters (refer to equation (2.121)):

  - A vector **b** containing the coefficients of the numerator polynomial, b_l, $l = 0, \ldots, M$;
  - A vector **a** containing the coefficients of the denominator polynomial, a_i, $i = 0, \ldots, N$;
  - The number **n** of points around the upper half of the unit circle, or, alternatively, the vector **w** of points in which the response is evaluated.

  Output parameters:

  - A vector **h** containing the frequency response;
  - A vector **w** containing the frequency points at which the frequency response is evaluated;
  - With no input arguments, `freqz` plots the magnitude and unwrapped phase of the transfer function.

  Example:
  ```
  a=[1 0.8 0.64]; b=[1 1]; [h,w]=freqz(b,a,200);
  plot(w,abs(h)); figure(2); plot(w,angle(h));
  ```

- `freqspace`: Generates a vector with equally spaced points on the unit circle.
  Input parameters:

    - The number `n` of points on the upper half of the unit circle;
    - If `'whole'` is given as a parameter, the `n` points are equally spaced on the whole unit circle.

  Output parameter: The vector of frequency points in Hz.
  Example:
    ```
    a=[1 0.8 0.64]; b=[1 1]; w=pi*freqspace(200);
    freqz(b,a,w);
    ```

- `unwrap`: Unwraps the phase so that phase jumps are always smaller than 2π.
  Input parameter: The vector `p` containing the phase;
  Output parameter: The vector `q` containing the unwrapped phase.
  Example:
    ```
    p1=[0:0.01:6*pi]; p=[p1 p1]; plot(p);
    q=unwrap(p); figure(2); plot(q);
    ```

- `grpdelay`: Gives the group delay of a linear discrete-time system specified by its transfer function.
  Input parameters: same as for the `freqz` command.
  Output parameters:

    - A vector `gd` containing the frequency response;
    - A vector `w` containing the frequency points at which the frequency response is evaluated;
    - With no arguments, `grpdelay` plots the group delay of the transfer function versus the normalized frequency ($\pi \rightarrow 1$).

  Example:
    ```
    a=[1 0.8 0.64]; b=[1 1]; [gd,w]=grpdelay(b,a,200);
    plot(w,gd);
    ```

- `tf2zp`: Given a transfer function, returns the zero and pole locations. See also Section 4.7.

- `zp2tf`: Given the zero and pole locations, returns the transfer function coefficients, thus reversing the operation of the `tf2zp` command. See also Section 4.7.

- `zplane`: Given a transfer function, plots the zero and pole locations.
  Input parameters:

    - A row vector `b` containing the coefficients of the numerator polynomial in decreasing powers of z;
    - A row vector `a` containing the coefficients of the denominator polynomial in decreasing powers of z;

– If we input column vectors instead, they are interpreted as the vectors containing the zeros and poles.

Example 1:
```
a=[1 0.8 0.64 0.3 0.02]; b=[1 1 0.3]; zplane(b,a);
```
Example 2:
```
a=[1 0.8 0.64 0.3 0.02]; b=[1 1 0.3];
[z p k]=tf2zp(b,a); zplane(z,p);
```

• `residuez`: Computes the partial-fraction expansion of a transfer function. See also Section 4.7.

• See also the commands `conv` and `impz` described in Section 1.6, and the command `filter` described in Section 4.7.

2.11 Summary

In this chapter, we have studied the z and Fourier transforms, two very important tools in the analysis and design of linear discrete-time systems. We first defined the z transform and its inverse, and then we presented some interesting z-transform properties. Next, we introduced the concept of the transfer function as a way of characterizing linear discrete-time systems using the z transform. Then a criterion for determining the stability of discrete-time linear systems was presented. Linked with the concept of transfer functions, we defined the frequency response of a system and the Fourier transform of discrete-time signals, presenting some of its properties. We closed the chapter by describing some MATLAB functions related to z transforms, Fourier transforms, and discrete-time transfer functions.

2.12 Exercises

2.1   Compute the z transform of the following sequences, indicating their regions of convergence:

(a) $x(n) = \sin(\omega n + \theta)u(n)$

(b) $x(n) = \cos(\omega n)u(n)$

(c) $x(n) = \begin{cases} n, & 0 \leq n \leq 4 \\ 0, & n < 0 \text{ and } n > 4 \end{cases}$

(d) $x(n) = a^n u(-n)$

(e) $x(n) = e^{-\alpha n} u(n)$

(f) $x(n) = e^{-\alpha n} \sin(\omega n)u(n)$

(g) $x(n) = n^2 u(n)$

2.2   Prove equation (2.35).

2.3   Suppose that the transfer function of a digital filter is given by

$$\frac{(z-1)^2}{z^2+(m_1-m_2)z+(1-m_1-m_2)}$$

Plot a graph specifying the region of the $m_2 \times m_1$ plane in which the digital filter is stable.

2.4   Compute the time response of the system described by the following transfer function:

$$H(z) = \frac{(z-1)^2}{z^2-0.32z+0.8}$$

when the input signal is the unit step.

2.5   Determine the inverse z transform of the following functions of the complex variable z:

(a) $\dfrac{z}{z-0.8}$

(b) $\dfrac{z^2}{z^2-z+0.5}$

(c) $\dfrac{z^2+2z+1}{z^2-z+0.5}$

(d) $\dfrac{z^2}{(z-a)(z-1)}$

2.6   Compute the inverse z transform of the functions below. Suppose that the sequences are right handed and one sided.

(a) $X(z) = \sin\left(\dfrac{1}{z}\right)$

(b) $X(z) = \sqrt{\dfrac{z}{1+z}}$

2.7   Given a sequence $x(n)$, form a new sequence consisting of only the even samples of $x(n)$, that is, $y(n) = x(2n)$. Determine the z transform of $y(n)$ as a function of the z transform of $x(n)$, using the auxiliary sequence $(-1)^n x(n)$.

2.8   Compute the frequency response of a system with the following impulse response

$$h(n) = \begin{cases} (-1)^n, & |n| < N-1 \\ 0, & \text{otherwise} \end{cases}$$

2.9   Compute and plot the magnitude and phase of the frequency response of the systems described by the following difference equations:

(a) $y(n) = x(n) + 2x(n-1) + 3x(n-2) + 2x(n-3) + x(n-4)$

(b) $y(n) = y(n-1) + x(n)$

(c) $y(n) = x(n) + 3x(n-1) + 2x(n-2)$

2.10   Plot the magnitude and phase of the frequency response of the digital filters characterized by the following transfer functions:

(a) $H(z) = z^{-4} + 2z^{-3} + 2z^{-1} + 1$

(b) $H(z) = \dfrac{z^2 - 1}{z^2 - 1.2z + 0.95}$

2.11   If a digital filter has transfer function $H(z)$, compute the steady-state response of this system for an input of the type $x(n) = \sin(\omega n)u(n)$.

2.12   Compute the Fourier transform of each of the sequences in Exercise 2.1.

2.13   Show that the unit step $u(n)$ does not have a Fourier transform, by computing the summation in equation (2.173) up to n, and taking the limit as $n \to \infty$.

2.14   Compute $h(n)$, the inverse Fourier transform of

$$H(e^{j\omega}) = \begin{cases} -j, & \text{for } 0 \le \omega < \pi \\ j, & \text{for } -\pi \le \omega < 0 \end{cases}$$

and show that:

(a) Equation (2.177) does not hold.

(b) The sequence $h(n)$ does not have a z transform.

2.15   Determine whether the polynomials below can be the denominator of a causal stable filter:

(a) $z^5 + 2z^4 + z^3 + 2z^2 + z + 0.5$

(b) $z^6 - z^5 + z^4 + 2z^3 + z^2 + z + 0.25$

(c) $z^4 + 0.5z^3 - 2z^2 + 1.75z + 0.5$

2.16   Prove the properties of the Fourier transform of real sequences given by equations (2.190)–(2.193).

2.17   State and prove the properties of the Fourier transform of imaginary sequences that correspond to the ones given by equations (2.190)–(2.193).

2.18   Show that the direct–inverse Fourier transform pair of the correlation of two sequences is

$$\sum_{n=-\infty}^{\infty} x_1(n)x_2(n+l) \longleftrightarrow X_1(e^{-j\omega})X_2(e^{j\omega})$$

2.19   Show that the Fourier transform of an imaginary and odd sequence is real and odd.

2.20   Show that the Fourier transform of a conjugate antisymmetric sequence is imaginary.

2.21   We define the even and odd parts of a complex sequence $x(n)$ as

$$\mathcal{E}\{x(n)\} = \frac{x(n) + x^*(-n)}{2}$$

$$\mathcal{O}\{x(n)\} = \frac{x(n) - x^*(-n)}{2}$$

respectively. Show that

$$\mathcal{F}\{\mathcal{E}\{x(n)\}\} = \mathrm{Re}\{X(e^{j\omega})\}$$
$$\mathcal{F}\{\mathcal{O}\{x(n)\}\} = j\,\mathrm{Im}\{X(e^{j\omega})\}$$

where $X(e^{j\omega}) = \mathcal{F}\{x(n)\}$.

2.22   Prove that

$$\mathcal{F}^{-1}\left\{ \sum_{k=-\infty}^{\infty} \delta\left(\omega - \frac{2\pi}{N}k\right) \right\} = \frac{N}{2\pi} \sum_{p=-\infty}^{\infty} \delta(n - Np)$$

by computing its left-hand side with the inverse Fourier transform in equation (2.173), and verifying that the resulting sequence is equal to its right-hand side.

3 Discrete transforms

3.1 Introduction

In Chapter 2, we saw that discrete-time signals and systems can be characterized in the frequency domain by their Fourier transform. Also, as seen in Chapter 2, one of the main advantages of discrete-time signals is that they can be processed and represented in digital computers. However, when we examine the definition of the Fourier transform in equation (2.173),

$$X(e^{j\omega}) = \sum_{n=-\infty}^{\infty} x(n)e^{-j\omega n} \tag{3.1}$$

we notice that such a characterization in the frequency domain depends on the continuous variable ω. This implies that the Fourier transform, as it is, is not suitable for the processing of discrete-time signals in digital computers. As a consequence of (3.1), we need a transform depending on a discrete-frequency variable. This can be obtained from the Fourier transform itself in a very simple way, by sampling uniformly the continuous-frequency variable ω. In this way, we obtain a mapping of a signal depending on a discrete-time variable n to a transform depending on a discrete-frequency variable k. Such a mapping is referred to as the discrete Fourier transform (DFT).

In the main part of this chapter, we will study the DFT. First, the expression for the direct and inverse DFTs will be derived. Then, the limitations of the DFT in the representation of generic discrete-time signals will be analyzed. Also, a useful matrix form of the DFT will be presented.

Next, the properties of the DFT will be analyzed, with special emphasis given to the convolution property, which allows the computation of a discrete convolution in the frequency domain. The overlap-and-add and overlap-and-save methods of performing convolution of long signals in the frequency domain will be presented.

A major limitation to the practical use of the DFT is the large number of arithmetic operations involved in its computation, especially for long sequences. This problem has been partially solved with the introduction of efficient algorithms for DFT computation, generally known as Fast Fourier Transforms (FFTs). The first FFT algorithms were proposed in Cooley & Tukey (1965). Since then, the DFT has been widely used in signal processing applications. In this chapter, some of the most commonly used FFT algorithms are studied.

After the FFT algorithms, discrete transforms other than the DFT will be dealt with. Among others, the discrete cosine transform and the Hartley transform will be analyzed.

Then a summary of discrete signal representations will be given, highlighting the relations between the continuous-time and discrete-time Fourier transforms, the Laplace transform, the z transform, the Fourier series, and the DFT.

Finally, a brief description of MATLAB commands which are useful for computing fast transforms and performing fast convolutions is given.

3.2   Discrete Fourier transform

The Fourier transform of a sequence $x(n)$ is given by equation (3.1). Since $X(e^{j\omega})$ is periodic with period 2π, it is convenient to sample it with a sampling frequency equal to an integer multiple of its period, that is, taking N uniformly spaced samples between 0 and 2π. Let us use the frequencies $\omega_k = \frac{2\pi}{N}k$, $k \in \mathbb{Z}$. This sampling process is equivalent to generating a Fourier transform $X'(e^{j\omega})$ such that

$$X'(e^{j\omega}) = X(e^{j\omega}) \sum_{k=-\infty}^{\infty} \delta\left(\omega - \frac{2\pi}{N}k\right) \tag{3.2}$$

Applying the convolution theorem seen in Chapter 2, the above equation becomes

$$x'(n) = \mathcal{F}^{-1}\left\{X'(e^{j\omega})\right\} = x(n) * \mathcal{F}^{-1}\left\{\sum_{k=-\infty}^{\infty} \delta\left(\omega - \frac{2\pi}{N}k\right)\right\} \tag{3.3}$$

and since (see Exercise 2.22)

$$\mathcal{F}^{-1}\left\{\sum_{k=-\infty}^{\infty} \delta\left(\omega - \frac{2\pi}{N}k\right)\right\} = \frac{N}{2\pi} \sum_{p=-\infty}^{\infty} \delta(n - Np) \tag{3.4}$$

equation (3.3) becomes

$$x'(n) = x(n) * \frac{N}{2\pi} \sum_{p=-\infty}^{\infty} \delta(n - Np) = \frac{N}{2\pi} \sum_{p=-\infty}^{\infty} x(n - Np) \tag{3.5}$$

Equation (3.5) indicates that, from equally spaced samples of the Fourier transform, we are able to recover a signal $x'(n)$ consisting of a sum of periodic repetitions of the original discrete signal $x(n)$. In this case, the period of the repetitions is equal to the number, N, of samples of the Fourier transform in one period. It is then easy to see that if the length, L, of $x(n)$ is larger than N, then $x(n)$ cannot be obtained from $x'(n)$. On the other hand, if $L \leq N$, then $x'(n)$ is a precise periodic repetition of $x(n)$, and therefore $x(n)$ can be obtained by isolating one complete period of N samples of $x'(n)$, as given by

$$x(n) = \frac{2\pi}{N}x'(n), \quad \text{for } 0 \le n \le N - 1 \tag{3.6}$$

From the above discussion, we can draw two important conclusions:

- The samples of the Fourier transform can provide an effective discrete representation, in frequency, of a finite-length discrete-time signal.
- This representation is useful only if the number of samples, N, of the Fourier transform in one period is greater than or equal to the original signal length L.

It is interesting to note that equation (3.5) is of the same form as equation (1.88), which gives the Fourier transform of a sampled continuous-time signal. In that case, equation (1.88) implies that, in order for a continuous-time signal to be recoverable from its samples, aliasing has to be avoided. This can be done by using a sampling frequency greater than two times the bandwidth of the analog signal. Similarly, equation (3.5) implies that one can recover a digital signal from the samples of its Fourier transform provided that the signal length, L, is smaller than or equal to the number of samples, N, taken in one complete period of its Fourier transform.

We can express $x(n)$ directly as a function of the samples of $X(e^{j\omega})$ by manipulating equation (3.2) in a different manner to that described above. We begin by noting that equation (3.2) is equivalent to

$$X'(e^{j\omega}) = \sum_{k=-\infty}^{\infty} X\left(e^{j\frac{2\pi}{N}k}\right) \delta\left(\omega - \frac{2\pi}{N}k\right) \tag{3.7}$$

By applying the inverse Fourier transform relation in equation (2.173), we have

$$\begin{aligned} x'(n) &= \frac{1}{2\pi} \int_0^{2\pi} X'(e^{j\omega}) e^{j\omega n} d\omega \\ &= \frac{1}{2\pi} \int_0^{2\pi} \sum_{k=-\infty}^{\infty} X\left(e^{j\frac{2\pi}{N}k}\right) \delta\left(\omega - \frac{2\pi}{N}k\right) e^{j\omega n} d\omega \\ &= \frac{1}{2\pi} \sum_{k=0}^{N-1} X\left(e^{j\frac{2\pi}{N}k}\right) e^{j\frac{2\pi}{N}kn} \end{aligned} \tag{3.8}$$

Substituting equation (3.8) in equation (3.6), $x(n)$ can be expressed as

$$x(n) = \frac{1}{N} \sum_{k=0}^{N-1} X\left(e^{j\frac{2\pi}{N}k}\right) e^{j\frac{2\pi}{N}kn}, \quad \text{for } 0 \le n \le N - 1 \tag{3.9}$$

Equation (3.9) shows how a discrete-time signal can be recovered from its discrete-frequency representation. This relation is known as the inverse discrete Fourier transform (IDFT).

An inverse expression to equation (3.9), relating the discrete-frequency representation to the discrete-time signal can be obtained by just rewriting equation (3.1) for the frequencies $\omega_k = \frac{2\pi}{N}k$, $k = 0, \ldots, N - 1$. Since $x(n)$ has finite duration, we assume that its nonzero samples are within the interval $0 \leq n \leq N - 1$, and then

$$X\left(e^{j\frac{2\pi}{N}k}\right) = \sum_{n=0}^{N-1} x(n)e^{-j\frac{2\pi}{N}kn}, \quad \text{for } 0 \leq k \leq N - 1 \tag{3.10}$$

Equation (3.10) is known as the discrete Fourier transform (DFT).

It is important to point out that, in equation (3.10), if $x(n)$ has length $L < N$, it has to be padded with zeros up to length N, to adapt the sequence length when calculating its corresponding DFT. Referring to Figure 3.1, we can see that the amount of zero-padding also determines the frequency resolution of the DFT. Figure 3.1a, shows a signal $x(n)$ consisting of $L = 6$ samples along with its Fourier transform, which distinctively presents two pairs of close peaks within $[0, 2\pi)$. In Figure 3.1b, we see that sampling the Fourier transform with 8 samples in the interval $[0, 2\pi)$ is equivalent to computing the DFT of $x(n)$ using equation (3.10) with $N = 8$ samples, which requires $x(n)$ to be padded with $N - L = 2$ zeros. From this figure, we observe that in such a case the two close peaks of the Fourier transform of $x(n)$ could not be resolved by the DFT coefficients. This fact indicates that the resolution of the DFT should be improved by increasing the number of samples, N, thus requiring $x(n)$ to be padded with even more zeros. In Figure 3.1c, we can see that the close peaks can be easily identified by the DFT coefficients when N has been increased to 32, corresponding to a zero-padding of $N - L = 32 - 6 = 26$ zeros in $x(n)$.

We can summarize the issues in the above discussion as follows:

- The larger the number of zeros padded on $x(n)$ for the calculation of the DFT, the more it resembles its Fourier transform. This happens because of the larger number of samples taken within $[0, 2\pi)$.
- The amount of zero-padding used depends on the arithmetic complexity allowed by a particular application, because the larger the amount of zero-padding the greater the computational and storage requirements involved in the DFT computation.

There is, however, an important observation that has to be made about the discussion related to Figure 3.1. We have seen in Figure 3.1b that for $N = 8$ the DFT could not resolve the two close peaks of the Fourier transform. However, since the signal duration is $N = 6$, $x(n)$ can be recovered from the samples of the Fourier transform in Figure 3.1b using equation (3.10). With $x(n)$, the Fourier transform $X(e^{j\omega})$, as in Figure 3.1a, can be computed using equation (3.1), and therefore the two close peaks could be fully recovered and identified. At this point one could ask why one would need to use a DFT of larger resolution if a DFT of size equal to the

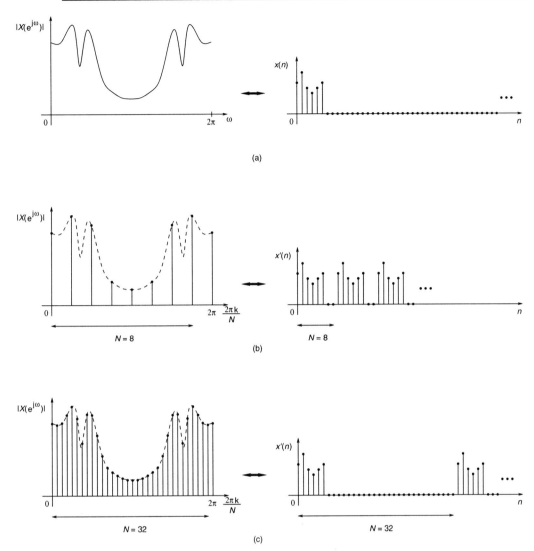

Figure 3.1 Equivalence between sampling the Fourier transform of a signal and its DFT. (a) Example of Fourier transform of a signal $x(n)$ with only $L = 6$ nonzero samples; (b) DFT with $N = 8$ samples, along with the corresponding zero-padded $x(n)$; (c) DFT with $N = 32$ samples, along with the corresponding zero-padded $x(n)$.

signal duration would be enough to fully recover the correct Fourier transform. A justification for the use of a DFT of a larger size than the signal duration is that, for example, for N large enough, one would not need to perform indirect calculations to identify the two close peaks in Figure 3.1, because they would be represented directly by the DFT coefficients, as depicted in Figure 3.1c. In what follows, we derive an expression relating the Fourier transform of a signal $x(n)$ to its DFT.

Equation (3.6), which relates the signal $x(n)$ to the signal $x'(n)$ obtainable from the samples of its Fourier transform, can be rewritten as

$$x(n) = \frac{2\pi}{N} x'(n)(u(n) - u(n - N)) \tag{3.11}$$

Using the fact that a multiplication in the time domain corresponds to a periodic convolution in the frequency domain, discussed in Chapter 2, we have

$$
\begin{aligned}
X(e^{j\omega}) &= \frac{1}{2\pi} \frac{2\pi}{N} X'(e^{j\omega}) \circledast \mathcal{F}\{u(n) - u(n - N)\} \\
&= \frac{1}{N} X'(e^{j\omega}) \circledast \left[\frac{\sin \frac{\omega N}{2}}{\sin \frac{\omega}{2}} e^{-j\frac{\omega(N-1)}{2}} \right]
\end{aligned}
\tag{3.12}
$$

Substituting equation (3.7) in equation (3.12), $X(e^{j\omega})$ becomes

$$
\begin{aligned}
X(e^{j\omega}) &= \frac{1}{N} \left[\sum_{k=-\infty}^{\infty} X\left(e^{j\frac{2\pi}{N}k}\right) \delta\left(\omega - \frac{2\pi}{N}k\right) \right] \circledast \left[\frac{\sin \frac{\omega N}{2}}{\sin \frac{\omega}{2}} e^{-j\frac{\omega(N-1)}{2}} \right] \\
&= \frac{1}{N} \sum_{k=-\infty}^{\infty} X\left(e^{j\frac{2\pi}{N}k}\right) \left[\frac{\sin \frac{(\omega - \frac{2\pi}{N}k)N}{2}}{\sin \frac{(\omega - \frac{2\pi}{N}k)}{2}} e^{-j\frac{(\omega - \frac{2\pi}{N}k)(N-1)}{2}} \right] \\
&= \frac{1}{N} \sum_{k=-\infty}^{\infty} X\left(e^{j\frac{2\pi}{N}k}\right) \left[\frac{\sin \left(\frac{\omega N}{2} - \pi k\right)}{\sin \left(\frac{\omega}{2} - \frac{\pi k}{N}\right)} e^{-j(\frac{\omega}{2} - \frac{\pi}{N}k)(N-1)} \right]
\end{aligned}
\tag{3.13}
$$

This equation corresponds to an interpolation formula that gives the Fourier transform of a signal as a function of its DFT. One should note, once again, that such a relationship only works when N is larger than the signal length L.

In order to simplify the notation, it is common practice to use $X(k)$ instead of $X\left(e^{j\frac{2\pi}{N}k}\right)$, and to define

$$W_N = e^{-j\frac{2\pi}{N}} \tag{3.14}$$

Using this notation, the definitions of the DFT and IDFT, as given in equations (3.10) and (3.9), become

$$X(k) = \sum_{n=0}^{N-1} x(n) W_N^{kn}, \quad \text{for } 0 \le k \le N - 1 \tag{3.15}$$

$$x(n) = \frac{1}{N} \sum_{k=0}^{N-1} X(k) W_N^{-kn}, \quad \text{for } 0 \le n \le N - 1 \tag{3.16}$$

respectively.

From the development represented by equations (3.7)–(3.10), it can be seen that equation (3.16) is the inverse of equation (3.15). This can be shown in an alternative way by substituting equation (3.16) into equation (3.15) directly (see Exercise 3.2).

One should note that if, in the above definitions, $x(n)$ and $X(k)$ are not restricted to being between 0 and $N-1$, then they should be interpreted as being periodic sequences with period N.

From the above, the discrete Fourier transform as expressed in equation (3.15) can be interpreted in two related ways:

- As a discrete-frequency representation of finite-length signals whereby a length-N signal is mapped into N discrete-frequency coefficients, which correspond to N samples of the Fourier transform of $x(n)$.

- As the Fourier transform of a periodic signal having period N. This periodic signal may correspond to the finite-length signal $x(n)$ repeated periodically, not restricting the index n in equation (3.16) to the interval $0 \leq n \leq N - 1$.

EXAMPLE 3.1
Compute the DFT of the following sequence:

$$x(n) = \begin{cases} 1, & 0 \leq n \leq 4 \\ -1, & 5 \leq n \leq 9 \end{cases} \tag{3.17}$$

SOLUTION

$$X(k) = \frac{1 - W_5^{5k}}{1 - W_5^k} - \frac{W_5^{5k} - W_5^{10k}}{1 - W_5^k} = \frac{(1 - W_5^{5k})^2}{1 - W_5^k} \tag{3.18}$$

$\triangle$

Equations (3.15) and (3.16) can be written in matrix notation as

$$
\begin{bmatrix} X(0) \\ X(1) \\ X(2) \\ \vdots \\ X(N-1) \end{bmatrix} =
\begin{bmatrix}
W_N^0 & W_N^0 & \cdots & W_N^0 \\
W_N^0 & W_N^1 & \cdots & W_N^{(N-1)} \\
W_N^0 & W_N^2 & \cdots & W_N^{2(N-1)} \\
\vdots & \vdots & \ddots & \vdots \\
W_N^0 & W_N^{(N-1)} & \cdots & W_N^{(N-1)^2}
\end{bmatrix}
\begin{bmatrix} x(0) \\ x(1) \\ x(2) \\ \vdots \\ x(N-1) \end{bmatrix} \tag{3.19}
$$

$$
\begin{bmatrix} x(0) \\ x(1) \\ x(2) \\ \vdots \\ x(N-1) \end{bmatrix} = \frac{1}{N}
\begin{bmatrix}
W_N^0 & W_N^0 & \cdots & W_N^0 \\
W_N^0 & W_N^{-1} & \cdots & W_N^{-(N-1)} \\
W_N^0 & W_N^{-2} & \cdots & W_N^{-2(N-1)} \\
\vdots & \vdots & \ddots & \vdots \\
W_N^0 & W_N^{-(N-1)} & \cdots & W_N^{-(N-1)^2}
\end{bmatrix}
\begin{bmatrix} X(0) \\ X(1) \\ X(2) \\ \vdots \\ X(N-1) \end{bmatrix} \tag{3.20}
$$

By defining

$$
\mathbf{x} = \begin{bmatrix} x(0) \\ x(1) \\ \vdots \\ x(N-1) \end{bmatrix}, \quad \mathbf{X} = \begin{bmatrix} X(0) \\ X(1) \\ \vdots \\ X(N-1) \end{bmatrix} \tag{3.21}
$$

and a matrix $\mathbf{W}_N$ such that

$$
\{\mathbf{W}_N\}_{ij} = W_N^{ij}, \quad \text{for } 0 \le i, j \le N-1 \tag{3.22}
$$

equations (3.19) and (3.20) can be rewritten more concisely as

$$
\mathbf{X} = \mathbf{W}_N \mathbf{x} \tag{3.23}
$$

$$
\mathbf{x} = \frac{1}{N} \mathbf{W}_N^* \mathbf{X} \tag{3.24}
$$

Note that the matrix $\mathbf{W}_N$ enjoys very special properties. Equation (3.22) above implies that it is symmetric, that is, $\mathbf{W}_N^T = \mathbf{W}_N$. Also, the direct and inverse DFT relations in equations (3.23) and (3.24) imply that $\mathbf{W}_N^{-1} = \frac{1}{N}\mathbf{W}_N^*$.

From either equations (3.15) and (3.16), or (3.19) and (3.20), one can easily conclude that a length-N DFT requires N^2 complex multiplications, including possible trivial multiplications by $W_N^0 = 1$. If we do not take into account these trivial cases, the total number of multiplications is equal to $(N^2 - 2N + 1)$.

3.3   Properties of the DFT

In this section, we describe the main properties of the direct and inverse DFTs, and provide proofs for some of them. Note that since the DFT corresponds to samples of the Fourier transform, its properties closely resemble the ones of the Fourier transform presented in Section 2.9. However, from N samples of the Fourier transform, one can only recover a signal corresponding to the periodic repetition of the signal $x(n)$ with period N, as given by equation (3.5). This makes the properties of the DFT slightly different from the ones of the Fourier transform.

One should bear in mind that, although the DFT can be interpreted as the Fourier transform of a periodic signal, it is just a mapping from a length-N signal into N frequency coefficients and vice versa. However, we often resort to this periodic signal interpretation, since some of the DFT properties follow naturally from it. Note that in this case the periodic signal is the one in equation (3.5), which is the same one as that obtained by allowing the index n in equation (3.16) to vary within $(-\infty, \infty)$.

3.3.1 Linearity

The DFT of a linear combination of two sequences is the linear combination of the DFT of the individual sequences, that is, if $x(n) = k_1 x_1(n) + k_2 x_2(n)$ then

$$X(k) = k_1 X_1(k) + k_2 X_2(k) \tag{3.25}$$

Note that the two DFTs must have the same length, and thus the two sequences, if necessary, should be zero-padded accordingly in order to reach the same length N.

3.3.2 Time-reversal

The DFT of $x(-n)$ is such that

$$x(-n) \longleftrightarrow X(-k) \tag{3.26}$$

It must be noted that, if both indexes n and k are constrained to be between 0 and $N-1$, then $-n$ and $-k$ are outside this interval. Therefore, consistency with equations (3.15) and (3.16) requires that $x(-n) = x(N - n)$ and $X(-k) = X(N - k)$. These relations can also be deduced from the fact that we can interpret both $x(n)$ and $X(k)$ as being periodic with period N.

3.3.3 Time-shift theorem

The DFT of a sequence shifted in time is such that

$$x(n + l) \longleftrightarrow W_N^{-lk} X(k) \tag{3.27}$$

In the definition of the IDFT given in equation (3.16), if the index n is allowed to vary outside the set $0, 1, \ldots, N - 1$, then $x(n)$ can be interpreted as being periodic with period N. This interpretation implies that the signal $x(n + l)$ obtained from the inverse DFT of $W_N^{-lk} X(k)$ corresponds to a circular shift of $x(n)$, that is, if

$$y(n) = x(n + l) \tag{3.28}$$

then

$$y(n) = \begin{cases} x(n + l), & \text{for } 0 \leq n \leq N - l - 1 \\ x(n + l - N), & \text{for } N - l \leq n \leq N - 1 \end{cases} \tag{3.29}$$

This result indicates that $y(n)$ is a sequence whose last l samples are equal to the first l samples of $x(n)$. An example of circular shift is illustrated in Figure 3.2. A formal proof of this property is provided below.

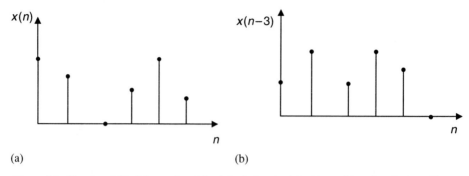

(a)   (b)

Figure 3.2 Circular shift of 3 samples: (a) original signal $x(n)$; (b) resulting signal $x(n-3)$.

PROOF

$$X'(k) = \sum_{n=0}^{N-1} x(n+l) W_N^{nk}$$

$$= W_N^{-lk} \sum_{n=0}^{N-1} x(n+l) W_N^{(n+l)k}$$

$$= W_N^{-lk} \sum_{m=l}^{N+l-1} x(m) W_N^{mk}$$

$$= W_N^{-lk} \left(\sum_{m=l}^{N-1} x(m) W_N^{mk} + \sum_{m=N}^{N+l-1} x(m) W_N^{mk} \right) \qquad (3.30)$$

Since W_N^k has period N and $x(n+l)$ is obtained from a circular shift of $x(n)$, then the summation from N to $(N+l-1)$ is equivalent to a summation from 0 to $(l-1)$. Therefore

$$X'(k) = W_N^{-lk} \left(\sum_{m=l}^{N-1} x(m) W_N^{mk} + \sum_{m=0}^{l-1} x(m) W_N^{mk} \right)$$

$$= W_N^{-lk} \sum_{m=0}^{N-1} x(m) W_N^{mk}$$

$$= W_N^{-lk} X(k) \qquad (3.31)$$

$\square$

3.3.4   Circular frequency-shift theorem (modulation theorem)

$$W_N^{ln} x(n) \longleftrightarrow X(k+l) \qquad (3.32)$$

The proof is analogous to the one of the time-shift theorem, and is left as an exercise to the reader.

Noting that

$$W_N^{ln} = \cos\left(\frac{2\pi}{N}ln\right) - j\sin\left(\frac{2\pi}{N}ln\right) \tag{3.33}$$

equation (3.32) also implies the following properties:

$$x(n)\sin\left(\frac{2\pi}{N}ln\right) \longleftrightarrow \frac{1}{2j}(X(k-l) - X(k+l)) \tag{3.34}$$

$$x(n)\cos\left(\frac{2\pi}{N}ln\right) \longleftrightarrow \frac{1}{2}(X(k-l) + X(k+l)) \tag{3.35}$$

3.3.5 Circular convolution in time

If $x(n)$ and $h(n)$ are periodic with period N, then

$$\sum_{l=0}^{N-1} x(l)h(n-l) = \sum_{l=0}^{N-1} x(n-l)h(l) \longleftrightarrow X(k)H(k) \tag{3.36}$$

where $X(k)$ and $H(k)$ are the DFTs of the length-N signals corresponding to one period of $x(n)$ and $h(n)$, respectively.

PROOF
If $Y(k) = X(k)H(k)$, then

$$
\begin{aligned}
y(n) &= \frac{1}{N}\sum_{k=0}^{N-1} H(k)X(k)W_N^{-kn} \\
&= \frac{1}{N}\sum_{k=0}^{N-1} H(k)\left(\sum_{l=0}^{N-1} x(l)W_N^{kl}\right) W_N^{-kn} \\
&= \frac{1}{N}\sum_{k=0}^{N-1}\sum_{l=0}^{N-1} H(k)x(l)W_N^{(l-n)k} \\
&= \sum_{l=0}^{N-1} x(l)\frac{1}{N}\sum_{k=0}^{N-1} H(k)W_N^{-(n-l)k} \\
&= \sum_{l=0}^{N-1} x(l)h(n-l) \tag{3.37}
\end{aligned}
$$

$\square$

This result is the basis of one of the most important applications of the DFT, which is the ability to compute a discrete convolution in time through the application of the inverse DFT to the product of the DFTs of the two sequences. However, one should bear in mind that when computing the DFT all the sequence shifts involved are circular, as depicted in Figure 3.2. Therefore, we say that the product of two DFTs actually corresponds to a circular convolution in the time domain. In fact, the circular convolution is equivalent to a linear convolution between the periodic versions of the original finite-length sequences. It is important to note that, as seen in Chapter 1, linear convolutions are usually the ones of interest in practice. In the next section, we will discuss how linear convolutions can be implemented through circular convolutions, and therefore through the computation of DFTs.

Since $x(n)$ and $h(n)$ in equation (3.36) have period N, then their convolution $y(n)$ also has period N, and needs to be specified only for $0 \leq n \leq N - 1$. Therefore, the circular convolution in equation (3.36) can be expressed as a function of only the samples of $x(n)$ and $h(n)$ between 0 and $N - 1$ as

$$y(n) = \sum_{l=0}^{N-1} x(l)h(n-l)$$

$$= \sum_{l=0}^{n} x(l)h(n-l) + \sum_{l=n+1}^{N-1} x(l)h(n-l+N), \quad \text{for } 0 \leq n \leq N-1 \tag{3.38}$$

which can be rewritten in a compact form as

$$y(n) = \sum_{l=0}^{N-1} x(l)h((n-l) \bmod N) = x(n) \otimes h(n) \tag{3.39}$$

where $(l \bmod N)$ represents the remainder of the integer division of l by N.

From equation (3.38) we can deduce the matrix form for the circular convolution:

$$\begin{bmatrix} y(0) \\ y(1) \\ y(2) \\ \vdots \\ y(N-1) \end{bmatrix} = \begin{bmatrix} h(0) & h(N-1) & h(N-2) & \cdots & h(1) \\ h(1) & h(0) & h(N-1) & \cdots & h(2) \\ h(2) & h(1) & h(0) & \cdots & h(3) \\ \vdots & \vdots & \vdots & \ddots & \vdots \\ h(N-1) & h(N-2) & h(N-3) & \cdots & h(0) \end{bmatrix} \begin{bmatrix} x(0) \\ x(1) \\ x(2) \\ \vdots \\ x(N-1) \end{bmatrix} \tag{3.40}$$

Note that each row of the matrix in the above equation is a circular right-shift of the previous row.

In the remainder of this chapter, unless otherwise stated, it will be assumed that all the sequences are periodic and all the convolutions are circular.

3.3.6   Correlation

The DFT of the correlation in time between two sequences is such that

$$\sum_{n=0}^{N-1} h(n)x(l+n) \longleftrightarrow H(-k)X(k) \tag{3.41}$$

3.3.7   Real and imaginary sequences

If $x(n)$ is a real sequence, then

$$\left.\begin{array}{l} \mathrm{Re}\{X(k)\} = \mathrm{Re}\{X(-k)\} \\ \mathrm{Im}\{X(k)\} = -\mathrm{Im}\{X(-k)\} \end{array}\right\} \tag{3.42}$$

The proof can be easily obtained from the definition of the DFT, by using the expression for W_N^{kn} in equation (3.33).

When $x(n)$ is imaginary, then

$$\left.\begin{array}{l} \mathrm{Re}\{X(k)\} = -\mathrm{Re}\{X(-k)\} \\ \mathrm{Im}\{X(k)\} = \mathrm{Im}\{X(-k)\} \end{array}\right\} \tag{3.43}$$

The properties above can be explored for the computation of two DFTs of real sequences $x_1(n)$ and $x_2(n)$ through the computation of just the DFT of the sequence $x_1(n) + jx_2(n)$. We begin by noting that, from the linearity of the DFT,

$$x_1(n) + jx_2(n) \longleftrightarrow Y(k) = X_1(k) + jX_2(k) \tag{3.44}$$

From this equation, we have that

$$\left.\begin{array}{l} \mathrm{Re}\{Y(k)\} = \mathrm{Re}\{X_1(k)\} - \mathrm{Im}\{X_2(k)\} \\ \mathrm{Im}\{Y(k)\} = \mathrm{Re}\{X_2(k)\} + \mathrm{Im}\{X_1(k)\} \end{array}\right\} \tag{3.45}$$

and substituting these results into equation (3.42), we get

$$\left.\begin{array}{l} \mathrm{Re}\{X_1(k)\} = \dfrac{1}{2}\left(\mathrm{Re}\{Y(k)\} + \mathrm{Re}\{Y(-k)\}\right) \\[2mm] \mathrm{Im}\{X_1(k)\} = \dfrac{1}{2}\left(\mathrm{Im}\{Y(k)\} - \mathrm{Im}\{Y(-k)\}\right) \\[2mm] \mathrm{Re}\{X_2(k)\} = \dfrac{1}{2}\left(\mathrm{Im}\{Y(k)\} + \mathrm{Im}\{Y(-k)\}\right) \\[2mm] \mathrm{Im}\{X_2(k)\} = \dfrac{1}{2}\left(\mathrm{Re}\{Y(-k)\} - \mathrm{Re}\{Y(k)\}\right) \end{array}\right\} \tag{3.46}$$

3.3.8   Symmetric and antisymmetric sequences

Symmetric and antisymmetric sequences are of special interest because their DFTs have some interesting properties. In Section 2.9 we gave the properties of the Fourier transform relating to symmetric and antisymmetric sequences. In this chapter, the meanings of symmetry and antisymmetry are slightly different. This happens because, unlike the Fourier and z transforms, that are applied to infinite-duration signals, the DFT is applied to finite-duration signals. In fact, the DFT can be interpreted as the Fourier transform of a periodic signal formed from the infinite repetition of the finite-duration signal. Therefore, before describing the properties of the DFT relating to symmetric and antisymmetric sequences, we give precise definitions for them in the context of the DFT.

- A sequence is called symmetric (or even) if $x(n) = x(-n)$. Since, for indexes outside the set $0, 1, \ldots, N - 1$, $x(n)$ can be interpreted as periodic with period N (see equation (3.16)), then $x(-n) = x(N - n)$. Therefore, symmetry is equivalent to $x(n) = x(N - n)$.
- A sequence is antisymmetric (or odd) if $x(n) = -x(-n) = -x(N - n)$.
- A complex sequence is said to be conjugate symmetric if $x(n) = x^*(-n) = x^*(N - n)$.
- A complex sequence is called conjugate antisymmetric if $x(n) = -x^*(-n) = -x^*(N - n)$.

Using such concepts, the following properties hold:

- If $x(n)$ is real and symmetric, $X(k)$ is also real and symmetric.

PROOF
From equation (3.15), $X(k)$ is given by

$$X(k) = \sum_{n=0}^{N-1} x(n) \cos\left(\frac{2\pi}{N}kn\right) - j \sum_{n=0}^{N-1} x(n) \sin\left(\frac{2\pi}{N}kn\right) \tag{3.47}$$

Since $x(n) = x(N - n)$, the imaginary part of the above summation is null, because, for N even, we get that

$$\sum_{n=0}^{N-1} x(n) \sin\left(\frac{2\pi}{N}kn\right) = \sum_{n=0}^{\frac{N}{2}-1} x(n) \sin\left(\frac{2\pi}{N}kn\right) + \sum_{n=\frac{N}{2}}^{N-1} x(n) \sin\left(\frac{2\pi}{N}kn\right)$$

$$= \sum_{n=1}^{\frac{N}{2}-1} x(n) \sin\left(\frac{2\pi}{N}kn\right) + \sum_{n=\frac{N}{2}+1}^{N-1} x(n) \sin\left(\frac{2\pi}{N}kn\right)$$

$$= \sum_{n=1}^{\frac{N}{2}-1} x(n) \sin\left(\frac{2\pi}{N}kn\right)$$

$$+ \sum_{m=1}^{\frac{N}{2}-1} x(N - m) \sin\left[\frac{2\pi}{N}k(N - m)\right]$$

$$= \sum_{n=1}^{\frac{N}{2}-1} x(n) \sin\left(\frac{2\pi}{N}kn\right) - \sum_{m=1}^{\frac{N}{2}-1} x(m) \sin\left(\frac{2\pi}{N}km\right)$$

$$= 0 \tag{3.48}$$

Therefore, we have that

$$X(k) = \sum_{n=0}^{N-1} x(n) \cos\left(\frac{2\pi}{N}kn\right) \tag{3.49}$$

which is real and symmetric (even). The proof for N odd is analogous, and is left as an exercise to the reader.

□

- If $x(n)$ is imaginary and even, then $X(k)$ is imaginary and even.
- If $x(n)$ is real and odd, then $X(k)$ is imaginary and odd.
- If $x(n)$ is imaginary and odd, then $X(k)$ is real and odd.
- If $x(n)$ is conjugate symmetric, then $X(k)$ is real.

PROOF

A conjugate symmetric sequence $x(n)$ can be expressed as

$$x(n) = x_e(n) + jx_o(n) \tag{3.50}$$

where $x_e(n)$ is real and even and $x_o(n)$ is real and odd. Therefore,

$$X(k) = X_e(k) + jX_o(k) \tag{3.51}$$

From the above properties, $X_e(k)$ is real and even, and $X_o(k)$ is imaginary and odd. Thus, $X(k) = X_e(k) + jX_o(k)$ is real.

□

- If $x(n)$ is conjugate antisymmetric, then $X(k)$ is imaginary.

The proofs of all other of these properties are left as exercises to the interested reader.

3.3.9   Parseval's theorem

$$\sum_{n=0}^{N-1} |x(n)|^2 = \frac{1}{N} \sum_{n=0}^{N-1} |X(k)|^2 \tag{3.52}$$

3.3.10   Relationship between the DFT and the z transform

In Section 3.2, we defined the DFT as samples of the Fourier transform, and showed, in equation (3.13), how to obtain the Fourier transform directly from the DFT. Since the Fourier transform corresponds to the z transform for $z = e^{j\omega}$, then clearly the DFT can be obtained by sampling the z transform at $\omega = \frac{2\pi}{N}k$.

Mathematically, since the z transform $X_z(z)$ of a length-N sequence $x(n)$ is

$$X_z(z) = \sum_{n=0}^{N-1} x(n) z^{-n} \tag{3.53}$$

if we make $z = e^{j\frac{2\pi}{N}k}$, then we have that

$$X_z\left(e^{j\frac{2\pi}{N}kn}\right) = \sum_{n=0}^{N-1} x(n) e^{-j\frac{2\pi}{N}kn} \tag{3.54}$$

This equation corresponds to samples of the z transform equally spaced on the unit circle, and is identical to the definition of the DFT in equation (3.15).

In order to obtain the z transform from the DFT coefficients $X(k)$, we substitute equation (3.16) into equation (3.53), obtaining

$$
\begin{aligned}
X_z(z) &= \sum_{n=0}^{N-1} x(n) z^{-n} \\
&= \sum_{n=0}^{N-1} \frac{1}{N} \sum_{k=0}^{N-1} X(k) W_N^{-kn} z^{-n} \\
&= \frac{1}{N} \sum_{k=0}^{N-1} X(k) \sum_{n=0}^{N-1} \left(W_N^{-k} z^{-1}\right)^n \\
&= \frac{1}{N} \sum_{k=0}^{N-1} X(k) \frac{1 - W_N^{-kN} z^{-N}}{1 - W_N^{-k} z^{-1}} \\
&= \frac{1 - z^{-N}}{N} \sum_{k=0}^{N-1} \frac{X(k)}{1 - W_N^{-k} z^{-1}}
\end{aligned}
\tag{3.55}
$$

which, similarly to equation (3.13) for the Fourier transform, relates the DFT to the z transform.

3.4   Digital filtering using the DFT

3.4.1   Linear and circular convolutions

A linear time-invariant system implements the linear convolution of the input signal with the impulse response of the system. Since the Fourier transform of the convolution of two sequences is the product of their Fourier transforms, it is natural to consider the computation of convolutions in the frequency domain. The DFT is the discrete version of the Fourier transform, and therefore should be the transform used for such computations. However, as is described in equation (3.5), the sampling process in the frequency domain forces the signal to be periodic in time. In Section 3.3, see equation (3.36), we saw that this implies that the IDFT of the product of the DFTs of two length-N signals corresponds to the convolution of two periodic signals. These periodic signals are obtained by repeating the two length-N signals with period N. As seen before, the convolution between the periodic signals is called the circular convolution between the two length-N signals. This means that, using the DFT, one can in principle compute only circular convolutions, and not the linear convolution necessary to implement a linear system. In this section, we describe techniques to circumvent this problem and allow us to implement discrete-time linear systems in the frequency domain.

These techniques are essentially based on a very simple trick. Supposing that the DFTs are of size N, we have that the circular convolution between two sequences $x(n)$ and $h(n)$ is given by equation (3.38), which is repeated here for the reader's convenience

$$
\begin{aligned}
y(n) &= \sum_{l=0}^{N-1} x(l)h(n-l) \\
&= \sum_{l=0}^{n} x(l)h(n-l) + \sum_{l=n+1}^{N-1} x(l)h(n-l+N), \quad \text{for } 0 \leq n \leq N-1
\end{aligned}
\tag{3.56}
$$

If we want the circular convolution to be equal to the linear convolution between $x(n)$ and $h(n)$, the second summation in the above equation must be null, that is

$$
c(n) = \sum_{l=n+1}^{N-1} x(l)h(n-l+N) = 0, \quad \text{for } 0 \leq n \leq N-1
\tag{3.57}
$$

Assuming that $x(n)$ has duration L and $h(n)$ has duration K, that is,

$$
x(n) = 0, \quad \text{for } n \geq L; \quad h(n) = 0, \quad \text{for } n \geq K
\tag{3.58}
$$

we have that the summation $c(n)$ in equation (3.57) is different from zero only if both $x(l)$ and $h(n-l+N)$ are nonzero, and this happens if

$$
l \leq L-1
\tag{3.59}
$$

$$
n - l + N \leq K - 1 \quad \Rightarrow \quad l \geq n + N - K + 1
\tag{3.60}
$$

If we then want the summation $c(n)$ to be null for $0 \leq n \leq N-1$, we must have, from equations (3.59) and (3.60), that

$$
n + N - K + 1 > L - 1
\tag{3.61}
$$

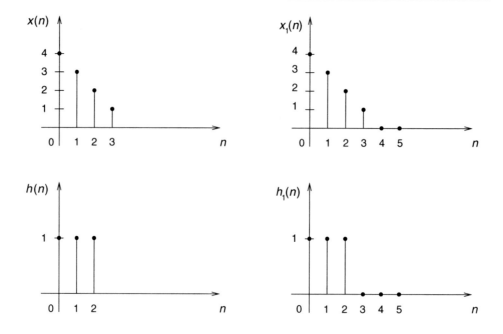

Figure 3.3 Zero-padding of two sequences in order to perform linear convolution using the DFT: $x_1(n)$ corresponds to $x(n)$, and $h_1(n)$ corresponds to $h(n)$ after appropriate padding.

and then

$$N > L + K - 2 - n, \quad \text{for } 0 \le n \le N - 1 \tag{3.62}$$

Since the most strict case of equation (3.62) is for $n = 0$, we have that the condition for $c(n) = 0$, and therefore for the circular convolution to be equivalent to the linear convolution, is

$$N \ge L + K - 1 \tag{3.63}$$

Thus, in order to perform a linear convolution using the inverse DFT of the product of the DFT of two sequences, we must choose a DFT size N satisfying equation (3.63). This is equivalent to padding $x(n)$ using at least $K - 1$ zeros and padding $h(n)$ using at least $L - 1$ zeros. This zero-padding process is illustrated in Figure 3.3 for $L = 4$ and $K = 3$, where, after zero-padding, the sequences $x(n)$ and $h(n)$ are denoted as $x_1(n)$ and $h_1(n)$, respectively.

The following example will help to clarify the above discussion.

EXAMPLE 3.2
Compute the linear convolution of the two sequences $x(n)$ and $h(n)$, as well as the circular convolution of $x(n)$ and $h(n)$, and also of $x_1(n)$ and $h_1(n)$.

SOLUTION

First, we compute the linear convolution of $x(n)$ and $h(n)$,

$$y_l(n) = x(n) * h(n) = \sum_{l=0}^{7} x(l)h(n-l) \tag{3.64}$$

such that

$$\left.\begin{aligned}
y_l(0) &= x(0)h(0) = 4 \\
y_l(1) &= x(0)h(1) + x(1)h(0) = 7 \\
y_l(2) &= x(0)h(2) + x(1)h(1) + x(2)h(0) = 9 \\
y_l(3) &= x(1)h(2) + x(2)h(1) + x(3)h(0) = 6 \\
y_l(4) &= x(2)h(2) + x(3)h(1) = 3 \\
y_l(5) &= x(3)h(2) = 1
\end{aligned}\right\} \tag{3.65}$$

The circular convolution of length $N = 4$ between $x(n)$ and $h(n)$, using equation (3.39), is equal to

$$y_{c_4}(n) = x(n) \otimes h(n) = \sum_{l=0}^{3} x(l)h((n-l) \bmod 4) \tag{3.66}$$

such that

$$\left.\begin{aligned}
y_{c_4}(0) &= x(0)h(0) + x(1)h(3) + x(2)h(2) + x(3)h(1) = 7 \\
y_{c_4}(1) &= x(0)h(1) + x(1)h(0) + x(2)h(3) + x(3)h(2) = 8 \\
y_{c_4}(2) &= x(0)h(2) + x(1)h(1) + x(2)h(0) + x(3)h(3) = 9 \\
y_{c_4}(3) &= x(0)h(3) + x(1)h(2) + x(2)h(1) + x(3)h(0) = 6
\end{aligned}\right\} \tag{3.67}$$

We can also use equation (3.39) to compute the circular convolution of length $N = 6$ of $x_1(n)$ and $h_1(n)$, obtaining

$$y_{c_6}(n) = x_1(n) \otimes h_1(n) = \sum_{l=0}^{5} x(l)h((n-l) \bmod 6) \tag{3.68}$$

such that

$$
\begin{aligned}
y_{c_6}(0) &= x_1(0)h_1(0) + x_1(1)h_1(5) + x_1(2)h_1(4) \\
&\quad + x_1(3)h_1(3) + x_1(4)h_1(2) + x_1(5)h_1(1) \\
&= x_1(0)h_1(0) \\
&= 4
\end{aligned}
$$

$$
\begin{aligned}
y_{c_6}(1) &= x_1(0)h_1(1) + x_1(1)h_1(0) + x_1(2)h_1(5) \\
&\quad + x_1(3)h_1(4) + x_1(4)h_1(3) + x_1(5)h_1(2) \\
&= x_1(0)h_1(1) + x(1)h(0) \\
&= 7
\end{aligned}
$$

$$
\begin{aligned}
y_{c_6}(2) &= x_1(0)h_1(2) + x_1(1)h_1(1) + x_1(2)h_1(0) \\
&\quad + x_1(3)h_1(5) + x_1(4)h_1(4) + x_1(5)h_1(3) \\
&= x_1(0)h_1(2) + x(1)h(1) + x(2)h(0) \\
&= 9
\end{aligned}
$$

$$
\begin{aligned}
y_{c_6}(3) &= x_1(0)h_1(3) + x_1(1)h_1(2) + x_1(2)h_1(1) \\
&\quad + x_1(3)h_1(0) + x_1(4)h_1(5) + x_1(5)h_1(4) \\
&= x_1(1)h_1(2) + x(2)h(1) + x(3)h(0) \\
&= 6
\end{aligned}
$$

$$
\begin{aligned}
y_{c_6}(4) &= x_1(0)h_1(4) + x_1(1)h_1(3) + x_1(2)h_1(2) \\
&\quad + x_1(3)h_1(1) + x_1(4)h_1(0) + x_1(5)h_1(5) \\
&= x_1(2)h_1(2) + x(3)h(1) \\
&= 3
\end{aligned}
$$

$$
\begin{aligned}
y_{c_6}(5) &= x_1(0)h_1(5) + x_1(1)h_1(4) + x_1(2)h_1(3) \\
&\quad + x_1(3)h_1(2) + x_1(4)h_1(1) + x_1(5)h_1(0) \\
&= x_1(3)h_1(2) \\
&= 1
\end{aligned}
$$

$$(3.69)$$

Comparing equations (3.65) and (3.69), it is easy to confirm that $y_{c_6}(n)$ corresponds exactly to the linear convolution between $x(n)$ and $h(n)$.

△

We have now seen how to implement the linear convolution between two finite-length signals using the DFT. However, in practice, it is frequently necessary to implement the convolution of a finite-length sequence with an infinite-length sequence, or even to convolve a short-duration sequence with a long-duration sequence. In both cases, it is not feasible to compute the DFT of a very long or infinite sequence. The solution adopted, in these cases, is to divide the long sequence into blocks of duration N, and perform the convolution of each block with the short sequence. In most practical cases, the long or infinite sequence corresponds to the system input and the short-length sequence to the system impulse response. However, the results of the convolution of each block must be properly combined so that the final result corresponds to the convolution of the long sequence with the short sequence. Two methods for performing this combination are the so-called overlap-and-add and overlap-and-save methods discussed below.

3.4.2 Overlap-and-add method

We can describe a signal $x(n)$ decomposed in non-overlapping blocks $x_m(n)$ of length N as

$$x(n) = \sum_{m=0}^{\infty} x_m(n - mN) \tag{3.70}$$

where

$$x_m(n) = \begin{cases} x(n + mN), & \text{for } 0 \leq n \leq N - 1 \\ 0, & \text{otherwise} \end{cases} \tag{3.71}$$

that is, each block $x_m(n)$ is nonzero between 0 and $N - 1$, and $x(n)$ is composed of the sum of the blocks $x_m(n)$ shifted to $n = mN$.

Using equation (3.70), the convolution of $x(n)$ with another signal $h(n)$ can be written as

$$y(n) = \sum_{m=0}^{\infty} y_m(n - mN) = x(n) * h(n) = \sum_{m=0}^{\infty} (x_m(n - mN) * h(n - mN)) \tag{3.72}$$

Note that $y_m(n - mN)$ is the block decomposition of the result of the convolution of $h(n)$ with the mth block $x_m(n)$.

As seen in Subsection 3.4.1, if we want to compute the linear convolutions leading to $y_m(n)$ using DFTs, their lengths must be at least $(N + K - 1)$, where K is the duration of $h(n)$. Thus, if $x_m(n)$ and $h(n)$ are zero-padded to length $(N + K - 1)$, then the linear convolutions in equation (3.72) can be implemented using circular convolutions. If $x'_m(n)$ and $h'(n)$ are the zero-padded versions of $x_m(n)$ and $h(n)$, we have that the filtered blocks $y_m(n)$ in equation (3.72) become

$$y_m(n) = \sum_{l=0}^{N+K-1} x'_m(l)h'(n - l), \quad \text{for } 0 \leq n \leq N + K - 1 \tag{3.73}$$

From equations (3.72) and (3.73), we see that in order to convolve the long $x(n)$ with the length-K $h(n)$ it suffices to:

(i)   Divide $x(n)$ into length-N blocks $x_m(n)$.
(ii) Zero-pad $h(n)$ and each block $x_m(n)$ to length $N + K - 1$.
(iii) Perform the circular convolution of each block using the length-$(N + K - 1)$ DFT.
(iv) Add the results according to equation (3.72).

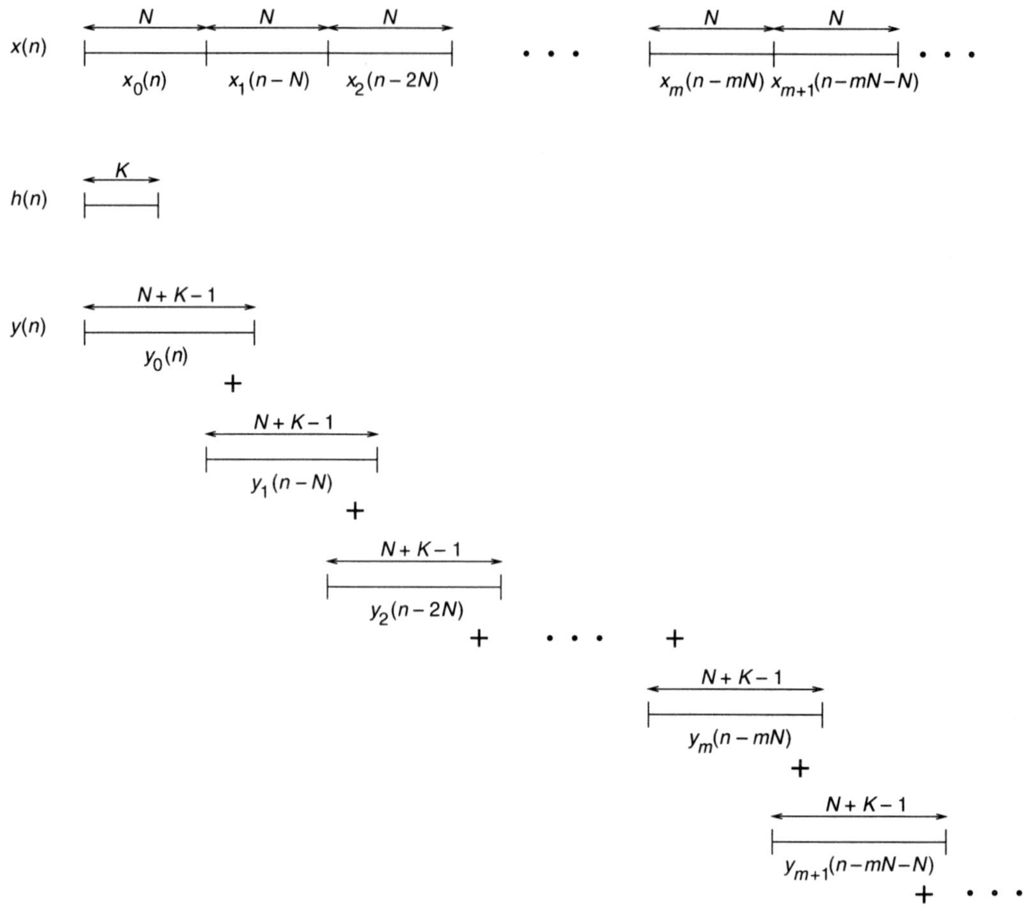

Figure 3.4 Illustration of the overlap-and-add method.

Note that in the addition performed in step (iv), there is an overlap of the $K - 1$ last samples of $y_m(n)$ and the $K - 1$ first samples of $y_{m+1}(n)$. This is why the above procedure is called the overlap-and-add method, which is illustrated schematically in Figure 3.4.

EXAMPLE 3.3

Compute graphically the linear convolution of $x(n)$ and $h(n)$ in Figure 3.5a by splitting $x(n)$ into blocks of two samples and using the overlap-and-add method.

SOLUTION

Figures 3.5b–e depict the solution.

$\triangle$

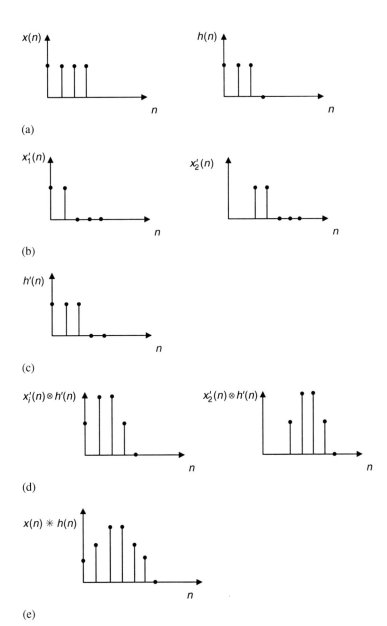

Figure 3.5 Example of the application of the overlap-and-add method. (a) Original sequences; (b) Input sequence divided into length-2 blocks. Each block has been zero-padded to length $N = 5$; (c) zero-padded $h(n)$; (d) circular convolution of each zero-padded block with zero-padded $h(n)$; (e) final result (normalized).

3.4.3   Overlap-and-save method

In the overlap-and-add method, one divides the signal into blocks of length N and computes the convolutions using DFTs of size $(N + K - 1)$, where K is the duration of $h(n)$. In the overlap-and-save method, one uses DFTs of length N instead. This poses a problem, because the length-N circular convolution of a length-N block $x_m(n)$ and a length-K $h(n)$ is not equal to their linear convolution. To circumvent this, one only uses the samples of the circular convolution that are equal to the ones of the linear convolution. These valid samples can be determined by referring to the expression of the circular convolution in Subsection 3.4.1, equation (3.56). From there, we see that the condition for computing a linear convolution using a circular convolution is given by equation (3.57), repeated here for convenience, with $x(n)$ replaced by $x_m(n)$:

$$c(n) = \sum_{l=n+1}^{N-1} x_m(l)h(n - l + N) = 0, \quad \text{for } 0 \le n \le N - 1 \tag{3.74}$$

This equation indicates that the length-N circular convolution, between a length-N block $x_m(n)$ and a length-K $h(n)$, is equal to their linear convolution whenever the above summation is null. Since, for $0 \le n \le N - 1$, we have that $x_m(n) \ne 0$, then $c(n)$ above is null only for n such that all the $h(n)$ in the summation are zero, that is for $n + 1 \le l \le N - 1$, then $h(n - l + N) = 0$. Since $h(n)$ has length K, then $h(n) = 0$, for $n \ge K$, and then $h(n - l + N) = 0$, for $n - l + N \ge K$, or accordingly $n \ge K - N + l$. The most strict case in this inequality is when $l = N - 1$. This implies that the condition in equation (3.74) is satisfied only when $n \ge K - 1$.

The conclusion is that the only samples of the length-N circular convolution that are equal to the ones of the linear convolution are for $n \ge K - 1$. Therefore, when computing the convolution of the blocks $x_m(n)$ with $h(n)$, the first $K - 1$ samples of the result have to be discarded. In order to compensate for the discarded samples, there must be an overlap of an extra $K - 1$ samples between adjacent blocks.

Thus, the signal $x(n)$ must be divided into blocks $x_m(n)$ of length N such that

$$x_m(n) = \begin{cases} x(n + m(N - K + 1) - K + 1), & \text{for } 0 \le n \le N - 1 \\ 0, & \text{otherwise} \end{cases} \tag{3.75}$$

We then take, for each block $x_m(n)$, only the samples of the circular convolution with $h(n)$ for which $n \ge K - 1$. Note that $x(n)$ has to be padded with $K - 1$ zeros at the beginning, since the first $K - 1$ samples of the output are discarded. Then, if $h'(n)$ is the version of $h(n)$ zero-padded to length N, the useful output $y_m(n)$ of each block can be expressed as

$$y_m(n) = \sum_{l=0}^{N} x'_m(l)h'((n - l) \bmod N), \quad \text{for } K - 1 \le n \le N - 1 \tag{3.76}$$

Discarding the $K - 1$ first samples of $y_m(n)$, the output $y(n) = x(n) * h(n)$ is built, for each m, as

$$y(n) = y_m(n - m(N - K + 1)) \tag{3.77}$$

for $m(N - K + 1) + K - 1 \le n \le m(N - K + 1) + N - 1$.

From equations (3.76) and (3.77), we see that, in order to convolve the long $x(n)$ with the length-K $h(n)$ using the overlap-and-save method, it suffices to:

(i)   Divide $x(n)$ into length-N blocks $x_m(n)$ with an overlap of $K - 1$ samples as in equation (3.75). The first block should be zero-padded with $K - 1$ zeros at its

:ach block using the length-N DFT.
equation (3.77).

e $K - 1$ last samples of block $y_m(n)$ being
$K - 1$ first samples of block $y_{m+1}(n)$, thus
ve method, which is illustrated schematically

volution of $x(n)$ and $h(n)$ in Figure 3.7a
ngth-6 blocks and using the overlap-and-save

ed from $x(n)$ and the circular convolutions of
added to length 6. The figure also shows the
(n) and $h(n)$.

$l = 6$, and each partial convolution generates

$\triangle$

3.5   Fast Fourier transform

In the previous section, we saw that the DFT is an effective discrete representation in frequency that can be used to compute linear convolutions between two discrete sequences. However, by examining the DFT and IDFT definitions in equations (3.15)

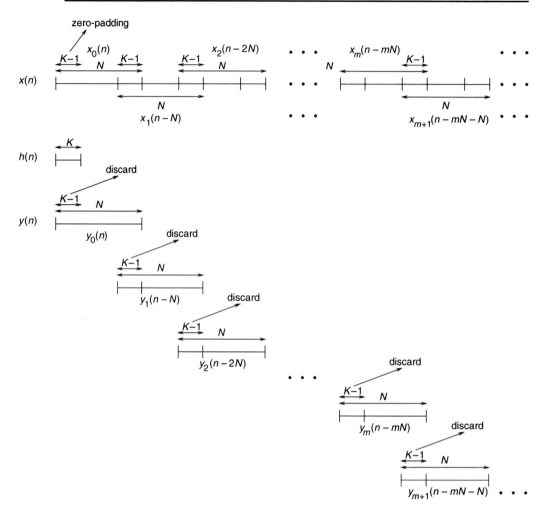

Figure 3.6 Illustration of the overlap-and-save method.

and (3.16), repeated here for convenience

$$X(k) = \sum_{n=0}^{N-1} x(n) W_N^{kn}, \quad \text{for } 0 \le k \le N - 1 \tag{3.78}$$

$$x(n) = \frac{1}{N} \sum_{k=0}^{N-1} X(k) W_N^{-kn}, \quad \text{for } 0 \le n \le N - 1 \tag{3.79}$$

we see that in order to compute the DFT and IDFT of a length-N sequence, one needs about N^2 complex multiplications, that is, the complexity of the DFT grows with the square of the signal length. This severely limits its practical use for lengthy signals. Fortunately, in 1965, Cooley and Tukey proposed an efficient algorithm to compute the DFT (Cooley & Tukey, 1965), which requires a number of complex multiplications of

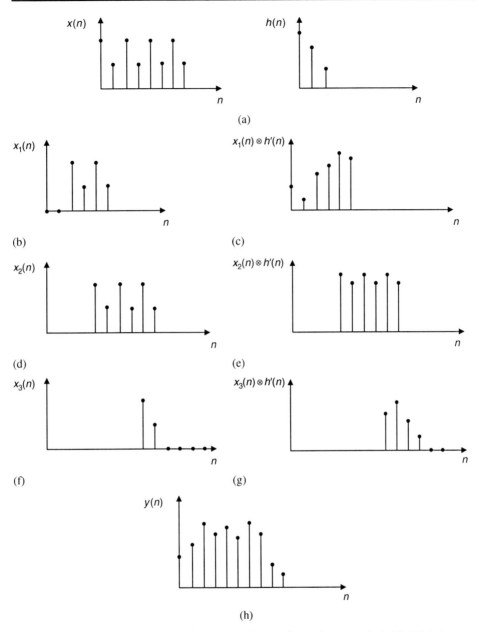

Figure 3.7 Linear convolution using the DFT and the overlap-and-save method. (a) Original sequences; (b) first block; (c) first partial convolution; (d) second block; (e) second partial convolution; (f) third block; (g) third partial convolution; (h) final result.

the order of $N \log_2 N$. This may represent a tremendous decrease in complexity. For example, even for signal lengths as low as 1024 samples, the decrease in complexity is of the order of 100 times, that is, two orders of magnitude. It is needless to say that the advent of this algorithm opened up an endless list of practical applications for the DFT, ranging from signal analysis to fast linear filtering. Today, there is an enormous number of fast algorithms for the computation of the DFT, and they are collectively known as FFT (Fast Fourier Transform) algorithms (Cochran et al., 1967). In this section we will study some of the most popular types of FFT algorithms.

3.5.1 Radix-2 algorithm with decimation in time

Suppose we have a sequence $x(n)$ whose length N is a power of two, that is, $N = 2^l$. Now, let us express the DFT relation in equation (3.78) by splitting the summation into two parts, one with the even-indexed $x(n)$ and another with the odd-indexed $x(n)$, obtaining

$$X(k) = \sum_{n=0}^{N-1} x(n) W_N^{nk}$$

$$= \sum_{n=0}^{\frac{N}{2}-1} x(2n) W_N^{2nk} + \sum_{n=0}^{\frac{N}{2}-1} x(2n+1) W_N^{(2n+1)k}$$

$$= \sum_{n=0}^{\frac{N}{2}-1} x(2n) W_N^{2nk} + W_N^k \sum_{n=0}^{\frac{N}{2}-1} x(2n+1) W_N^{2nk} \tag{3.80}$$

If we note that for N even, we have that

$$W_N^{2nk} = e^{-j\frac{2\pi}{N} 2nk} = e^{-j\frac{2\pi}{\frac{N}{2}} nk} = W_{\frac{N}{2}}^{nk} \tag{3.81}$$

then equation (3.80) becomes

$$X(k) = \sum_{n=0}^{\frac{N}{2}-1} x(2n) W_{\frac{N}{2}}^{nk} + W_N^k \sum_{n=0}^{\frac{N}{2}-1} x(2n+1) W_{\frac{N}{2}}^{nk} \tag{3.82}$$

and we can see that each summation may represent a distinct DFT of size $N/2$. Therefore, a DFT of size N can be computed through two DFTs of size $N/2$, in addition to the multiplications by W_N^k. Note that each new DFT has only $N/2$ coefficients, and for the computation of its coefficients we now need solely $(N/2)^2$ complex multiplications. In addition, since we have one distinct coefficient W_N^k for each k between 0 and $N - 1$, then we need to perform N multiplications by W_N^k. Therefore, the computation of the DFT according to equation (3.82) requires

$$2 \left(\frac{N}{2} \right)^2 + N = \frac{N^2}{2} + N \tag{3.83}$$

complex multiplications. Since $(N + \frac{N^2}{2})$ is smaller than N^2, for $N > 2$, equation (3.82) provides a decrease in complexity when compared to the usual DFT computation.

We should also compare the number of complex additions of the two forms of computation. The usual length-N DFT computation needs a total of $N(N-1) = N^2 - N$ additions. In equation (3.82), we need to compute two DFTs of length $N/2$, and then perform the N additions of the two DFTs after multiplication by W_N^k. Therefore, the total number of complex additions in equation (3.82) is

$$2\left[\left(\frac{N}{2}\right)^2 - \frac{N}{2}\right] + N = \frac{N^2}{2} \tag{3.84}$$

which also corresponds to a reduction in complexity.

From the above, exploiting the fact that N is a power of 2, it is easy to see that if the procedure shown in equation (3.82) is recursively applied to each of the resulting DFTs, we may get a very significant reduction in complexity until all the remaining DFTs are of length 2. The overall procedure is formalized in an algorithm by first rewriting equation (3.82) as

$$X(k) = X_e(k) + W_N^k X_o(k) \tag{3.85}$$

where $X_e(k)$ and $X_o(k)$ are, respectively, the DFTs of length $\frac{N}{2}$ of the even- and odd-indexed samples of $x(n)$, that is

$$\left.\begin{array}{l} X_e(k) = \displaystyle\sum_{n=0}^{\frac{N}{2}-1} x(2n) W_{\frac{N}{2}}^{nk} = \sum_{n=0}^{\frac{N}{2}-1} x_e(n) W_{\frac{N}{2}}^{nk} \\[4ex] X_o(k) = \displaystyle\sum_{n=0}^{\frac{N}{2}-1} x(2n+1) W_{\frac{N}{2}}^{nk} = \sum_{n=0}^{\frac{N}{2}-1} x_o(n) W_{\frac{N}{2}}^{nk} \end{array}\right\} \tag{3.86}$$

The DFTs above can be computed by separating $x_e(n)$ and $x_o(n)$ into their even- and odd-indexed samples, as follows

$$\left.\begin{array}{l} X_e(k) = \displaystyle\sum_{n=0}^{\frac{N}{4}-1} x_e(2n) W_{\frac{N}{4}}^{nk} + W_{\frac{N}{2}}^k \sum_{n=0}^{\frac{N}{4}-1} x_e(2n+1) W_{\frac{N}{4}}^{nk} \\[4ex] X_o(k) = \displaystyle\sum_{n=0}^{\frac{N}{4}-1} x_o(2n) W_{\frac{N}{4}}^{nk} + W_{\frac{N}{2}}^k \sum_{n=0}^{\frac{N}{4}-1} x_o(2n+1) W_{\frac{N}{4}}^{nk} \end{array}\right\} \tag{3.87}$$

such that

$$\left.\begin{array}{l} X_e(k) = X_{ee}(k) + W_{\frac{N}{2}}^k X_{eo}(k) \\[2ex] X_o(k) = X_{oe}(k) + W_{\frac{N}{2}}^k X_{oo}(k) \end{array}\right\} \tag{3.88}$$

where $X_{ee}(k)$, $X_{eo}(k)$, $X_{oe}(k)$, and $X_{oo}(k)$ correspond now to DFTs of length $\frac{N}{4}$.

Generically, in each step we compute DFTs of length L using DFTs of length $\frac{L}{2}$ as follows

$$X_i(k) = X_{ie}(k) + W_L^k X_{io}(k) \tag{3.89}$$

A recursive application of the above procedure can convert the computation of a DFT of length $N = 2^l$ in l steps to the computation of 2^l DFTs of length 1, because each step converts a DFT of length L into two DFTs of length $\frac{L}{2}$ plus a complex product by W_L^k and a complex sum. Therefore, supposing that $\mathcal{M}(N)$ and $\mathcal{A}(N)$ are respectively the number of complex multiplications and additions to compute a DFT of length N the following relations hold:

$$\mathcal{M}(N) = 2\mathcal{M}\left(\frac{N}{2}\right) + N \tag{3.90}$$

$$\mathcal{A}(N) = 2\mathcal{A}\left(\frac{N}{2}\right) + N \tag{3.91}$$

In order to compute the values of $\mathcal{M}(N)$ and $\mathcal{A}(N)$, we must solve the recursive equations above. The initial conditions are $\mathcal{M}(1) = \mathcal{A}(1) = 0$, since a length-1 DFT is equal to the identity. Note that, since equations (3.90) and (3.91) are the same, then it is clear that $\mathcal{M}(N) = \mathcal{A}(N)$.

Using the change of variables $N = 2^l$ and $T(l) = \frac{\mathcal{M}(N)}{N}$, equation (3.90) becomes

$$T(l) = T(l-1) + 1 \tag{3.92}$$

Since $T(0) = \mathcal{M}(1) = 0$, then $T(l) = l$. Therefore, one concludes that

$$\mathcal{M}(N) = \mathcal{A}(N) = N \log_2 N \tag{3.93}$$

that is, the Fast Fourier Transform (FFT) can be computed using $N \log_2 N$ complex multiplications and additions, which comprises an economy of the order of $\frac{N}{\log_2 N}$, when compared to the direct implementation in equation (3.78).

This FFT algorithm is known as a decimation-in-time algorithm because it recursively divides the sequence $x(n)$ into subsequences. We can devise a graphical representation of the above procedure by noting that in the operation in equation (3.89) each value of either $X_{ie}(k)$ or $X_{io}(k)$ is used twice. This is so because, if $X(k)$ has period L, then $X_{ie}(k)$ and $X_{io}(k)$ have period $\frac{L}{2}$. In other words,

$$X_i(k) = X_{ie}(k) + W_L^k X_{io}(k) \tag{3.94}$$

$$X_i\left(k + \frac{L}{2}\right) = X_{ie}\left(k + \frac{L}{2}\right) + W_L^{k+\frac{L}{2}} X_{io}\left(k + \frac{L}{2}\right) = X_{ie}(k) + W_L^{k+\frac{L}{2}} X_{io}(k) \tag{3.95}$$

Therefore, if we have the DFTs of length $\frac{L}{2}$, $X_{ie}(k)$ and $X_{io}(k)$, we can compute the length-L DFT $X(k)$ by applying equations (3.94) and (3.95), for $k = 0, \ldots, \frac{L}{2} - 1$. This

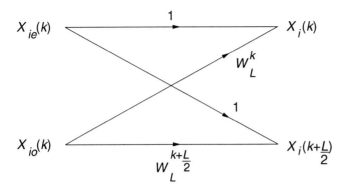

Figure 3.8 Basic cell of the decimation-in-time FFT algorithm.

process is illustrated by the graph in Figure 3.8. Since the FFT algorithm is composed of repetitions of this procedure, it is often called the basic cell of the algorithm. Due to the appearance of its graph, the basic cell is sometimes called a butterfly.

A graph for the complete FFT algorithm above is obtained by repeating the basic cell in Figure 3.8, for $L = N, \frac{N}{2}, \ldots, 2$ and for $k = 0, \ldots, \frac{L}{2} - 1$. Figure 3.9 illustrates the case when $N = 8$.

The graph in Figure 3.9 has an interesting property. The nodes of one section of the graph depend only on nodes of the previous section of the graph. For example, $X_i(k)$ depends only on $X_{ie}(k)$ and $X_{io}(k)$. In addition, $X_{ie}(k)$ and $X_{io}(k)$ are only used to compute $X_i(k)$ and $X_i(k + \frac{L}{2})$. Therefore, once computed, the values of $X_i(k)$ and $X_i(k + \frac{L}{2})$ can be stored in the same place as $X_{ie}(k)$ and $X_{io}(k)$. This implies that the intermediate results of the length-N FFT computation can be stored in a single size-N vector, that is, the results of section l can be stored in the same place as the results of section $l - 1$. This is why the computation of the intermediate results of the FFT is often said to be 'in place'.

Another important aspect of the FFT algorithm relates to the ordering of the input vector. As seen in Figure 3.9, the output vector is ordered sequentially while the input vector is not. A general rule for the ordering of the input vector can be devised by referring to the FFT graph in Figure 3.9. Moving from right to left, we note that in the second section, the upper half corresponds to the DFT of the even-indexed samples and the lower half to the DFT of the odd-indexed samples. Since the even-indexed samples have the least significant bit (LSB) equal to 0 and the odd-indexed samples have the LSB equal to 1, then the DFT of the upper half corresponds to samples with an index having LSB = 0 and the lower half to the DFT of samples with an index having LSB = 1. Likewise, for each half, the upper half corresponds to the even-indexed samples belonging to this half, that is, the ones with the second LSB = 0. Similarly, the lower half corresponds to the second LSB = 1. If we proceed until we reach the input signal on the left, we end up with the uppermost sample having an index with all bits equal

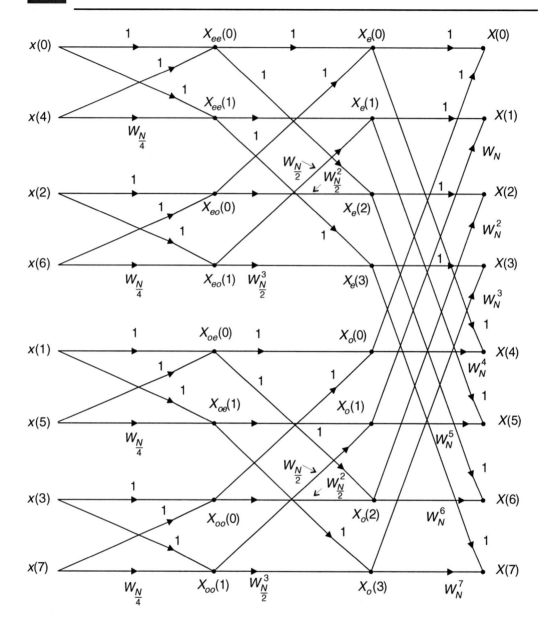

Figure 3.9 Graph of the decimation-in-time 8-point FFT algorithm.

to zero, the second one having the first bit equal to one and the others equal to zero, and so on. We note, then, that the ordering of the input vector is the ascending order of the bit-reversed indexes. For example, $x(3) = x(011)$ will be put in position 110 of the input, that is, position 6.

An additional economy in the number of complex multiplications of the algorithm can be obtained by noting that, in equation (3.95)

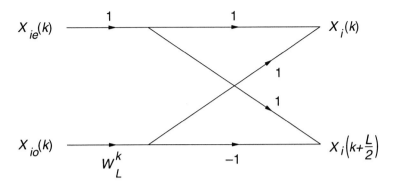

Figure 3.10 More efficient basic cell of the decimation-in-time FFT algorithm.

$$W_L^{k+\frac{L}{2}} = W_L^k W_L^{\frac{L}{2}} = W_L^k W_2 = -W_L^k \tag{3.96}$$

Then, equations (3.94) and (3.95) can be rewritten as

$$X_i(k) = X_{ie}(k) + W_L^k X_{io}(k) \tag{3.97}$$

$$X_i\left(k + \frac{L}{2}\right) = X_{ie}(k) - W_L^k X_{io}(k) \tag{3.98}$$

This allows a more efficient implementation of the basic cell in Figure 3.8, using one complex multiplication instead of two. The resulting cell is depicted in Figure 3.10. Substituting the basic cells corresponding to Figure 3.8 by the ones corresponding to Figure 3.10, we have the more efficient graph for the FFT shown in Figure 3.11.

With this new basic cell, the number of complex multiplications has dropped to half, that is, there is a total of $\frac{N}{2} \log_2 N$ complex multiplications. In order to perform a more accurate calculation, we have to discount the trivial multiplications. One set of them are by $W_L^0 = 1$. Since, when the DFTs have length L, we have $\frac{N}{L}$ DFTs and the term W_L^0 appears $\frac{N}{L}$ times. Then, we have $\frac{N}{2} + \frac{N}{4} + \cdots + \frac{N}{N} = N - 1$ multiplications by 1. The other set of trivial multiplications are the ones by $W_L^{\frac{L}{4}} = -j$. Since, in the first stage, there are no terms equal to $-j$ and, from the second stage on, the number of times a term $-j$ appears is the same as the number of times a term 1 appears, then we have $N - 1 - \frac{N}{2} = \frac{N}{2} - 1$ multiplications by $-j$. This gives an overall number of nontrivial complex multiplications equal to

$$\mathcal{M}(N) = \frac{N}{2} \log_2 N - N + 1 - \frac{N}{2} + 1 = \frac{N}{2} \log_2 N - \frac{3}{2}N + 2 \tag{3.99}$$

Note that the number of complex additions remains $\mathcal{A}(N) = N \log_2 N$.

If we shuffle the horizontal branches of the graph in Figure 3.11 so that the input signal is in normal ordering, then the output of the graph comes in bit-reversed

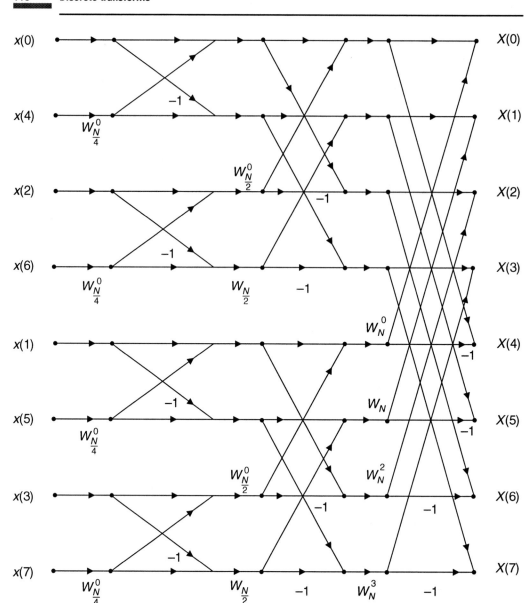

Figure 3.11 More efficient graph of the decimation-in-time 8-point FFT algorithm. The unmarked branches are equal to 1.

ordering. In this case, we have another 'in place' algorithm. In fact, there are a plethora of forms for the FFT. For example, there is an algorithm in which the input and output appear in normal ordering, but where there is no possibility of 'in place' computation (Cochran et al., 1967; Oppenheim & Schafer, 1975).

It is clear from equations (3.78) and (3.79) that, in order to compute the inverse DFT it suffices to change, in the graphs of Figures 3.9 or 3.11, the terms W_N^k to W_N^{-k} and divide the output of the graph by N.

An interesting interpretation of the FFT algorithms can be found if we look at the matrix form of the DFT in equation (3.23),

$$\mathbf{X} = \mathbf{W}_N \mathbf{x} \tag{3.100}$$

The matrix form is widely used for describing fast transform algorithms (Elliott & Rao, 1982). For instance, the decimation-in-time FFT algorithm can be represented in matrix form if we notice that equations (3.94) and (3.95), corresponding to the basic cell in Figure 3.8, can be expressed in matrix form as

$$\left[\begin{array}{c} X_i(k) \\ X_i(k + \frac{L}{2}) \end{array} \right] = \left[\begin{array}{cc} 1 & W_L^k \\ 1 & W_L^{k+\frac{L}{2}} \end{array} \right] \left[\begin{array}{c} X_{ie}(k) \\ X_{io}(k) \end{array} \right] \tag{3.101}$$

Then the graph in Figure 3.9 can be expressed as

$$\mathbf{X} = \mathbf{F}_8^{(8)} \mathbf{F}_8^{(4)} \mathbf{F}_8^{(2)} \mathbf{P}_8 \mathbf{x} \tag{3.102}$$

where

$$\mathbf{P}_8 = \left[\begin{array}{cccccccc} 1 & 0 & 0 & 0 & 0 & 0 & 0 & 0 \\ 0 & 0 & 0 & 0 & 1 & 0 & 0 & 0 \\ 0 & 0 & 1 & 0 & 0 & 0 & 0 & 0 \\ 0 & 0 & 0 & 0 & 0 & 0 & 1 & 0 \\ 0 & 1 & 0 & 0 & 0 & 0 & 0 & 0 \\ 0 & 0 & 0 & 0 & 0 & 1 & 0 & 0 \\ 0 & 0 & 0 & 1 & 0 & 0 & 0 & 0 \\ 0 & 0 & 0 & 0 & 0 & 0 & 0 & 1 \end{array} \right] \tag{3.103}$$

corresponds to the bit-reversal operation onto the indexes of the input vector $\mathbf{x}$,

$$\mathbf{F}_8^{(2)} = \left[\begin{array}{cccccccc} 1 & W_2^0 & 0 & 0 & 0 & 0 & 0 & 0 \\ 1 & W_2^1 & 0 & 0 & 0 & 0 & 0 & 0 \\ 0 & 0 & 1 & W_2^0 & 0 & 0 & 0 & 0 \\ 0 & 0 & 1 & W_2^1 & 0 & 0 & 0 & 0 \\ 0 & 0 & 0 & 0 & 1 & W_2^0 & 0 & 0 \\ 0 & 0 & 0 & 0 & 1 & W_2^1 & 0 & 0 \\ 0 & 0 & 0 & 0 & 0 & 0 & 1 & W_2^0 \\ 0 & 0 & 0 & 0 & 0 & 0 & 1 & W_2^1 \end{array} \right] \tag{3.104}$$

corresponds to the basic cells of the first stage,

$$\mathbf{F}_8^{(4)} = \begin{bmatrix} 1 & 0 & W_4^0 & 0 & 0 & 0 & 0 & 0 \\ 0 & 1 & 0 & W_4^1 & 0 & 0 & 0 & 0 \\ 1 & 0 & W_4^2 & 0 & 0 & 0 & 0 & 0 \\ 0 & 1 & 0 & W_4^3 & 0 & 0 & 0 & 0 \\ 0 & 0 & 0 & 0 & 1 & 0 & W_4^0 & 0 \\ 0 & 0 & 0 & 0 & 0 & 1 & 0 & W_4^1 \\ 0 & 0 & 0 & 0 & 1 & 0 & W_4^2 & 0 \\ 0 & 0 & 0 & 0 & 0 & 1 & 0 & W_4^3 \end{bmatrix} \tag{3.105}$$

corresponds to the basic cells of the second stage, and

$$\mathbf{F}_8^{(8)} = \begin{bmatrix} 1 & 0 & 0 & 0 & W_8^0 & 0 & 0 & 0 \\ 0 & 1 & 0 & 0 & 0 & W_8^1 & 0 & 0 \\ 0 & 0 & 1 & 0 & 0 & 0 & W_8^2 & 0 \\ 0 & 0 & 0 & 1 & 0 & 0 & 0 & W_8^3 \\ 1 & 0 & 0 & 0 & W_8^4 & 0 & 0 & 0 \\ 0 & 1 & 0 & 0 & 0 & W_8^5 & 0 & 0 \\ 0 & 0 & 1 & 0 & 0 & 0 & W_8^6 & 0 \\ 0 & 0 & 0 & 1 & 0 & 0 & 0 & W_8^7 \end{bmatrix} \tag{3.106}$$

corresponds to the basic cells of the third stage.

Comparing equations (3.100) and (3.102), one can see the FFT algorithm in Figure 3.9 as a factorization of the DFT matrix $\mathbf{W}_8$ such that

$$\mathbf{W}_8 = \mathbf{F}_8^{(8)} \mathbf{F}_8^{(4)} \mathbf{F}_8^{(2)} \mathbf{P}_8 \tag{3.107}$$

This factorization corresponds to a fast algorithm because the matrices $\mathbf{F}_8^{(8)}, \mathbf{F}_8^{(4)}, \mathbf{F}_8^{(2)}$, and $\mathbf{P}_8$ have most elements equal to zero. Because of that, the product by each matrix above can be effected at the expense of at most 8 complex multiplications, except for $\mathbf{P}_8$, which is just a permutation, thus the product by it requires no multiplications at all.

Equations (3.102)–(3.107) exemplify the general fact that each different FFT algorithm corresponds to a different factorization of the DFT matrix $\mathbf{W}_N$ into sparse matrices. For example, the reduction in complexity achieved by replacing Figure 3.8 (equations (3.94) and (3.95)) by Figure 3.10 (equations (3.97) and (3.98)) is equivalent to factoring the matrix in equation (3.101) as

$$\begin{bmatrix} 1 & W_L^k \\ 1 & W_L^{k+\frac{L}{2}} \end{bmatrix} = \begin{bmatrix} 1 & 1 \\ 1 & -1 \end{bmatrix} \begin{bmatrix} 1 & 0 \\ 0 & W_L^k \end{bmatrix} \tag{3.108}$$

3.5.2   Decimation in frequency

An alternative algorithm for the fast computation of the DFT can be obtained using decimation in frequency, that is, division of $X(k)$ into subsequences. The algorithm is generated as follows:

$$
\begin{aligned}
X(k) &= \sum_{n=0}^{N-1} x(n) W_N^{nk} \\
&= \sum_{n=0}^{\frac{N}{2}-1} x(n) W_N^{nk} + \sum_{n=\frac{N}{2}}^{N-1} x(n) W_N^{nk} \\
&= \sum_{n=0}^{\frac{N}{2}-1} x(n) W_N^{nk} + \sum_{n=0}^{\frac{N}{2}-1} x\left(n + \frac{N}{2}\right) W_N^{\frac{N}{2}k} W_N^{nk} \\
&= \sum_{n=0}^{\frac{N}{2}-1} \left(x(n) + W_N^{\frac{N}{2}k} x\left(n + \frac{N}{2}\right)\right) W_N^{nk}
\end{aligned}
\tag{3.109}
$$

We can now compute the even and odd samples of $X(k)$ separately, that is

$$
\begin{aligned}
X(2l) &= \sum_{n=0}^{\frac{N}{2}-1} \left(x(n) + W_N^{Nl} x\left(n + \frac{N}{2}\right)\right) W_N^{2nl} \\
&= \sum_{n=0}^{\frac{N}{2}-1} \left(x(n) + x\left(n + \frac{N}{2}\right)\right) W_N^{2nl}
\end{aligned}
\tag{3.110}
$$

for $l = 0, 1, 2, \ldots, (\frac{N}{2} - 1)$, and

$$
\begin{aligned}
X(2l+1) &= \sum_{n=0}^{\frac{N}{2}-1} \left(x(n) + W_N^{(2l+1)\frac{N}{2}} x\left(n + \frac{N}{2}\right)\right) W_N^{(2l+1)n} \\
&= \sum_{n=0}^{\frac{N}{2}-1} \left(x(n) - x\left(n + \frac{N}{2}\right)\right) W_N^{(2l+1)n} \\
&= \sum_{n=0}^{\frac{N}{2}-1} \left[\left(x(n) - x\left(n + \frac{N}{2}\right)\right) W_N^{n}\right] W_N^{2ln}
\end{aligned}
\tag{3.111}
$$

for $l = 0, 1, 2, \ldots, (\frac{N}{2} - 1)$.

Equations (3.110) and (3.111) can be recognized as DFTs of length $\frac{N}{2}$, since $W_N^{2ln} = W_{\frac{N}{2}}^{ln}$. Before computing these DFTs, we have to compute the two intermediate

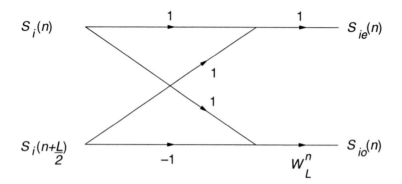

Figure 3.12 Basic cell of the decimation-in-frequency FFT algorithm.

signals $S_e(n)$ and $S_o(n)$, of length $\frac{N}{2}$, given by

$$
\left.
\begin{aligned}
S_e(n) &= x(n) + x\left(n + \frac{N}{2}\right) \\
S_o(n) &= \left(x(n) - x\left(n + \frac{N}{2}\right)\right) W_N^n
\end{aligned}
\right\}
\tag{3.112}
$$

Naturally, the above procedure can be repeated for each of the DFTs of length $\frac{N}{2}$, thus generating DFTs of length $\frac{N}{4}, \frac{N}{8}$, and so on. Therefore, the basic cell used in the decimation-in-frequency computation of the DFT is characterized by

$$
\left.
\begin{aligned}
S_{ie}(n) &= S_i(n) + S_i\left(n + \frac{L}{2}\right) \\
S_{io}(n) &= \left(S_i(n) - S_i\left(n + \frac{L}{2}\right)\right) W_L^n
\end{aligned}
\right\}
\tag{3.113}
$$

where $S_{ie}(n)$ and $S_{io}(n)$ have length $\frac{N}{2}$, and $S_i(n)$ has length L. The graph of such a basic cell is depicted in Figure 3.12.

Figure 3.13 shows the complete decimation-in-frequency FFT algorithm for $N = 8$. It is important to note that, in this algorithm, the input sequence $x(n)$ is in normal ordering and the output sequence $X(k)$ is in bit-reversed ordering. Also, comparing Figures 3.11 and 3.13, it is interesting to see that one graph is the transpose of the other.

3.5.3 Radix-4 algorithm

If $N = 2^{2l}$, instead of using radix-2 algorithms, we can use radix-4 FFT algorithms, which can give us additional economy in the required number of complex multiplications.

The derivation of the radix-4 algorithms parallels those of the radix-2 ones. If we use decimation-in-time, a length-N sequence is divided into 4 sequences of length $\frac{N}{4}$

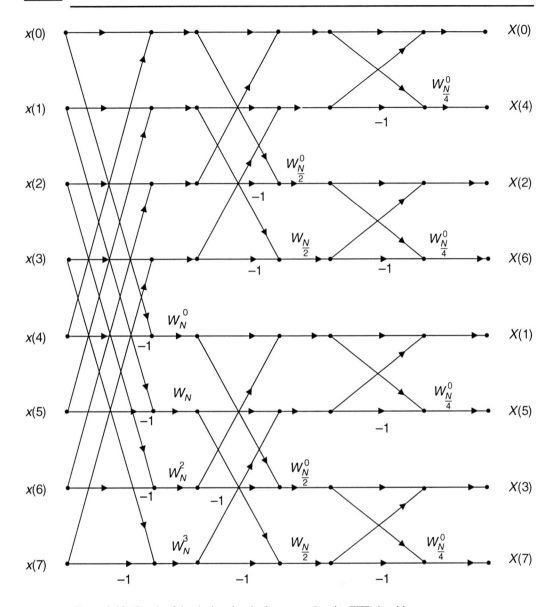

Figure 3.13 Graph of the decimation-in-frequency 8-point FFT algorithm.

such that

$$X(k) = \sum_{m=0}^{\frac{N}{4}-1} x(4m) W_N^{4mk} + \sum_{m=0}^{\frac{N}{4}-1} x(4m+1) W_N^{(4m+1)k}$$

$$+ \sum_{m=0}^{\frac{N}{4}-1} x(4m+2) W_N^{(4m+2)k} + \sum_{m=0}^{\frac{N}{4}-1} x(4m+3) W_N^{(4m+3)k}$$

$$
= \sum_{m=0}^{\frac{N}{4}-1} x(4m) W_{\frac{N}{4}}^{mk} + W_N^k \sum_{m=0}^{\frac{N}{4}-1} x(4m+1) W_{\frac{N}{4}}^{mk}
$$

$$
+ W_N^{2k} \sum_{m=0}^{\frac{N}{4}-1} x(4m+2) W_{\frac{N}{4}}^{mk} + W_N^{3k} \sum_{m=0}^{\frac{N}{4}-1} x(4m+3) W_{\frac{N}{4}}^{mk}
$$

$$
= \sum_{l=0}^{3} W_N^{lk} \sum_{m=0}^{\frac{N}{4}-1} x(4m+l) W_{\frac{N}{4}}^{mk} \tag{3.114}
$$

We can rewrite the above equation as

$$
X(k) = \sum_{l=0}^{3} W_N^{lk} F_l(k) \tag{3.115}
$$

where each $F_l(k)$ can be computed using 4 DFTs of length $\frac{N}{16}$, as shown below

$$
F_l(k) = \sum_{m=0}^{\frac{N}{4}-1} x(4m+l) W_{\frac{N}{4}}^{mk}
$$

$$
= \sum_{q=0}^{\frac{N}{16}-1} x(16q+l) W_{\frac{N}{4}}^{4qk} + W_{\frac{N}{4}}^{k} \sum_{q=0}^{\frac{N}{16}-1} x(16q+l+1) W_{\frac{N}{4}}^{4qk}
$$

$$
+ W_{\frac{N}{4}}^{2k} \sum_{q=0}^{\frac{N}{16}-1} x(16q+l+2) W_{\frac{N}{4}}^{4qk} + W_{\frac{N}{4}}^{3k} \sum_{q=0}^{\frac{N}{16}-1} x(16q+l+3) W_{\frac{N}{4}}^{4qk}
$$

$$
= \sum_{r=0}^{3} W_{\frac{N}{4}}^{rk} \sum_{q=0}^{\frac{N}{16}-1} x(16q+l+r) W_{\frac{N}{16}}^{qk} \tag{3.116}
$$

This process is recursively applied until we have to compute $\frac{N}{4}$ DFTs of length 4. From equation (3.115), we can see that the basic cell implements the equations below when computing a DFT $S(k)$ of length L using 4 DFTs of length $\frac{L}{4}$, $S_l(k)$, $l = 0, 1, 2, 3$.

$$
\left.
\begin{aligned}
S(k) &= \sum_{l=0}^{3} W_L^{lk} S_l(k) \\[2mm]
S\left(k+\tfrac{L}{4}\right) &= \sum_{l=0}^{3} W_L^{l\left(k+\frac{L}{4}\right)} S_l(k) = \sum_{l=0}^{3} W_L^{lk}(-\mathrm{j})^l S_l(k) \\[2mm]
S\left(k+\tfrac{L}{2}\right) &= \sum_{l=0}^{3} W_L^{l\left(k+\frac{L}{2}\right)} S_l(k) = \sum_{l=0}^{3} W_L^{lk}(-1)^l S_l(k) \\[2mm]
S\left(k+\tfrac{3L}{4}\right) &= \sum_{l=0}^{3} W_L^{l\left(k+\frac{3L}{4}\right)} S_l(k) = \sum_{l=0}^{3} W_L^{lk}(\mathrm{j})^l S_l(k)
\end{aligned}
\right\} \tag{3.117}
$$

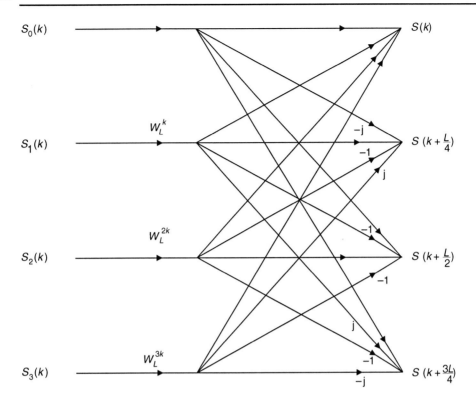

Figure 3.14 Basic cell of the radix-4 FFT algorithm.

The corresponding graph for the radix-4 butterfly is shown in Figure 3.14. As an illustration of the radix-4 algorithm, Figure 3.15 shows a sketch of the computation of a DFT of length 64 using a radix-4 FFT.

As can be deduced from Figure 3.14 and equation (3.116), at each stage of the application of the radix-4 FFT algorithm we need N complex multiplications and $3N$ complex additions, giving a total number of complex operations of

$$\mathcal{M}(N) = N \log_4 N = \frac{N}{2} \log_2 N \qquad (3.118)$$

$$\mathcal{A}(N) = 3N \log_4 N = \frac{3N}{2} \log_2 N \qquad (3.119)$$

Apparently, the radix-4 algorithms do not present any advantage when compared to radix-2 algorithms. However, the number of additions in the radix-4 basic cell

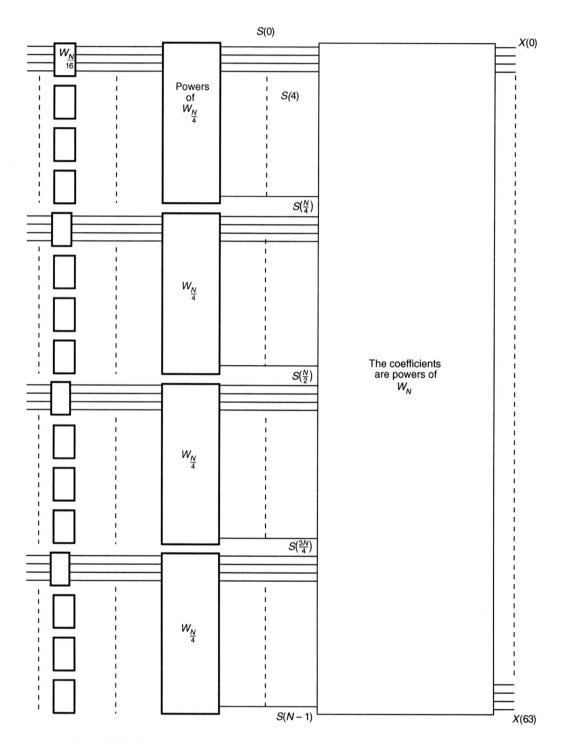

Figure 3.15　Sketch of a length-64 DFT using a radix-4 FFT algorithm.

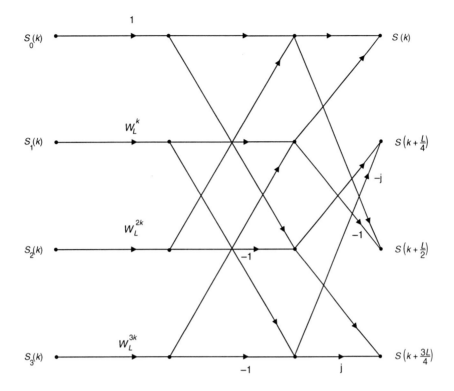

Figure 3.16 More efficient basic cell of the radix-4 FFT algorithm.

can be decreased if we note from Figure 3.14 and equation (3.117) that the quantities

$$
\left.
\begin{aligned}
&W_L^0 S_0(k) + W_L^{2k} S_2(k) \\[4pt]
&W_L^0 S_0(k) - W_L^{2k} S_2(k) \\[4pt]
&W_L^k S_1(k) + W_L^{3k} S_3(k) \\[4pt]
&W_L^k S_1(k) - W_L^{3k} S_3(k)
\end{aligned}
\right\}
\tag{3.120}
$$

are each computed twice, unnecessarily. By exploiting this fact, the more economical basic cell for the radix-4 algorithm shown in Figure 3.16 results, and we decrease the number of complex additions to $2N$ per stage, instead of $3N$.

The number of multiplications can be further decreased if we do not consider the multiplications by W_N^0. There is one of them in each basic cell, and three more of them in the basic cells corresponding to the index $k = 0$. Since there are $\log_4 N$ stages, we have, in the former case, $\frac{N}{4} \log_4 N$ elements W_N^0, while the number of elements corresponding to $k = 0$ is $3(1 + 4 + 16 + \cdots + \frac{N}{4}) = 3(N - 1)$. Therefore, the total number of multiplications is given by

$$\mathcal{M}(N) = N \log_4 N - \frac{N}{4} \log_4 N - N + 1 = \frac{3}{8} N \log_2 N - N + 1 \qquad (3.121)$$

Some additional trivial multiplications can also be detected, as shown in Nussbaumer (1982). If we then compare equations (3.99) and (3.121), we note that the radix-4 algorithm can be more economical, in terms of the number of overall complex multiplications, than the radix-2 algorithm.

It is important to point out that, in general, the shorter the length of the DFTs of the basic cell of a FFT algorithm, the more efficient it is. The exceptions to this rule are the radix-4, -8, -16,..., algorithms, where we can obtain a smaller number of multiplications than in radix-2 algorithms.

Although the radix-4 algorithm introduced here was based on the decimation-in-time approach, a similar algorithm based on the decimation-in-frequency method could be readily obtained.

3.5.4 Algorithms for arbitrary values of N

Efficient algorithms for the computation of DFTs having a generic length N are possible provided that N is not prime (Singleton, 1969; Rabiner, 1979). In such cases, we can decompose N as a product of factors

$$N = N_1 N_2 N_3 \cdots N_l = N_1 N_{2l} \qquad (3.122)$$

where $N_{2l} = N_2 N_3 \cdots N_l$. We can then initially divide the input sequence into N_1 sequences of length N_{2l}, thus writing the DFT of $x(n)$ as

$$X(k) = \sum_{n=0}^{N-1} x(n) W_N^{nk}$$

$$= \sum_{m=0}^{N_{2l}-1} x(N_1 m) W_N^{m N_1 k} + \sum_{m=0}^{N_{2l}-1} x(N_1 m + 1) W_N^{m N_1 k + k} + \cdots$$

$$+ \sum_{m=0}^{N_{2l}-1} x(N_1 m + N_1 - 1) W_N^{m N_1 k + (N_1 - 1)k}$$

$$= \sum_{m=0}^{N_{2l}-1} x(N_1 m) W_{N_{2l}}^{mk} + W_N^k \sum_{m=0}^{N_{2l}-1} x(N_1 m + 1) W_{N_{2l}}^{mk}$$

$$+ W_N^{2k} \sum_{m=0}^{N_{2l}-1} x(N_1 m + 2) W_{N_{2l}}^{mk} + \cdots + W_N^{(N_1 - 1)k} \sum_{m=0}^{N_{2l}-1} x(N_1 m + N_1 - 1) W_{N_{2l}}^{mk}$$

$$= \sum_{r=0}^{N_1 - 1} W_N^{rk} \sum_{m=0}^{N_{2l}-1} x(N_1 m + r) W_{N_{2l}}^{mk} \qquad (3.123)$$

This equation can be interpreted as being the computation of a length-N DFT using N_1 DFTs of length N_{2l}. We then need $N_1 N_{2l}^2$ complex multiplications to compute the N_1 mentioned DFTs, plus $N(N_1 - 1)$ complex multiplications to compute the products of W_N^{rk} with the N_1 DFTs. We can continue this process by computing each of the N_1 DFTs of length N_{2l} using N_2 DFTs of length N_{3l}, where $N_{3l} = N_3 N_4 \cdots N_l$, and so on, until all the DFTs have length N_l.

It can be shown that in such a case the total number of complex multiplications is given by (see Exercise 3.16)

$$\mathcal{M}(N) = N(N_1 + N_2 + \cdots + N_{l-1} + N_l - l) \tag{3.124}$$

For example, if $N = 63 = 3 \times 3 \times 7$ we have that $\mathcal{M}(N) = 63(3+3+7-3) = 630$. It is interesting to note that in order to compute a length-64 FFT we need only 384 complex multiplications if we use a radix-2 algorithm. This example reinforces the idea that we should, as a rule of thumb, divide N into factors as small as possible. In practice, whenever possible, the input sequences are zero-padded to force N to be a power of 2. As seen in the example above, this usually leads to an overall economy in the number of multiplications.

3.5.5   Alternative techniques for determining the DFT

The algorithms for efficient computation of the DFT presented in the previous subsections are generically called FFT algorithms. They were, for a long time, the only known methods to enable the efficient computation of long DFTs. In 1976, however, Winograd showed that there are algorithms with smaller complexity than the FFTs. These algorithms are based on convolution calculations exploring 'number-theoretic' properties of the addresses of the data (McClellan & Rader, 1979). These algorithms are denominated Winograd Fourier transform (WFT) algorithms.

In terms of practical implementations, the WFT has a smaller number of complex multiplications than the FFT, at the expense of a more complex algorithm. This gives an advantage to the WFT in many cases. However, the FFTs are more modular, which is an advantage in hardware implementations, especially in very large scale integration (VLSI). The main disadvantage of the WFT in this case is the complexity of the control path. Therefore, when implemented in special-purpose digital signal processors (DSPs) the FFTs have a clear advantage. This is so because multiplications are not a problem in such processors, and the more complex algorithms of the WFT make it slower than the FFTs in most cases.

Another class of techniques for the computation of convolutions and DFTs is given by the number-theoretic transform (NTT), which explores number-theoretic properties of the data. NTT techniques have practical implementations in machines with modular arithmetic. Also, they are useful for computations using hardware based on residue arithmetic (McClellan & Rader, 1979; Nussbaumer, 1982; Elliott & Rao, 1982).

3.6   Other discrete transforms

As seen in the previous sections, the DFT is a natural discrete-frequency representation for finite-length discrete signals, consisting of uniformly spaced samples of the Fourier transform. The direct and inverse DFTs can be expressed in matrix form as given by equations (3.23) and (3.24), repeated here for convenience

$$\mathbf{X} = \mathbf{W}_N \mathbf{x} \tag{3.125}$$

$$\mathbf{x} = \frac{1}{N} \mathbf{W}_N^* \mathbf{X} \tag{3.126}$$

where $\{\mathbf{W}_N\}_{ij} = W_N^{ij}$.

By using a different matrix, $\mathbf{C}_N$, instead of $\mathbf{W}_N$, we can generalize the above definition to

$$\mathbf{X} = \mathbf{C}_N \mathbf{x} \tag{3.127}$$

$$\mathbf{x} = \gamma \mathbf{C}_N^{*T} \mathbf{X} \tag{3.128}$$

where the superscript T denotes matrix transposition and γ is a constant. For matrices $\mathbf{C}_N$ such that $\mathbf{C}_N^{-1} = \gamma \mathbf{C}_N^{*T}$, equations (3.127) and (3.128) represent several discrete transforms which are frequently employed in signal processing applications. In the remainder of this section, we describe some of the most commonly employed discrete transforms.

3.6.1   Discrete cosine transform

The length-N discrete cosine transform (DCT) of a signal $x(n)$ can be defined (Ahmed et al., 1974) as

$$C(k) = \alpha(k) \sum_{n=0}^{N-1} x(n) \cos \left[\frac{\pi (n + \frac{1}{2})k}{N} \right], \quad \text{for } 0 \leq k \leq N - 1 \tag{3.129}$$

where

$$\alpha(k) = \begin{cases} \sqrt{\frac{1}{N}}, & \text{for } k = 0 \\ \\ \sqrt{\frac{2}{N}}, & \text{for } 1 \leq k \leq N - 1 \end{cases} \tag{3.130}$$

Accordingly, the inverse DCT is given by

$$x(n) = \sum_{k=0}^{N-1} \alpha(k) C(k) \cos \left[\frac{\pi (n + \frac{1}{2})k}{N} \right], \quad \text{for } 0 \leq n \leq N - 1 \tag{3.131}$$

Note that the DCT is a real transform, that is, it maps a real signal into real DCT coefficients. From equations (3.129)–(3.131), we can define the DCT matrix $\mathbf{C}_N$ by

$$\{\mathbf{C}_N\}_{kn} = \alpha(k) \cos\left[\frac{\pi(n+\frac{1}{2})k}{N}\right] \tag{3.132}$$

and the matrix form of the DCT becomes

$$\mathbf{c} = \mathbf{C}_N \mathbf{x} \tag{3.133}$$

$$\mathbf{x} = \mathbf{C}_N^{\mathrm{T}} \mathbf{c} \tag{3.134}$$

Note that for the above equations to be valid, $\mathbf{C}_N^{-1} = \mathbf{C}_N^{\mathrm{T}}$, which implies that the matrix $\mathbf{C}_N$ is unitary. Due to this fact, it is easy to show that Parseval's relation holds, that is, the energy of the signal is equal to the energy of the transform coefficients,

$$\sum_{n=0}^{N-1} C^2(k) = \mathbf{c}^{\mathrm{T}}\mathbf{c} = (\mathbf{C}_N\mathbf{x})^{\mathrm{T}}\mathbf{C}_N\mathbf{x} = \mathbf{x}^{\mathrm{T}}\mathbf{C}_N^{\mathrm{T}}\mathbf{C}_N\mathbf{x} = \mathbf{x}^{\mathrm{T}}\mathbf{x} = \sum_{n=0}^{N-1} x^2(n) \tag{3.135}$$

The DCT enjoys a very important property: when applied to signals such as voice and video, most of the energy is concentrated in few transform coefficients. For example, in Figure 3.17a, we can see a digital signal $x(n)$ corresponding to one line of a digitized television signal, and in Figure 3.17b, we show its DCT coefficients $C(k)$. It can be seen that the energy is mostly concentrated in the first transform coefficients. Due to this property, the DCT is widely used in video compression schemes, because the coefficients with lowest energy can be discarded during transmission without introducing significant distortion in the original signal (Bhaskaran & Konstantinides, 1997). In fact, the DCT is part of most of the digital video broadcasting systems in operation in the world (Whitaker, 1999).

Since the DCT is based on sinusoids and $\cos(x) = \frac{1}{2}(e^{jx} + e^{-jx})$, then in the worst case the DCT of $x(n)$ can be computed using one length-$2N$ DFT. This can be deduced by observing that

$$\cos\left[\frac{\pi(n+\frac{1}{2})k}{N}\right] = \cos\left(\frac{2\pi}{2N}kn + \frac{\pi}{2N}k\right)$$

$$= \frac{1}{2}\left[e^{\left(j\frac{2\pi}{2N}kn + \frac{\pi}{2N}k\right)} + e^{-\left(j\frac{2\pi}{2N}kn + \frac{\pi}{2N}k\right)}\right]$$

$$= \frac{1}{2}\left(W_{2N}^{\frac{k}{2}} W_{2N}^{kn} + W_{2N}^{-\frac{k}{2}} W_{2N}^{-kn}\right) \tag{3.136}$$

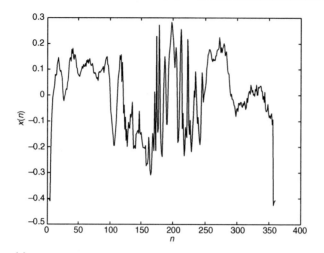

(a)

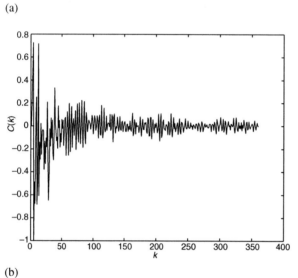

(b)

Figure 3.17 The DCT of a video signal: (a) discrete-time video signal $x(n)$; (b) the DCT of $x(n)$.

which implies that

$$C(k) = \alpha(k) \sum_{n=0}^{N-1} x(n) \cos\left[\frac{\pi(n + \frac{1}{2})k}{N}\right]$$

$$= \frac{1}{2}\left(\alpha(k)W_{2N}^{\frac{k}{2}} \sum_{n=0}^{N-1} x(n)W_{2N}^{kn} + \alpha(k)W_{2N}^{-\frac{k}{2}} \sum_{n=0}^{N-1} x(n)W_{2N}^{-kn}\right)$$

$$= \frac{1}{2}\alpha(k)\left(W_{2N}^{\frac{k}{2}} \mathrm{DFT}_{2N}\left\{\hat{x}(n)\right\}(k) + W_{2N}^{-\frac{k}{2}} \mathrm{DFT}_{2N}\left\{\hat{x}(n)\right\}(-k)\right) \tag{3.137}$$

for $0 \leq k \leq N - 1$, where $\hat{x}(n)$ is equal to $x(n)$ zero-padded to length $2N$. Note that

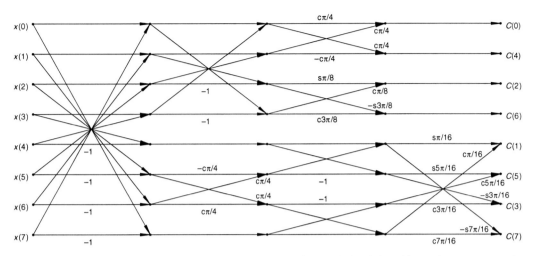

Figure 3.18 A fast algorithm for the computation of a length-8 DCT. In this graph, c*x* corresponds to cos *x* and s*x* corresponds to sin *x*.

the second term in the above equation is equivalent to the first one computed at the index $-k$. Therefore, we actually need to compute only one length-$2N$ DFT.

One can see that the algorithm in equation (3.137) has a complexity of one length-$2N$ DFT, plus the $2N$ multiplications by W_{2N}^k, which gives a total of $(2N + 2N \log_2 2N)$ complex multiplications.

However, there are other fast algorithms for the DCT which give a complexity of the order of $N \log_2 N$ real multiplications. One popular fast DCT algorithm is given in Chen et al. (1977). Its graph for $N = 8$ is shown in Figure 3.18.

This graph corresponds to the following factorization of $\mathbf{C}_8$:

$$\mathbf{C}_8 = \mathbf{P}_8 \mathbf{A}_4 \mathbf{A}_3 \mathbf{A}_2 \mathbf{A}_1 \tag{3.138}$$

where

$$\mathbf{A}_1 = \begin{bmatrix} 1 & 0 & 0 & 0 & 0 & 0 & 0 & 1 \\ 0 & 1 & 0 & 0 & 0 & 0 & 1 & 0 \\ 0 & 0 & 1 & 0 & 0 & 1 & 0 & 0 \\ 0 & 0 & 0 & 1 & 1 & 0 & 0 & 0 \\ 0 & 0 & 0 & 1 & -1 & 0 & 0 & 0 \\ 0 & 0 & 1 & 0 & 0 & -1 & 0 & 0 \\ 0 & 1 & 0 & 0 & 0 & 0 & -1 & 0 \\ 1 & 0 & 0 & 0 & 0 & 0 & 0 & -1 \end{bmatrix} \tag{3.139}$$

$$
\mathbf{A}_2 =
\begin{bmatrix}
1 & 0 & 0 & 1 & 0 & 0 & 0 & 0 \\
0 & 1 & 1 & 0 & 0 & 0 & 0 & 0 \\
0 & 1 & -1 & 0 & 0 & 0 & 0 & 0 \\
1 & 0 & 0 & -1 & 0 & 0 & 0 & 0 \\
0 & 0 & 0 & 0 & 1 & 0 & 0 & 0 \\
0 & 0 & 0 & 0 & 0 & -\cos\frac{\pi}{4} & \cos\frac{\pi}{4} & 0 \\
0 & 0 & 0 & 0 & 0 & \cos\frac{\pi}{4} & \cos\frac{\pi}{4} & 0 \\
0 & 0 & 0 & 0 & 0 & 0 & 0 & 1
\end{bmatrix}
\tag{3.140}
$$

$$
\mathbf{A}_3 =
\begin{bmatrix}
\cos\frac{\pi}{4} & \cos\frac{\pi}{4} & 0 & 1 & 0 & 0 & 0 & 0 \\
\cos\frac{\pi}{4} & -\cos\frac{\pi}{4} & 0 & 0 & 0 & 0 & 0 & 0 \\
0 & 0 & \sin\frac{\pi}{8} & \cos\frac{\pi}{8} & 0 & 0 & 0 & 0 \\
0 & 0 & -\sin\frac{3\pi}{8} & \cos\frac{3\pi}{8} & 0 & 0 & 0 & 0 \\
0 & 0 & 0 & 0 & 1 & 1 & 0 & 0 \\
0 & 0 & 0 & 0 & 1 & -1 & 0 & 0 \\
0 & 0 & 0 & 0 & 0 & 0 & -1 & 1 \\
0 & 0 & 0 & 0 & 0 & 0 & 1 & 1
\end{bmatrix}
\tag{3.141}
$$

$$
\mathbf{A}_4 =
\begin{bmatrix}
1 & 0 & 0 & 1 & 0 & 0 & 0 & 0 \\
0 & 1 & 0 & 0 & 0 & 0 & 0 & 0 \\
0 & 0 & 1 & 0 & 0 & 0 & 0 & 0 \\
0 & 0 & 0 & 1 & 0 & 0 & 0 & 0 \\
0 & 0 & 0 & 0 & \sin\frac{\pi}{16} & 0 & 0 & \cos\frac{\pi}{16} \\
0 & 0 & 0 & 0 & 0 & \sin\frac{5\pi}{16} & \cos\frac{5\pi}{16} & 0 \\
0 & 0 & 0 & 0 & 0 & -\sin\frac{3\pi}{16} & \cos\frac{3\pi}{16} & 0 \\
0 & 0 & 0 & 0 & -\sin\frac{7\pi}{16} & 0 & 0 & \cos\frac{7\pi}{16}
\end{bmatrix}
\tag{3.142}
$$

and $\mathbf{P}_8$ is given by equation (3.103), corresponding to the placement in normal ordering of the bit-reversed indexes of the output vector $\mathbf{c}$. Note that the operation corresponding to $\mathbf{P}_8$ is not shown in Figure 3.18.

After discounting the trivial multiplications, we have that, for a generic $N = 2^l$, the numbers of real multiplications $\mathcal{M}(N)$ and real additions $\mathcal{A}(N)$ in this fast implementation of the DCT are (Chen et al., 1977)

$$
\mathcal{M}(N) = N \log_2 N - \frac{3N}{2} + 4
\tag{3.143}
$$

$$
\mathcal{A}(N) = \frac{3N}{2}(\log_2 N - 1) + 2
\tag{3.144}
$$

3.6.2 A family of sine and cosine transforms

The DCT is a particular case of a more general class of transforms, whose transform matrix has its rows composed of sines or cosines of increasing frequency. Such transforms have a number of applications including the design of filter banks (Section 9.8).

Table 3.1. *Definition of the even cosine and sine transforms.*

	ϵ_1	ϵ_2	ϵ_3	ϵ_4	ϵ_5
C_{kn}^{I}	-1	1	1	0	0
C_{kn}^{II}	0	1	0	0	$\frac{1}{2}$
C_{kn}^{III}	0	0	1	$\frac{1}{2}$	0
C_{kn}^{IV}	0	0	0	$\frac{1}{2}$	$\frac{1}{2}$
S_{kn}^{I}	1	0	0	0	0
S_{kn}^{II}	0	1	0	0	$-\frac{1}{2}$
S_{kn}^{III}	0	0	1	$-\frac{1}{2}$	0
S_{kn}^{IV}	0	0	0	$\frac{1}{2}$	$\frac{1}{2}$

In what follows, we present a brief description of the so-called even forms of the cosine and sine transforms.

There are four types of even cosine transforms and four types of even sine transforms. Their transform matrices are referred to as $\mathbf{C}_N^{I-IV}$ for the cosine transforms and $\mathbf{S}_N^{I-IV}$ for the sine transforms. Their (k, n) elements are given by

$$\{\mathbf{C}_N^x\}_{kn} = \sqrt{\frac{2}{N + \epsilon_1}} \left(\alpha_{N+\epsilon_1}(k)\right)^{\epsilon_2} \left(\alpha_{N+\epsilon_1}(n)\right)^{\epsilon_3} \cos\left[\frac{\pi(k + \epsilon_4)(n + \epsilon_5)}{N + \epsilon_1}\right] \qquad (3.145)$$

$$\{\mathbf{S}_N^x\}_{kn} = \sqrt{\frac{2}{N + \epsilon_1}} \left(\alpha_{N+\epsilon_1}(k)\right)^{\epsilon_2} \left(\alpha_{N+\epsilon_1}(n)\right)^{\epsilon_3} \sin\left[\frac{\pi(k + \epsilon_4)(n + \epsilon_5)}{N + \epsilon_1}\right] \qquad (3.146)$$

where

$$\alpha_\gamma(k) = \begin{cases} \frac{1}{\sqrt{2}}, & \text{for } k = 0 \text{ or } k = \gamma \\ 1, & \text{for } 1 \le k \le \gamma - 1 \end{cases} \qquad (3.147)$$

The set $(\epsilon_1, \epsilon_2, \epsilon_3, \epsilon_4, \epsilon_5)$ defines the transform. Table 3.1 gives those values for all even cosine and sine transforms. One should note that all these transforms have fast algorithms and are unitary, that is $\mathbf{A}^{-1} = \mathbf{A}^{*T}$. For instance, the DCT defined in Subsection 3.6.1 is the same as $\mathbf{C}_N^{II}$.

3.6.3   Discrete Hartley transform

The discrete Hartley transform (DHT) can be viewed as a real counterpart of the DFT when it is applied to real signals. The definition of a length-N direct DHT is (Olejniczak & Heydt, 1994; Bracewell, 1994)

$$H(k) = \sum_{n=0}^{N-1} x(n) \operatorname{cas}\left(\frac{2\pi}{N}kn\right), \quad \text{for } 0 \leq k \leq N - 1 \tag{3.148}$$

where $\operatorname{cas} x = \cos x + \sin x$. Similarly, the inverse DHT is determined by

$$x(n) = \frac{1}{N} \sum_{k=0}^{N-1} H(k) \operatorname{cas}\left(\frac{2\pi}{N}kn\right), \quad \text{for } 0 \leq k \leq N - 1 \tag{3.149}$$

The Hartley transform is attractive due to the following properties:

- If the input signal is real, the DHT is real.
- The DFT can be easily derived from the DHT and vice versa.
- It has efficient fast algorithms (Bracewell, 1984).
- It has a convolution-multiplication property.

The first property follows trivially from the definition of the DHT. The other properties can be derived from the relations between the DFT $X(k)$ and the DHT $H(k)$ of a real sequence $x(n)$, which are

$$H(k) = \operatorname{Re}\{X(k)\} - \operatorname{Im}\{X(k)\} \tag{3.150}$$
$$X(k) = \mathcal{E}\{H(k)\} - j\mathcal{O}\{H(k)\} \tag{3.151}$$

where the operators $\mathcal{E}\{\cdot\}$ and $\mathcal{O}\{\cdot\}$ correspond to the even and odd parts, respectively, that is

$$\mathcal{E}\{H(k)\} = \frac{H(k) + H(-k)}{2} \tag{3.152}$$

$$\mathcal{O}\{H(k)\} = \frac{H(k) - H(-k)}{2} \tag{3.153}$$

The convolution property (the fourth property above) can be stated more precisely as, given two arbitrary length-N sequences $x_1(n)$ and $x_2(n)$, the DHT of their length-N circular convolution $y(n)$ is given by (from equations (3.150) and (3.36))

$$Y(k) = H_1(k)\mathcal{E}\{H_2(k)\} + H_1(-k)\mathcal{O}\{H_2(k)\} \tag{3.154}$$

where $H_1(k)$ is the DHT of $x_1(n)$ and $H_2(k)$ is the DHT of $x_2(n)$. This result is especially useful when the sequence $x_2(n)$ is even, when it yields

$$Y(k) = H_1(k)H_2(k) \tag{3.155}$$

This equation only involves real functions while equation (3.36) relates complex functions to determine $Y(k)$. Therefore, it is advantageous to use equation (3.155) instead of equation (3.36) for performing convolutions.

To conclude, we can say that the DHT, as a signal representation, is not as widely used in practice as the DFT or the DCT. However, it has been used in a large range of applications as a tool to perform fast convolutions (Bracewell, 1984), as well as a tool to compute FFTs and fast DCTs (Malvar, 1986, 1987).

3.6.4   Hadamard transform

The Hadamard transform, also known as the Walsh–Hadamard transform, is a transform that is not based on sinusoidal functions, unlike the other transforms seen so far. Instead, the elements of its transform matrix are either 1 or -1. When $N = 2^n$, its transform matrix is defined by the following recursion (Harmuth, 1970; Elliott & Rao, 1982):

$$\mathbf{H}_1 = \frac{1}{\sqrt{2}} \begin{bmatrix} 1 & 1 \\ 1 & -1 \end{bmatrix}$$

$$\mathbf{H}_N = \frac{1}{\sqrt{2}} \begin{bmatrix} \mathbf{H}_{\frac{N}{2}} & \mathbf{H}_{\frac{N}{2}} \\ \mathbf{H}_{\frac{N}{2}} & -\mathbf{H}_{\frac{N}{2}} \end{bmatrix} \tag{3.156}$$

From the above equations, it is easy to see that the Hadamard transform is unitary. For example, a length-8 Hadamard transform matrix is

$$\mathbf{H}_8 = \frac{1}{\sqrt{8}} \begin{bmatrix} 1 & 1 & 1 & 1 & 1 & 1 & 1 & 1 \\ 1 & -1 & 1 & -1 & 1 & -1 & 1 & -1 \\ 1 & 1 & -1 & -1 & 1 & 1 & -1 & -1 \\ 1 & -1 & -1 & 1 & 1 & -1 & -1 & 1 \\ 1 & 1 & 1 & 1 & -1 & -1 & -1 & -1 \\ 1 & -1 & 1 & -1 & -1 & 1 & -1 & 1 \\ 1 & 1 & -1 & -1 & -1 & -1 & 1 & 1 \\ 1 & -1 & -1 & 1 & -1 & 1 & 1 & -1 \end{bmatrix} \tag{3.157}$$

An important aspect of the Hadamard transform is that, since the elements of its transform matrix are only either 1 or -1, then the Hadamard transform needs no multiplications for its computation, leading to simple hardware implementations. Because of this fact, the Hadamard transform has been used in digital video schemes in the past, although its compression performance is not as high as that of the DCT (Jain, 1989). Nowadays, with the advent of specialized hardware to compute the DCT, the Hadamard transform is seldom used in digital video. However, one important area of application of the Hadamard transform today is in carrier division multiple access (CDMA) systems for mobile communications, where it is employed as channelization code in synchronous communications systems (Stüber, 1996).

The Hadamard transform also has a fast algorithm. For example, the matrix $\mathbf{H}_8$ can be factored as (Jain, 1989)

$$\mathbf{H}_8 = \mathbf{A}_8^3 \tag{3.158}$$

where

$$\mathbf{A}_8 = \frac{1}{\sqrt{2}} \begin{bmatrix} 1 & 1 & 0 & 0 & 0 & 0 & 0 & 0 \\ 0 & 0 & 1 & 1 & 0 & 0 & 0 & 0 \\ 0 & 0 & 0 & 0 & 1 & 1 & 0 & 0 \\ 0 & 0 & 0 & 0 & 0 & 0 & 1 & 1 \\ 1 & -1 & 0 & 0 & 0 & 0 & 0 & 0 \\ 0 & 0 & 1 & -1 & 0 & 0 & 0 & 0 \\ 0 & 0 & 0 & 0 & 1 & -1 & 0 & 0 \\ 0 & 0 & 0 & 0 & 0 & 0 & 1 & -1 \end{bmatrix} \qquad (3.159)$$

and, therefore, the number of additions of a fast Hadamard transform algorithm is of the order of $N \log_2 N$.

3.6.5   Other important transforms

Wavelet transforms

Wavelet transforms constitute a different class of transforms which have become very popular in the past years. They come from recent developments in functional analysis, and are usually studied under the discipline of multirate signal processing. An introduction to wavelet transforms is given in Chapter 9 of this book.

Karhunen–Loève transform

As seen in Subsection 3.6.1, the DCT is widely used in image and video compression because it has the ability to concentrate the energy of a signal in a few transform coefficients. A question that naturally arises is which is the transform that maximizes this energy concentration. Given a statistical distribution of an ensemble of signals, the optimum transform in terms of this energy compaction capability is the Karhunen–Loève transform (KLT) (Jain, 1989). It is defined as the transform that diagonalizes the autocovariance matrix of a discrete random process.

From the above, we see that there is a different KLT for each of the different signal statistics. However, it can be shown that the DCT approximates the KLT when the signals can be modeled as Gauss–Markov processes with correlation coefficients near to 1 (Jain, 1989). This is a reasonably good model for several useful signals, video, for instance. For those signals, the DCT is indeed approximately optimum in terms of energy compaction capability, which explains why it is widely used.

3.7   Signal representations

In Chapters 1–3, we have dealt with several forms of signal representations, both continuous and discrete. We now summarize the main characteristics of those representations. We classify them in terms of both the time and frequency variables as

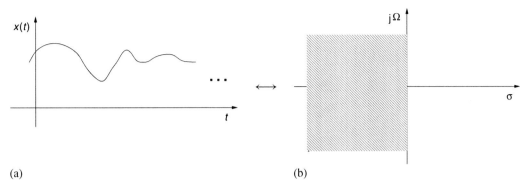

Figure 3.19 Laplace transform: (a) continuous-time signal; (b) domain of the corresponding Laplace transform.

continuous or discrete, as well as real, imaginary, or complex. Also, we classify the representations as being periodic or nonperiodic. The time–frequency relationship for each transform is shown in Figures 3.19–3.24.

Laplace transform

$$X(s) = \int_{-\infty}^{\infty} x(t)e^{-st}\,dt \quad \longleftrightarrow \quad x(t) = \frac{1}{2\pi}e^{\sigma t}\int_{-\infty}^{\infty} X(\sigma + j\omega)e^{j\omega t}\,d\omega$$

(3.160)

- Time domain: nonperiodic function of a continuous and real-time variable.
- Frequency domain: nonperiodic function of a continuous and complex frequency variable.

z transform

$$X(z) = \sum_{n=-\infty}^{\infty} x(n)z^{-n} \quad \longleftrightarrow \quad x(n) = \frac{1}{2\pi j}\oint_C X(z)z^{n-1}\,dz$$

(3.161)

- Time domain: nonperiodic function of a discrete and integer time variable.
- Frequency domain: nonperiodic function of a continuous and complex frequency variable.

Fourier transform (continuous time)

$$X(\Omega) = \int_{-\infty}^{\infty} x(t)e^{-j\Omega t}\,dt \quad \longleftrightarrow \quad x(t) = \frac{1}{2\pi}\int_{-\infty}^{\infty} X(\Omega)e^{j\Omega t}\,d\Omega$$

(3.162)

- Time domain: nonperiodic function of a continuous and real-time variable.
- Frequency domain: nonperiodic function of a continuous and imaginary frequency variable.

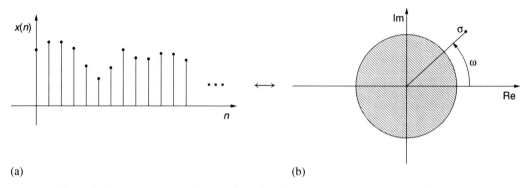

(a)   (b)

Figure 3.20 z transform: (a) discrete-time signal; (b) domain of the corresponding z transform.

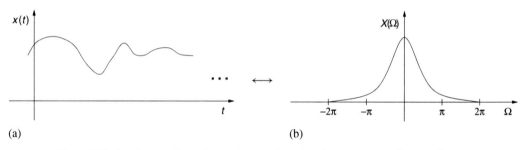

(a)   (b)

Figure 3.21 Fourier transform: (a) continuous-time signal; (b) corresponding Fourier transform.

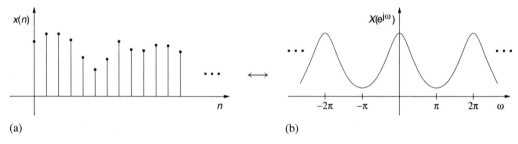

(a)   (b)

Figure 3.22 Discrete-time Fourier transform: (a) discrete-time signal; (b) corresponding discrete-time Fourier transform.

Fourier transform (discrete time)

$$X(\mathrm{e}^{\mathrm{j}\omega}) = \sum_{n=-\infty}^{\infty} x(n)\mathrm{e}^{-\mathrm{j}\omega n} \quad \longleftrightarrow \quad x(n) = \frac{1}{2\pi} \int_{-\infty}^{\infty} X(\mathrm{e}^{\mathrm{j}\omega})\mathrm{e}^{\mathrm{j}\omega n}\,\mathrm{d}\omega \qquad (3.163)$$

- Time domain: nonperiodic function of a discrete and integer time variable.
- Frequency domain: periodic function of a continuous and imaginary frequency variable.

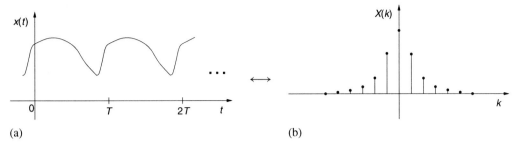

(a)   (b)

Figure 3.23 Fourier series: (a) continuous-time periodic signal; (b) corresponding Fourier series.

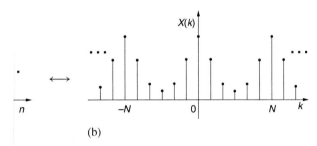

(b)

ısform: (a) discrete-time signal; (b) corresponding discrete Fourier

$$t \quad \longleftrightarrow \quad x(t) = \sum_{k=-\infty}^{\infty} X(k) e^{j\frac{2\pi}{T} kt} \qquad (3.164)$$

nction of a continuous and real-time variable.

eriodic function of a discrete and integer frequency vari-

$$\longleftrightarrow \quad x(n) = \frac{1}{N} \sum_{k=0}^{N-1} X(k) e^{j\frac{2\pi}{N} kn} \qquad (3.165)$$

ınction of a discrete and integer time variable.

dic function of a discrete and integer frequency variable.

3.8   Discrete transforms with MATLAB

The functions described below are part of the MATLAB Signal Processing toolbox.

- `fft`: Computes the DFT of a vector.
  Input parameter: The input vector x.
  Output parameter: The DFT of the input vector x.
  Example:

  ```
  t=0:0.001:0.25; x=sin(2*pi*50*t)+sin(2*pi*120*t);
  y=fft(x); plot(abs(y));
  ```

- `ifft`: Computes the inverse DFT of a vector.
  Input parameter: The complex input vector y having the DFT coefficients.
  Output parameter: The inverse DFT of the input vector y.
  Example:

  ```
  w=[1:256]; y=zeros(size(w));
  y(1:10)=1; y(248:256)=1;
  x=ifft(y); plot(real(x));
  ```

- `fftshift`: Swaps the left and right halves of a vector. When applied to a DFT, shows it from $-\frac{N}{2} + 1$ to $\frac{N}{2}$.
  Input parameter: The DFT vector y.
  Output parameter: The swapped DFT vector z.
  Example:

  ```
  t=0:0.001:0.25; x=sin(2*pi*50*t)+sin(2*pi*120*t);
  y=fft(x); z=fftshift(y); plot(abs(z));
  ```

- `dftmtx`: Generates a DFT matrix.
  Input parameter: The size of the DFT n.
  Output parameter: The n×n DFT matrix.
  Example:

  ```
  t=0:0.001:0.25; x=sin(2*pi*50*t)+sin(2*pi*120*t);
  F=dftmtx(251); y=F*x'; z=fftshift(y); plot(abs(z));
  ```

- `fftfilt`: Performs linear filtering using the overlap-and-add method.
  Input parameters:

  - The vector h containing the filter coefficients;
  - The vector x containing the input signal;
  - The length n of the non-overlapping blocks into which x is divided. If n is not provided, the function uses its own optimized n (recommended).

  Output parameter: The filtered signal y.
  Example:

  ```
  h=[1 2 3 4 4 3 2 1]; t=0:0.05:1;
  x=sin(2*pi*3*t);
  y=fftfilt(h); plot(y)
  ```

- dct: Computes the DCT of a vector.
  Input parameter: The input vector x.
  Output parameter: The DCT of the input vector x.
  Example:
  ```
  t=0:0.05:1; x=sin(2*pi*3*t);
  y=dct(x); plot(y);
  ```

- idct: Computes the inverse DCT of a vector of coefficients.
  Input parameter: The coefficient vector y.
  Output parameter: The inverse DCT of the coefficient vector y.
  Example:
  ```
  y=1:32; x=idct(y); plot(x);
  ```

- dctmtx: Generates a DCT matrix.
  Input parameter: The size of the DCT n.
  Output parameter: The n × n DCT matrix.
  Example (plots the first and sixth basis functions of a length-8 DCT):
  ```
  C=dctmtx(8);
  subplot(2,1,1),stem(C(1,:));
  subplot(2,1,2),stem(C(6,:));
  ```

3.9   Summary

In this chapter, we have thoroughly studied discrete transforms. We began with the discrete Fourier transform (DFT), which is a discrete-frequency representation of a finite discrete-time signal. We have shown that the DFT is a powerful tool for the computation of discrete-time convolutions. Several algorithms for the efficient implementation of the DFT (FFT algorithms) have been studied. In particular, we emphasized the radix-2 algorithms because they are the most widely used. We have also dealt with discrete transforms other than the DFT, such as the discrete cosine transform, widely used in signal compression, and the Hartley transform, with many applications in fast convolutions and computation of the FFT. We then presented an overview of signal representations, where we compared the transforms studied in Chapters 1–3. The chapter also included a brief description of MATLAB basic functions related to all the discrete transforms previously discussed.

3.10   Exercises

3.1   Show that

$$\sum_{n=0}^{N-1} W_N^{nk} = \begin{cases} N, & \text{for } k = 0, \pm N, \pm 2N, \ldots \\ 0, & \text{otherwise} \end{cases}$$

3.2   Show, by substituting equation (3.16) into equation (3.15), and using the result given in Exercise 3.1, that equation (3.16) gives the IDFT of a sequence $x(n)$.

3.3   Prove the following properties of the DFT:

(a) Circular frequency-shift theorem: equations (3.32)–(3.35).
(b) Correlation: equation (3.41).
(c) Parseval's theorem: equation (3.52).
(d) DFT of real and imaginary sequences: equations (3.42) and (3.43).

3.4   Show how to compute the DFT of two even complex length-N sequences performing only one length-N transform calculation. Follow the steps below:

(i)   Build the auxiliary sequence $y(n) = W_N^n x_1(n) + x_2(n)$.
(ii)   Show that $Y(k) = X_1(k+1) + X_2(k)$.
(iii) Using properties of symmetric sequences, show that
       $Y(-k-1) = X_1(k) + X_2(k+1)$.
(iv)   Use the results of (ii) and (iii) to create a recursion to compute $X_1(k)$ and $X_2(k)$.
       Note that $X(0) = \sum_{n=0}^{N-1} x(n)$.

3.5   Repeat Exercise 3.4 for the case of two complex antisymmetric sequences.

3.6   Show how to compute the DFT of four real, even, length-N sequences using only one length-N transform, using the results of Exercise 3.4.

3.7   Compute the coefficients of the Fourier series of the periodic sequences below using the DFT.

(a) $x'(n) = \sin\left(2\pi\dfrac{n}{N}\right), \quad$ for $N = 20$.

(b) $x'(n) = \begin{cases} 1, & \text{for } n \text{ even} \\ -1, & \text{for } n \text{ odd} \end{cases}$

3.8   Compute and plot the magnitude and phase of the DFT of the following finite-length sequences:

(a) $x(n) = 2\cos\left(\pi\dfrac{n}{N}\right) + \sin^2\left(\pi\dfrac{n}{N}\right), \quad$ for $0 \le n \le 10$ and $N = 11$.
(b) $x(n) = e^{-2n}, \quad$ for $0 \le n \le 20$.
(c) $x(n) = \delta(n-1), \quad$ for $0 \le n \le 2$.

(d) $x(n) = n$, for $0 \leq n \leq 5$.

3.9   We want to compute the linear convolution of a long sequence $x(n)$, of length L, with a short sequence $h(n)$, of length K. If we use the overlap-and-save method to compute the convolution, determine the block length that minimizes the number of arithmetic operations involved in the convolution.

3.10   Repeat Exercise 3.9 for the case when the overlap-and-add method is used.

3.11   Express the algorithm described in the graph for the decimation-in-time FFT in Figure 3.11 in matrix form.

3.12   Express the algorithm described in the graph of the decimation-in-frequency FFT in Figure 3.13 in matrix form.

3.13   Determine the graph of a decimation-in-time length-6 DFT, and express its algorithm in matrix form.

3.14   Determine the basic cell of a radix-5 algorithm. Analyze the possible simplifications in the graph of the cell.

3.15   Repeat Exercise 3.14 for the radix-8 case, determining the complexity of a generic radix-8 algorithm.

3.16   Show that the number of complex multiplications of the FFT algorithm for generic N is given by equation (3.124).

3.17   Compute, using an FFT algorithm, the linear convolution of the sequences (a) and (b), as well as (b) and (c) in Exercise 3.8.

3.18   Show that the DCT of a length-N sequence $x(n)$ can be computed from the length-$2N$ DFT of a sequence $\tilde{x}(n)$ consisting of $x(n)$ extended symmetrically, that is

$$\tilde{x}(n) = \begin{cases} x(n), & \text{for } 0 \leq n \leq N - 1 \\ x(2N - n - 1), & \text{for } N \leq n \leq 2N - 1 \end{cases}$$

3.19   Show that the discrete cosine transform of a length-N sequence $x(n)$ can be computed from the length-N DFT of a sequence $\hat{x}(n)$ consisting of the following reordering of the even and odd elements of $x(n)$

$$\hat{x}(n) = x(2n)$$
$$\hat{x}(N - 1 - n) = x(2n + 1)$$

for $0 \leq n \leq \frac{N}{2} - 1$.

3.20   Prove the relations between the DHT and the DFT given in equations (3.150) and (3.151).

3.21 Prove the convolution-multiplication property of the DHT in equation (3.154).

3.22 Show that the Hadamard transform matrix is unitary, using equation (3.156).

3.23 The spectrum of a signal is given by its Fourier transform. In order to compute it, we need all the samples of the signal. Therefore, the spectrum is a characteristic of the whole signal. However, in many applications (e.g., in speech processing) one needs to find the spectrum of a short section of the signal. In order to do it, we define a length-N sliding window $x_i(n)$ of the signal $x(n)$ as

$$x_i(n) = x(n + i - N + 1), \quad \text{for } -\infty < i < \infty$$

that is, we take N samples of $x(n)$ starting from position i backwards. We then define the short-time spectrum of $x(n)$ at the position i as the DFT of $x_i(n)$, that is,

$$X(k, i) = \sum_{n=0}^{N-1} x_i(n) W_N^{kn}, \quad \text{for } 0 \le k \le N - 1, \quad \text{for } -\infty < i < \infty$$

Therefore, we have, for each frequency k, a signal $X(k, i)$, $-\infty < i < \infty$, which is the short-time spectrum of $x(n)$ at frequency k.

(a) Show that the short-time spectrum can be efficiently computed with a bank of N IIR filters having the following transfer function for frequency k

$$H_k(z) = \frac{1 - z^{-N}}{W_N^k - z^{-1}}, \quad 0 \le k \le N - 1$$

(b) Compare, in terms of the number of complex additions and multiplications, the complexity of the computations of the short-time spectrum using the above formula for $H_k(z)$ with that using the FFT algorithm.

(c) Discuss whether it is advantageous or not, in terms of arithmetic operations, to use the above formula for $H_k(z)$ to compute a linear convolution using the overlap-and-add method. Repeat this item for the overlap-and-save method.

A good coverage of recursive computations of sinusoidal transforms can be found in Liu et al. (1994).

3.24 Linear convolution using `fft` in MATLAB.

(a) Use the `fft` command to determine the linear convolution between two given signals $x(n)$ and $h(n)$.

(b) Compare the function you have created in (a) with the **conv** and **filter** commands with respect to the output signal and to the total number of flops required to convolve the two signals of orders N and K, respectively.

(c) Verify your results experimentally (using the command **flops**) for general values of N and K.

(d) Repeat (c) considering solely the values of N and K such that $(N + K - 1)$ is a power of two.

3.25   Given the signal

$$x(n) = \sin\left(\frac{\omega_s}{10}n\right) + \sin\left[\left(\frac{\omega_s}{10} + \frac{\omega_s}{l}\right)n\right]$$

(a) For $l = 100$, compute the DFT of $x(n)$ using 64 samples. Can you observe the presence of both sinusoids?

(b) Increase the length of the DFT to 128 samples by padding 64 zeros to the original samples of $x(n)$. Comment on the results.

(c) Compute the DFT of $x(n)$ using 128 samples. Can you now observe the presence of both sinusoids?

(d) Increase the length of the DFT of $x(n)$ by 128 samples by padding 128 zeros to the samples of $x(n)$. Repeat for a padding of 384 zeros. Comment on the results.

4 Digital filters

4.1 Introduction

In the previous chapters we studied different ways of describing discrete-time systems that are linear and time invariant. It was verified that the z transform greatly simplifies the analysis of discrete-time systems, especially those initially described by a difference equation.

In this chapter, we study several structures used to realize a given transfer function associated with a specific difference equation through the use of the z transform. The transfer functions considered here will be of the polynomial form (nonrecursive filters) and of the rational-polynomial form (recursive filters). We also analyze the properties of the given generic digital filter structures associated with practical discrete-time systems.

4.2 Basic structures of nonrecursive digital filters

Nonrecursive filters are characterized by a difference equation in the form

$$y(n) = \sum_{l=0}^{M} b_l x(n-l) \tag{4.1}$$

where the b_l coefficients are directly related to the system impulse response, that is, $b_l = h(l)$. Due to the finite length of their impulse responses, nonrecursive filters are also referred to as finite-duration impulse-response (FIR) filters. We can rewrite equation (4.1) as

$$y(n) = \sum_{l=0}^{M} h(l) x(n-l) \tag{4.2}$$

Applying the z transform to equation (4.2), we end up with the following input–output relationship

$$H(z) = \frac{Y(z)}{X(z)} = \sum_{l=0}^{M} b_l z^{-l} = \sum_{l=0}^{M} h(l) z^{-l} \tag{4.3}$$

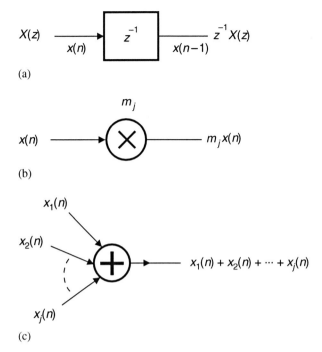

Figure 4.1 Classic representation of basic elements of digital filters: (a) delay; (b) multiplier; (c) adder.

In practical terms, equation (4.3) can be implemented in several distinct forms, using as basic elements the delay, the multiplier, and the adder blocks. These basic elements of digital filters and their corresponding standard symbols are depicted in Figure 4.1. An alternative way of representing such elements is the so-called signal flowgraph shown in Figure 4.2. These two sets of symbolisms representing the delay, multiplier, and adder elements, are used throughout this book interchangeably.

4.2.1 Direct form

The simplest realization of an FIR digital filter is derived from equation (4.3). The resulting structure, seen in Figure 4.3, is called the direct-form realization, as the multiplier coefficients are obtained directly from the filter transfer function. Such a structure is also referred to as the canonic direct form, where for a canonic form we understand any structure that realizes a given transfer function with the minimum number of delays, multipliers, and adders. More specifically, a structure that utilizes the minimum number of delays is said to be canonic with respect to the delay element, and so on.

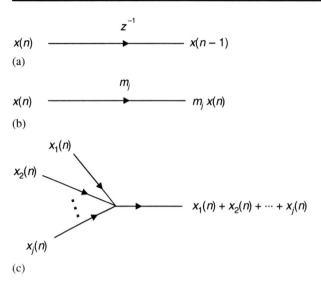

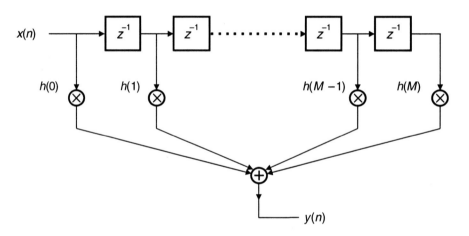

Figure 4.2 Signal-flowgraph representation of basic elements of digital filters: (a) delay; (b) multiplier; (c) adder.

Figure 4.3 Direct form for FIR digital filters.

An alternative canonic direct form for equation (4.3) can be derived by expressing $H(z)$ as

$$H(z) = \sum_{l=0}^{M} h(l)z^{-l} = \left(h(0) + z^{-1}\left[h(1) + z^{-1}\left(h(2) + \cdots + h(M) \right) \right] + \cdots \right) \quad (4.4)$$

The implementation of this form is shown in Figure 4.4.

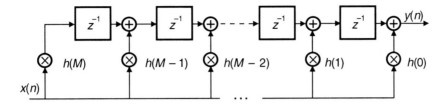

Figure 4.4 Alternative direct form for FIR digital filters.

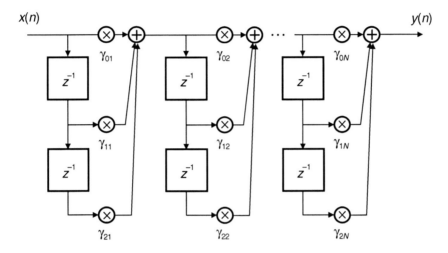

Figure 4.5 Cascade form for FIR digital filters.

4.2.2 Cascade form

Equation (4.3) can be realized through a series of equivalent structures. However, the coefficients of such distinct realizations are not explicitly the filter impulse response or the corresponding transfer function. An important example of such a realization is the so-called cascade form which consists of a series of second-order FIR filters connected in cascade, thus the name of the resulting structure, as seen in Figure 4.5.

The transfer function associated with such a realization is of the form

$$H(z) = \prod_{k=1}^{N} (\gamma_{0k} + \gamma_{1k}z^{-1} + \gamma_{2k}z^{-2}) \tag{4.5}$$

where if M is the filter order, then $N = M/2$ when M is even, and $N = (M + 1)/2$ when M is odd. In the latter case, one of the γ_{2k} becomes zero.

4.2.3 Linear-phase forms

An important subclass of FIR digital filters is the one that includes linear-phase filters. Such filters are characterized by a constant group delay τ, and therefore they must present a frequency response of the following form:

$$H(e^{j\omega}) = B(\omega)e^{-j\omega\tau + j\phi} \tag{4.6}$$

where $B(\omega)$ is real, and τ and ϕ are constant. Hence, the impulse response $h(n)$ of linear-phase filters satisfies

$$
\begin{aligned}
h(n) &= \frac{1}{2\pi} \int_{-\infty}^{\infty} H(e^{j\omega}) e^{j\omega n} d\omega \\
&= \frac{1}{2\pi} \int_{-\infty}^{\infty} B(\omega) e^{-j\omega\tau + j\phi} e^{j\omega n} d\omega \\
&= \frac{e^{j\phi}}{2\pi} \int_{-\infty}^{\infty} B(\omega) e^{j\omega(n-\tau)} d\omega \\
&= e^{j\phi} b(n - \tau) \tag{4.7}
\end{aligned}
$$

where $b(n)$ is the inverse Fourier transform of $B(\omega)$.

Since $B(\omega)$ is real,

$$b(n) = b^*(-n) \tag{4.8}$$

and from equation (4.7), we have that

$$b(n) = e^{-j\phi} h(n + \tau) \tag{4.9}$$

We can then deduce that for a filter to be linear phase with group delay τ, its impulse response must satisfy

$$e^{-j\phi} h(n + \tau) = e^{j\phi} h^*(-n + \tau) \tag{4.10}$$

and then

$$h(n + \tau) = e^{2j\phi} h^*(-n + \tau) \tag{4.11}$$

We now proceed to show that linear-phase FIR filters present impulse responses of very particular forms. In fact, equation (4.11) implies that $h(0) = e^{2j\phi} h^*(2\tau)$. Hence, if $h(n)$ is causal and of finite duration, for $0 \leq n \leq M$, we must necessarily have that

$$\tau = \frac{M}{2} \tag{4.12}$$

and then, equation (4.11) becomes

$$h(n) = e^{2j\phi} h^*(M - n) \tag{4.13}$$

This is the general equation that the coefficients of a linear-phase FIR filter must satisfy.

In the common case where all the filter coefficients are real, then $h(n) = h^*(n)$, and equation (4.13) implies that $e^{2j\phi}$ must be real. Thus

$$\phi = \frac{k\pi}{2}, \; k \in \mathbb{Z} \tag{4.14}$$

and equation (4.13) becomes

$$h(n) = (-1)^k h(M - n), \; k \in \mathbb{Z} \tag{4.15}$$

That is, the filter impulse response must be either symmetric or antisymmetric.

From equation (4.6), the frequency response of linear-phase FIR filters with real coefficients becomes

$$H(e^{j\omega}) = B(\omega)e^{-j\omega \frac{M}{2} + j\frac{k\pi}{2}} \tag{4.16}$$

For all practical purposes, we only need to consider the cases when $k = 0, 1, 2, 3$, as all other values of k will be equivalent to one of these four cases. Furthermore, as $B(\omega)$ can be either positive or negative, the cases $k = 2$ and $k = 3$ are obtained from cases $k = 0$ and $k = 1$ respectively, by making $B(\omega) \leftarrow -B(\omega)$.

Therefore, we consider solely the four distinct cases described by equations (4.13) and (4.16). They are referred to as follows:

- Type I: $k = 0$ and M even.
- Type II: $k = 0$ and M odd.
- Type III: $k = 1$ and M even.
- Type IV: $k = 1$ and M odd.

We now proceed to demonstrate that $h(n) = (-1)^k h(M - n)$ is a sufficient condition for an FIR filter with real coefficients to have a linear phase. The four types above are considered separately.

- Type I: $k = 0$ implies that the filter has symmetric impulse response, that is, $h(M - n) = h(n)$. Since the filter order M is even, equation (4.3) may be rewritten as

$$H(z) = \sum_{n=0}^{\frac{M}{2}-1} h(n)z^{-n} + h\left(\frac{M}{2}\right)z^{-\frac{M}{2}} + \sum_{n=\frac{M}{2}+1}^{M} h(n)z^{-n}$$

$$= \sum_{n=0}^{\frac{M}{2}-1} h(n)\left[z^{-n} + z^{-(M-n)}\right] + h\left(\frac{M}{2}\right)z^{-\frac{M}{2}} \tag{4.17}$$

Evaluating this equation over the unit circle, that is, using the variable transformation $z \rightarrow e^{j\omega}$, one obtains

$$H(e^{j\omega}) = \sum_{n=0}^{\frac{M}{2}-1} h(n) \left[e^{-j\omega n} + e^{(-j\omega M + j\omega n)} \right] + h\left(\frac{M}{2}\right) e^{-j\omega \frac{M}{2}}$$

$$= e^{-j\omega \frac{M}{2}} \left\{ h\left(\frac{M}{2}\right) + \sum_{n=0}^{\frac{M}{2}-1} 2h(n) \cos \left[\omega \left(n - \frac{M}{2} \right) \right] \right\} \tag{4.18}$$

Substituting n by $(\frac{M}{2} - m)$, we get

$$H(e^{j\omega}) = e^{-j\omega \frac{M}{2}} \left[h\left(\frac{M}{2}\right) + \sum_{m=1}^{\frac{M}{2}} 2h\left(\frac{M}{2} - m\right) \cos(\omega m) \right]$$

$$= e^{-j\omega \frac{M}{2}} \sum_{m=0}^{\frac{M}{2}} a(m) \cos(\omega m) \tag{4.19}$$

with $a(0) = h(\frac{M}{2})$ and $a(m) = 2h(\frac{M}{2} - m)$, for $m = 1, \ldots, \frac{M}{2}$.

Since this equation is in the form of equation (4.16), this completes the sufficiency proof for Type-I filters.

- Type II: $k = 0$ implies that the filter has a symmetric impulse response, that is, $h(M - n) = h(n)$. Since the filter order M is odd, equation (4.3) may be rewritten as

$$H(z) = \sum_{n=0}^{\frac{M-1}{2}} h(n) z^{-n} + \sum_{n=\frac{M+1}{2}}^{M} h(n) z^{-n}$$

$$= \sum_{n=0}^{\frac{M-1}{2}} h(n)[z^{-n} + z^{-(M-n)}] \tag{4.20}$$

Evaluating this equation over the unit circle, one obtains

$$H(e^{j\omega}) = \sum_{n=0}^{\frac{M-1}{2}} h(n) \left[e^{-j\omega n} + e^{(-j\omega M + j\omega n)} \right]$$

$$= e^{-j\omega \frac{M}{2}} \sum_{n=0}^{\frac{M-1}{2}} h(n) \left[e^{-j\omega(n-\frac{M}{2})} + e^{j\omega(n-\frac{M}{2})} \right]$$

$$= e^{-j\omega \frac{M}{2}} \sum_{n=0}^{\frac{M-1}{2}} 2h(n) \cos \left[\omega \left(n - \frac{M}{2} \right) \right] \tag{4.21}$$

Substituting n with $(\frac{M+1}{2} - m)$

$$H(e^{j\omega}) = e^{-j\omega\frac{M}{2}} \sum_{m=1}^{\frac{M+1}{2}} 2h\left(\frac{M+1}{2} - m\right) \cos\left[\omega\left(m - \frac{1}{2}\right)\right]$$

$$= e^{-j\omega\frac{M}{2}} \sum_{m=1}^{\frac{M+1}{2}} b(m) \cos\left[\omega\left(m - \frac{1}{2}\right)\right] \tag{4.22}$$

with $b(m) = 2h(\frac{M+1}{2} - m)$, for $m = 1, \ldots, (\frac{M+1}{2})$.

Since this equation is in the form of equation (4.16), this completes the sufficiency proof for Type-II filters.

Notice that at $\omega = \pi$, $H(e^{j\omega}) = 0$, as it consists of a summation of cosine functions evaluated at $\pm\frac{\pi}{2}$, which are obviously null. Therefore, highpass and bandstop filters cannot be approximated as Type-II filters.

- Type III: $k = 1$ implies that the filter has an antisymmetric impulse response, that is, $h(M - n) = -h(n)$. In this case, $h(\frac{M}{2})$ is necessarily null. Since the filter order M is even, equation (4.3) may be rewritten as

$$H(z) = \sum_{n=0}^{\frac{M}{2}-1} h(n)z^{-n} - \sum_{n=\frac{M}{2}+1}^{M} h(n)z^{-n}$$

$$= \sum_{n=0}^{\frac{M}{2}-1} h(n)\left[z^{-n} - z^{-(M-n)}\right] \tag{4.23}$$

which, when evaluated over the unit circle, yields

$$H(e^{j\omega}) = \sum_{n=0}^{\frac{M}{2}-1} h(n)\left[e^{-j\omega n} - e^{(-j\omega M + j\omega n)}\right]$$

$$= e^{-j\omega\frac{M}{2}} \sum_{n=0}^{\frac{M}{2}-1} h(n)\left[e^{-j\omega(n-\frac{M}{2})} - e^{j\omega(n-\frac{M}{2})}\right]$$

$$= e^{-j\omega\frac{M}{2}} \sum_{n=0}^{\frac{M}{2}-1} -2jh(n) \sin\left[\omega\left(n - \frac{M}{2}\right)\right]$$

$$= e^{-j(\omega\frac{M}{2}-\frac{\pi}{2})} \sum_{n=0}^{\frac{M}{2}-1} -2h(n) \sin\left[\omega\left(n - \frac{M}{2}\right)\right] \tag{4.24}$$

Substituting n by $(\frac{M}{2} - m)$

$$H(e^{j\omega}) = e^{-j(\omega\frac{M}{2} - \frac{\pi}{2})} \sum_{m=1}^{\frac{M}{2}} -2h\left(\frac{M}{2} - m\right) \sin[\omega(-m)]$$

$$= e^{-j(\omega\frac{M}{2} - \frac{\pi}{2})} \sum_{m=1}^{\frac{M}{2}} c(m) \sin(\omega m) \tag{4.25}$$

with $c(m) = 2h(\frac{M}{2} - m)$, for $m = 1, \ldots, \frac{M}{2}$.

Since this equation is in the form of equation (4.16), this completes the sufficiency proof for Type-III filters.

Notice, in this case, that the frequency response becomes null at $\omega = 0$ and at $\omega = \pi$. That makes this type of realization suitable for bandpass filters, differentiators, and Hilbert transformers, these last two due to the phase shift of $\frac{\pi}{2}$, as it will be seen in Chapter 5.

- Type IV: $k = 1$ implies that the filter has an antisymmetric impulse response, that is, $h(M - n) = -h(n)$. Since the filter order M is odd, equation (4.3) may be rewritten as

$$H(z) = \sum_{n=0}^{\frac{M-1}{2}} h(n)z^{-n} - \sum_{n=\frac{M+1}{2}}^{M} h(n)z^{-n}$$

$$= \sum_{n=0}^{\frac{M-1}{2}} h(n)\left[z^{-n} - z^{-(M-n)}\right] \tag{4.26}$$

Evaluating this equation over the unit circle

$$H(e^{j\omega}) = \sum_{n=0}^{\frac{M-1}{2}} h(n)\left[e^{-j\omega n} - e^{(-j\omega M + j\omega n)}\right]$$

$$= e^{-j\omega\frac{M}{2}} \sum_{n=0}^{\frac{M-1}{2}} h(n)\left[e^{-j\omega(n-\frac{M}{2})} - e^{j\omega(n-\frac{M}{2})}\right]$$

$$= e^{-j\omega\frac{M}{2}} \sum_{n=0}^{\frac{M-1}{2}} -2jh(n) \sin\left[\omega\left(n - \frac{M}{2}\right)\right]$$

$$= e^{-j(\omega\frac{M}{2} - \frac{\pi}{2})} \sum_{n=0}^{\frac{M-1}{2}} -2h(n) \sin\left[\omega\left(n - \frac{M}{2}\right)\right] \tag{4.27}$$

Table 4.1. *Main characteristics of linear-phase FIR filters: order, impulse response, frequency response, phase response, and group delay.*

Type	M	$h(n)$		$H(e^{j\omega})$	$\Theta(\omega)$	τ
I	Even	Symmetric		$e^{-j\omega\frac{M}{2}}\displaystyle\sum_{m=0}^{\frac{M}{2}} a(m)\cos(\omega m)$	$-\omega\dfrac{M}{2}$	$\dfrac{M}{2}$
				$a(0) = h\left(\frac{M}{2}\right); a(m) = 2h\left(\frac{M}{2} - m\right)$		
II	Odd	Symmetric		$e^{-j\omega\frac{M}{2}}\displaystyle\sum_{m=1}^{\frac{M+1}{2}} b(m)\cos\left[\omega\left(m-\frac{1}{2}\right)\right]$	$-\omega\dfrac{M}{2}$	$\dfrac{M}{2}$
				$b(m) = 2h\left(\frac{M+1}{2} - m\right)$		
III	Even	Antisymmetric		$e^{-j(\omega\frac{M}{2} - \frac{\pi}{2})}\displaystyle\sum_{m=1}^{\frac{M}{2}} c(m)\sin(\omega m)$	$-\omega\dfrac{M}{2} + \dfrac{\pi}{2}$	$\dfrac{M}{2}$
				$c(m) = 2h\left(\frac{M}{2} - m\right)$		
IV	Odd	Antisymmetric		$e^{-j(\omega\frac{M}{2} - \frac{\pi}{2})}\displaystyle\sum_{m=1}^{\frac{M+1}{2}} d(m)\sin\left[\omega\left(m-\frac{1}{2}\right)\right]$	$-\omega\dfrac{M}{2} + \dfrac{\pi}{2}$	$\dfrac{M}{2}$
				$d(m) = 2h\left(\frac{M+1}{2} - m\right)$		

Substituting n by $(\frac{M+1}{2} - m)$

$$H(e^{j\omega}) = e^{-j(\omega\frac{M}{2} - \frac{\pi}{2})}\sum_{m=1}^{\frac{M+1}{2}} -2h\left(\frac{M+1}{2} - m\right)\sin\left[\omega\left(\frac{1}{2} - m\right)\right]$$

$$= e^{-j(\omega\frac{M}{2} - \frac{\pi}{2})}\sum_{m=1}^{\frac{M+1}{2}} d(m)\sin\left[\omega\left(m - \frac{1}{2}\right)\right] \tag{4.28}$$

with $d(m) = 2h(\frac{M+1}{2} - m)$, for $m = 1, \ldots, (\frac{M+1}{2})$.

Since this equation is in the form of equation (4.16), this completes the sufficiency proof for Type-IV filters, thus finishing the whole proof.

Notice that $H(e^{j\omega}) = 0$, at $\omega = 0$. Hence, lowpass filters cannot be approximated as Type-IV filters, although they are still suitable for differentiators and Hilbert transformers, like the Type-III form.

Typical impulse responses of the four cases of linear-phase FIR digital filters are depicted in Figure 4.6. The properties of all four cases are summarized in Table 4.1.

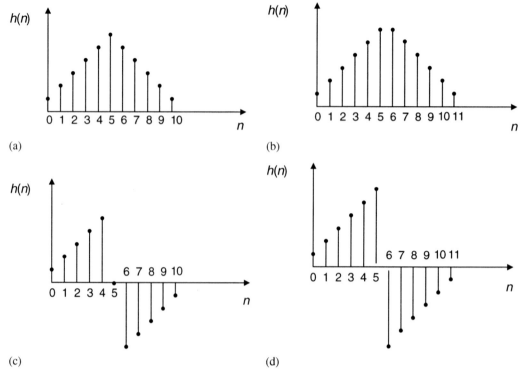

Figure 4.6 Example of impulse responses of linear-phase FIR digital filters: (a) Type I; (b) Type II; (c) Type III; (d) Type IV.

One can derive important properties of linear-phase FIR filters by representing equations (4.17), (4.20), (4.23), and (4.26) in a single framework as

$$H(z) = z^{-\frac{M}{2}} \sum_{k=0}^{K} f_k (z^k \pm z^{-k}) \tag{4.29}$$

where the f_k coefficients are derived from the filter's impulse response $h(n)$, and $K = \frac{M}{2}$, if M is even; or $K = \frac{M-1}{2}$, if M is odd. From equation (4.29), it is easy to observe that, if z_γ is a zero of $H(z)$, so is z_γ^{-1}. This implies that all zeros of $H(z)$ occur in reciprocal pairs. Considering that if the coefficients $h(n)$ are real, all complex zeros occur in conjugate pairs, and then one can infer that the zeros of $H(z)$ must satisfy the following relationships:

- All complex zeros which are not on the unit circle occur in conjugate and reciprocal quadruples. In other words, if z_γ is complex, then z_γ^{-1}, z_γ^*, and $(z_\gamma^{-1})^*$ are also zeros of $H(z)$.
- There can be any given number of zeros over the unit circle, in conjugate pairs, since in this case we automatically have that $z_\gamma^{-1} = z_\gamma^*$.
- All real zeros outside the unit circle occur in reciprocal pairs.

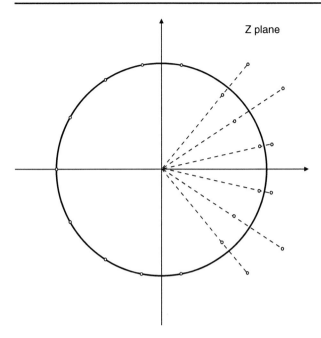

Figure 4.7 Typical zero plot of a linear-phase FIR digital filter.

- There can be any given number of zeros at $z = z_\gamma = \pm 1$, since in this case we necessarily have that $z_\gamma^{-1} = \pm 1$.

A typical zero plot for a linear-phase lowpass FIR filter is shown in Figure 4.7.

An interesting property of linear-phase FIR digital filters is that they can be realized with efficient structures that exploit their symmetric or antisymmetric impulse-response characteristics. In fact, when M is even, these efficient structures require $\frac{M}{2} + 1$ multiplications, while when M is odd, only $\frac{M+1}{2}$ multiplications are necessary. Figure 4.8 depicts two of these efficient structures for linear-phase FIR filters when the impulse response is symmetric.

4.3   Basic structures of recursive digital filters

4.3.1   Direct forms

The transfer function of a recursive filter is given by

$$H(z) = \frac{N(z)}{D(z)} = \frac{\displaystyle\sum_{i=0}^{M} b_i z^{-i}}{1 + \displaystyle\sum_{i=1}^{N} a_i z^{-i}} \tag{4.30}$$

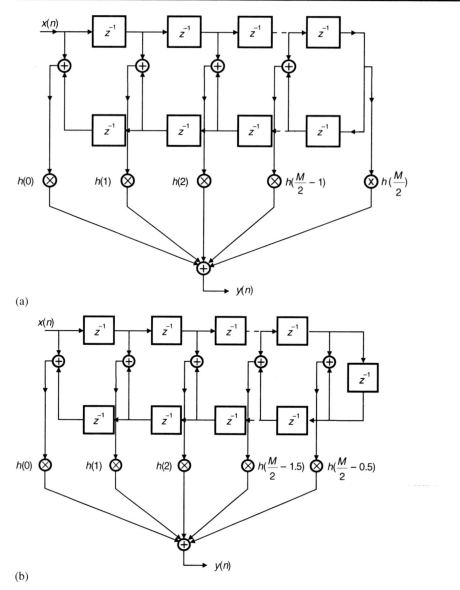

(a)

(b)

Figure 4.8 Realizations of linear-phase filters with symmetric impulse response: (a) even order; (b) odd order.

Since, in most cases, such transfer functions give rise to filters with impulse responses having infinite durations, recursive filters are also referred to as infinite-duration impulse-response (IIR) filters.[1]

We can consider that $H(z)$ as above results from the cascading of two separate filters of transfer functions $N(z)$ and $\frac{1}{D(z)}$. The $N(z)$ polynomial can be realized with the FIR

---

[1] It is important to notice that in the cases where $D(z)$ divides $N(z)$, the filter $H(z)$ turns out to have a finite-duration impulse response, and is actually an FIR filter.

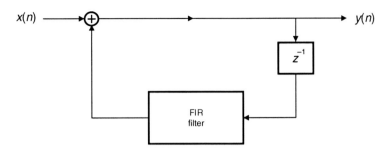

Figure 4.9 Block diagram realization of $\frac{1}{D(z)}$.

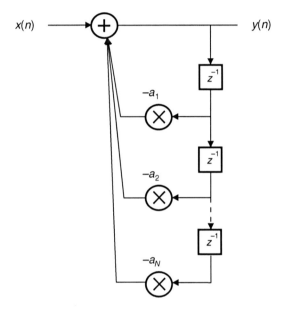

Figure 4.10 Detailed realization of $\frac{1}{D(z)}$.

direct form, as shown in the previous section. The realization of $\frac{1}{D(z)}$ can be performed as depicted in Figure 4.9, where the FIR filter shown will be an $(N-1)$th-order filter with transfer function

$$D'(z) = z(1 - D(z)) = -z \sum_{i=1}^{N} a_i z^{-i} \qquad (4.31)$$

which can be realized as in Figure 4.3. The direct form for realizing $\frac{1}{D(z)}$ is then shown in Figure 4.10.

The complete realization of $H(z)$, as a cascade of $N(z)$ and $\frac{1}{D(z)}$, is shown in Figure 4.11. Such a structure is not canonic with respect to the delays, since for a (M, N)th-order filter, with $N \geq M$ in practice, this realization requires $(N + M)$ delays.

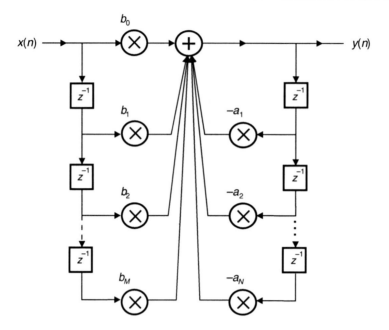

Figure 4.11 Noncanonic IIR direct-form realization.

Clearly, in the general case we can change the order in which we cascade the two separate filters, that is, $H(z)$ can be realized as $N(z) \times \frac{1}{D(z)}$ or $\frac{1}{D(z)} \times N(z)$. In the second option, all delays employed start from the same node, which allows us to eliminate the consequent redundant delays. In that manner, the resulting structure, usually referred to as the Type 1 canonic direct form, is the one depicted in Figure 4.12, for the special case when $N = M$.

An alternative structure, the so-called Type 2 canonic direct form, is shown in Figure 4.13. Such a realization is generated from the nonrecursive form in Figure 4.4. The majority of IIR filter transfer functions used in practice present a numerator degree, M, smaller than or equal to the denominator degree, N. In general, one can consider, without much loss of generality, that $M = N$. In the case where $M < N$, we just make the coefficients $b_{M+1}, \ldots, b_N$ in Figures 4.12 and 4.13 equal to zero.

4.3.2   Cascade form

In the same way as their FIR counterparts, the IIR digital filters present a large variety of possible alternative realizations. An important one, referred to as the cascade realization, is depicted in Figure 4.14a, where the basic blocks represent simple transfer functions of orders 2 or 1. In fact, the cascade form, based on second-order

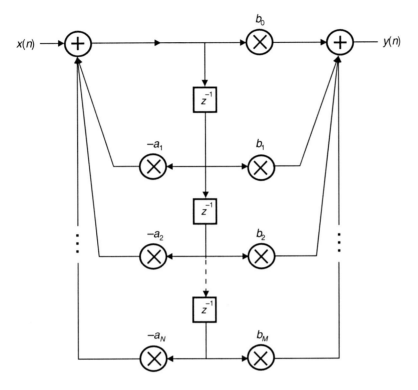

Figure 4.12 Type 1 canonic direct form for IIR filters.

blocks, is associated to the following transfer function decomposition:

$$
\begin{aligned}
H(z) &= \prod_{k=1}^{m} \frac{\gamma_{0k} + \gamma_{1k}z^{-1} + \gamma_{2k}z^{-2}}{1 + m_{1k}z^{-1} + m_{2k}z^{-2}} \\
&= \prod_{k=1}^{m} \frac{\gamma_{0k}z^2 + \gamma_{1k}z + \gamma_{2k}}{z^2 + m_{1k}z + m_{2k}} \\
&= H_0 \prod_{k=1}^{m} \frac{z^2 + \gamma'_{1k}z + \gamma'_{2k}}{z^2 + m_{1k}z + m_{2k}}
\end{aligned}
\tag{4.32}
$$

4.3.3   Parallel form

Another important realization for recursive digital filters is the parallel form represented in Figure 4.14b. Using second-order blocks, which are the most commonly used in practice, the parallel realization corresponds to the following transfer function decomposition:

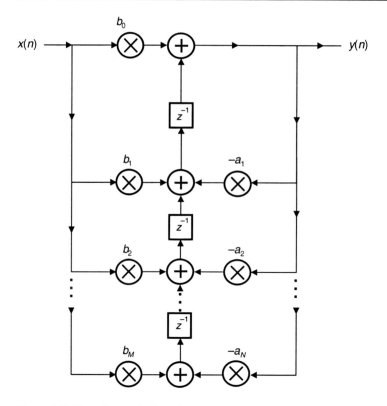

Figure 4.13 Type 2 canonic direct form for IIR filters.

$$H(z) = \sum_{k=1}^{m} \frac{\gamma_{0k}^{p} z^2 + \gamma_{1k}^{p} z + \gamma_{2k}^{p}}{z^2 + m_{1k} z + m_{2k}}$$

$$= h_0 + \sum_{k=1}^{m} \frac{\gamma_{1k}^{p'} z + \gamma_{2k}^{p'}}{z^2 + m_{1k} z + m_{2k}}$$

$$= h_0' + \sum_{k=1}^{m} \frac{\gamma_{0k}^{p''} z^2 + \gamma_{1k}^{p''} z}{z^2 + m_{1k} z + m_{2k}} \qquad (4.33)$$

also known as the partial-fraction decomposition. This equation indicates three alternative forms of the parallel realization, where the last two are canonic with respect to the number of multiplier elements.

It should be mentioned that each second-order block in the cascade and parallel forms can be realized by any of the existing distinct structures, as, for instance, one of the direct forms shown in Figure 4.15.

As it will be seen in future chapters, all these digital filter realizations present different properties when one considers practical finite-precision implementations, that is, the quantization of the coefficients and the finite precision of the arithmetic operations,

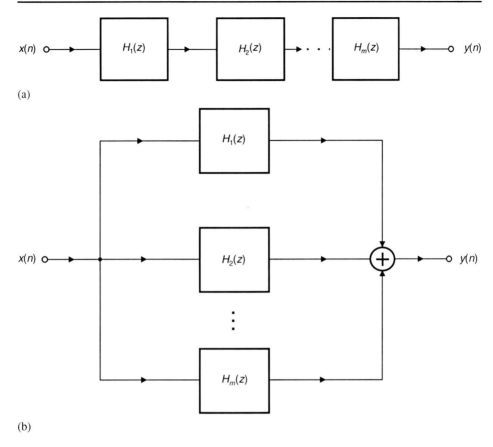

(a)

(b)

Figure 4.14 Block diagrams of: (a) cascade form; (b) parallel form.

such as additions and multiplications (Jackson, 1969; Oppenheim & Schafer, 1975; Antoniou, 1993; Jackson, 1996). In fact, the analysis of the finite-precision effects in the distinct realizations is a fundamental step in the overall process of designing any digital filter, as will be discussed in detail in Chapter 7.

4.4   Digital network analysis

The signal-flowgraph representation greatly simplifies the analysis of digital networks composed of delays, multipliers, and adder elements (Crochiere & Oppenheim, 1975). In practice, the analysis of such devices is implemented by first numbering all nodes of the graph of interest. Then, one determines the relationship between the output signal of each node with respect to the output signals of all other nodes. The connections between two nodes, referred to as branches, consist of combinations of delays and/or multipliers. Branches that inject external signals into the graph are called source branches. Such branches have a transmission coefficient equal to 1. Following this

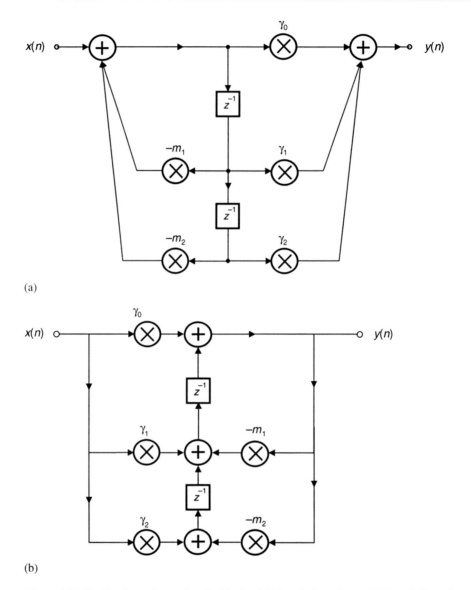

Figure 4.15 Realizations of second-order blocks: (a) Type 1 direct form; (b) Type 2 direct form.

framework, we can describe the output signal of each node as a combination of the signals of all other nodes and possibly an external signal, that is

$$Y_j(z) = X_j(z) + \sum_{k=1}^{N} (a_{kj} Y_k(z) + z^{-1} b_{kj} Y_k(z)) \tag{4.34}$$

for $j = 1, \ldots, N$, where N is the number of nodes, a_{kj} and $z^{-1} b_{kj}$ are the transmission coefficients of the branch connecting node k to node j, $Y_j(z)$ is the z transform of the

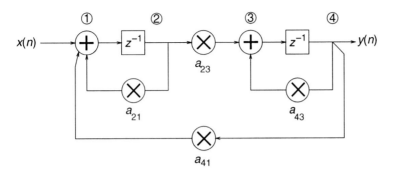

Figure 4.16 Second-order digital filter.

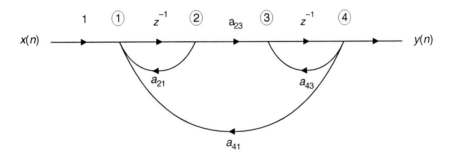

Figure 4.17 Signal-flowgraph representation of a digital filter.

output signal of node j, and $X_j(z)$ is the z transform of the external signal injected in node j. We can express equation (4.34) in a more compact form as

$$\mathbf{y}(z) = \mathbf{x}(z) + \mathbf{A}^\mathrm{T}\mathbf{y}(z) + \mathbf{B}^\mathrm{T}\mathbf{y}(z)z^{-1} \tag{4.35}$$

where $\mathbf{y}(z)$ is the output signal $N \times 1$ vector and $\mathbf{x}(z)$ is the external input signal $N \times 1$ vector for all nodes in the given graph. Also, $\mathbf{A}^\mathrm{T}$ is an $N \times N$ matrix formed by the multiplier coefficients of the delayless branches of the circuit, while $\mathbf{B}^\mathrm{T}$ is the $N \times N$ matrix of multiplier coefficients of the branches with a delay.

EXAMPLE 4.1
Describe the digital filter seen in Figure 4.16 using the compact representation given in equation (4.35).

SOLUTION
In order to carry out the description of the filter as in equations (4.34) and (4.35), it is more convenient to represent it in the signal-flowgraph form, as in Figure 4.17.

Following the procedure described above, one can easily write that

$$
\begin{bmatrix} Y_1(z) \\ Y_2(z) \\ Y_3(z) \\ Y_4(z) \end{bmatrix} = \begin{bmatrix} X_1(z) \\ 0 \\ 0 \\ 0 \end{bmatrix} + \begin{bmatrix} 0 & a_{21} & 0 & a_{41} \\ 0 & 0 & 0 & 0 \\ 0 & a_{23} & 0 & a_{43} \\ 0 & 0 & 0 & 0 \end{bmatrix} \begin{bmatrix} Y_1(z) \\ Y_2(z) \\ Y_3(z) \\ Y_4(z) \end{bmatrix}
$$

$$
+ z^{-1} \begin{bmatrix} 0 & 0 & 0 & 0 \\ 1 & 0 & 0 & 0 \\ 0 & 0 & 0 & 0 \\ 0 & 0 & 1 & 0 \end{bmatrix} \begin{bmatrix} Y_1(z) \\ Y_2(z) \\ Y_3(z) \\ Y_4(z) \end{bmatrix} \tag{4.36}
$$

$\triangle$

The z-domain signals $Y_j(z)$ associated with the network nodes can be determined as

$$
\mathbf{y}(z) = \mathbf{T}^{\mathrm{T}}(z)\mathbf{x}(z) \tag{4.37}
$$

with

$$
\mathbf{T}^{\mathrm{T}}(z) = \left(\mathbf{I} - \mathbf{A}^{\mathrm{T}} - \mathbf{B}^{\mathrm{T}} z^{-1} \right)^{-1} \tag{4.38}
$$

where $\mathbf{I}$ is the Nth-order identity matrix. In the above equation, $\mathbf{T}(z)$ is the so-called transfer matrix, whose entry $T_{ij}(z)$ describes the transfer function from node i to node j, that is

$$
T_{ij}(z) = \left. \frac{Y_j(z)}{X_i(z)} \right|_{X_k(z)=0,\, k=1,\ldots,N,\, k \neq j} \tag{4.39}
$$

and then $T_{ij}(z)$ gives the response at node j when the only nonzero input is applied to node i. If one is interested in the output signal of a particular node when several signals are injected into the network, equation (4.37) can be used to give

$$
Y_j(z) = \sum_{i=1}^{N} T_{ij}(z) X_i(z) \tag{4.40}
$$

Equation (4.35) can be expressed in the time domain as

$$
\mathbf{y}(n) = \mathbf{x}(n) + \mathbf{A}^{\mathrm{T}}\mathbf{y}(n) + \mathbf{B}^{\mathrm{T}}\mathbf{y}(n-1) \tag{4.41}
$$

If the above equation is used as a recurrence relation to determine the signal at a particular node given the input signal and initial conditions, there is no guarantee that the signal at a particular node does not depend on the signal at a different node whose output is yet to be determined. This undesirable situation can be avoided by the use of special node orderings, such as the one used by the algorithm below:

(i)   Enumerate the nodes only connected to either source branches or branches with a delay element. Note that for computing the outputs of these nodes, we need only the current values of the external input signals or values of the internal signals at instant $(n - 1)$.

(ii) Enumerate the nodes only connected to source branches, branches with a delay element, or branches connected to the nodes whose outputs were already computed in step (i). For this new group of nodes, their corresponding outputs depend on external signals, signals at instant $(n - 1)$, or signals previously determined in step (i).

(iii) Repeat the procedure above until the outputs of all nodes have been enumerated. The only case in which this is not achievable (notice that at each step, at least the output of one new node should be enumerated), occurs when the given network presents a delayless loop, which is of no practical use.

EXAMPLE 4.2

Analyze the network given in the previous example, using the algorithm described above.

SOLUTION

In the given example, the first group consists of nodes 2 and 4, and the second group consists of nodes 1 and 3. If we reorder the nodes 2, 4, 1, and 3 as 1, 2, 3, and 4, respectively, we end up with the network shown in Figure 4.18, which corresponds to

$$
\begin{bmatrix} Y_1(z) \\ Y_2(z) \\ Y_3(z) \\ Y_4(z) \end{bmatrix} = \begin{bmatrix} 0 \\ 0 \\ X_3(z) \\ 0 \end{bmatrix} + \begin{bmatrix} 0 & 0 & 0 & 0 \\ 0 & 0 & 0 & 0 \\ a_{21} & a_{41} & 0 & 0 \\ a_{23} & a_{43} & 0 & 0 \end{bmatrix} \begin{bmatrix} Y_1(z) \\ Y_2(z) \\ Y_3(z) \\ Y_4(z) \end{bmatrix}
$$

$$
+ z^{-1} \begin{bmatrix} 0 & 0 & 1 & 0 \\ 0 & 0 & 0 & 1 \\ 0 & 0 & 0 & 0 \\ 0 & 0 & 0 & 0 \end{bmatrix} \begin{bmatrix} Y_1(z) \\ Y_2(z) \\ Y_3(z) \\ Y_4(z) \end{bmatrix} \tag{4.42}
$$

$\triangle$

In general, after reordering, $\mathbf{A}^{\mathrm{T}}$ can be put in the following form:

$$
\mathbf{A}^{\mathrm{T}} = \begin{bmatrix} 0 & \cdots & 0 & \cdots & 0 & 0 \\ 0 & \cdots & 0 & \cdots & 0 & 0 \\ \vdots & \ddots & \vdots & \ddots & \vdots & \vdots \\ a_{1j} & \cdots & a_{kj} & \cdots & \vdots & \vdots \\ \vdots & \ddots & \vdots & \ddots & \vdots & \vdots \\ a_{1N} & \cdots & a_{kN} & \cdots & 0 & 0 \end{bmatrix} \tag{4.43}
$$

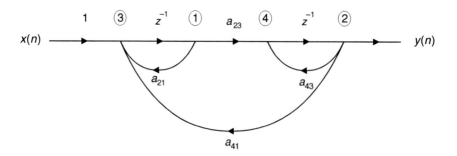

Figure 4.18 Reordering the nodes in the signal flowgraph.

This means that $\mathbf{A}^T$ and $\mathbf{B}^T$ tend to be sparse matrices. Therefore, efficient algorithms can be employed to solve the time- or z-domain analyses described in this section.

4.5   State-space description

An alternative form of representing digital filters is to use what is called the state-space representation. In such a description, the outputs of the memory elements (delays) are considered the system states. Once all the values of the external and state signals are known, we can determine the future values of the system states (the delay inputs) and the system output signals as follows:

$$\left.\begin{array}{l} \mathbf{x}(n+1) = \mathbf{A}\mathbf{x}(n) + \mathbf{B}\mathbf{u}(n) \\ \mathbf{y}(n) = \mathbf{C}^T\mathbf{x}(n) + \mathbf{D}\mathbf{u}(n) \end{array}\right\} \tag{4.44}$$

where $\mathbf{x}(n)$ is the $N \times 1$ vector of the state variables. If M is the number of system inputs and M' is the number of system outputs, we have that $\mathbf{A}$ is $N \times N$, $\mathbf{B}$ is $N \times M$, $\mathbf{C}$ is $N \times M'$, and $\mathbf{D}$ is $M' \times M$. In general, we work with single-input and single-output systems. In such cases, $\mathbf{B}$ is $N \times 1$, $\mathbf{C}$ is $N \times 1$, and $\mathbf{D}$ is 1×1, that is, $\mathbf{D} = d$ is a scalar.

Note that this representation is essentially different from the one given in equation (4.41), because in that equation the variables are the outputs of each node, whereas in the state-space approach the variables are just the outputs of the delays.

The impulse response for a system as described in equation (4.44) is given by

$$h(n) = \begin{cases} d, & \text{for } n = 0 \\ \mathbf{C}^T\mathbf{A}^{n-1}\mathbf{B}, & \text{for } n > 0 \end{cases} \tag{4.45}$$

To determine the corresponding transfer function, we first apply the z transform to equation (4.44), obtaining

$$\left.\begin{array}{l} z\mathbf{X}(z) = \mathbf{A}\mathbf{X}(z) + \mathbf{B}\mathbf{U}(z) \\ \mathbf{Y}(z) = \mathbf{C}^T\mathbf{X}(z) + \mathbf{D}\mathbf{U}(z) \end{array}\right\} \tag{4.46}$$

and then

$$H(z) = \frac{Y(z)}{U(z)} = \mathbf{C}^{\mathrm{T}} (z\mathbf{I} - \mathbf{A})^{-1} \mathbf{B} + d \tag{4.47}$$

From equation (4.47), it should be noticed that the poles of $H(z)$ are the eigenvalues of $\mathbf{A}$, as the denominator of $H(z)$ will be given by the determinant of $(z\mathbf{I} - \mathbf{A})$.

By applying a linear transformation $\mathbf{T}$ to the state vector such that

$$\mathbf{x}(n) = \mathbf{T}\mathbf{x}'(n) \tag{4.48}$$

where $\mathbf{T}$ is any $N \times N$ nonsingular matrix, we end up with a system characterized by

$$\mathbf{A}' = \mathbf{T}^{-1}\mathbf{A}\mathbf{T}; \quad \mathbf{B}' = \mathbf{T}^{-1}\mathbf{B}; \quad \mathbf{C}' = \mathbf{T}^{\mathrm{T}}\mathbf{C}; \quad d' = d \tag{4.49}$$

Such a system will present the same transfer function as the original system, and consequently the same poles and zeros. The proof of this fact is left to the interested reader as an exercise.

EXAMPLE 4.3
Determine the state-space equations of the filters given in Figures 4.15a and 4.16.

SOLUTION
Following the procedure described above, for the filter shown in Figure 4.15a, we have that

$$\left.\begin{array}{l} \begin{bmatrix} x_1(n+1) \\ x_2(n+1) \end{bmatrix} = \begin{bmatrix} -m_1 & -m_2 \\ 1 & 0 \end{bmatrix} \begin{bmatrix} x_1(n) \\ x_2(n) \end{bmatrix} + \begin{bmatrix} 1 \\ 0 \end{bmatrix} u(n) \\[20pt] y(n) = \begin{bmatrix} (\gamma_1 - m_1\gamma_0) & (\gamma_2 - m_2\gamma_0) \end{bmatrix} \begin{bmatrix} x_1(n) \\ x_2(n) \end{bmatrix} + \gamma_0 u(n) \end{array}\right\} \tag{4.50}$$

and for the filter in Figure 4.16, we get

$$\left.\begin{array}{l} \begin{bmatrix} x_1(n+1) \\ x_2(n+1) \end{bmatrix} = \begin{bmatrix} a_{21} & a_{41} \\ a_{23} & a_{43} \end{bmatrix} \begin{bmatrix} x_1(n) \\ x_2(n) \end{bmatrix} + \begin{bmatrix} 1 \\ 0 \end{bmatrix} u(n) \\[20pt] y(n) = \begin{bmatrix} 0 & 1 \end{bmatrix} \begin{bmatrix} x_1(n) \\ x_2(n) \end{bmatrix} + 0u(n) \end{array}\right\} \tag{4.51}$$

$\triangle$

4.6　Basic properties of digital networks

In this section, we introduce some network properties that are very useful for designing and analyzing digital filters. The material covered in this section is mainly based on (Fettweis, 1971b).

4.6.1   Tellegen's theorem

Consider a digital network represented by the corresponding signal flowgraph, in which the signal that reaches node j from node i is denoted by x_{ij}. We can use this notation to represent a branch that leaves from and arrives at the same node. Such a branch is called a loop. Among other things, loops are used to represent source branches entering a node. In fact, every source branch will be represented as a loop having the value of its source. In this case, x_{ii} includes the external signal and any other loop connecting node i to itself. Following this framework, for each node of a given graph, we can write that

$$y_j = \sum_{i=1}^{N} x_{ij} \tag{4.52}$$

where N in this case is the total number of nodes. Consider now the following result.

THEOREM 4.1 (TELLEGEN'S THEOREM)
All corresponding signals, (x_{ij}, y_j) and (x'_{ij}, y'_j), of two distinct networks with equal signal-flowgraph representations satisfy

$$\sum_{i=1}^{N} \sum_{j=1}^{N} (y_j x'_{ij} - y'_i x_{ji}) = 0 \tag{4.53}$$

where both sums include all nodes of both networks.

$\diamond$

PROOF
Equation (4.53) can be rewritten as

$$\sum_{j=1}^{N} \left(y_j \sum_{i=1}^{N} x'_{ij} \right) - \sum_{i=1}^{N} \left(y'_i \sum_{j=1}^{N} x_{ji} \right) = \sum_{j=1}^{N} y_j y'_j - \sum_{i=1}^{N} y'_i y_i = 0 \tag{4.54}$$

which completes the proof.

$\square$

Tellegen's theorem can be generalized for the frequency domain, since

$$Y_j = \sum_{i=1}^{N} X_{ij} \tag{4.55}$$

and then

$$\sum_{i=1}^{N} \sum_{j=1}^{N} \left(Y_j X'_{ij} - Y'_i X_{ji} \right) = 0 \tag{4.56}$$

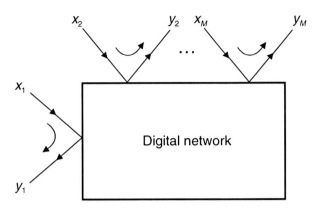

Figure 4.19 General digital network with *M* ports.

Notice that in Tellegen's theorem x_{ij} is actually the sum of all signals departing from node i and arriving at node j. Therefore, in the most general case where the two graphs have different topologies, Tellegen's theorem can still be applied, making the two topologies equal by adding as many nodes and branches with null transmission values as necessary.

4.6.2 Reciprocity

Consider a particular network in which *M* of its nodes each have two branches connecting them to the outside world, as depicted in Figure 4.19. The first branch in each of these nodes is a source branch through which an external signal is injected into the node. The second branch makes the signal of each node available as an output signal. Naturally, in nodes where there are neither external input nor output signals, one must consider the corresponding branch to have a null transmission value. For generality, if a node does not have either external or output signals, both corresponding branches are considered to have a null transmission value.

Suppose that we apply a set of signals X_i to this network, and collect as output the signals Y_i. Alternatively, we could apply the signals X_i', and observe the Y_i' signals. The particular network is said to be reciprocal if

$$\sum_{i=1}^{M} \left(X_i Y_i' - X_i' Y_i \right) = 0 \tag{4.57}$$

In such cases, if the *M*-port network is described by

$$Y_i = \sum_{j=1}^{M} T_{ji} X_j \tag{4.58}$$

where T_{ji} is the transfer function from port j to port i, equation (4.57) is then equivalent to

$$T_{ij} = T_{ji} \tag{4.59}$$

The proof for this statement is based on the substitution of equation (4.58) into equation (4.57), which yields

$$\sum_{i=1}^{M} \left(X_i \sum_{j=1}^{M} T_{ji} X_j' - X_i' \sum_{j=1}^{M} T_{ji} X_j \right) = \sum_{i=1}^{M} \sum_{j=1}^{M} \left(X_i T_{ji} X_j' \right) - \sum_{i=1}^{M} \sum_{j=1}^{M} \left(X_i' T_{ji} X_j \right)$$

$$= \sum_{i=1}^{M} \sum_{j=1}^{M} \left(X_i T_{ji} X_j' \right) - \sum_{i=1}^{M} \sum_{j=1}^{M} \left(X_j' T_{ij} X_i \right)$$

$$= \sum_{i=1}^{M} \sum_{j=1}^{M} \left(T_{ji} - T_{ij} \right) \left(X_i X_j' \right)$$

$$= 0 \tag{4.60}$$

and thus

$$T_{ij} = T_{ji} \tag{4.61}$$

$\square$

4.6.3 Interreciprocity

The vast majority of the digital networks associated with digital filters are not reciprocal. However, such a concept is crucial in some cases. Fortunately, there is another related property, called interreciprocity, between two networks which is very common and useful. Consider two networks with the same number of nodes, and also consider that X_i and Y_i are respectively input and output signals of the first network. Correspondingly, X_i' and Y_i' represent input and output signals of the second network. Such networks are considered interreciprocal if equation (4.57) holds for (X_i, Y_i) and (X_i', Y_i'), $i = 1, \ldots, M$.

If two networks are described by

$$Y_i = \sum_{j=1}^{M} T_{ji} X_j \tag{4.62}$$

and

$$Y_i' = \sum_{j=1}^{M} T_{ji}' X_j' \tag{4.63}$$

it can be easily shown that these two networks are interreciprocal if

$$T_{ji} = T_{ij}' \tag{4.64}$$

Once again, the proof is left as an exercise for the interested reader.

4.6.4 Transposition

Given any signal-flowgraph representation of a digital network, we can generate another network by reversing the directions of all branches. In such a procedure, all addition nodes turn into distribution nodes, and vice versa. Also, if in the original network the branch from node i to node j is F_{ij}, the transpose network will have a branch from node j to node i with transmission F'_{ji} such that

$$F_{ij} = F'_{ji} \tag{4.65}$$

Using Tellegen's theorem, one can easily show that the original network and its corresponding transpose network are interreciprocal. In fact, applying Tellegen's theorem to all signals of both networks, one obtains

$$\sum_{i=1}^{N} \sum_{\substack{j=1 \\ j \neq i, \text{ if } i < M+1}}^{N} \left(Y_j X'_{ij} - Y'_i X_{ji} \right) + \sum_{i=1}^{M} \left(Y_i X'_{ii} - Y'_i X_{ii} \right)$$

$$= \sum_{i=1}^{N} \sum_{\substack{j=1 \\ j \neq i, \text{ if } i < M+1}}^{N} \left(Y_j F'_{ij} Y'_i - Y'_i F_{ji} Y_j \right) + \sum_{i=1}^{M} \left(Y_i X'_{ii} - Y'_i X_{ii} \right)$$

$$= 0 + \sum_{i=1}^{M} \left(Y_i X'_{ii} - Y'_i X_{ii} \right)$$

$$= 0 \tag{4.66}$$

where all external signals are considered to be injected at the first M nodes. Naturally, $\sum_{i=1}^{M} (Y_i X'_{ii} - Y'_i X_{ii}) = 0$ is equivalent to interreciprocity (see equation (4.57) applied to the interreciprocity case), which implies that (equation (4.64))

$$T_{ij} = T'_{ji} \tag{4.67}$$

This is a very important result, because it indicates that a network and its transpose must have the same transfer function. For instance, the equivalence of the networks in Figures 4.3 and 4.4 can be deduced from the fact that one is the transpose of the other. The same can be said about the networks in Figures 4.12 and 4.13.

4.6.5 Sensitivity

Sensitivity is a measure of the degree of variation of a network's overall transfer function with respect to small fluctuations in the value of one of its elements. In the specific case of digital filters, one is often interested in the sensitivity with respect to variations of the multiplier coefficients, that is

$$S_{m_i}^{H(z)} = \frac{\partial H(z)}{\partial m_i} \tag{4.68}$$

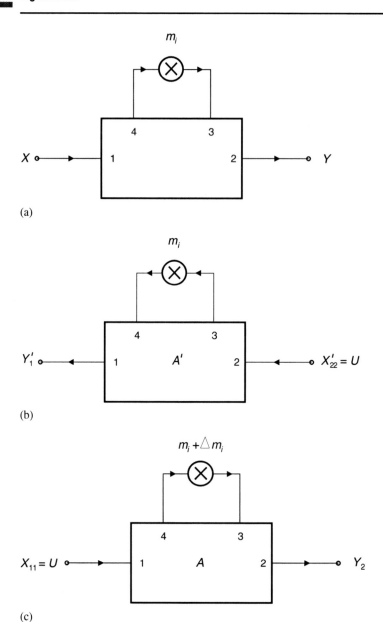

Figure 4.20 Digital networks: (a) original; (b) transpose; (c) original with modified coefficient.

for $i = 1, \ldots, L$, where L is the total number of multipliers in the particular network.

Using the concept of transposition, we can determine the sensitivity of $H(z)$ with respect to a given coefficient m_i in a very efficient way. To understand how, consider a network, its transpose, and also the original network with a specific coefficient slightly modified, as depicted in Figure 4.20.

Using Tellegen's theorem on the networks shown in Figures 4.20b and 4.20c, one obtains

$$\sum_{i=1}^{N}\sum_{j=1}^{N}\left(Y_j X'_{ij} - Y'_i X_{ji}\right) = \sum_{j=1}^{N}\left(Y_j X'_{1j} - Y'_1 X_{j1}\right) + \sum_{j=1}^{N}\left(Y_j X'_{2j} - Y'_2 X_{j2}\right)$$

$$+ \sum_{j=1}^{N}\left(Y_j X'_{3j} - Y'_3 X_{j3}\right) + \sum_{i=4}^{N}\sum_{j=1}^{N}\left(Y_j X'_{ij} - Y'_i X_{ji}\right)$$

$$= A_1 + A_2 + A_3 + A_4$$

$$= 0 \tag{4.69}$$

where A_1, A_2, A_3, and A_4 are separately determined below.

$$A_1 = Y_1 X'_{11} - Y'_1 X_{11} + \sum_{j=2}^{N}\left(Y_j X'_{1j} - Y'_1 X_{j1}\right)$$

$$= -U Y'_1 + \sum_{j=2}^{N}\left(Y_j F'_{1j} Y'_1 - Y'_1 F_{j1} Y_j\right)$$

$$= -U Y'_1 \tag{4.70}$$

since $F'_{1j} = F_{j1}$, for all j. In addition,

$$A_2 = Y_2 X'_{22} - Y'_2 X_{22} + \sum_{\substack{j=1 \\ j \neq 2}}^{N}\left(Y_j X'_{2j} - Y'_2 X_{j2}\right)$$

$$= U Y_2 + \sum_{\substack{j=1 \\ j \neq 2}}^{N}\left(Y_j F'_{2j} Y'_2 - Y'_2 F_{j2} Y_j\right)$$

$$= U Y_2 \tag{4.71}$$

$$A_3 = Y_4 X'_{34} - Y'_3 X_{43} + \sum_{\substack{j=1 \\ j \neq 4}}^{N}\left(Y_j F'_{3j} Y'_3 - Y'_3 F_{j3} Y_j\right)$$

$$= Y_4 m_i Y'_3 - Y'_3 (m_i + \Delta m_i) Y_4$$

$$= -\Delta m_i Y_4 Y'_3 \tag{4.72}$$

$$A_4 = \sum_{i=4}^{N}\sum_{j=1}^{N}\left(Y_j X'_{ij} - Y'_i X_{ji}\right)$$

$$= \sum_{i=4}^{N}\sum_{j=1}^{N}\left(Y_j F'_{ij} Y'_i - Y'_i F_{ji} Y_j\right)$$

$$= 0 \tag{4.73}$$

Hence, one has that

$$-UY'_1 + UY_2 - \Delta m_i Y_4 Y'_3 = 0 \tag{4.74}$$

Thus

$$U\left(Y_2 - Y'_1\right) = \Delta m_i Y_4 Y'_3 \tag{4.75}$$

Defining

$$\Delta H_{12} = \left(H_{12} - H'_{21}\right) = \left(\frac{Y_2}{U} - \frac{Y'_1}{U}\right) \tag{4.76}$$

one gets, from equation (4.75), that

$$U^2\left(H_{12} - H'_{21}\right) = U^2 \Delta H_{12} = \Delta m_i Y_4 Y'_3 \tag{4.77}$$

If we now let Δm_i converge to zero, H_{12} tends to H'_{21}, and consequently

$$\frac{\partial H_{12}}{\partial m_i} = \frac{Y_4 Y'_3}{U^2} = H'_{23} H_{14} = H_{32} H_{14} \tag{4.78}$$

This equation indicates that the sensitivity of the transfer function of the original network, H_{12}, with respect to variations of one of its coefficients, can be determined based on transfer functions between the system input and the node before the multiplier, H_{14}, and between the multiplier output node and the system output, H_{32}.

EXAMPLE 4.4
Determine the sensitivity of $H(z)$ with respect to the coefficients a_{11}, a_{22}, a_{12}, and a_{21} in the network of Figure 4.21.

SOLUTION
The necessary transfer functions are

$$F_1(z) = \frac{X_1(z)}{X(z)} = \frac{b_1 z + (b_2 a_{12} - b_1 a_{22})}{D(z)} \tag{4.79}$$

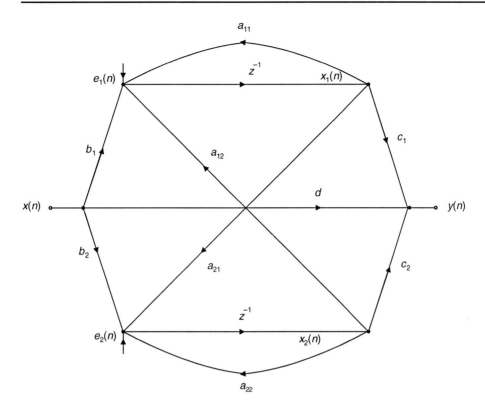

Figure 4.21 State variable network.

$$F_2(z) = \frac{X_2(z)}{X(z)} = \frac{b_2 z + (b_1 a_{21} - b_2 a_{11})}{D(z)} \tag{4.80}$$

$$G_1(z) = \frac{Y(z)}{E_1(z)} = \frac{c_1 z + (c_2 a_{21} - c_1 a_{22})}{D(z)} \tag{4.81}$$

$$G_2(z) = \frac{Y(z)}{E_2(z)} = \frac{c_2 z + (c_1 a_{12} - c_2 a_{11})}{D(z)} \tag{4.82}$$

where the common denominator $D(z)$ is given by

$$D(z) = z^2 - (a_{11} + a_{22})z + (a_{11} a_{22} - a_{12} a_{21}) \tag{4.83}$$

The required sensitivities are then

$$S_{a_{11}}^{H(z)} = F_1(z) G_1(z) \tag{4.84}$$

$$S_{a_{22}}^{H(z)} = F_2(z) G_2(z) \tag{4.85}$$

$$S_{a_{12}}^{H(z)} = F_2(z) G_1(z) \tag{4.86}$$

$$S_{a_{21}}^{H(z)} = F_1(z) G_2(z) \tag{4.87}$$

$\triangle$

Table 4.2. *List of MATLAB commands for transforming digital filter representations.*

	Direct	Zero-pole	Cascade	Parallel	State-space
Direct		`tf2zp` `roots`		`residuez`	`tf2ss`
Zero-pole	`zp2tf` `poly`		`zp2sos`		`zp2ss`
Cascade	`sos2tf`	`sos2zp`			`sos2ss`
Parallel	`residuez`				
State-space	`ss2tf`	`ss2zp`	`ss2sos`		

4.7   Digital filter forms with MATLAB

As has been seen in this chapter, there are various different forms for representing a given transfer function. These include the direct form, the cascade form, the parallel form, and the state-space form. The MATLAB Signal Processing toolbox has a series of commands that are suitable for transforming a given representation into another one of interest. Such commands are summarized in Table 4.2 and explained in detail below.

- `tf2zp`: Converts the direct form into the zero-pole-gain form, inverting the operation of `zp2tf`. The zero-pole-gain form is described by

$$H(z) = k \frac{[z - Z(1)][z - Z(2)] \cdots [z - Z(M)]}{[z - P(1)][z - P(2)] \cdots [z - P(N)]} \tag{4.88}$$

where k is a gain factor, $Z(1), Z(2), \ldots, Z(M)$ is the set of filter zeros, and $P(1), P(2), \ldots, P(N)$ is the set of filter poles.
Input parameters: Vectors with the numerator, b, and denominator coefficients, a.
Output parameters: Column vectors of filter zeros, z, and poles, p, and the filter gain factor, k.
Example:
```
b=[1 0.6 -0.16]; a=[1 0.7 0.12];
[z,p,k]=tf2zp(b,a);
```

- `zp2tf`: Converts the zero-pole-gain form (see `tf2zp` command) into the direct form, inverting the operation of `tf2zp`.
Input parameters: Column vectors of filter zeros, z, and poles, p, and the filter gain factor, k.
Output parameters: Vectors with the numerator, b, and denominator coefficients, a.

Example:
```
z=[-0.8 0.2]'; p=[-0.4 -0.3]'; k=1;
[num,den]=zp2tf(z,p,k);
```

- `roots`: Determines the roots of a polynomial. This command also can be used to decompose a given transfer function into the zero-pole-gain format.
  Input parameter: A vector of polynomial coefficients.
  Output parameter: A vector of roots.
  Example:
  ```
  r=roots([1 0.6 -0.16]);
  ```

- `poly`: Inverts the operation of `roots`, that is, given a set of roots, this command determines the monic polynomial associated with these roots.
  Input parameter: A vector of roots.
  Output parameter: A vector of polynomial coefficients.
  Example:
  ```
  pol=poly([-0.8 0.2]);
  ```

- `sos2tf`: Converts the cascade form into the direct form. Notice that there is no direct command that reverses this operation.
  Input parameter: An $L \times 6$ matrix, `sos`, whose rows contain the coefficients of each second-order section of the form

$$H_k(z) = \frac{b_{0k} + b_{1k}z^{-1} + b_{2k}z^{-2}}{a_{0k} + a_{1k}z^{-1} + a_{2k}z^{-2}} \tag{4.89}$$

such that

$$\mathsf{sos} = \begin{bmatrix} b_{01} & b_{11} & b_{21} & a_{01} & a_{11} & a_{21} \\ b_{02} & b_{12} & b_{22} & a_{02} & a_{12} & a_{22} \\ \vdots & \vdots & \vdots & \vdots & \vdots & \vdots \\ b_{0L} & b_{1L} & b_{2L} & a_{0L} & a_{1L} & a_{2L} \end{bmatrix} \tag{4.90}$$

Output parameters: Vectors `num` and `den` containing the numerator and denominator coefficients.
  Example:
  ```
  sos=[1 1 1 1 10 1; -2 3 1 1 0 -1];
  [num,den]=sos2tf(sos);
  ```

- `residuez`: Performs the partial-fraction expansion in the z domain, if there are two input parameters. This command considers complex roots. To obtain a parallel expansion of a given transfer function, one should combine such roots in complex conjugate pairs with the `cplxpair` command to form second-order sections with solely real coefficients. The `residuez` command also converts the partial-fraction expansion back to original direct form, if there are three input parameters.

Input parameters: Vectors of numerator coefficients, b, and denominator coefficients, a.

Output parameters: Vectors of residues, r, and poles, p, and gain factor, k.

Example:

```
b=[1 0.6 -0.16]; a=[1 0.7 0.12];
[r,p,k]=residuez(b,a);
```

Input parameters: Vectors of residues, r, poles, p, and gain factor k.

Output parameters: Vectors of numerator coefficients, b, and denominator coefficients, a.

Example:

```
r=[14 -43/3]; p=[-0.4 -0.3]; k=4/3;
[b,a]=residuez(r,p,k);
```

- cplxpair: Rearranges the elements of a vector into complex conjugate pairs. The pairs are ordered by increasing real part. Real elements are placed after all complex pairs.

  Input parameter: A vector of complex numbers.

  Output parameter: A vector containing the ordered complex numbers.

  Example:

  ```
  Xord=cplxpair(roots([ 1 4 2 1 3 1 4]));
  ```

- tf2ss: Converts the direct form into the state-space form, inverting the operation of ss2tf.

  Input parameters: As for command tf2zp.

  Output parameters: The state-space matrix, A, the input column vector, B, the state row vector, C, and the scalar, D.

  Example:

  ```
  b=[1 0.6 -0.16]; a=[1 0.7 0.12];
  [A,B,C,D]=tf2ss(b,a);
  ```

- ss2tf: Converts the state-space form into the direct form, inverting the operation of tf2ss.

  Input parameters: The state-space matrix, A, the input column vector, B, the state row vector, C, and the scalar, D.

  Output parameters: As for command zp2tf.

  Example:

  ```
  A=[-0.7 -0.12; 1 0]; b=[1 0]'; c=[-0.1 -0.28]; d=1;
  [num,den]=ss2tf(A,b,c,d);
  ```

- zp2sos: Converts the zero-pole-gain form into the cascade form, inverting the operation of sos2zp. The zp2sos command can also order the resulting sections according to the pole positions with respect to the unit circle.

  Input parameters: The same as for zp2tf, with the addition of a string, up (default) or `down`, that indicates the desired section ordering starts with the poles further away from or closer to the unit circle, respectively.

Output parameter: As for the input parameter of the command `sos2tf`.

Example:
```
z=[1 1 j -j]; p=[0.9 0.8 0.7 0.6];
sos=zp2sos(z,p);
```

- `sos2zp`: Converts the cascade form into the zero-pole-gain form, inverting the operation of `zp2sos`.

  Input parameter: As for command `sos2tf`.

  Output parameter: As for command `tf2zp`.

  Example:
  ```
  sos=[1 1 1 1 10 1; -2 3 1 1 0 -1];
  [z,p,k]=sos2zp(sos);
  ```

- `zp2ss`: Converts the zero-pole-gain form into the state-space form, inverting the operation of `ss2zp`.

  Input parameters: As for command `zp2tf`, but with no restriction on the format of the zero and pole vectors.

  Output parameters: As for command `tf2ss`.

  Example:
  ```
  z=[-0.8 0.2]; p=[-0.4 -0.3]; k=1;
  [A,B,C,D]=zp2ss(z,p,k);
  ```

- `ss2zp`: Converts the state-space form into the zero-pole-gain form, inverting the operation of `zp2ss`.

  Input parameters: As for command `ss2tf`.

  Output parameters: As for command `tf2zp`.

  Example:
  ```
  A=[-0.7 -0.12; 1 0]; b=[1 0]'; c=[-0.1 -0.28]; d=1;
  [z,p,k]=ss2zp(A,b,c,d);
  ```

- `sos2ss`: Converts the cascade form into the state-space form, inverting the operation of `ss2sos`.

  Input parameter: As for command `sos2tf`.

  Output parameters: As for command `tf2ss`.

  Example:
  ```
  sos=[1 1 1 1 10 1; -2 3 1 1 0 -1];
  [A,B,C,D]=sos2ss(sos);
  ```

- `ss2sos`: Converts the state-space form into the cascade form, inverting the operation of `sos2ss`.

  Input parameters: As for command `ss2tf`.

Output parameter: As for input parameter of command `sos2tf`.
Example:

```
A=[-0.7 -0.12; 1 0]; b=[1 0]'; c=[-0.1 -0.28]; d=1;
sos=ss2sos(A,b,c,d);
```

- `filter`: Performs signal filtering using the Type 2 canonic direct form for IIR filters (see Figure 4.13).

  Input parameters: A vector b of numerator coefficients; a vector a of denominator coefficients; a vector x holding the signal to be filtered; a vector zi holding the initial conditions of the delays.

  Output parameters: A vector y holding the filtered signal; a vector zf holding the initial conditions of the delays.

  Example:

```
a=[1 -0.22 -0.21 0.017 0.01];
b=[1 2 1]; zi=[0 0 0 0];
x=randn(100,1); [y,zf]=filter(b,a,x,zi); plot(y);
```

4.8   Summary

In this chapter, some basic realizations for digital filters with finite-duration impulse responses (FIR) or infinite-duration impulse responses (IIR) were presented. In particular, FIR filters with linear phase were introduced, and their practical importance was highlighted. Other more advanced FIR and IIR structures will be discussed later in this book.

A procedure for analyzing digital networks in the time and frequency domains was devised. Following this, the state-space description was introduced.

The digital version of Tellegen's theorem was presented, as well as the reciprocity, interreciprocity, and transposition network properties. Then, with the help of Tellegen's theorem, a simple formulation for calculating the sensitivity of a given transfer function with respect to coefficient variations was provided.

Finally, the MATLAB functions related to basic digital filter representations were discussed.

4.9   Exercises

4.1   Give two distinct realizations for the transfer functions below.

(a) $H(z) = 0.0034 + 0.0106z^{-2} + 0.0025z^{-4} + 0.0149z^{-6}$

(b) $H(z) = \left[\frac{z^2 - 1.349z + 1}{z^2 - 1.919z + 0.923} \right] \left[\frac{z^2 - 1.889z + 1}{z^2 - 1.937z + 0.952} \right]$

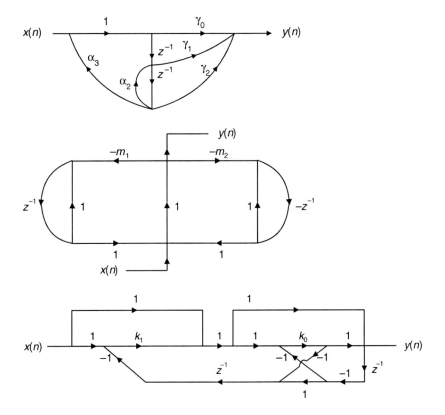

Figure 4.22 Signal flowgraphs of three digital filters.

4.2   Some FIR filters present a rational transfer function. Show that the transfer function

$$H(z) = \frac{z^{-(M+1)} - 1}{z^{-1} - 1}$$

corresponds to an FIR filter. Determine the operation performed by such a filter. Discuss when a rational transfer function corresponds to an FIR filter.

4.3   Write the equations that describe the networks in Figure 4.22, by numbering the nodes appropriately.

4.4   Determine the transfer functions of the digital filters in Figure 4.22.

4.5   Show that the space-state description of a given digital filter is invariant with respect to a linear transformation of the state vector

$$\mathbf{x}(n) = \mathbf{T}\mathbf{x}'(n)$$

where $\mathbf{T}$ is any $N \times N$ nonsingular matrix.

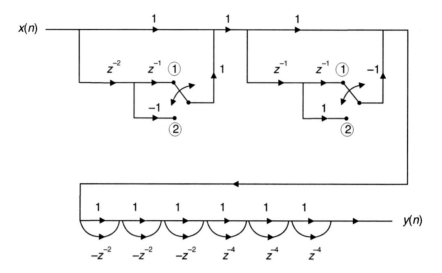

Figure 4.23 Signal flowgraph of a digital filter.

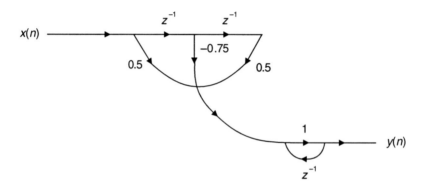

Figure 4.24 Signal flowgraph of a digital filter.

4.6 Describe the networks in Figure 4.22 using state variables.

4.7 Find the transpose for each network in Figure 4.22.

4.8 Determine the sensitivity of the transfer functions determined in Exercise 4.4, for the filters given in Figure 4.22, with respect to each filter coefficient.

4.9 Determine the transfer function of the filter in Figure 4.23, considering the two possible positions for the switches.

4.10 Determine and plot the frequency response of the filter shown in Figure 4.23, considering the two possible positions for the switches.

4.11 Determine the impulse response of the filter shown in Figure 4.24.

4.12   Show that if two given networks are described by $Y_i = \sum_{j=1}^{M} T_{ij} X_j$ and $Y_i' = \sum_{j=1}^{M} T_{ij}' X_j'$, then these networks are interreciprocal if $T_{ji} = T_{ij}'$.

4.13   Create MATLAB commands to fill in the blanks of Table 4.2.

5 FIR filter approximations

5.1 Introduction

In this chapter, we will study the approximation schemes for digital filters with finite-duration impulse response (FIR) and we will present the methods for determining the multiplier coefficients and the filter order, in such a way that the resulting frequency response satisfies a set of prescribed specifications.

In some cases, FIR filters are considered inefficient in the sense that they require a high-order transfer function to satisfy the system requirements when compared to the order required by digital filters with infinite-duration impulse response. However, FIR digital filters do possess a few implementation advantages such as a possible exact linear-phase characteristic and intrinsically stable implementations, when using non-recursive realizations. In addition, the computational complexity of FIR digital filters can be reduced if they are implemented using fast numerical algorithms such as the fast Fourier transform.

We start by discussing the ideal frequency response characteristics of commonly used FIR filters, as well as their corresponding impulse responses. We include in the discussion lowpass, highpass, bandpass, and bandstop filters, and also treat two other important filters, namely differentiators and Hilbert transformers.

We go on to discuss the frequency sampling and the window methods for approximating FIR digital filters, focusing on the rectangular, triangular, Bartlett, Hamming, Blackman, Kaiser, and Dolph–Chebyshev windows. In addition, the design of maximally flat FIR filters is addressed.

Following this, numerical methods for designing FIR filters are discussed. A unified framework for the general approximation problem is provided. The weighted-least-squares (WLS) method is presented as a generalization of the rectangular window approach. We then introduce the Chebyshev (or minimax) approach as the most efficient form, with respect to the resulting filter order, to approximate FIR filters which minimize the maximum passband and stopband ripples. We also discuss the WLS–Chebyshev approach, which is able to combine the desired characteristics of high attenuation of the Chebyshev scheme with the low energy level of the WLS scheme in the filter stopband.

We conclude the chapter by discussing the use of MATLAB for designing FIR filters.

5.2   Ideal characteristics of standard filters

In this section, we analyze the time and frequency response characteristics of commonly used FIR filters. First, we deal with lowpass, highpass, bandpass, and bandstop filters. Then, two types of filters widely used in the field of digital signal processing, the differentiators and Hilbert transformers (Oppenheim & Schafer, 1975; Antoniou, 1993), are analyzed, and their implementations as special cases of FIR digital filters are studied.

The behavior of a filter is usually best characterized by its frequency response $H(e^{j\omega})$. As seen in Chapter 4, a filter implementation is based on its transfer function $H(z)$ of the form

$$H(z) = \sum_{n=-\infty}^{\infty} h(n)z^{-n} \tag{5.1}$$

The FIR filter design starts by calculating the coefficients $h(n)$ which will be used in one of the structures discussed in Section 4.2.

As seen in Section 2.8, the relationship between $H(e^{j\omega})$ and $h(n)$ is given by the following pair of equations:

$$H(e^{j\omega}) = \sum_{n=-\infty}^{\infty} h(n)e^{-j\omega n} \tag{5.2}$$

$$h(n) = \frac{1}{2\pi} \int_{-\pi}^{\pi} H(e^{j\omega})e^{j\omega n} d\omega \tag{5.3}$$

In what follows, we determine $H(e^{j\omega})$ and $h(n)$ related to ideal standard filters.

5.2.1   Lowpass, highpass, bandpass, and bandstop filters

The ideal magnitude responses of some standard digital filters are depicted in Figure 5.1.

For instance, the lowpass filter, as seen in Figure 5.1a, is described by

$$|H(e^{j\omega})| = \begin{cases} 1, & \text{for } |\omega| \le \omega_c \\ 0, & \text{for } \omega_c < |\omega| \le \pi \end{cases} \tag{5.4}$$

Using (5.3), the impulse response for the ideal lowpass filter is

$$h(n) = \frac{1}{2\pi} \int_{-\omega_c}^{\omega_c} e^{j\omega n} d\omega = \begin{cases} \dfrac{\omega_c}{\pi}, & \text{for } n = 0 \\[2mm] \dfrac{\sin(\omega_c n)}{\pi n}, & \text{for } n \ne 0 \end{cases} \tag{5.5}$$

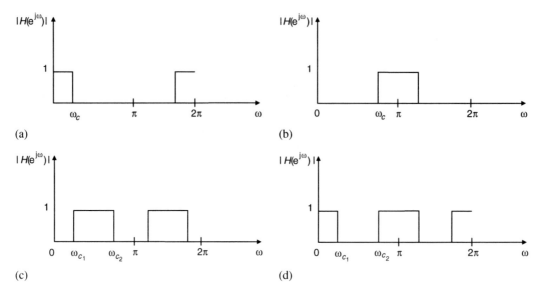

Figure 5.1 Ideal magnitude responses: (a) lowpass; (b) highpass; (c) bandpass; (d) bandstop filters.

Likewise, for bandstop filters, the ideal magnitude response, depicted in Figure 5.1d, is given by

$$|H(e^{j\omega})| = \begin{cases} 1, & \text{for} \quad 0 \le |\omega| \le \omega_{c_1} \\ 0, & \text{for} \quad \omega_{c_1} < |\omega| < \omega_{c_2} \\ 1, & \text{for} \quad \omega_{c_2} \le |\omega| \le \pi \end{cases} \tag{5.6}$$

Then, using (5.3), the impulse response for such an ideal filter is

$$h(n) = \frac{1}{2\pi} \left[\int_{-\omega_{c_1}}^{\omega_{c_1}} e^{j\omega n} d\omega + \int_{\omega_{c_2}}^{\pi} e^{j\omega n} d\omega + \int_{-\pi}^{-\omega_{c_2}} e^{j\omega n} d\omega \right]$$

$$= \begin{cases} 1 + \dfrac{\omega_{c_1} - \omega_{c_2}}{\pi}, & \text{for } n = 0 \\ \dfrac{1}{\pi n} \left[\sin(\omega_{c_1} n) - \sin(\omega_{c_2} n) \right], & \text{for } n \ne 0 \end{cases} \tag{5.7}$$

Following an analogous reasoning, one can easily find the magnitude responses of the ideal highpass and bandpass filters, depicted in Figures 5.1b and 5.1c, respectively. Table 5.1 (Subsection 5.2.4) includes the ideal magnitude responses and their respective impulse responses for the ideal lowpass, highpass, bandpass, and bandstop filters.

5.2.2   Differentiators

An ideal discrete-time differentiator is a linear system that, when samples of a band-limited continuous signal are used as input, the output samples represent the derivative

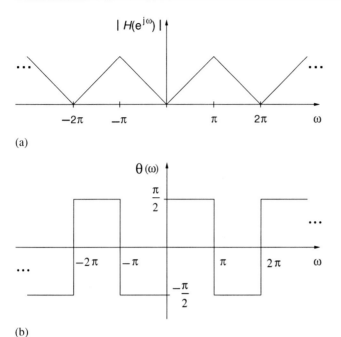

(a)

(b)

Figure 5.2 Characteristics of an ideal discrete-time differentiator: (a) magnitude response; (b) phase response.

of the continuous signal. More precisely, given a continuous-time signal $x_a(t)$ band-limited to $[-\frac{\pi}{T}, \frac{\pi}{T}]$, when its corresponding sampled version $x(n) = x_a(nT)$ is input to an ideal differentiator, it produces the output signal, $y(n)$, such that

$$y(n) = \frac{\mathrm{d}x_a(t)}{\mathrm{d}t}\bigg|_{t=nT} \tag{5.8}$$

If the Fourier transform of the continuous-time signal is denoted by $X_a(\mathrm{j}\Omega)$, we have that the Fourier transform of its derivative is $\mathrm{j}\Omega X_a(\mathrm{j}\Omega)$, as can be deduced from equation (2.172). Therefore, an ideal discrete-time differentiator is characterized by a frequency response of the form

$$H(\mathrm{e}^{\mathrm{j}\omega}) = \mathrm{j}\omega, \text{ for } -\pi < \omega < \pi \tag{5.9}$$

The magnitude and phase responses of a differentiator are depicted in Figure 5.2.

Using equation (5.3), the corresponding impulse response is given by

$$h(n) = \frac{1}{2\pi}\int_{-\pi}^{\pi} \mathrm{j}\omega \mathrm{e}^{\mathrm{j}\omega n}\mathrm{d}\omega = \begin{cases} 0, & \text{for } n = 0 \\ \left[\mathrm{e}^{\mathrm{j}\omega n}\left(\frac{\omega}{n} - \frac{1}{\mathrm{j}n^2}\right)\right]\bigg|_{-\pi}^{\pi} = \frac{(-1)^n}{n}, & \text{for } n \neq 0 \end{cases} \tag{5.10}$$

One should note that, comparing equation (5.9) with equation (4.16), if a differentiator is to be approximated by a linear-phase FIR filter, one should necessarily use either a Type-III or a Type-IV form.

5.2.3 Hilbert transformers

The Hilbert transformer is a system that, when fed with the real part of a complex signal whose Fourier transform is null for $-\pi \leq \omega < 0$, produces at its output the imaginary part of the complex signal. In other words, let $x(n)$ be the inverse Fourier transform of $X(e^{j\omega})$ such that $X(e^{j\omega}) = 0, -\pi \leq \omega < 0$. The real and imaginary parts of $x(n)$, $x_R(n)$ and $x_I(n)$, are defined as

$$\left. \begin{aligned} \text{Re}\{x(n)\} &= \frac{x(n) + x^*(n)}{2} \\ \text{Im}\{x(n)\} &= \frac{x(n) - x^*(n)}{2j} \end{aligned} \right\} \tag{5.11}$$

Hence, their Fourier transforms, $X_R(e^{j\omega}) = \mathcal{F}\{\text{Re}\{x(n)\}\}$ and $X_I(e^{j\omega}) = \mathcal{F}\{\text{Im}\{x(n)\}\}$, are

$$\left. \begin{aligned} X_R(e^{j\omega}) &= \frac{X(e^{j\omega}) + X^*(e^{-j\omega})}{2} \\ X_I(e^{j\omega}) &= \frac{X(e^{j\omega}) - X^*(e^{-j\omega})}{2j} \end{aligned} \right\} \tag{5.12}$$

Since $X(e^{j\omega}) = 0$, we have that, for $-\pi \leq \omega < 0$

$$\left. \begin{aligned} X_R(e^{j\omega}) &= \frac{X^*(e^{-j\omega})}{2} \\ X_I(e^{j\omega}) &= j\frac{X^*(e^{-j\omega})}{2} \end{aligned} \right\} \tag{5.13}$$

and, for $0 \leq \omega < \pi$, since $X^*(e^{-j\omega}) = 0$, we also have that

$$\left. \begin{aligned} X_R(e^{j\omega}) &= \frac{X(e^{j\omega})}{2} \\ X_I(e^{j\omega}) &= -j\frac{X(e^{j\omega})}{2} \end{aligned} \right\} \tag{5.14}$$

From equations (5.13) and (5.14), we can easily conclude that

$$\left. \begin{aligned} X_I(e^{j\omega}) &= -jX_R(e^{j\omega}), \quad \text{for } 0 \leq \omega < \pi \\ X_I(e^{j\omega}) &= jX_R(e^{j\omega}), \quad \text{for } -\pi \leq \omega < 0 \end{aligned} \right\} \tag{5.15}$$

These equations provide a relation between the Fourier transforms of the real and imaginary parts of a signal whose Fourier transform is null for $-\pi \leq \omega < 0$. It thus implies that the ideal Hilbert transformer has the following transfer function:

$$H(e^{j\omega}) = \begin{cases} -j, & \text{for } 0 \leq \omega < \pi \\ j, & \text{for } -\pi \leq \omega < 0 \end{cases} \tag{5.16}$$

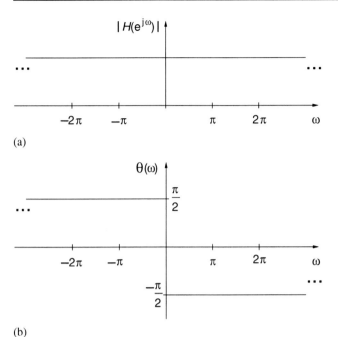

(a)

(b)

Figure 5.3 Characteristics of an ideal Hilbert transformer: (a) magnitude response; (b) phase response.

The magnitude and phase components of such a frequency response are depicted in Figure 5.3.

Using equation (5.3), the corresponding impulse response for the ideal Hilbert transformer is given by

$$h(n) = \frac{1}{2\pi} \left[\int_0^\pi -\mathrm{j}e^{\mathrm{j}\omega n}\,d\omega + \int_{-\pi}^0 \mathrm{j}e^{\mathrm{j}\omega n}\,d\omega \right] = \begin{cases} 0, & \text{for } n = 0 \\ \dfrac{1}{\pi n}\left[1 - (-1)^n\right], & \text{for } n \neq 0 \end{cases}$$

$$(5.17)$$

By examining equation (5.16) and comparing it with equation (4.16), we conclude, as in the case of the differentiator, that a Hilbert transformer must be approximated, when using a linear-phase FIR filter, by either a Type-III or Type-IV structure.

An interesting interpretation of Hilbert transformers comes from the observation that equation (5.16) implies that every positive-frequency sinusoid $e^{\mathrm{j}\omega_0}$ input to a Hilbert transformer has its phase shifted by $-\frac{\pi}{2}$ at the output, whereas every negative-frequency sinusoid $e^{-\mathrm{j}\omega_0}$ has its phase shifted by $+\frac{\pi}{2}$ at the output, as seen in Figure 5.3b. This is equivalent to shifting the phase of every sine or cosine function by $-\frac{\pi}{2}$. Therefore, an ideal Hilbert transformer transforms every 'cosine' component of a signal into a 'sine' and every 'sine' component into a 'cosine'.

5.2.4   Summary

Table 5.1 summarizes the ideal frequency responses and corresponding impulse responses for the basic lowpass, highpass, bandpass, and bandstop filters, as well as for differentiators and Hilbert transformers. By examining this table, we note that the impulse responses corresponding to all these ideal filters are not directly realizable, since they have infinite duration and are noncausal. In the remainder of this chapter, we deal with the problem of approximating ideal frequency responses, as the ones seen in this section, by a digital filter with a finite-duration impulse response.

5.3   FIR filter approximation by frequency sampling

In general, the problem of FIR filter design is to find a finite-length impulse response $h(n)$, whose Fourier transform $H(e^{j\omega})$ approximates a given frequency response well enough. As seen in Section 3.2, one way of achieving such a goal is by noting that the DFT of a length-N sequence $h(n)$ corresponds to samples of its Fourier transform at the frequencies $\omega = \frac{2\pi k}{N}$, that is

$$H(e^{j\omega}) = \sum_{n=0}^{N-1} h(n)e^{-j\omega n} \tag{5.18}$$

and then

$$H(e^{j\frac{2\pi k}{N}}) = \sum_{n=0}^{N-1} h(n)e^{-j\frac{2\pi kn}{N}}, \text{ for } k = 0, \ldots, N-1 \tag{5.19}$$

It is then natural to consider designing a length-N FIR filter by finding an $h(n)$ whose DFT corresponds exactly to samples of the desired frequency response. In other words, $h(n)$ can be determined by sampling the desired frequency response at the N points $e^{j\frac{2\pi}{N}k}$ and finding its inverse DFT, given in equation (3.16). This method is generally referred to as the frequency sampling approach (Gold & Jordan, 1969; Rabiner et al., 1970; Rabiner & Gold, 1975).

More precisely, if the desired frequency response is given by $D(\omega)$, one must first find

$$A(k)e^{j\theta(k)} = D\left(\frac{\omega_s k}{N}\right), \text{ for } k = 0, \ldots, N-1 \tag{5.20}$$

where $A(k)$ and $\theta(k)$ are samples of the desired amplitude and phase responses, respectively. If we want the resulting filter to have linear phase, $h(n)$ must be of one of the forms given in Subsection 4.2.3. For each form, the functions $A(k)$ and $\theta(k)$ present particular properties. We then summarize the results given in Antoniou (1993) for these four cases separately.[1]

---

[1] To maintain consistency with the notation in Subsection 4.2.3, in the following discussion we will use the filter order $M = N - 1$ instead of the filter length N.

Table 5.1. *Ideal frequency characteristics and corresponding impulse responses for lowpass, highpass, bandpass, and bandstop filters, as well as for differentiators and Hilbert transformers.*

Filter type	Magnitude response $|H(e^{j\omega})|$	Impulse response $h(n)$
Lowpass	$\begin{cases} 1, & \text{for } 0 \le |\omega| \le \omega_c \\ 0, & \text{for } \omega_c < |\omega| \le \pi \end{cases}$	$\begin{cases} \dfrac{\omega_c}{\pi}, & \text{for } n = 0 \\ \dfrac{1}{\pi n}\sin(\omega_c n), & \text{for } n \ne 0 \end{cases}$
Highpass	$\begin{cases} 0, & \text{for } 0 \le |\omega| < \omega_c \\ 1, & \text{for } \omega_c \le |\omega| \le \pi \end{cases}$	$\begin{cases} 1 - \dfrac{\omega_c}{\pi}, & \text{for } n = 0 \\ -\dfrac{1}{\pi n}\sin(\omega_c n), & \text{for } n \ne 0 \end{cases}$
Bandpass	$\begin{cases} 0, & \text{for } 0 \le |\omega| < \omega_{c_1} \\ 1, & \text{for } \omega_{c_1} \le |\omega| \le \omega_{c_2} \\ 0, & \text{for } \omega_{c_2} < |\omega| \le \pi \end{cases}$	$\begin{cases} \dfrac{(\omega_{c_2} - \omega_{c_1})}{\pi}, & \text{for } n = 0 \\ \dfrac{1}{\pi n}\left[\sin(\omega_{c_2} n) - \sin(\omega_{c_1} n)\right], & \text{for } n \ne 0 \end{cases}$
Bandstop	$\begin{cases} 1, & \text{for } 0 \le |\omega| \le \omega_{c_1} \\ 0, & \text{for } \omega_{c_1} < |\omega| < \omega_{c_2} \\ 1, & \text{for } \omega_{c_2} \le |\omega| \le \pi \end{cases}$	$\begin{cases} 1 - \dfrac{(\omega_{c_2} - \omega_{c_1})}{\pi}, & \text{for } n = 0 \\ \dfrac{1}{\pi n}\left[\sin(\omega_{c_1} n) - \sin(\omega_{c_2} n)\right], & \text{for } n \ne 0 \end{cases}$

Filter type	Frequency response $H(e^{j\omega})$	Impulse response $h(n)$
Differentiator	$j\omega, \quad \text{for } -\pi \le \omega \le \pi$	$\begin{cases} 0, & \text{for } n = 0 \\ \dfrac{(-1)^n}{n}, & \text{for } n \ne 0 \end{cases}$
Hilbert transformer	$\begin{cases} -j, & \text{for } 0 \le \omega < \pi \\ j, & \text{for } -\pi \le \omega < 0 \end{cases}$	$\begin{cases} 0, & \text{for } n = 0 \\ \dfrac{1}{\pi n}\left[1 - (-1)^n\right], & \text{for } n \ne 0 \end{cases}$

- Type I: Even order M and symmetrical impulse response. In this case, the phase and amplitude responses must satisfy

$$\theta(k) = -\frac{\pi k M}{M + 1}, \text{ for } 0 \le k \le M \tag{5.21}$$

$$A(k) = A(M - k + 1), \text{ for } 1 \le k \le \frac{M}{2} \tag{5.22}$$

and then, the impulse response is given by

$$h(n) = \frac{1}{N} \left[A(0) + 2 \sum_{k=1}^{\frac{M}{2}} (-1)^k A(k) \cos \frac{\pi k(1 + 2n)}{M + 1} \right] \tag{5.23}$$

for $n = 0, \ldots, M$.

- Type II: Odd order M and symmetrical impulse response. The phase and amplitude responses, in this case, become

$$\theta(k) = \begin{cases} -\dfrac{\pi k M}{M + 1}, & \text{for } 0 \le k \le \frac{M-1}{2} \\[2mm] \pi - \dfrac{\pi k M}{M + 1}, & \text{for } \frac{M+3}{2} \le k \le M \end{cases} \tag{5.24}$$

$$A(k) = A(M - k + 1), \text{ for } 1 \le k \le \frac{M+1}{2} \tag{5.25}$$

$$A\left(\frac{M + 1}{2}\right) = 0 \tag{5.26}$$

and the impulse response is

$$h(n) = \frac{1}{N} \left[A(0) + 2 \sum_{k=1}^{\frac{M-1}{2}} (-1)^k A(k) \cos \frac{\pi k(1 + 2n)}{M + 1} \right] \tag{5.27}$$

for $n = 0, \ldots, M$.

- Type III: Even order M and antisymmetric impulse response. The phase and amplitude responses are such that

$$\theta(k) = \frac{(1 + 2r)\pi}{2} - \frac{\pi k M}{M + 1}, \text{ for } r \in \mathbb{Z} \text{ and } 0 \le k \le M \tag{5.28}$$

$$A(k) = A(M - k + 1), \text{ for } 1 \le k \le \frac{M}{2} \tag{5.29}$$

$$A(0) = 0 \tag{5.30}$$

and the impulse response is given by

$$h(n) = \frac{2}{N} \sum_{k=1}^{\frac{M}{2}} (-1)^{k+1} A(k) \sin \frac{\pi k(1 + 2n)}{M + 1} \tag{5.31}$$

for $n = 0, \ldots, M$.

- Type IV: Odd order M and antisymmetric impulse response. In this case, the phase and amplitude responses are of the form

$$\theta(k) = \begin{cases} \dfrac{\pi}{2} - \dfrac{\pi k M}{M + 1}, & \text{for } 1 \le k \le \frac{M-1}{2} \\[2mm] -\dfrac{\pi}{2} - \dfrac{\pi k M}{M + 1}, & \text{for } \frac{M+1}{2} \le k \le M \end{cases} \tag{5.32}$$

$$A(k) = A(M - k + 1), \qquad \text{for } 1 \le k \le M \tag{5.33}$$

$$A(0) = 0 \tag{5.34}$$

Table 5.2. *Impulse responses for linear-phase FIR filters using the frequency sampling approach.*

Filter type	Impulse response $h(n)$, for $n = 0, \ldots, M$	Condition
Type I	$\dfrac{1}{N}\left[A(0) + 2 \displaystyle\sum_{k=1}^{\frac{M}{2}} (-1)^k A(k) \cos \dfrac{\pi k(1 + 2n)}{M + 1} \right]$	
Type II	$\dfrac{1}{N}\left[A(0) + 2 \displaystyle\sum_{k=1}^{\frac{M-1}{2}} (-1)^k A(k) \cos \dfrac{\pi k(1 + 2n)}{M + 1} \right]$	$A\left(\dfrac{M+1}{2}\right) = 0$
Type III	$\dfrac{2}{N} \displaystyle\sum_{k=1}^{\frac{M}{2}} (-1)^{k+1} A(k) \sin \dfrac{\pi k(1 + 2n)}{M + 1}$	$A(0) = 0$
Type IV	$\dfrac{1}{N}\left[(-1)^{\frac{M+1}{2}+n} A\left(\dfrac{M+1}{2}\right) + 2 \displaystyle\sum_{k=1}^{\frac{M-1}{2}} (-1)^k A(k) \sin \dfrac{\pi k(1+2n)}{M+1} \right]$	$A(0) = 0$

and the impulse response then becomes

$$h(n) = \frac{1}{N}\left[(-1)^{\frac{M+1}{2}+n} A\left(\frac{M + 1}{2}\right) + 2 \sum_{k=1}^{\frac{M-1}{2}} (-1)^k A(k) \sin \frac{\pi k(1 + 2n)}{M + 1} \right] \quad (5.35)$$

for $n = 0, \ldots, M$.

The results given above are summarized in Table 5.2.

EXAMPLE 5.1

Design a lowpass filter satisfying the specification below using the frequency sampling method:[2]

$$\left.\begin{array}{rcl} M &=& 52 \\ \Omega_p &=& 4.0 \text{ rad/s} \\ \Omega_r &=& 4.2 \text{ rad/s} \\ \Omega_s &=& 10.0 \text{ rad/s} \end{array}\right\} \quad (5.36)$$

SOLUTION

We have that the sampling frequency Ω_s corresponds to $k = M + 1 = 53$. Therefore,

---

[2] Note that in this text, in general, the variable Ω represents an analog frequency, and the variable ω a digital frequency.

Table 5.3. *Coefficients of the lowpass filter designed with the frequency sampling method.*

h(0) to h(26)		
$h(0) = -5.512\,926\,18\mathrm{E}{-}03$	$h(9) = -1.353\,963\,28\mathrm{E}{-}02$	$h(18) = 4.130\,186\,35\mathrm{E}{-}02$
$h(1) = 1.470\,949\,85\mathrm{E}{-}02$	$h(10) = 1.375\,711\,69\mathrm{E}{-}03$	$h(19) = -3.957\,755\,97\mathrm{E}{-}02$
$h(2) = -1.900\,164\,78\mathrm{E}{-}02$	$h(11) = 1.235\,625\,49\mathrm{E}{-}02$	$h(20) = 2.184\,395\,95\mathrm{E}{-}02$
$h(3) = 1.688\,538\,26\mathrm{E}{-}02$	$h(12) = -2.308\,640\,96\mathrm{E}{-}02$	$h(21) = 1.142\,775\,21\mathrm{E}{-}02$
$h(4) = -8.931\,503\,41\mathrm{E}{-}03$	$h(13) = 2.679\,076\,60\mathrm{E}{-}02$	$h(22) = -5.595\,254\,59\mathrm{E}{-}02$
$h(5) = -2.355\,649\,93\mathrm{E}{-}03$	$h(14) = -2.132\,492\,49\mathrm{E}{-}02$	$h(23) = 1.043\,773\,04\mathrm{E}{-}01$
$h(6) = 1.329\,182\,76\mathrm{E}{-}02$	$h(15) = 7.303\,577\,98\mathrm{E}{-}03$	$h(24) = -1.478\,332\,70\mathrm{E}{-}01$
$h(7) = -2.016\,029\,59\mathrm{E}{-}02$	$h(16) = 1.176\,027\,96\mathrm{E}{-}02$	$h(25) = 1.779\,269\,63\mathrm{E}{-}01$
$h(8) = 2.041\,037\,91\mathrm{E}{-}02$	$h(17) = -3.014\,553\,20\mathrm{E}{-}02$	$h(26) = 8.113\,207\,55\mathrm{E}{-}01$

according to the prescribed specifications, we have that Ω_p and Ω_r correspond, respectively, to:

$$k_p = \left\lfloor 53 \times \frac{4}{10} \right\rfloor = 21 \tag{5.37}$$

$$k_r = \left\lfloor 53 \times \frac{4.2}{10} \right\rfloor = 22 \tag{5.38}$$

Thus, we assign

$$A(k) = \begin{cases} 1, & \text{for } 0 \le k \le 21 \\ 0, & \text{for } 22 \le k \le 26 \end{cases} \tag{5.39}$$

and then, by using the first row in Table 5.2, we end up with the set of coefficients shown in Table 5.3, where only half of the filter coefficients are given, as the other half can be obtained from symmetry as $h(n) = h(52 - n)$.

The corresponding magnitude response is shown in Figure 5.4.

$\triangle$

By examining the magnitude response shown in Figure 5.4, one notices that there is a great deal of ripple both at the passband and at the stopband. This is the main reason why this method has not found widespread use in filter design. This is not a surprising result, because the equations derived in this section guarantee only that the Fourier transform of $h(n)$ and the desired frequency response $D(\omega)$ (expressed as a function of the digital frequencies, that is, $\omega = 2\pi \frac{\Omega}{\Omega_s} = \Omega T$) coincide at the $M + 1$ distinct frequencies $\frac{2\pi k}{M+1}$, $k = 0, \ldots, M + 1$, where M is the filter order. At the other frequencies, as is illustrated in Figure 5.5, there is no constraint on the magnitude response, and, as a consequence, no control over the ripple δ.

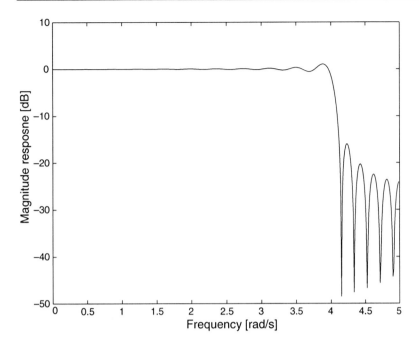

Figure 5.4 Magnitude response of the lowpass filter designed with the frequency sampling method.

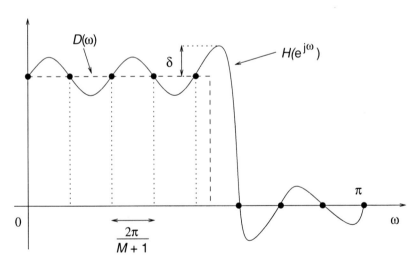

Figure 5.5 The desired magnitude response and the Fourier transform of $h(n)$ coincide only at the frequencies $\frac{2\pi k}{M+1}$, when using the frequency sampling approximation method.

An interesting explanation of this fact comes from the expression of the inverse Fourier transform of the desired frequency response $D(\omega)$, which, from equation (5.3), is given by

$$h(n) = \frac{1}{2\pi} \int_{-\pi}^{\pi} D(\omega) e^{j\omega n} d\omega = \frac{1}{2\pi} \int_{0}^{2\pi} D(\omega) e^{j\omega n} d\omega \tag{5.40}$$

If we try to approximate the above integral as a summation over the discrete frequencies $\frac{2\pi k}{N}$, by substituting $\omega \to \frac{2\pi k}{N}$ and $d\omega \to \frac{2\pi}{N}$, we end up with an approximation, $d(n)$, of $h(n)$ given by

$$d(n) = \frac{1}{2\pi} \sum_{n=0}^{N-1} D\left(\frac{2\pi k}{N}\right) e^{-j\frac{2\pi kn}{N}} \frac{2\pi}{N} = \frac{1}{N} \sum_{n=0}^{N-1} D\left(\frac{2\pi k}{N}\right) e^{-j\frac{2\pi kn}{N}} \tag{5.41}$$

From equation (3.16), we see that $d(n)$ represents the IDFT of the sequence $D(\frac{2\pi k}{N})$, for $k = 0, \ldots, N - 1$. However, considering that the argument of the integral in equation (5.40) is $D(\omega) e^{j\omega n}$, the resolution of a good sampling grid for approximating it would have to be of the order of 10% of the period of the sinusoid $e^{j\omega n}$. This would require a sampling grid with a resolution of the order of $\frac{2\pi}{10N}$. In equation (5.41) we are approximating the integral using a sampling grid with resolution $\frac{2\pi}{N}$ and such approximation would only be valid for values of $n \leq \frac{N}{10}$. This is clearly not sufficient in most practical cases, which explains the large values of ripple depicted in Figure 5.4.

One important situation, however, in which the frequency sampling method gives exact results is when the desired frequency response $D(\omega)$ is composed of a sum of sinusoids equally spaced in frequency. Such a result is formally stated in the following theorem.

THEOREM 5.1
If the desired frequency response $D(\omega)$ is a finite sum of complex sinusoids equally spaced in frequency, that is

$$D(\omega) = \sum_{n=N_0}^{N_1} a(n) e^{-j\omega n} \tag{5.42}$$

then the frequency sampling method yields exact results, except for a constant group-delay term, provided that the length of the impulse response, N, satisfies $N \geq N_1 - N_0 + 1$.

◇

PROOF
The theorem essentially states that the Fourier transform of the impulse response, $h(n)$, given by the frequency sampling method is identical to the desired frequency response $D(\omega)$, except for a constant group-delay term. That is, the above theorem is equivalent to

$$\mathcal{F}\left\{\text{IDFT}\left[D\left(\frac{2\pi k}{N}\right)\right]\right\} = D(\omega) \qquad (5.43)$$

where $\mathcal{F}\{\cdot\}$ is the Fourier transform. The proof becomes simpler if we rewrite equation (5.43) as

$$D\left(\frac{2\pi k}{N}\right) = \text{DFT}\left\{\mathcal{F}^{-1}\left[D(\omega)\right]\right\} \qquad (5.44)$$

for $k = 0, \ldots, N - 1$.

For a desired frequency response in the form of equation (5.42), the corresponding inverse Fourier transform is given by

$$d(n) = \begin{cases} 0, & \text{for } n < N_0 \\ a(n), & \text{for } n = N_0, \ldots, N_1 \\ 0, & \text{for } n > N_1 \end{cases} \qquad (5.45)$$

The length-N DFT of $d(n)$, $H(k)$, is then equal to the length-N DFT of $a(n)$, adequately shifted in time to the interval $n \in [0, N - 1]$. Therefore, if $N \geq N_1 - N_0 + 1$, we get that

$$
\begin{aligned}
H(k) = \text{DFT}\left[a(n' + N_0)\right] &= \sum_{n'=0}^{N-1} a(n' + N_0) e^{-j\frac{2\pi k}{N} n'} \\
&= \sum_{n'=0}^{N-1} d(n' + N_0) e^{-j\frac{2\pi k}{N} n'} \\
&= \sum_{n=N_0}^{N+N_0-1} d(n) e^{-j\frac{2\pi k}{N}(n-N_0)} \\
&= e^{j\frac{2\pi k}{N} N_0} \sum_{n=N_0}^{N_1} d(n) e^{-j\frac{2\pi k}{N} n} \\
&= e^{j\frac{2\pi k}{N} N_0} D\left(\frac{2\pi k}{N}\right) \qquad (5.46)
\end{aligned}
$$

for $k = 0, \ldots, N - 1$, and this completes the proof.

□

This result is very useful whenever the desired frequency response $D(\omega)$ is of the form described by equation (5.42), as is the case in the approximation methods discussed in Section 5.5 and Subsection 5.6.2.

5.4   FIR filter approximation with window functions

For all ideal filters analyzed in Section 5.2, the impulse responses obtained from equation (5.3) have infinite duration, which leads to non-realizable FIR filters. A straightforward way to overcome this limitation is to define a finite-length auxiliary sequence $h'(n)$, yielding a filter of order M, as

$$h'(n) = \begin{cases} h(n), & \text{for } |n| \leq \frac{M}{2} \\ 0, & \text{for } |n| > \frac{M}{2} \end{cases} \tag{5.47}$$

assuming that M is even. The resulting transfer function is written as

$$H'(z) = h(0) + \sum_{n=1}^{\frac{M}{2}} \left(h(-n)z^n + h(n)z^{-n} \right) \tag{5.48}$$

This is still a noncausal function which we can make causal by multiplying it by $z^{\frac{-M}{2}}$, without either distorting the filter magnitude response or destroying the linear-phase property. The example below highlights some of the impacts that the truncation of the impulse response in equations (5.47) and (5.48) has on the filter frequency response.

EXAMPLE 5.2
Design a bandstop filter satisfying the specification below:

$$\left. \begin{array}{rcl} M & = & 50 \\ \Omega_{c_1} & = & \frac{\pi}{4} \text{ rad/s} \\ \Omega_{c_2} & = & \frac{\pi}{2} \text{ rad/s} \\ \Omega_s & = & 2\pi \text{ rad/s} \end{array} \right\} \tag{5.49}$$

SOLUTION
Applying equations (5.47) and (5.48) to the corresponding bandstop equations in Table 5.1, we obtain the filter coefficients listed in Table 5.4 (only half of them are listed as the others can be found using $h(n) = h(50 - n)$). The resulting magnitude response is depicted in Figure 5.6.

$\triangle$

The ripple seen in Figure 5.6 close to the band edges is due to the slow convergence of the Fourier series $h(n)$ when approximating functions presenting discontinuities, such as the ideal responses seen in Figure 5.1. This implies that large-amplitude ripples in the magnitude response appear close to the edges whenever an infinite-length $h(n)$ is truncated to generate a finite-length filter. These ripples are commonly referred to as Gibbs' oscillations. It can be shown that Gibbs' oscillations possess the property

Table 5.4. *Bandstop filter coefficients.*

$h(0)$ to $h(25)$		
−3.729 243 54E−03	−1.591 503 65E−08	1.331 867 55E−02
−1.591 465 05E−08	6.215 375 89E−03	−5.305 164 77E−02
4.053 502 10E−03	−2.273 642 04E−02	−1.086 777 82E−01
−1.446 863 12E−02	−4.179 913 99E−02	1.591 706 30E−08
−2.587 565 38E−02	1.591 831 75E−08	1.811 296 66E−01
1.591 694 72E−08	4.939 900 81E−02	1.591 549 43E−01
2.859 943 05E−02	3.183 098 86E−02	−9.323 081 84E−02
1.768 388 26E−02	−1.035 898 98E−02	7.499 999 84E−01
−5.484 176 38E−03	−1.591 474 70E−08	

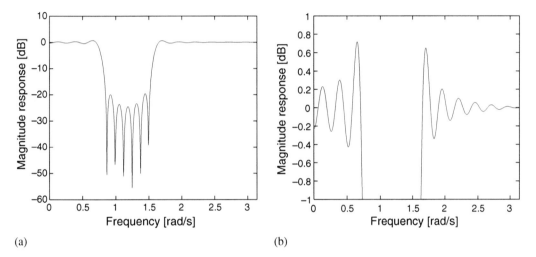

(a)   (b)

Figure 5.6 Bandstop filter: (a) magnitude response; (b) passband detail.

that their amplitudes do not decrease even when the filter order M is increased dramatically (Oppenheim et al., 1983; Kreider et al., 1966). This severely limits the practical usefulness of equations (5.47) and (5.48) in FIR design, because the maximum deviation from the ideal magnitude response cannot be minimized by increasing the filter length.

Although we cannot remove the ripples introduced by the poor convergence of the Fourier series, we can still attempt to control their amplitude by multiplying the impulse response $h(n)$ by a window function $w(n)$. The window $w(n)$ must be designed such that it introduces minimum deviation from the ideal frequency response. The coefficients of the resulting impulse response $h'(n)$ become

$$h'(n) = h(n)w(n) \tag{5.50}$$

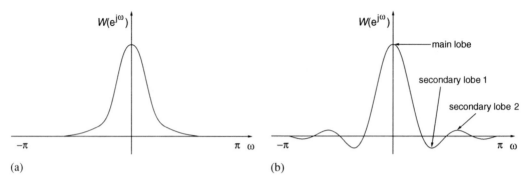

Figure 5.7 Magnitude responses of a window function: (a) ideal case; (b) practical case.

In the frequency domain, such a multiplication corresponds to a periodic convolution operation between the frequency responses of the ideal filter, $H(e^{j\omega})$, and of the window function, $W(e^{j\omega})$, that is

$$H'(e^{j\omega}) = \frac{1}{2\pi} \int_{-\pi}^{\pi} H(e^{j\omega'}) W(e^{j(\omega-\omega')}) d\omega' \qquad (5.51)$$

We can then infer that a good window is a finite-length sequence, whose frequency response, when convolved with an ideal frequency response, produces the minimum distortion possible. This minimum distortion would occur when the frequency response of the window has an impulse-like shape, concentrated around $\omega = 0$, as depicted in Figure 5.7a. However, band-limited signals in frequency are not limited in time, and therefore such a window sequence would have to be infinite, which contradicts our main requirement. This means that we must find a finite-length window which has a frequency response that has most of its energy concentrated around $\omega = 0$. Also, in order to avoid the oscillations in the filter magnitude response, the sidelobes of the window magnitude response should quickly decay as $|\omega|$ is increased (Kaiser, 1974; Rabiner & Gold, 1975; Rabiner et al., 1975; Antoniou, 1993).

A practical window function is in general as shown in Figure 5.7b. The effect of the secondary lobe is to introduce the largest ripple close to the band edges. From equation (5.51), we see that the main-lobe width determines the transition bandwidth of the resulting filter. Based on these facts, a practical window function must present a magnitude response characterized by:

- The ratio of the main-lobe amplitude to the secondary-lobe amplitude must be as large as possible.
- The energy must decay rapidly when $|\omega|$ increases from 0 to π.

We can now proceed to perform a thorough study of the more widely used window functions in FIR filter design.

5.4.1 Rectangular window

The simple truncation of the Fourier series as described in equation (5.47) can be interpreted as the product between the ideal $h(n)$ and a rectangular window defined by[3]

$$
w_r(n) = \begin{cases} 1, & \text{for } |n| \le \frac{M}{2} \\ 0, & \text{for } |n| > \frac{M}{2} \end{cases} \tag{5.52}
$$

Accordingly, the frequency response of the rectangular window is given by

$$
\begin{aligned}
W_r(e^{j\omega}) &= \sum_{n=-\frac{M}{2}}^{\frac{M}{2}} e^{-j\omega n} \\
&= \frac{e^{j\omega \frac{M}{2}} - e^{-j\omega \frac{M}{2}} e^{-j\omega}}{1 - e^{-j\omega}} \\
&= e^{-j\frac{\omega}{2}} \frac{\left[e^{j\omega\left(\frac{M+1}{2}\right)} - e^{-j\omega\left(\frac{M+1}{2}\right)} \right]}{1 - e^{-j\omega}} \\
&= \frac{\sin\left[\omega\left(\frac{M+1}{2}\right) \right]}{\sin\left(\frac{\omega}{2}\right)}
\end{aligned} \tag{5.53}
$$

5.4.2 Triangular windows

The main problem associated with the rectangular window is the presence of ripples near the band edges of the resulting filter, which are caused by the existence of sidelobes in the frequency response of the window. Such a problem is due to the inherent discontinuity of the rectangular window in the time domain. One way to reduce such a discontinuity is to employ a triangular-shaped window, which will present only small discontinuities near its edges. The standard triangular window is defined as

$$
w_t(n) = \begin{cases} -\dfrac{2|n|}{M+2} + 1, & \text{for } |n| \le \frac{M}{2} \\ 0, & \text{for } |n| > \frac{M}{2} \end{cases} \tag{5.54}
$$

---

[3] Note that if we want to truncate the impulse responses in Table 5.1 and still keep the linear-phase property, the resulting truncated sequences would have to be either symmetric or antisymmetric. This implies that, for those cases, M would have to be even (Type-I and Type-III filters, as seen in Subsection 4.2.3). For the case of M odd (Type-II and Type-IV filters), the solution would be to shift $h(n)$ so that it is causal and apply a window different from zero from $n = 0$ to $n = M - 1$. This solution, however, it is not commonly used in practice.

A small variation of such a window is called the Bartlett window and is defined by

$$
w_{tB}(n) = \begin{cases} -\dfrac{2|n|}{M} + 1, & \text{for } |n| \le \dfrac{M}{2} \\ 0, & \text{for } |n| > \dfrac{M}{2} \end{cases} \tag{5.55}
$$

Clearly, these two triangular-type window functions are closely related. Their main difference lies in the fact that the Bartlett window presents one null element at each of its extremities. In that manner, an Mth-order Bartlett window can be obtained by juxtaposing one zero at each extremity of the $(M - 2)$th-order standard triangular window.

In some cases, an even greater reduction of the sidelobes is necessary, and then more complex window functions should be used, such as the ones described in the following subsections.

5.4.3   Hamming and Hanning windows

The generalized Hamming window is defined as

$$
w_H(n) = \begin{cases} \alpha + (1 - \alpha)\cos\left(\dfrac{2\pi n}{M}\right), & \text{for } |n| \le \dfrac{M}{2} \\ 0, & \text{for } |n| > \dfrac{M}{2} \end{cases} \tag{5.56}
$$

with $0 \le \alpha \le 1$. This generalized window is referred to as the Hamming window when $\alpha = 0.54$, and for $\alpha = 0.5$, it is known as the Hanning window.

The frequency response for the general Hamming window can be expressed based on the frequency response of the rectangular window. We first write equation (5.56) as

$$
w_H(n) = w_r(n)\left[\alpha + (1 - \alpha)\cos\left(\frac{2\pi n}{M}\right)\right] \tag{5.57}
$$

By transforming the above equation to the frequency domain, clearly the frequency response of the generalized Hamming window results from the periodic convolution between $W_r(e^{j\omega})$ and three impulse functions as

$$
W_H(e^{j\omega}) = W_r(e^{j\omega}) * \left[\alpha\delta(\omega) + \left(\frac{1-\alpha}{2}\right)\delta\left(\omega - \frac{2\pi}{M}\right) + \left(\frac{1-\alpha}{2}\right)\delta\left(\omega + \frac{2\pi}{M}\right)\right] \tag{5.58}
$$

and then

$$
W_H(e^{j\omega}) = \alpha W_r(e^{j\omega}) + \left(\frac{1-\alpha}{2}\right)W_r\left(e^{j(\omega - \frac{2\pi}{M})}\right) + \left(\frac{1-\alpha}{2}\right)W_r\left(e^{j(\omega + \frac{2\pi}{M})}\right) \tag{5.59}
$$

From this equation, one clearly sees that $W_H(e^{j\omega})$ is composed of three versions of the rectangular spectrum $W_r(e^{j\omega})$: The main component, $\alpha W_r(e^{j\omega})$, centered at $\omega = 0$,

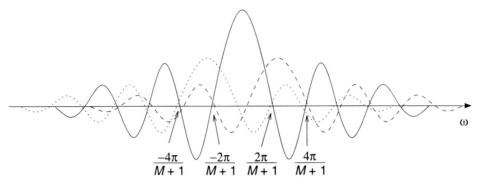

Figure 5.8 The three components of the generalized Hamming window combine to reduce the resulting secondary lobes. (Solid line – $\alpha W_r(e^{j\omega})$; dashed line – $\frac{1-\alpha}{2} W_r\left(e^{j(\omega-\frac{\pi}{M})}\right)$; dotted line – $\frac{1-\alpha}{2} W_r\left(e^{j(\omega+\frac{\pi}{M})}\right)$.)

and two additional ones with smaller amplitudes, centered at $\omega = \pm 2\pi/M$, that reduce the secondary lobe of the main component. This is illustrated in Figure 5.8.

The main characteristics of the generalized Hamming window are:

- All three $W_r(e^{j\omega})$ components have zeros close to $\omega = \pm\frac{4\pi}{M+1}$. Hence, the main-lobe total width is $\frac{8\pi}{M+1}$.
- When $\alpha = 0.54$, the main-lobe total energy is approximately 99.96% of the window total energy.
- The transition band of the Hamming window is larger than the transition band of the rectangular window, due to its wider main lobe.
- The ratio between the amplitudes of the main and secondary lobes of the Hamming window is much larger than for the rectangular window.
- The stopband attenuation for the Hamming window is larger than the attenuation for the rectangular window.

5.4.4   Blackman window

The Blackman window is defined as

$$
w_B(n) = \begin{cases} 0.42 + 0.5\cos\left(\dfrac{2\pi n}{M}\right) + 0.08\cos\left(\dfrac{4\pi n}{M}\right), & \text{for } |n| \leq \dfrac{M}{2} \\[2mm] 0, & \text{for } |n| > \dfrac{M}{2} \end{cases} \tag{5.60}
$$

Compared to the Hamming window function, the Blackman window introduces a second cosine term to further reduce the effects of the secondary lobes of $W_r(e^{j\omega})$. The Blackman window is characterized by:

- The main-lobe width is approximately $\frac{12\pi}{M+1}$, which is wider than that for the previous windows.

Table 5.5. *Filter coefficients using the rectangular window.*

$h(0)$ to $h(40)$		
7.527 030 03E−13	4.447 416 72E−03	4.081 904 57E−02
−2.964 944 48E−03	1.204 324 80E−12	1.051 207 59E−02
−1.289 022 50E−02	−4.818 034 78E−03	6.948 027 72E−13
1.323 860 94E−02	−2.129 689 34E−02	−1.284 809 27E−02
3.212 023 18E−03	2.226 493 40E−02	−6.122 856 85E−02
8.602 320 03E−13	5.506 325 46E−03	6.997 550 69E−02
−3.400 965 73E−03	6.948 027 72E−13	1.927 213 91E−02
−1.484 328 93E−02	−6.085 938 67E−03	1.389 605 54E−12
1.530 714 21E−02	−2.721 269 71E−02	−2.890 820 87E−02
3.730 091 44E−03	2.881 344 40E−02	−1.632 761 83E−01
1.466 805 85E−12	7.227 052 17E−03	2.449 142 74E−01
−3.987 339 13E−03	1.389 605 54E−12	1.156 328 35E−01
−1.749 387 67E−02	−8.259 488 20E−03	6.000 000 00E−01
1.814 179 81E−02	−3.767 911 91E−02	

- The passband ripples are smaller than in the previous windows.
- The stopband attenuation is larger than in the previous windows.

EXAMPLE 5.3

Design a bandstop filter satisfying the specification below using the rectangular, Hamming, Hanning, and Blackman windows:

$$\left. \begin{array}{l} M = 80 \\ \Omega_{p_1} = 2000 \text{ rad/s} \\ \Omega_{p_2} = 4000 \text{ rad/s} \\ \Omega_s = 10\,000 \text{ rad/s} \end{array} \right\} \tag{5.61}$$

SOLUTION

The results are shown in Figure 5.9 and Tables 5.5–5.8. Due to the symmetry inherent to all these window functions, Tables 5.5–5.8 only show the filter coefficients for $0 \leq n \leq 40$. The remaining coefficients are obtained as $h(n) = h(80 - n)$. The reader should notice the compromise between the width of the transition bands and ripple in the passband and stopband when going from the rectangular to the Blackman window, that is, as the ripple decreases, the width of the transition band increases accordingly.

$\triangle$

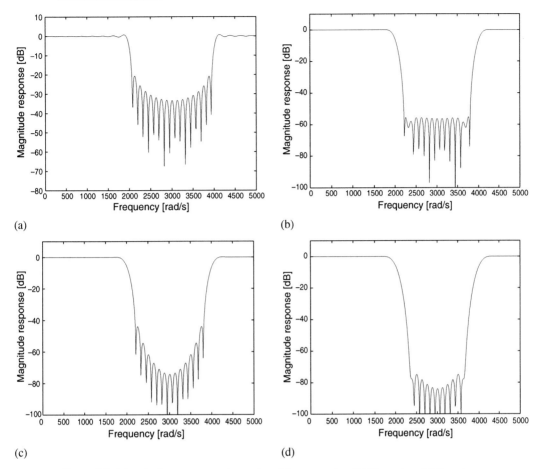

Figure 5.9 Magnitude responses when using: (a) rectangular; (b) Hamming; (c) Hanning; (d) Blackman windows.

5.4.5 Kaiser window

All the window functions seen so far allow us to control the transition band through a proper choice of the filter order M. However, no control can be achieved over the passband and stopband ripples, which makes these windows of little use when designing filters with prescribed frequency specifications, such as that depicted in Figure 5.10. Such problems are overcome with the Kaiser and Dolph–Chebyshev windows, presented in this and in the next subsections.

As seen earlier in this section, the ideal window should be a finite-duration function such that most of its spectral energy is concentrated around $|\omega| = 0$, quickly decaying when $|\omega|$ increases. There is a family of continuous-time functions, called the prolate spheroidal functions (Kaiser, 1974), which are optimal for achieving these properties. Such functions, although very difficult to implement in practice, can be effectively

Table 5.6. *Filter coefficients using the Hamming window.*

$h(0)$ to $h(40)$		
6.021 624 02E−14	1.472 825 96E−03	3.307 898 79E−02
−2.413 999 28E−04	4.383 328 26E−13	8.816 962 67E−03
−1.104 220 02E−03	−1.916 865 65E−03	6.011 913 83E−13
1.227 349 32E−03	−9.213 358 36E−03	−1.143 206 26E−02
3.292 773 53E−04	1.042 088 30E−02	−5.584 950 51E−02
9.893 994 09E−14	2.774 685 94E−03	6.523 218 05E−02
−4.425 914 73E−04	3.751 934 97E−13	1.830 589 19E−02
−2.193 623 30E−03	−3.506 055 61E−03	1.340 947 93E−12
2.569 337 22E−03	−1.665 307 82E−02	−2.825 736 92E−02
7.095 128 37E−04	1.865 338 76E−02	−1.612 009 69E−01
3.149 685 12E−13	4.929 917 86E−03	2.435 272 36E−01
−9.619 611 07E−04	9.950 053 43E−13	1.154 688 64E−01
−4.716 677 77E−03	−6.184 999 05E−03	6.000 000 00E−01
5.436 201 77E−03	−2.940 287 57E−02	

Table 5.7. *Filter coefficients using the Hanning window.*

$h(0)$ to $h(40)$		
2.053 738 18E−24	1.214 165 89E−03	3.240 593 94E−02
−4.569 967 17E−06	3.717 248 27E−13	8.669 561 52E−03
−7.935 002 96E−05	−1.664 590 08E−03	5.930 512 62E−13
1.828 919 16E−04	−8.162 616 19E−03	−1.130 892 96E−02
7.860 380 22E−05	9.390 965 49E−03	−5.538 176 05E−02
3.274 063 11E−14	2.537 152 07E−03	6.481 971 73E−02
−1.853 415 38E−04	3.474 013 86E−13	1.822 187 04E−02
−1.093 652 34E−03	−3.281 717 96E−03	1.336 716 83E−12
1.461 702 01E−03	−1.573 485 04E−02	−2.820 077 45E−02
4.468 538 28E−04	1.776 990 45E−02	−1.610 205 16E−01
2.148 087 44E−13	4.730 167 05E−03	2.434 066 24E−01
−6.988 847 57E−04	9.606 922 82E−13	1.154 546 06E−01
−3.605 616 99E−03	−6.004 608 68E−03	6.000 000 00E−01
4.331 367 31E−03	−2.868 320 24E−02	

approximated with the hyperbolic-sine $I_0(\cdot)$ functions as

$$
w(t) = \begin{cases} \dfrac{I_0\left[\beta\sqrt{1-(\frac{t}{\tau})^2}\right]}{I_0(\beta)}, & \text{for } |t| \le \tau \\ 0, & \text{for } |t| > \tau \end{cases}
\tag{5.62}
$$

Table 5.8. *Filter coefficients using the Blackman window.*

$h(0)$ to $h(40)$		
1.026 869 10E−24	6.492 424 76E−04	2.813 131 34E−02
−1.649 696 24E−06	2.072 519 44E−13	7.697 039 37E−03
−2.887 862 84E−05	−9.673 175 33E−04	5.374 670 40E−13
6.745 815 16E−05	−4.940 811 04E−03	−1.044 187 29E−02
2.952 845 48E−05	5.915 753 92E−03	−5.199 712 85E−02
1.258 414 15E−14	1.661 563 36E−03	6.176 313 40E−02
−7.318 727 95E−05	2.362 329 42E−13	1.758 632 96E−02
−4.452 861 79E−04	−2.313 962 01E−03	1.304 156 35E−12
6.155 440 00E−04	−1.148 736 96E−02	−2.775 909 63E−02
1.951 277 03E−04	1.341 099 16E−02	−1.595 968 33E−01
9.746 427 55E−14	3.684 258 24E−03	2.424 476 67E−01
−3.299 970 46E−04	7.709 158 78E−13	1.153 407 15E−01
−1.773 634 44E−03	−4.955 465 20E−03	6.000 000 00E−01
2.221 127 15E−03	−2.430 039 59E−02	

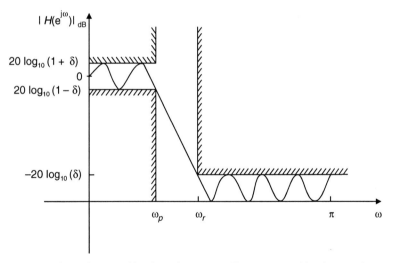

Figure 5.10 Typical specification of a lowpass filter. The specifications are in terms of the digital frequency $\omega = 2\pi \frac{\Omega}{\Omega_s} = \Omega T$.

where $\beta = \Omega_a \tau$ and $I_0(x)$ is the zeroth-order modified Bessel function of the first kind, which can be efficiently determined through its series expansion given by

$$I_0(x) = 1 + \sum_{k=1}^{\infty} \left[\frac{\left(\frac{x}{2}\right)^k}{k!} \right]^2 \qquad (5.63)$$

The Fourier transform of $w(t)$ is given by

$$W(\Omega) = \frac{2\tau \sin\left[\beta\sqrt{(\frac{\Omega}{\Omega_a})^2 - 1}\right]}{\beta I_0(\beta)\sqrt{(\frac{\Omega}{\Omega_a})^2 - 1}} \tag{5.64}$$

The Kaiser window is derived from equation (5.62) by making the transformation to the discrete-time domain given by $\tau \to \frac{M}{2}T$ and $t \to nT$. The window is then described by

$$w_K(n) = \begin{cases} \dfrac{I_0\left[\beta\sqrt{1 - (\frac{2n}{M})^2}\right]}{I_0(\beta)}, & \text{for } |n| \leq \frac{M}{2} \\ 0, & \text{for } |n| > \frac{M}{2} \end{cases} \tag{5.65}$$

Since the functions given by equation (5.64) tend to be highly concentrated around $|\Omega| = 0$, we can assume that $W(\Omega) \approx 0$, for $|\Omega| \geq \frac{\Omega_s}{2}$. Therefore, from equation (2.180), we can approximate the frequency response for the Kaiser window by

$$W_K(e^{j\omega}) \approx \frac{1}{T} W\left(\frac{\omega}{T}\right) \tag{5.66}$$

where $W(\Omega)$ is given by equation (5.64) when τ is replaced by $\frac{M}{2}T$. This yields

$$W_K(e^{j\omega}) \approx \frac{M \sin\left[\beta\sqrt{(\frac{\omega}{\omega_a})^2 - 1}\right]}{\beta I_0(\beta)\sqrt{(\frac{\omega}{\omega_a})^2 - 1}} \tag{5.67}$$

where $\omega_a = \Omega_a T$ and $\beta = \Omega_a \tau = \frac{\omega_a}{T}\frac{M}{2}T = \omega_a \frac{M}{2}$.

The main advantage of the Kaiser window appears in the design of FIR digital filters with prescribed specifications, such as that depicted in Figure 5.10. In such an application, the parameter β is used to control both the main-lobe width and the ratio of the main to the secondary lobes.

The overall procedure for designing FIR filters using the Kaiser window is as follows:

(i) From the ideal frequency response that the filter is supposed to approximate, determine the impulse response $h(n)$ using Table 5.1. If the filter is either lowpass or highpass, one should make $\Omega_c = \frac{\Omega_p + \Omega_r}{2}$. The case of bandpass and bandstop filters is dealt with later in this subsection.

(ii) Given the maximum passband ripple in dB, A_p, and the minimum stopband attenuation in dB, A_r, determine the corresponding ripples

$$\delta_p = \frac{10^{0.05A_p} - 1}{10^{0.05A_p} + 1} \tag{5.68}$$

$$\delta_r = 10^{-0.05A_r} \tag{5.69}$$

(iii)   As with all other window functions, the Kaiser window can only be used to design filters that present the same passband and stopband ripples. Therefore, in order to satisfy the prescribed specifications, one should use $\delta = \min\{\delta_p, \delta_r\}$.

(iv)   Compute the resulting passband ripple and stopband attenuation in dB using

$$A_p = 20 \log \frac{1 + \delta}{1 - \delta} \tag{5.70}$$

$$A_r = -20 \log \delta \tag{5.71}$$

(v)   Given the passband and stopband edges, Ω_p and Ω_r, respectively, compute the transition bandwidth $T_r = (\Omega_r - \Omega_p)$.

(vi)   Compute β using

$$\beta = \begin{cases} 0, & \text{for } A_r \leq 21 \\ 0.5842(A_r - 21)^{0.4} + 0.078\,86(A_r - 21), & \text{for } 21 < A_r \leq 50 \\ 0.1102(A_r - 8.7), & \text{for } 50 < A_r \end{cases} \tag{5.72}$$

This empirical formula was devised by (Kaiser, 1974) based on the behavior of the function $W(\Omega)$ in equation (5.64).

(vii)   Defining the normalized window length D as

$$D = \frac{T_r M}{\Omega_s} \tag{5.73}$$

where Ω_s is the sampling frequency, we have that D is related to A_r by the following empirical formula:

$$D = \begin{cases} 0.9222, & \text{for } A_r \leq 21 \\ \dfrac{(A_r - 7.95)}{14.36}, & \text{for } 21 < A_r \end{cases} \tag{5.74}$$

(viii)   Having computed D using equation (5.74), we can use equation (5.73) to determine the filter order M as the smallest even number that satisfies

$$M \geq \frac{\Omega_s D}{T_r} \tag{5.75}$$

One should remember that T_r must be in the same units as Ω_s.

(ix)   With M and β determined, we compute the window $w_K(n)$ using equation (5.65). We are now ready to form the sequence $h'(n) = w_K(n)h(n)$, where $h(n)$ is the ideal filter impulse response computed in step (i).

(x)   The desired transfer function is then given by

$$H(z) = z^{-\frac{M}{2}} \mathcal{Z}\{h'(n)\} \tag{5.76}$$

The above procedure applies to lowpass filters (see Figure 5.10) as well as highpass filters. If the filter is either bandpass or bandstop, we must include the following reasoning in step (i) above:

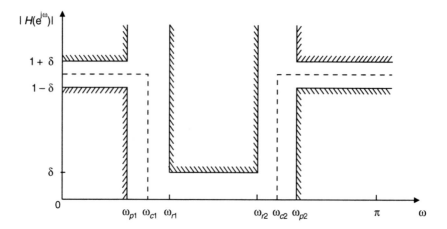

Figure 5.11 Typical specification of a bandstop filter.

1. Compute the narrower transition band

$$T_r = \pm \min\{|\Omega_{r_1} - \Omega_{p_1}|, |\Omega_{p_2} - \Omega_{r_2}|\} \tag{5.77}$$

Notice that T_r is negative for bandpass filters and positive for bandstop filters.

2. Determine the two central frequencies as

$$\Omega_{c_1} = \left(\Omega_{p_1} + \frac{T_r}{2}\right) \tag{5.78}$$

$$\Omega_{c_2} = \left(\Omega_{p_2} - \frac{T_r}{2}\right) \tag{5.79}$$

A typical magnitude specification for a bandstop filter is depicted in Figure 5.11.

EXAMPLE 5.4

Design a bandstop filter satisfying the specification below using the Kaiser window:

$$\left.\begin{aligned}
A_p &= 1.0\,\text{dB} \\
A_r &= 45\,\text{dB} \\
\Omega_{p_1} &= 800\,\text{Hz} \\
\Omega_{r_1} &= 950\,\text{Hz} \\
\Omega_{r_2} &= 1050\,\text{Hz} \\
\Omega_{p_2} &= 1200\,\text{Hz} \\
\Omega_s &= 6000\,\text{Hz}
\end{aligned}\right\} \tag{5.80}$$

SOLUTION

Following the procedure described above, the resulting filter is obtained as follows (note that in the FIR design procedure described above, the parameters of the Kaiser window depend only on the ratio of the analog frequencies in the filter specification to the sampling frequency; therefore, the frequencies can be entered in the formula in hertz, as long as the sampling frequency Ω_s is also in hertz):

(i)   From equations (5.77)–(5.79), we have that

$$T_r \ = + \min\{(950 - 800), (1200 - 1050)\} = 150 \text{ Hz} \tag{5.81}$$

$$\Omega_{c_1} = 800 + 75 = 875 \text{ Hz} \tag{5.82}$$

$$\Omega_{c_2} = 1200 - 75 = 1125 \text{ Hz} \tag{5.83}$$

(ii)   From equations (5.68) and (5.69),

$$\delta_p = \frac{10^{0.05} - 1}{10^{0.05} + 1} = 0.0575 \tag{5.84}$$

$$\delta_r = 10^{-0.05 \times 45} = 0.005\,62 \tag{5.85}$$

$$\tag{5.86}$$

(iii) $\delta = \min\{0.0575, 0.005\,62\} = 0.005\,62$

(iv)   From equations (5.70) and (5.71),

$$A_p = 20 \log \frac{1 + 0.005\,62}{1 - 0.005\,62} = 0.0977 \text{ dB} \tag{5.87}$$

$$A_r = -20 \log 0.005\,62 = 45 \text{ dB} \tag{5.88}$$

(v) T_r has already been computed as 150 Hz in step (i).

(vi)   From equation (5.72), since $A_r = 45$ dB, then

$$\beta = 0.5842(45 - 21)^{0.4} + 0.078\,86(45 - 21) = 3.975\,4327 \tag{5.89}$$

(vii) From equation (5.74), since $A_r = 45$ dB, then

$$D = \frac{(45 - 7.95)}{14.36} = 2.580\,0835 \tag{5.90}$$

(viii) Since the sampling period is $T = \frac{1}{6000}$ s, we have, from equation (5.75),

$$M \geq \frac{6000 \times 2.580\,0835}{150} = 103.203\,34 \quad \Rightarrow \quad M = 104 \tag{5.91}$$

The designed filter characteristics are summarized in Table 5.9.

(ix)   Plots of the window coefficients and magnitude response are shown in Figure 5.12.

The filter coefficients $h'(n)$ are given in Table 5.10 and Figure 5.13a. Once again, due to the symmetry inherent to the Kaiser window function, Table 5.10 only shows half of the filter coefficients, as the remaining coefficients are obtained as $h(n) = h(M - n)$.

Table 5.9. *Characteristics of the designed filter.*

Ω_{c_1}	875 Hz
Ω_{c_2}	1125 Hz
Ω_{p_1}	800 Hz
Ω_{r_1}	950 Hz
Ω_{r_2}	1050 Hz
Ω_{p_2}	1200 Hz
δ_p	0.0575
δ_r	0.005 62
T_r	150 Hz
D	2.580 0835
β	3.975 4327
M	104

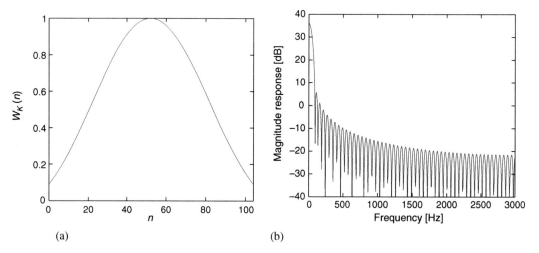

(a)   (b)

Figure 5.12 Kaiser window: (a) window function; (b) magnitude response.

(x) The filter magnitude response, along with the filter impulse response, are shown in Figure 5.13.

△

5.4.6   Dolph–Chebyshev window

Based on the Mth-order Chebyshev polynomial given by

$$C_M(x) = \begin{cases} \cos\left[M\cos^{-1}(x)\right], & \text{for } |x| \leq 1 \\ \cosh\left[M\cosh^{-1}(x)\right], & \text{for } |x| > 1 \end{cases} \tag{5.92}$$

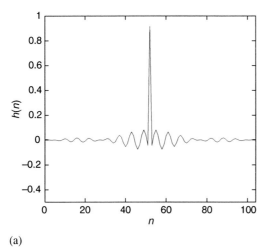

(a)

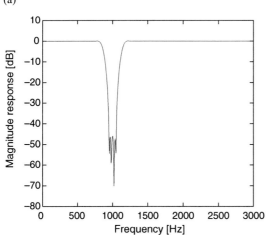

(b)

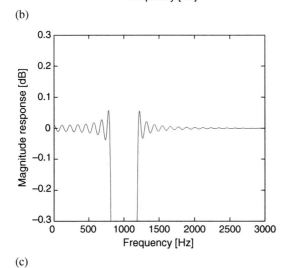

(c)

Figure 5.13 Resulting bandstop filter: (a) impulse response; (b) magnitude response; (c) passband detail.

Table 5.10. *Filter coefficients using the Kaiser window.*

	$h(0)$ to $h(52)$	
9.166 666 67E−01	−2.025 637 09E−02	7.018 040 81E−03
−4.152 143 92E−02	−8.058 394 12E−03	3.187 744 57E−03
4.108 795 45E−02	6.124 562 79E−03	−2.839 675 94E−03
8.074 552 18E−02	8.683 533 46E−03	−4.955 289 88E−03
3.938 662 38E−02	2.721 742 99E−03	−2.113 199 09E−03
−3.814 430 73E−02	−1.272 833 01E−03	1.756 456 66E−03
−7.332 855 03E−02	−2.303 350 15E−13	2.832 015 82E−03
−3.496 831 96E−02	1.095 082 11E−03	1.098 863 32E−03
3.308 112 52E−02	−2.013 873 12E−03	−8.104 467 84E−04
6.205 956 05E−02	−5.521 501 33E−03	−1.109 261 53E−03
2.884 324 52E−02	−3.342 741 76E−03	−3.338 094 03E−04
−2.655 179 21E−02	3.769 210 42E−03	1.490 046 61E−04
−4.837 285 43E−02	8.103 021 70E−03	5.114 812 86E−14
−2.177 831 01E−02	4.202 616 19E−03	−1.145 042 70E−04
1.935 818 38E−02	−4.236 733 00E−03	1.967 201 89E−04
3.391 165 07E−02	−8.337 823 08E−03	4.991 841 06E−04
1.459 953 30E−02	−4.014 661 65E−03	2.766 130 86E−04
−1.231 565 69E−02	3.789 583 33E−03	

the Dolph–Chebyshev window is defined as

$$
w_{DC}(n) =
\begin{cases}
\dfrac{1}{M+1}\left\{ \dfrac{1}{r} + 2 \sum_{i=1}^{\frac{M}{2}} C_M\left[x_0 \cos\left(\dfrac{i\pi}{M+1} \right) \right] \cos\left(\dfrac{2ni\pi}{M+1} \right) \right\}, & \text{for } |n| \le \dfrac{M}{2} \\
0, & \text{for } |n| > \dfrac{M}{2}
\end{cases}
$$

(5.93)

where r is the ripple ratio defined as

$$
r = \frac{\delta_r}{\delta_p}
\tag{5.94}
$$

and x_0 is given by

$$
x_0 = \cosh\left[\frac{1}{M} \cosh^{-1}\left(\frac{1}{r} \right) \right]
\tag{5.95}
$$

The procedure for designing FIR filters using the Dolph–Chebyshev window is very similar to the one for the Kaiser window, as follows:

(i) Perform steps (i) and (ii) of the Kaiser procedure.
(ii) Determine r from equation (5.94).

(iii) Perform steps (iii)–(v) and (vii)–(viii) of the Kaiser procedure, to determine the filter order M. In step (vii), however, as the stopband attenuation achieved with the Dolph–Chebyshev window is typically 1 to 4 dB higher than that obtained using the Kaiser window, one should compute D for the Dolph–Chebyshev window using equation (5.74) with A_r replaced by $A_r + 2.5$ (Saramäki, 1993). With this approximation, the resulting order value M may not be precise, and small corrections may be necessary at the end of the design routine to completely satisfy the prescribed specifications.

(iv) With r and M determined, compute x_0 from equation (5.95), and then compute the window coefficients from equation (5.93).

(v) We are now ready to form the sequence $h'(n) = w_{DC}(n)h(n)$, where $h(n)$ is the ideal filter impulse response computed in step (5.4.5).

(vi) Perform step (x) of the Kaiser procedure to determine the resulting FIR filter.

Overall, the Dolph–Chebyshev window is characterized by:

- The main-lobe width, and consequently the resulting filter transition band, can be controlled by varying M.

- The ripple ratio is controlled through an independent parameter r.

- All secondary lobes have the same amplitude. Therefore, the stopband of the resulting filter is equiripple.

5.5   Maximally flat FIR filter approximation

Maximally flat approximations should be employed when a signal must be preserved with minimal error around the zero frequency or when a monotone frequency response is necessary. FIR filters with a maximally flat frequency response at $\omega = 0$ and $\omega = \pi$ were introduced in Herrmann (1971). We consider here, following the standard literature on the subject (Herrmann, 1971; Vaidyanathan, 1984, 1985), the lowpass Type-I FIR filter of even order M and symmetric impulse response.

In this case, the frequency response of a maximally flat FIR filter is determined in such a way that $H(e^{j\omega}) - 1$ has $2L$ zeros at $\omega = 0$, and $H(e^{j\omega})$ has $2K$ zeros at $\omega = \pi$. To achieve a maximally flat response, the filter order M must satisfy $M = (2K + 2L - 2)$. Thus, the first $2L - 1$ derivatives of $H(e^{j\omega})$ are zero at $\omega = 0$, and the first $2K - 1$ derivatives of $H(e^{j\omega})$ are zero at $\omega = \pi$. If the above two conditions are satisfied, $H(e^{j\omega})$ can be written (Herrmann, 1971) either as

$$H(e^{j\omega}) = \left(\cos\frac{\omega}{2}\right)^{2K} \sum_{n=0}^{L-1} d(n) \left(\sin\frac{\omega}{2}\right)^{2n}$$

$$= \left(\frac{1+\cos\omega}{2}\right)^K \sum_{n=0}^{L-1} d(n) \left(\frac{1-\cos\omega}{2}\right)^n \tag{5.96}$$

or as

$$H(e^{j\omega}) = 1 - \left(\sin\frac{\omega}{2}\right)^{2L} \sum_{n=0}^{K-1} \hat{d}(n) \left(\cos\frac{\omega}{2}\right)^{2n}$$

$$= 1 - \left(\frac{1-\cos\omega}{2}\right)^L \sum_{n=0}^{K-1} \hat{d}(n) \left(\frac{1+\cos\omega}{2}\right)^n \tag{5.97}$$

where the coefficients $d(n)$ or $\hat{d}(n)$ are respectively given by

$$d(n) = \frac{(K-1+n)!}{(K-1)!n!} \tag{5.98}$$

$$\hat{d}(n) = \frac{(L-1+n)!}{(L-1)!n!} \tag{5.99}$$

From the above equations it is easy to see that $H(e^{j\omega})$ can be expressed as a sum of complex exponentials in ω having frequencies ranging from $(-K-L+1)$ to $(K+L-1)$, with an increment of 1. Therefore, from Theorem 5.1, one can see that $h(n)$ can be exactly recovered by sampling $H(e^{j\omega})$ at $(2K+2L-1) = M+1$ equally spaced points located at frequencies $\omega = 2\pi n/(M+1)$, for $n = 0, \ldots, M$, and taking the IDFT, as in the frequency sampling approach.

A more efficient implementation, however, results from writing the transfer function as (Vaidyanathan, 1984)

$$H(z) = \left(\frac{1+z^{-1}}{2}\right)^{2K} \sum_{n=0}^{L-1} (-1)^n d(n) z^{-(L-1-n)} \left(\frac{1-z^{-1}}{2}\right)^{2n} \tag{5.100}$$

or

$$H(z) = z^{-\frac{M}{2}} - (-1)^L \left(\frac{1-z^{-1}}{2}\right)^{2L} \sum_{n=0}^{K-1} \hat{d}(n) z^{-(K-1-n)} \left(\frac{1+z^{-1}}{2}\right)^{2n} \tag{5.101}$$

These equations require significantly fewer multipliers than the direct-form implementation. The drawback with these designs is the large dynamic range necessary to represent the sequences $d(n)$ and $\hat{d}(n)$. This can be avoided by an efficient cascade implementation of these coefficients, as discussed in Vaidyanathan (1984), utilizing the following relationships:

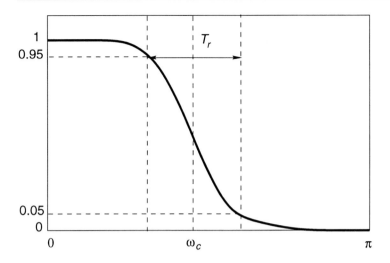

Figure 5.14 Typical specification of a maximally flat lowpass FIR filter.

$$d(n+1) = d(n)\frac{K+n}{n+1} \tag{5.102}$$

$$\hat{d}(n+1) = \hat{d}(n)\frac{L+n}{n+1} \tag{5.103}$$

In the procedure described above, the sole design parameters for the maximally flat FIR filters are the values of K and L. Given a desired magnitude response, as depicted in Figure 5.14, the transition band is defined as the region where the magnitude response varies from 0.95 to 0.05, and the normalized center frequency is the center of this band ($\omega_c = \frac{\omega_p + \omega_r}{2}$). If the transition band in rad/sample is T_r, we need to compute the following parameters:

$$M_1 = \left(\frac{\pi}{T_r}\right)^2 \tag{5.104}$$

$$\rho = \frac{1 + \cos \omega_c}{2} \tag{5.105}$$

Then, for all integer values of M_p in the range $M_1 \le M_p \le 2M_1$, we compute K_p as the nearest integer to ρM_p. We then choose K_p^* and M_p^* as the values of K_p and M_p for which the ratio $\frac{K_p}{M_p}$ is closest to ρ. The desired values of K, L, and M, are given by

$$K = K_p^* \tag{5.106}$$

$$L = M_p^* - K_p^* \tag{5.107}$$

$$M = 2K + 2L - 2 = 2M_p^* - 2 \tag{5.108}$$

Table 5.11. *Coefficients of the lowpass filter designed with the maximally flat method.*

	$d(0)$ to $d(7)$	
1	4495	1244 904
29	35 960	6724 520
435	237 336	

EXAMPLE 5.5

Design a maximally flat lowpass filter satisfying the specification below:

$$\left.\begin{aligned}\Omega_c &= 0.3\pi \text{ rad/s} \\ T_r &= 0.15\pi \text{ rad/s} \\ \Omega_s &= 2\pi \text{ rad/s}\end{aligned}\right\} \tag{5.109}$$

SOLUTION

For the given specification, it is easy to verify that the required values of K and L are $K = 29$ and $L = 8$. Therefore, the filter order is $M = 72$. The resulting coefficients $d(n)$ are given in Table 5.11 and the filter magnitude response is shown in Figure 5.15. Notice that in this case, even though the filter order is $M = 72$, there are only $L = 8$ coefficients $d(n)$.

$\triangle$

5.6　FIR filter approximation by optimization

The window method seen in Section 5.4 has a very straightforward design procedure for approximating the desired magnitude response. However, the window method is not efficient for designing, for example, FIR filters with different ripples in the passband and stopband, or nonsymmetric bandpass or bandstop filters. To fill this gap, in this section we present several numerical algorithms for designing more general FIR digital filters.

In many signal processing systems, filters with linear or zero phase are required. Unfortunately, filters designed to have zero phase are not causal; this can be a problem in applications where very little processing delay is permissible. Also, nonlinear phase causes distortion in the processed signal, which can be very perceptible in applications like data transmission, image processing, and so on. One of the major advantages of using an FIR system instead of a causal IIR system is that FIR systems can be designed with exact linear phase. As seen in Subsection 4.2.3, there are four distinct cases where an FIR filter presents linear phase. To present general algorithms for designing linear-phase FIR filters, a unified representation of these four cases is necessary. We define

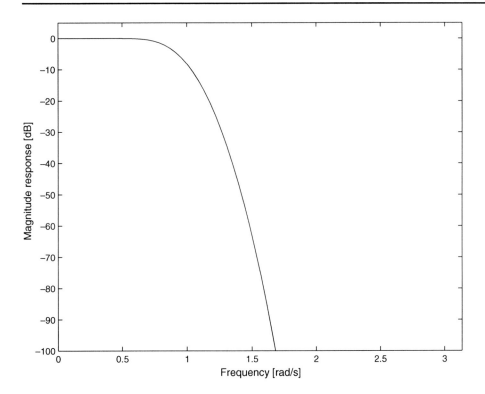

Figure 5.15 Magnitude response of maximally flat lowpass FIR filter.

an auxiliary function $P(\omega)$ as

$$P(\omega) = \sum_{l=0}^{L} p(l)\cos(\omega l) \tag{5.110}$$

where $L + 1$ is the number of cosine functions in the expression of $H(\mathrm{e}^{\mathrm{j}\omega})$. Based on this function, we can express the frequency response of the four types of linear-phase FIR filters as (McClellan & Parks, 1973):

- Type I: Even order M and symmetric impulse response. From Table 4.1, we can write that

$$H(\mathrm{e}^{\mathrm{j}\omega}) = \mathrm{e}^{-\mathrm{j}\omega\frac{M}{2}} \sum_{m=0}^{\frac{M}{2}} a(m)\cos(\omega m)$$

$$= \mathrm{e}^{-\mathrm{j}\omega\frac{M}{2}} \sum_{l=0}^{\frac{M}{2}} p(l)\cos(\omega l)$$

$$= \mathrm{e}^{-\mathrm{j}\omega\frac{M}{2}} P(\omega) \tag{5.111}$$

with

$$a(m) = p(m), \text{ for } m = 0, \ldots, L \tag{5.112}$$

where $L = \frac{M}{2}$.

- Type II: Odd order M and symmetric impulse response. In this case, from Table 4.1, we have

$$H(e^{j\omega}) = e^{-j\omega\frac{M}{2}} \sum_{m=1}^{\frac{M+1}{2}} b(m) \cos\left[\omega\left(m - \frac{1}{2}\right)\right] \tag{5.113}$$

Using

$$b(m) = \begin{cases} p(0) + \dfrac{1}{2}p(1), & \text{for } m = 1 \\[2mm] \dfrac{1}{2}\left(p(m-1) + p(m)\right), & \text{for } m = 2, \ldots, L \\[2mm] \dfrac{1}{2}p(L), & \text{for } m = L + 1 \end{cases} \tag{5.114}$$

with $L = \frac{M-1}{2}$, then $H(e^{j\omega})$ can be written in the form

$$H(e^{j\omega}) = e^{-j\omega\frac{M}{2}} \cos\left(\frac{\omega}{2}\right) P(\omega) \tag{5.115}$$

using the trigonometric identity

$$2\cos\left(\frac{\omega}{2}\right)\cos(\omega m) = \cos\left[\omega\left(m + \frac{1}{2}\right)\right] + \cos\left[\omega\left(m - \frac{1}{2}\right)\right] \tag{5.116}$$

The complete algebraic development is left as an exercise to the interested reader.

- Type III: Even order M and antisymmetric impulse response. In this case, from Table 4.1, we have

$$H(e^{j\omega}) = e^{-j(\omega\frac{M}{2} - \frac{\pi}{2})} \sum_{m=1}^{\frac{M}{2}} c(m) \sin(\omega m) \tag{5.117}$$

and then, by substituting

$$c(m) = \begin{cases} p(0) - \dfrac{1}{2}p(2), & \text{for } m = 1 \\[2mm] \dfrac{1}{2}\left(p(m-1) - p(m+1)\right), & \text{for } m = 2, \ldots, L - 1 \\[2mm] \dfrac{1}{2}p(m-1), & \text{for } m = L, L + 1 \end{cases} \tag{5.118}$$

with $L = \frac{M}{2} - 1$, equation (5.117) can be written as

$$H(e^{j\omega}) = e^{-j\left(\omega\frac{M}{2} - \frac{\pi}{2}\right)} \sin\omega P(\omega) \tag{5.119}$$

using, in this case, the identity

$$\sin \omega \cos (\omega m) = \sin [\omega (m + 1)] - \sin [\omega (m - 1)] \tag{5.120}$$

Once again, the algebraic proof is left as an exercise at the end of this chapter.

- Type IV: Odd order M and antisymmetric impulse response. We have, from Table 4.1, that

$$H(e^{j\omega}) = e^{-j(\omega \frac{M}{2} - \frac{\pi}{2})} \sum_{m=1}^{\frac{M+1}{2}} d(m) \sin \left[\omega \left(m - \frac{1}{2} \right) \right] \tag{5.121}$$

By substituting

$$d(m) = \begin{cases} p(0) - \dfrac{1}{2} p(1), & \text{for } m = 1 \\ \dfrac{1}{2} (p(m - 1) - p(m)), & \text{for } m = 2, \ldots, L \\ \dfrac{1}{2} p(L), & \text{for } m = L + 1 \end{cases} \tag{5.122}$$

with $L = \frac{M-1}{2}$, then $H(e^{j\omega})$ can be written as

$$H(e^{j\omega}) = e^{-j\left(\omega \frac{M}{2} - \frac{\pi}{2}\right)} \sin \left(\frac{\omega}{2} \right) P(\omega) \tag{5.123}$$

using the identity

$$2 \sin \left(\frac{\omega}{2} \right) \cos(\omega m) = \sin \left[\omega \left(m + \frac{1}{2} \right) \right] - \sin \left[\omega \left(m - \frac{1}{2} \right) \right] \tag{5.124}$$

Once more, the entire algebraic development is left as an exercise to the reader.

Equations (5.111), (5.115), (5.119), and (5.123) indicate that we can write the frequency response for any linear-phase FIR filter as

$$H(e^{j\omega}) = e^{-j(\alpha\omega - \beta)} Q(\omega) P(\omega) = e^{-j(\alpha\omega - \beta)} A(\omega) \tag{5.125}$$

where $A(\omega) = Q(\omega) P(\omega)$, $\alpha = \frac{M}{2}$, and for each case we have that:

- Type I: $\beta = 0$ and $Q(\omega) = 1$
- Type II: $\beta = 0$ and $Q(\omega) = \cos(\frac{\omega}{2})$
- Type III: $\beta = \frac{\pi}{2}$ and $Q(\omega) = \sin(\omega)$
- Type IV: $\beta = \frac{\pi}{2}$ and $Q(\omega) = \sin(\frac{\omega}{2})$.

Let $D(\omega)$ be the desired amplitude response. We define the weighted error function as

$$E(\omega) = W(\omega)(D(\omega) - A(\omega)) \tag{5.126}$$

We can then write $E(\omega)$ as

$$
\begin{aligned}
E(\omega) &= W(\omega)(D(\omega) - Q(\omega)P(\omega)) \\
&= W(\omega)Q(\omega)\left(\frac{D(\omega)}{Q(\omega)} - P(\omega)\right)
\end{aligned}
\tag{5.127}
$$

for all $0 \le \omega \le \pi$, as $Q(\omega)$ is independent of the coefficients for each ω. Defining

$$
W_q(\omega) = W(\omega)Q(\omega)
\tag{5.128}
$$

$$
D_q(\omega) = \frac{D(\omega)}{Q(\omega)}
\tag{5.129}
$$

the error function can be rewritten as

$$
E(\omega) = W_q(\omega)(D_q(\omega) - P(\omega))
\tag{5.130}
$$

and one can formulate the optimization problem for approximating linear-phase FIR filters as: *determine the set of coefficients $p(l)$ that minimizes some objective function of the weighted error $E(\omega)$ over a set of prescribed frequency bands.*

To solve such a problem numerically, we evaluate the weighted error function on a dense frequency grid with $0 \le \omega_i \le \pi$, for $i = 1, \ldots, KM$, where M is the filter order, obtaining a good discrete approximation of $E(\omega)$. For most practical purposes, using $8 \le K \le 16$ is recommended. Points associated with the transition bands can be disregarded, and the remaining frequencies should be linearly redistributed in the passbands and stopbands to include their corresponding edges. Thus, the following equation results

$$
\mathbf{e} = \mathbf{W}_q\left(\mathbf{d}_q - \mathbf{U}\mathbf{p}\right)
\tag{5.131}
$$

where

$$
\mathbf{e} = \begin{bmatrix} E(\omega_1) & E(\omega_2) & \cdots & E(\omega_{\bar{K}\bar{M}}) \end{bmatrix}^{\mathsf{T}}
\tag{5.132}
$$

$$
\mathbf{W}_q = \mathrm{diag}\begin{bmatrix} W_q(\omega_1) & W_q(\omega_2) & \cdots & W_q(\omega_{\bar{K}\bar{M}}) \end{bmatrix}
\tag{5.133}
$$

$$
\mathbf{d}_q = \begin{bmatrix} D_q(\omega_1) & D_q(\omega_2) & \cdots & D_q(\omega_{\bar{K}\bar{M}}) \end{bmatrix}^{\mathsf{T}}
\tag{5.134}
$$

$$
\mathbf{U} = \begin{bmatrix}
1 & \cos(\omega_1) & \cos(2\omega_1) & \cdots & \cos(L\omega_1) \\
1 & \cos(\omega_2) & \cos(2\omega_2) & \cdots & \cos(L\omega_2) \\
\vdots & \vdots & \vdots & \ddots & \vdots \\
1 & \cos(\omega_{\bar{K}\bar{M}}) & \cos(2\omega_{\bar{K}\bar{M}}) & \cdots & \cos(L\omega_{\bar{K}\bar{M}})
\end{bmatrix}
\tag{5.135}
$$

$$
\mathbf{p} = [p(0)\ p(1)\ \cdots\ p(L)]^{\mathsf{T}}
\tag{5.136}
$$

with $\bar{K}\bar{M} \le KM$, as the original frequencies in the transition band were discarded.

Table 5.12. *Weight functions and ideal magnitude responses for basic lowpass, highpass, bandpass, and bandstop filters, as well as differentiators and Hilbert transformers.*

Filter type	Weight function $W(\omega)$	Ideal amplitude response $D(\omega)$
Lowpass	$\begin{cases} 1, & \text{for } 0 \leq \omega \leq \omega_p \\ \dfrac{\delta_p}{\delta_r}, & \text{for } \omega_r \leq \omega \leq \pi \end{cases}$	$\begin{cases} 1, & \text{for } 0 \leq \omega \leq \omega_p \\ 0, & \text{for } \omega_r \leq \omega \leq \pi \end{cases}$
Highpass	$\begin{cases} \dfrac{\delta_p}{\delta_r}, & \text{for } 0 \leq \omega \leq \omega_r \\ 1, & \text{for } \omega_p \leq \omega \leq \pi \end{cases}$	$\begin{cases} 0, & \text{for } 0 \leq \omega \leq \omega_r \\ 1, & \text{for } \omega_p \leq \omega \leq \pi \end{cases}$
Bandpass	$\begin{cases} \dfrac{\delta_p}{\delta_r}, & \text{for } 0 \leq \omega \leq \omega_{r_1} \\ 1, & \text{for } \omega_{p_1} \leq \omega \leq \omega_{p_2} \\ \dfrac{\delta_p}{\delta_r}, & \text{for } \omega_{r_2} \leq \omega \leq \pi \end{cases}$	$\begin{cases} 0, & \text{for } 0 \leq \omega \leq \omega_{r_1} \\ 1, & \text{for } \omega_{p_1} \leq \omega \leq \omega_{p_2} \\ 0, & \text{for } \omega_{r_2} \leq \omega \leq \pi \end{cases}$
Bandstop	$\begin{cases} 1, & \text{for } 0 \leq \omega \leq \omega_{p_1} \\ \dfrac{\delta_p}{\delta_r}, & \text{for } \omega_{r_1} \leq \omega \leq \omega_{r_2} \\ 1, & \text{for } \omega_{p_2} \leq \omega \leq \pi \end{cases}$	$\begin{cases} 1, & \text{for } 0 \leq \omega \leq \omega_{p_1} \\ 0, & \text{for } \omega_{r_1} \leq \omega \leq \omega_{r_2} \\ 1, & \text{for } \omega_{p_2} \leq \omega \leq \pi \end{cases}$
Differentiator	$\begin{cases} \dfrac{1}{\omega}, & \text{for } 0 < \omega \leq \omega_p \\ 0, & \text{for } \omega_p < \omega \leq \pi \end{cases}$	$\omega, \quad \text{for } 0 \leq \omega \leq \pi$
Hilbert transformer	$\begin{cases} 0, & \text{for } 0 \leq \omega < \omega_{p_1} \\ 1, & \text{for } \omega_{p_1} \leq \omega \leq \omega_{p_2} \\ 0, & \text{for } \omega_{p_2} < \omega \leq \pi \end{cases}$	$1, \quad \text{for } 0 \leq \omega \leq \pi$

For the four standard types of filters, namely lowpass, highpass, bandpass, and bandstop, as well as differentiators and Hilbert transformers, the definitions of $W(\omega)$ and $D(\omega)$ are summarized in Table 5.12.

It is important to remember all design constraints, due to the magnitude and characteristics of the four linear-phase filter types. Such constraints are summarized in Table 5.13, where a 'Yes' entry indicates that the corresponding filter structure is suitable to implement the given filter type.

Table 5.13. *Suitability of linear-phase FIR structures for basic lowpass, highpass, bandpass, and bandstop filters, as well as differentiators and Hilbert transformers.*

Filter type	Type I	Type II	Type III	Type IV
Lowpass	Yes	Yes	No	No
Highpass	Yes	No	No	Yes
Bandpass	Yes	Yes	Yes	Yes
Bandstop	Yes	No	No	No
Differentiator	No	No	Yes	Yes
Hilbert transformer	No	No	Yes	Yes

5.6.1   Weighted-least-squares method

In the weighted-least-squares (WLS) approach, the idea is to minimize the square of the energy of the error function $E(\omega)$, that is

$$\min_{\mathbf{p}}\left\{\|E(\omega)\|_2^2\right\} = \min_{\mathbf{p}}\left\{\int_0^\pi |E(\omega)|^2\,d\omega\right\} \tag{5.137}$$

For a discrete set of frequencies, this objective function is approximated by (see equations (5.131)–(5.136))

$$\|E(\omega)\|_2^2 \approx \frac{1}{\bar{K}\bar{M}}\sum_{k=1}^{\bar{K}\bar{M}}|E(\omega_k)|^2 = \frac{1}{\bar{K}\bar{M}}\mathbf{e}^T\mathbf{e} \tag{5.138}$$

Using equation (5.131), and noting that $\mathbf{W}_q$ is diagonal, we can write that

$$\begin{aligned}
\mathbf{e}^T\mathbf{e} &= (\mathbf{d}_q^T - \mathbf{p}^T\mathbf{U}^T)\mathbf{W}_q^T\mathbf{W}_q(\mathbf{d}_q - \mathbf{U}\mathbf{p})\\
&= (\mathbf{d}_q^T - \mathbf{p}^T\mathbf{U}^T)\mathbf{W}_q^2(\mathbf{d}_q - \mathbf{U}\mathbf{p})\\
&= \mathbf{d}_q^T\mathbf{W}_q^2\mathbf{d}_q - \mathbf{d}_q^T\mathbf{W}_q^2\mathbf{U}\mathbf{p} - \mathbf{p}^T\mathbf{U}^T\mathbf{W}_q^2\mathbf{d}_q + \mathbf{p}^T\mathbf{U}^T\mathbf{W}_q^2\mathbf{U}\mathbf{p}\\
&= \mathbf{d}_q^T\mathbf{W}_q^2\mathbf{d}_q - 2\mathbf{p}^T\mathbf{U}^T\mathbf{W}_q^2\mathbf{d}_q + \mathbf{p}^T\mathbf{U}^T\mathbf{W}_q^2\mathbf{U}\mathbf{p}
\end{aligned} \tag{5.139}$$

because $\mathbf{d}_q^T\mathbf{W}_q^2\mathbf{U}\mathbf{p} = \mathbf{p}^T\mathbf{U}^T\mathbf{W}_q^2\mathbf{d}_q$, since these two terms are scalar. The minimization of such a functional is achieved by calculating its gradient vector with respect to the coefficient vector and equating it to zero, thus yielding

$$\nabla_{\mathbf{p}}\left\{\mathbf{e}^T\mathbf{e}\right\} = -2\mathbf{U}^T\mathbf{W}_q^2\mathbf{d}_q + 2\mathbf{U}^T\mathbf{W}_q^2\mathbf{U}\mathbf{p}^* = \mathbf{0} \tag{5.140}$$

which implies that

$$\mathbf{p}^* = \left(\mathbf{U}^T\mathbf{W}_q^2\mathbf{U}\right)^{-1}\mathbf{U}^T\mathbf{W}_q^2\mathbf{d}_q \tag{5.141}$$

It can be shown that when the weight function $W(\omega)$ is made constant, the WLS approach is equivalent to the rectangular window presented in the previous section, and so suffers from the same problem of Gibbs' oscillations near the band edges. When $W(\omega)$ is not constant, the oscillations still occur but their energies will vary from band to band.

Amongst the several extensions and generalizations of the WLS approach, we refer here to the constrained-WLS and eigenfilter methods. The constrained-WLS method was presented in Selesnick et al. (1996, 1998). In this method, the designer specifies the maximum and minimum values allowed for the desired magnitude response in each band. In the proposed algorithm, the transition bands are not completely specified, as only their central frequencies need be provided. The transition bands are then automatically adjusted to satisfy the constraints. The overall method consists of an iterative procedure where in each step a modified WLS design is performed, using Lagrange multipliers, and the constraints are subsequently tested and updated. Such a procedure involves verification of the Kuhn–Tucker conditions (Winston, 1991), that is, it checks if all the resulting multipliers are non-negative, followed by a search routine that finds the positions of all local extremals in each band and tests if all the constraints are satisfied. For the eigenfilter method presented in Vaidyanathan (1987) and used in Nguyen et al. (1994), the objective function in equation (5.130) is rewritten in a distinct form, and results from linear algebra are used to find the optimal filter for the resulting equation. With such a procedure, the eigenfilter method enables linear-phase FIR filters with different characteristics to be designed, and the WLS scheme appears as a special case of the eigenfilter approach.

5.6.2 Chebyshev method

In the Chebyshev optimization design approach, the idea is to minimize the maximum absolute value of the error function $E(\omega)$. Mathematically, such scheme is described by

$$\min_{\mathbf{p}} \{\|E(\omega)\|_\infty\} = \min_{\mathbf{p}} \left\{ \max_{\omega \in F} \{|E(\omega)|\} \right\} \tag{5.142}$$

where F is the set of prescribed frequency bands. This problem can be solved with the help of the following important theorem:

THEOREM 5.2 (ALTERNATION THEOREM)
If $P(\omega)$ is a linear combination of $(L + 1)$ cosine functions, that is

$$P(\omega) = \sum_{l=0}^{L} p(l) \cos(\omega l) \tag{5.143}$$

the necessary and sufficient condition for $P(\omega)$ to be the Chebyshev approximation of a continuous function $D(\omega)$ in F, a compact subset of $[0, \pi]$, is that the error function

$E(\omega)$ *must present at least* $(L + 2)$ *extreme frequencies in F. That is, there must be at least* $(L + 2)$ *points* ω_k *in F, where* $\omega_0 < \omega_1 < \cdots < \omega_{L+1}$, *such that*

$$E(\omega_k) = -E(\omega_{k+1}), \text{ for } k = 0, \ldots, L \tag{5.144}$$

and

$$|E(\omega_k)| = \max_{\omega \in F}\{|E(\omega)|\}, \text{ for } k = 0, \ldots, (L + 1) \tag{5.145}$$

◇

A proof of this theorem can be found in (Cheney, 1966).

The extremes of $E(\omega)$ are related to the extremes of $A(\omega)$ defined in equation (5.126). The values of ω for which $\frac{\partial A(\omega)}{\partial \omega} = 0$ allow us to determine that the number N_k of extremes of $A(\omega)$ is such that:

- Type I: $N_k \le \frac{M+2}{2}$
- Type II: $N_k \le \frac{M+1}{2}$
- Type III: $N_k \le \frac{M}{2}$
- Type IV: $N_k \le \frac{M+1}{2}$

In general, the extremes of $A(\omega)$ are also extremes of $E(\omega)$. However, $E(\omega)$ presents more extremes than $A(\omega)$, as $E(\omega)$ may present extremes at the band edges, which are, in general, not extremes of $A(\omega)$. The only exception to this rule occurs at the band edges $\omega = 0$ or $\omega = \pi$, where $A(\omega)$ also presents an extreme. For instance, for a Type-I bandstop filter, as depicted in Figure 5.16, $E(\omega)$ will have up to $(\frac{M}{2} + 5)$ extremes, where $(\frac{M}{2} + 1)$ are extremes of $A(\omega)$, and the other four are band edges.

To solve the Chebyshev approximation problem, we briefly describe the Remez exchange algorithm that searches for the extreme frequencies of $E(\omega)$ through the following steps:

(i) Initialize an estimate for the extreme frequencies $\omega_0, \omega_1, \ldots, \omega_{L+1}$ by selecting $(L + 2)$ equally spaced frequencies at the bands specified for the desired filter.

(ii) Find $P(\omega_k)$ and δ such that

$$W_q(\omega_k)(D_q(\omega_k) - P(\omega_k)) = (-1)^k \delta, \text{ for } k = 0, \ldots, (L + 1) \tag{5.146}$$

This equation can be written in a matrix form and have its solution analytically calculated. Such a procedure, however, is computationally intensive (Rabiner et al., 1975). An alternative and more efficient approach computes δ by

$$\delta = \frac{a_0 D_q(\omega_0) + a_1 D_q(\omega_1) + \cdots + a_{L+1} D_q(\omega_{L+1})}{\dfrac{a_0}{W_q(\omega_0)} - \dfrac{a_1}{W_q(\omega_1)} + \cdots + \dfrac{(-1)^{L+1} a_{L+1}}{W_q(\omega_{L+1})}} \tag{5.147}$$

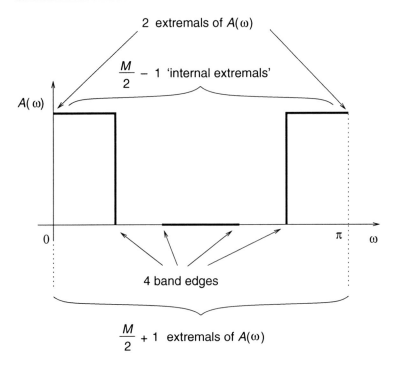

Figure 5.16 Extremes of $A(\omega)$ for a bandstop filter.

where

$$a_k = \prod_{i=0, i \neq k}^{L+1} \frac{1}{\cos \omega_k - \cos \omega_i} \qquad (5.148)$$

(iii) Use the barycentric-form Lagrange interpolator for $P(\omega)$, that is

$$P(\omega) = \begin{cases} c_k, & \text{for } \omega = \omega_0, \omega_1, \ldots, \omega_L \\[2em] \dfrac{\displaystyle\sum_{k=0}^{L} \dfrac{\beta_k}{\cos \omega - \cos \omega_k} c_k}{\displaystyle\sum_{k=0}^{L} \dfrac{\beta_k}{\cos \omega - \cos \omega_k}}, & \text{for } \omega \neq \omega_0, \omega_1, \ldots, \omega_L \end{cases} \qquad (5.149)$$

where

$$c_k = D_q(\omega_k) - (-1)^k \frac{\delta}{W_q(\omega_k)} \qquad (5.150)$$

$$\beta_k = \prod_{i=0, i \neq k}^{L} \frac{1}{\cos \omega_k - \cos \omega_i} = a_k(\cos \omega_k - \cos \omega_{L+1}) \qquad (5.151)$$

for $k = 0, \ldots, (L + 1)$.

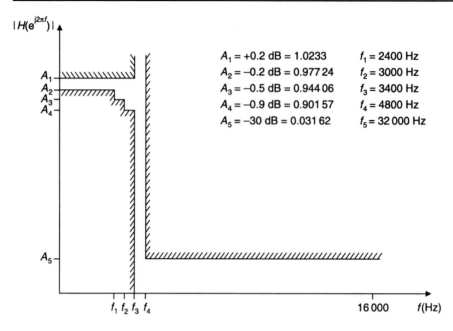

$A_1 = +0.2 \text{ dB} = 1.0233$ $f_1 = 2400 \text{ Hz}$
$A_2 = -0.2 \text{ dB} = 0.977\,24$ $f_2 = 3000 \text{ Hz}$
$A_3 = -0.5 \text{ dB} = 0.944\,06$ $f_3 = 3400 \text{ Hz}$
$A_4 = -0.9 \text{ dB} = 0.901\,57$ $f_4 = 4800 \text{ Hz}$
$A_5 = -30 \text{ dB} = 0.031\,62$ $f_5 = 32\,000 \text{ Hz}$

Figure 5.17 Lowpass PCM filter specifications.

(iv) Evaluate $|E(\omega)|$ in a dense set of frequencies. If $|E(\omega)| \leq |\delta|$ for all frequencies in the set, the optimal solution has been found, go to the next step. If $|E(\omega)| > |\delta|$ for some frequencies, a new set of candidate extremes must be chosen as the peaks of $E(\omega)$. In that manner, we force δ to grow and to converge to its upper limit. If there are more than $(L + 2)$ extremes in $E(\omega)$, keep the locations of the $(L + 2)$ largest values of the extremes of $|E(\omega)|$, making sure that the band edges are always kept, and return to step (ii).

(v) Since $P(\omega)$ is a sum of $(L + 1)$ cosines, with frequencies from zero to L, then it is also a sum of $(2L + 1)$ complex exponentials, with frequencies from $-L$ to L. Then, from Theorem 4.1, $p(l)$ can be exactly recovered by sampling $P(\omega)$ at $(2L + 1)$ equally spaced frequencies $\omega = \frac{2\pi n}{2L+1}$, $n = 0, \ldots, 2L$ and taking the inverse DFT. The resulting impulse response follows from equations (5.112), (5.114), (5.118), or (5.122), depending on the filter type.

Such an algorithm was coded in Fortran and published in McClellan et al. (1973). A few improvements to speed up the overall convergence routine are given in Antoniou (1982, 1983).

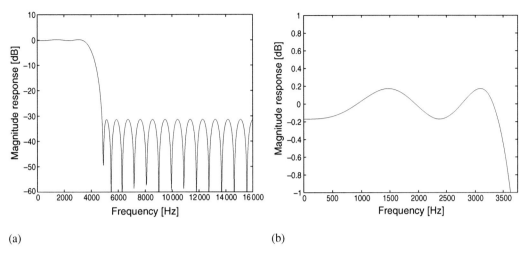

Figure 5.18 Approach 1: (a) magnitude response (b) passband detail.

EXAMPLE 5.6
Design the PCM filter specified as in Figure 5.17.

SOLUTION
Two approaches were employed:

- In the first approach, we simplify the specifications and consider a single passband with constant weight and ideal magnitude response. In this case, the specifications employed were

$$\left.\begin{array}{l} \Omega_p = 3400 \text{ Hz} \\[4pt] D(\Omega) = (1.0233 + 0.977\,24)/2 = 1.000\,27, \text{ for } 0 \leq \Omega \leq \Omega_p \\[4pt] W(\Omega) = 1, \text{ for } 0 \leq \Omega \leq \Omega_p \\[4pt] \Omega_r = 4800 \text{ Hz} \\[4pt] D(\Omega) = 0, \text{ for } \Omega_r \leq \Omega \leq \dfrac{\Omega_s}{2} \\[4pt] W(\Omega) = (1.0233 - 0.977\,24)/(2 \times 0.031\,62) \\[4pt] \qquad = 0.728\,3365, \text{ for } \Omega_r \leq \Omega \leq \dfrac{\Omega_s}{2} \\[4pt] \Omega_s = 32\,000 \text{ Hz} \end{array}\right\} \qquad (5.152)$$

The lowest filter order obtained was 34. The filter magnitude response is shown in Figure 5.18 and half of its coefficients are listed in Table 5.14. The other coefficients are obtained as $h(n) = h(34 - n)$.

Table 5.14. *Optimal PCM filter obtained with approach 1.*

h(0) to h(17)		
1.526 640 13E−02	9.259 159 97E−03	−4.738 738 01E−02
6.207 208 68E−04	2.257 899 58E−02	−7.788 436 55E−03
−6.647 366 97E−03	2.313 166 21E−02	6.739 316 76E−02
−1.384 779 84E−02	5.723 602 89E−03	1.569 395 70E−01
−1.488 200 29E−02	−2.369 204 01E−02	2.294 610 92E−01
−6.233 005 35E−03	−4.825 165 19E−02	2.572 998 25E−01

Table 5.15. *Optimal PCM filter obtained with approach 2.*

h(0) to h(14)		
−1.774 588 32E−02	1.709 248 70E−02	3.150 067 10E−04
−1.195 012 19E−02	2.797 583 84E−03	7.057 987 11E−02
2.325 175 87E−03	−2.784 688 03E−02	1.582 143 52E−01
9.884 162 09E−03	−4.495 420 64E−02	2.238 566 27E−01
2.407 499 07E−02	−4.332 983 75E−02	2.523 984 4E−01

- Considering a linear approximation for the given step-like specifications, we end up with a set of specifications described by

$$
\left.
\begin{aligned}
&\Omega_p = 3400 \text{ Hz} \\
&D(\Omega) = (1.0233 + 0.977\,24)/2 = 1.000\,27, \text{ for } 0 \leq \Omega \leq 2400 \\
&D(\Omega) = [1.000\,27 - 2.765 \times 10^{-5}(\Omega - 2400)], \\
&\qquad \text{for } 2400 \leq \Omega \leq 3000 \\
&D(\Omega) = [0.9837 - 5.3125 \times 10^{-5}(\Omega - 3000)], \\
&\qquad \text{for } 3000 \leq \Omega \leq 3400 \\
&W(\Omega) = (2.303 \times 10^{-5})/(1.0233 - D(\Omega)), \text{ for } 0 \leq \Omega \leq \Omega_p \\
&\Omega_r = 4800 \text{ Hz} \\
&D(\Omega) = 0, \text{ for } \Omega_r \leq \Omega \leq \frac{\Omega_s}{2} \\
&W(\Omega) = (1.0233 - 0.977\,24)/(2 \times 0.031\,62) \\
&\qquad = 0.728\,3365, \text{ for } \Omega_r \leq \Omega \leq \frac{\Omega_s}{2} \\
&\Omega_s = 32\,000 \text{ Hz}
\end{aligned}
\right\}
\qquad (5.153)
$$

In this case, the simplest filter obtained was of order 28. The filter magnitude response is shown in Figure 5.19, and its coefficients are listed in Table 5.15, where only half of the coefficients are shown due to filter symmetry. The remaining coefficients are determined by $h(n) = h(28 - n)$.

$\triangle$

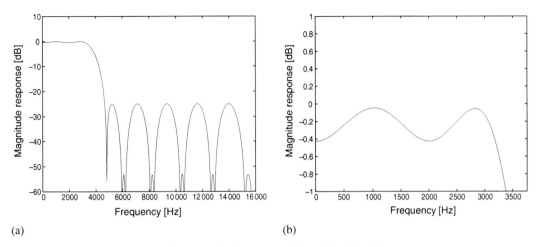

Figure 5.19 Approach 2: (a) magnitude response (b) passband detail.

5.6.3   WLS–Chebyshev method

In the standard literature, the design of FIR filters is dominated by the Chebyshev and the weighted-least-squares (WLS) approaches. Some applications that use narrowband filters, like frequency-division multiplexing for communications, do require both the minimum stopband attenuation and the total stopband energy to be considered simultaneously. For these cases, it can be shown that both the Chebyshev and WLS approaches are unsuitable as they completely disregard one of these two measurements in their objective function (Adams, 1991a,b). A solution to this problem is to combine the positive aspects of the WLS and Chebyshev methods to obtain a design procedure with good characteristics with respect to both the minimum attenuation and the total energy in the stopband.

In Lawson (1968) a scheme is derived that performs Chebyshev approximation as a limit of a special sequence of weighted-least-p (L_p) approximations, with p fixed. The particular case with $p = 2$ thus relates the Chebyshev approximation to the WLS method. The L_2 Lawson algorithm is implemented by a series of WLS approximations using a varying weight matrix $\mathbf{W}_k$, the elements of which are calculated by (Rice & Usow, 1968)

$$W_{k+1}^2(\omega) = W_k^2(\omega) B_k(\omega) \tag{5.154}$$

where

$$B_k(\omega) = |E_k(\omega)| \tag{5.155}$$

Convergence of the Lawson algorithm is slow, in practice usually 10 to 15 WLS iterations are required to approximate the Chebyshev solution. An efficiently accelerated version of the Lawson algorithm was presented in Lim et al. (1992). The

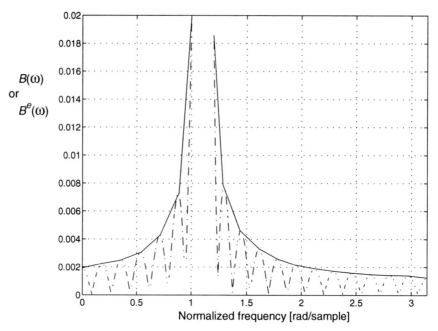

Figure 5.20 Typical absolute error function $B(\omega)$ (dash-dotted line) and corresponding envelope $B^e(\omega)$ (solid curve).

Lim–Lee–Chen–Yang (LLCY) approach is characterized by the weight matrix $\mathbf{W}_k$ recurrently updated by

$$W_{k+1}^2(\omega) = W_k^2(\omega)B_k^e(\omega) \tag{5.156}$$

where $B_k^e(\omega)$ is the envelope function of $B_k(\omega)$ composed by a set of piecewise linear segments that start and end at consecutive extremes of $B_k(\omega)$. Band edges are considered extreme frequencies, and edges from different bands are not connected. In that manner, labeling the extreme frequencies at a particular iteration k as ω_j^*, for $J \in \mathbb{N}$, the envelope function is formed as (Lim et al., 1992)

$$B_k^e(\omega) = \frac{(\omega - \omega_j^*)B_k(\omega_{J+1}^*) + (\omega_{J+1}^* - \omega)B_k(\omega_j^*)}{(\omega_{J+1}^* - \omega_j^*)}; \quad \omega_j^* \le \omega \le \omega_{J+1}^* \tag{5.157}$$

Figure 5.20 depicts a typical format of the absolute value of the error function (dash-dotted curve), at any particular iteration, used by the Lawson algorithm to update the weighting function, and its corresponding envelope (solid curve) used by the LLCY algorithm.

Comparing the adjustments used by the Lawson and LLCY algorithms, described in equations (5.154)–(5.157), and seen in Figure 5.20, with the piecewise-constant weight function used by the WLS method, one can devise a very simple approach for designing digital filters that compromises the minimax and WLS constraints. The approach consists of a modification of the weight-function updating procedure so that it becomes constant after a particular extreme of the stopband of $B_k(\omega)$, that is, (Diniz & Netto, 1999)

$$W_{k+1}^2(\omega) = W_k^2(\omega)\beta_k(\omega) \tag{5.158}$$

where, for the modified-Lawson algorithm, $\beta_k(\omega)$ is defined as

$$\beta_k(\omega) \equiv \tilde{B}_k(\omega) = \begin{cases} B_k(\omega), & 0 \le \omega \le \omega_J^* \\ B_k(\omega_J^*), & \omega_J^* < \omega \le \pi \end{cases} \tag{5.159}$$

and for the modified-LLCY algorithm, $\beta_k(\omega)$ is given by

$$\beta_k(\omega) \equiv \tilde{B}_k^e(\omega) = \begin{cases} B_k^e(\omega), & 0 \le \omega \le \omega_J^* \\ B_k^e(\omega_J^*), & \omega_J^* < \omega \le \pi \end{cases} \tag{5.160}$$

where ω_J^* is the Jth extreme value of the stopband of $B(\omega) = |E(\omega)|$. The passband values of $B(\omega)$ and $B^e(\omega)$ are left unchanged in equations (5.159) and (5.160) to preserve the equiripple property of the minimax method. The parameter J is the single design parameter for the WLS–Chebyshev scheme. Choosing $J = 1$, makes the new scheme similar to an equiripple-passband WLS design. On the other hand, choosing J as large as possible, that is, making $\omega_J^* = \pi$, turns the design method into the Lawson or the LLCY schemes.

An example of the new approach being applied to the generic functions seen in Figure 5.20 is depicted in Figure 5.21, where ω_J^* was chosen as the fifth extreme in the filter stopband.

The computational complexity of WLS-based algorithms, like the algorithms described here, is of the order of N^3, where N is the length of the filter. This burden, however, can be greatly reduced by taking advantage of the Toeplitz-plus-Hankel internal structure of the matrix $(\mathbf{U}^T\mathbf{W}^2\mathbf{U})$, as discussed in Merchant & Parks (1982), and by using an efficient grid scheme to minimize the number of frequency values, as described in Yang & Lim (1991, 1993, 1996). These simplifications make the computational complexity of WLS-based algorithms comparable to that for the minimax approach. The WLS-based methods, however, do have the additional advantage of being easily coded into computer routines.

The overall implementation of the WLS–Chebyshev algorithm is as follows:

(i)   Estimate the order M, select $8 \le K \le 16$, the maximum number of iterations $k_{\max}$, the value of J, and some small error tolerance $\epsilon > 0$.

(ii)   Create a frequency grid of KM points within $[0, \pi]$. For linear-phase filters, points in the transition band should be discarded and the remaining points redistributed in the interval $\omega \in [0, \omega_p] \cup [\omega_r, \pi]$.

(iii)   Set $k = 0$ and form $\mathbf{W}_q$, $\mathbf{d}_q$, and $\mathbf{U}$, as defined in equations (5.133)–(5.135), based on Table 5.12.

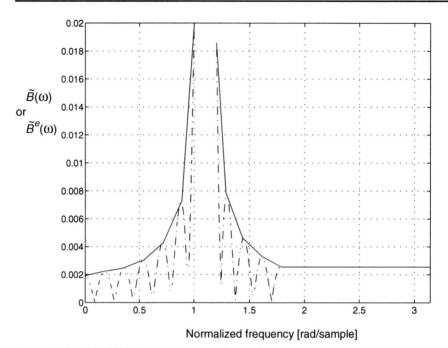

Figure 5.21 WLS–Chebyshev approach applied to the functions in Figure 5.20. Modified-Lawson algorithm $\tilde{B}(\omega)$ (dash-dotted curve) and modified-LLCY algorithm $\tilde{B}^e(\omega)$ (solid curve). The curves coincide for $\omega \geq \omega_5^*$.

(iv) Set $k = k + 1$, and determine $\mathbf{p}^*(k)$ from equation (5.141).

(v) Determine the error vector $\mathbf{e}(k)$ as given in equation (5.131), always using $\mathbf{W}_q$ corresponding to $k = 0$.

(vi) Check if $k > k_{\max}$ or if convergence has been achieved via, for instance, the criterion $\| \mathbf{e}(k) \| - \| \mathbf{e}(k-1) \| \leq \epsilon$. If so, then go to step (x).

(vii) Compute $B_k = |\mathbf{e}(k)|$ and B_k^e as the envelope of B_k.

(viii) Find the Jth stopband extreme of B_k^e. For a lowpass filter, consider the interval $\omega \in [\omega_r, \pi]$, starting at ω_r. For a highpass filter, search in the interval $\omega \in [0, \omega_r]$, starting at ω_r. For bandpass filters, consider the intervals $\omega \in [0, \omega_{r_1}]$, starting at ω_{r_1}, and $\omega \in [\omega_{r_2}, \pi]$, starting at ω_{r_2}. For the bandstop filter, look for the extreme in the interval $\omega \in [\omega_{r_1}, \omega_{r_2}]$, starting at both ω_{r_1} and ω_{r_2}.

(ix) Update $\mathbf{W}_q^2$ using either equation (5.159) or equation (5.160), and go back to step (iv).

(x) Determine the set of coefficients $h(n)$ of the linear-phase filter, and verify that the specifications are satisfied. If so, then decrease the filter order M and repeat the above procedure starting from step (ii). The best filter would be the one obtained at the iteration just before the specifications are not met. If the specifications are not satisfied at the first attempt, then increase the value of M

Table 5.16. *Coefficients of the bandpass filter designed with the WLS–Chebyshev method with* $J = 3$.

	$h(0)$ to $h(20)$	
$-4.529\,844\,99\text{E}-03$	$4.229\,044\,35\text{E}-18$	$-9.365\,213\,97\text{E}-02$
$-3.020\,444\,35\text{E}-18$	$1.561\,620\,96\text{E}-02$	$2.266\,071\,92\text{E}-17$
$6.228\,798\,87\text{E}-03$	$-4.017\,125\,23\text{E}-18$	$1.203\,500\,35\text{E}-01$
$2.826\,902\,86\text{E}-18$	$-3.692\,099\,90\text{E}-02$	$-1.782\,720\,69\text{E}-17$
$-4.871\,530\,56\text{E}-03$	$1.334\,835\,29\text{E}-17$	$-1.389\,891\,75\text{E}-01$
$-2.966\,004\,22\text{E}-18$	$6.417\,020\,81\text{E}-02$	$4.345\,069\,66\text{E}-18$
$-1.801\,344\,68\text{E}-03$	$-1.859\,104\,66\text{E}-17$	$1.457\,035\,85\text{E}-01$

and repeat the above procedure once again starting at step (ii). In this case, the best filter would be the one obtained when the specifications are first met.

EXAMPLE 5.7
Design a bandpass filter satisfying the specification below using the WLS and Chebyshev methods and discuss the results using the WLS–Chebyshev approach.

$$\left.\begin{array}{l} M = 40 \\ A_p = 1.0 \text{ dB} \\ \Omega_{r_1} = \frac{\pi}{2} - 0.4 \text{ rad/s} \\ \Omega_{p_1} = \frac{\pi}{2} - 0.1 \text{ rad/s} \\ \Omega_{p_2} = \frac{\pi}{2} + 0.1 \text{ rad/s} \\ \Omega_{r_2} = \frac{\pi}{2} + 0.4 \text{ rad/s} \\ \Omega_s = 2\pi \text{ rad/s} \end{array}\right\} \qquad (5.161)$$

SOLUTION
The magnitude responses for $J = 10$ (Chebyshev), $J = 5$, $J = 3$, and $J = 1$ (WLS) are shown in Figure 5.22. Table 5.16 shows half of the filter coefficients for the case when $J = 3$. The remaining coefficients are obtained as $h(n) = h(40 - n)$. Figure 5.23 shows the tradeoff between the stopband minimum attenuation and total stopband energy when J varies from 1 to 10. Notice how the two extremes correspond to optimal values for the attenuation and energy figures of merit, respectively. On the other hand, the same extremes are also the worst case scenarios for the energy and the attenuation in the stopband. In this example, a good compromise between the two measures can be obtained when $J = 3$.

$\triangle$

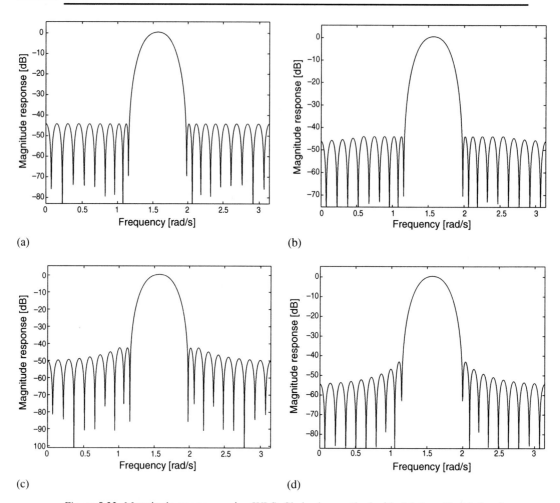

(a)

(b)

(c)

(d)

Figure 5.22 Magnitude responses using WLS–Chebyshev method with: (a) $J = 10$; (b) $J = 5$; (c) $J = 3$; (d) $J = 1$.

5.7   FIR filter approximation with MATLAB

MATLAB has the two specific functions described below to perform the discrete-time differentiation and the Hilbert transformation. However, if real-time processing is required, such operations must be implemented as a digital filter, whose approximation should be performed as described in this chapter.

- `diff`: Performs the difference between consecutive entries of a vector. It can be used to approximate the differentiation.

  Input parameter: A data vector `x`.

  Output parameter: A vector `h` with the differences.

  Example: `x=sin(0:0.01:pi); h=diff(x)/0.01;`

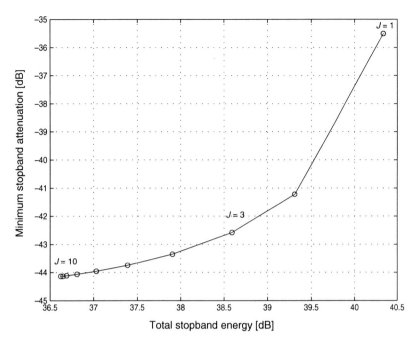

Figure 5.23 Tradeoff between the minimum stopband attenuation and total stopband energy using the WLS–Chebyshev method.

- `hilbert`: Performs the Hilbert transform on the real part of a vector. The result is a complex vector, whose real part is the original data, and whose imaginary part is the resulting transform.

  Input parameter: A data vector `x`.

  Output parameter: A vector `H` with the Hilbert transform.

  Example:
  ```
  x=rand(30:1); H=hilbert(x);
  ```

MATLAB also has a series of commands that are useful for solving the approximation problem of FIR digital filters. More specifically, the window-based, the weighted-least-squares, and the Chebyshev methods are easily implemented in MATLAB with the help of the following commands.

MATLAB commands related to the window method

- `fir1`: Designs standard (lowpass, highpass, bandpass, and bandstop) FIR filters using the window method.

  Input parameters:

  - The filter order `M`. For highpass and bandstop filters this value must be even. In such cases, if an odd value is provided, MATLAB increments it by 1;
  - A vector `f` of band edges. If `f` has one element, the filter is lowpass or highpass, and the given value becomes the passband edge. If `f` has two elements, the

filter is either bandpass, where the two values become the passband edges, or bandstop, where the two values become the stopband edges. A larger **f** corresponds to multiband filters;

– A string specifying the standard filter type. The default value is associated with lowpass and bandpass filters, according to the size of **f**. **'high'** indicates that the filter is highpass; **'stop'** indicates that the filter is bandstop; **'DC-1'** indicates that the first band of a multiband filter is a passband; **'DC-0'** indicates that the first band of a multiband filter is a stopband;

– The default window type used by **fir1** is the Hamming window. The user can change it, if desired, by using one of the window commands seen below.

Output parameter: A vector **h** containing the filter coefficients.
Example:

```
M=40; f=0.2;
h=fir1(M,f,'high',chebwin(M+1,30));
```

- **fir2**: Designs arbitrary-response FIR filters using the window method.
  Input parameters:

  – The filter order **M**;
  – A frequency vector **f** specifying the filter bands;
  – A magnitude response vector **m**, the same size as **f**;
  – A string specifying the window type, as in the case of the **fir1** command.

  Output parameter: A vector **h** containing the filter coefficients.

- **boxcar**: Determines the rectangular window function.
  Input parameter: The window length **N=M+1**.
  Output parameter: A vector **wr** containing the window.
  Example:

  ```
  N=11; wr=boxcar(N);
  ```

- **triang**: Determines the triangular window function.
  Input parameter: The window length **N=M+1**. If this value is even, the direct relationship between the triangular and Bartlett windows disappears.
  Output parameter: A vector **wt** containing the window.
  Example:

  ```
  N=20; wt=triang(N);
  ```

- **bartlett**: Determines the Bartlett window function.
  Input parameter: The window length **N=M+1**.
  Output parameter: A vector **wtB** containing the window.
  Example:

  ```
  N=10; wtB=bartlett(N);
  ```

- `hamming`: Determines the Hamming window function.
  Input parameter: The window length N=M+1.
  Output parameter: A vector wH containing the window.
  Example:
      N=31; wH=hamming(N);

- `hanning`: Determines the Hanning window function.
  Input parameter: The window length N=M+1.
  Output parameter: A vector wHn containing the window.
  Example:
      N=18; wHn=hanning(N);

- `blackman`: Determines the Blackman window function.
  Input parameter: The window length N=M+1.
  Output parameter: A vector wB containing the window.
  Example:
      N=49; wB=blackman(N);

- `kaiser`: Determines the Kaiser window function.
  Input parameters: The window length N=M+1 and the auxiliary parameter beta, as determined in equation (5.72).
  Output parameter: A vector wK containing the window.
  Example:
      N=23; beta=4.1;
      wK=kaiser(N,beta);

- `kaiserord`: Estimates the order of the filter designed with the Kaiser window (see Exercise 5.19). This command is suited for use with the firl command with the Kaiser window.
  Input parameters:

  - A vector f of band edges;
  - A vector a of desired amplitudes in the bands defined by f;
  - A vector, the same size as f and a, that specifies the maximum error in each band between the desired amplitude and the resulting amplitude of the designed filter;
  - The sampling frequency Fs.

  Output parameter: The order of the Kaiser window.
  Example:
      f=[100 200]; a=[1 0]; err=[0.1 0.01]; Fs=800;
      M=kaiserord(f,a,err,Fs);

- `chebwin`: Determines the Dolph–Chebyshev window function.
  Input parameters: The window length N=M+1 and the ripple ratio r in dB,

$20 \log(\frac{\delta_p}{\delta_r})$.

Output parameter: A vector wDC containing the window.

Example:

```
N=51; r=20;
wDC=chebwin(N,r);
```

MATLAB commands related to the weighted-least-squares method

- firls: Designs linear-phase FIR filters using the weighted-least-squares method.
  Input parameters:

  - The filter order M;
  - A vector f of pairs of normalized frequency points between 0 and 1;
  - A vector a containing the desired amplitudes of the points specified in f;
  - A vector w, half of the size of f and a, of weights for each band specified in f;
  - A string specifying the filter type other than standard ones. The options 'dif-ferentiator' and 'hilbert' can be used to design a differentiator or a Hilbert transformer, respectively.

  Output parameter: A vector h containing the filter coefficients.
  Example:

  ```
  M=40; f=[0 0.4 0.6 0.9]; a=[0 1 0.5 0.5]; w=[1 2];
  h=firls(M,f,a,w);
  ```

- fircls: Designs multiband FIR filters with the constrained-least-squares method.
  Input parameters:

  - The filter order M;
  - A vector f of normalized band edges starting with 0 and ending with 1, necessarily;
  - A vector a describing the piecewise constant desired amplitude response. The length of a is the number of prescribed bands, that is, length(f) − 1;
  - Two vectors, up and lo, with the same length as a, defining the upper and lower bounds for the magnitude response in each band.

  Output parameter: A vector h containing the filter coefficients.
  Example:

  ```
  M=40; f=[0 0.4 0.6 1]; a=[1 0 1];
  up=[1.02 0.02 1.02]; lo=[0.98 -0.02 0.98];
  h=fircls(M,f,a,up,lo);
  ```

- fircls1: Designs lowpass and highpass linear-phase FIR filters with the constrained-least-squares method.
  Input parameters:

- The filter order M;
- The passband normalized frequency wp;
- The passband ripple dp;
- The stopband ripple dr;
- The stopband normalized frequency wr;
- A string 'high' to indicate that the filter is highpass, if this is the case.

Output parameter: A vector h containing the filter coefficients.

Example:

```
M=30; wp=0.4; dp=0.1; dr=0.01; wr=0.5;
h=fircls1(M,wp,dp,dr,wr);
```

MATLAB commands related to the Chebyshev optimal method

- remez: Designs a linear-phase FIR filter using the Parks–McClellan algorithm (McClellan et al., 1973).
  Input parameters: As for the command firls.
  Output parameter: A vector h containing the filter coefficients.
  Example:

  ```
  M=40; f=[0 0.4 0.6 0.9]; a=[0 1 0.5 0.5]; w=[1 2];
  h=remez(M,f,a,w);
  ```

- cremez: Generalizes the remez command for complex and nonlinear-phase FIR filters.
  Input parameters: There are several possibilities for using the cremez command. One of them is similar to the one for the remez and firls commands (see example below), where f, for the cremez command, should be specified in the range −1 to 1.
  Output parameter: A vector h containing the filter coefficients.
  Example:

  ```
  M=30; f=[-1 -0.4 0.6 0.9]; a=[0 1 0.5 0.5]; w=[2 1];
  h=cremez(M,f,a,w);
  ```

- remezord: Estimates the order of the filter designed with the Chebyshev method (see Exercise 5.20). This command is perfectly suited to the remez command.
  Input parameters:

  - A vector f of band edges;
  - A vector a of desired amplitudes in the bands defined by f. The length of f in this case is twice the length of a, minus 2;
  - A vector, the same size as a, that specifies the maximum error in each band between the desired amplitude and the resulting amplitude of the designed filter;
  - The sampling frequency Fs.

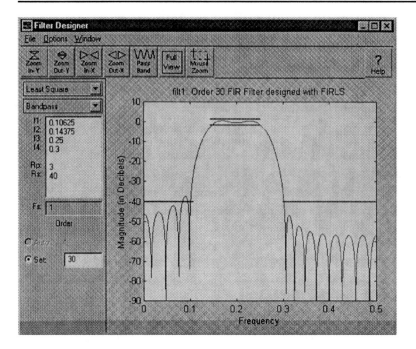

Figure 5.24 Filter specification using `sptool` in MATLAB.

Output parameter: The order M of the filter.

Example:

```
f=[400 500]; a=[1 0]; err=[0.1 0.01]; Fs=2000;
M=remezord(f,a,err,Fs);
```

In versions 5.0 or higher, MATLAB provides a filter design tool along with the Signal Processing toolbox. Such a tool is invoked with the command `sptool` presenting a very friendly graphic user's interface for filter design. In that manner, defining the filter specifications becomes a very easy task, as depicted in Figure 5.24, and analyzing the resulting filter characteristics also becomes straightforward.

Possible designs can be based on the window approach, the Chebyshev method, or the least-squares approach. Filter types include solely the standard lowpass, highpass, bandpass, and bandstop types. Possible analyses are the magnitude, phase, and group delay responses, in the frequency domain, and the impulse and step response, in the time domain, as well as a clear visualization of the filter's zero-pole diagram, as seen, for instance, in Figure 5.25.

5.8 Summary

The subject of FIR filter design is very extensive and could be, in its own right, the subject of a complete textbook. Among the methods studied in the present chapter

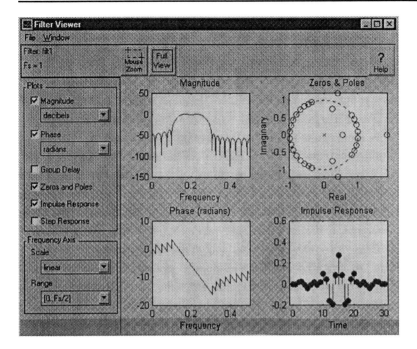

Figure 5.25 Filter analyses using `sptool` in MATLAB.

for designing FIR filters, we focused on the frequency sampling, window-based, maximally flat, weighted-least-squares (WLS), Chebyshev, and WLS–Chebyshev approaches.

The frequency sampling method consists of performing an inverse DFT over a set of samples of a desired frequency response. Its implementation is very simple, but the results tend to be poor, especially when very sharp transition bands are involved. Despite this drawback, the frequency sampling method is very useful when the desired frequency response is composed of a sum of complex sinusoids.

The window method consists of truncating the Fourier series of a given desired magnitude response through the use of a so-called window function. Simple window functions, the rectangular, triangular, Blackman, and so on, do not allow designs satisfying prescribed specifications, in general. More sophisticated window functions, such as the Dolph–Chebyshev and Kaiser windows, are able to control the passband and stopband ripples simultaneously, thus enabling FIR filters to be designed that satisfy prescribed specifications. The subject of window functions is very rich. Other window functions, distinct from the ones seen here, can be found for instance in Nuttall (1981); Webster (1985); Ha & Pearce (1989); Adams (1991b); Yang & Ke (1992); Saramäki (1993); Kay & Smith (1999).

The maximally flat filter design generates lowpass and highpass FIR filters with very flat passband and stopband. The method is extremely simple, although it is only suitable for low-order filters, because the filter coefficients tend to present a very large dynamic range as the order increases.

In addition, a unified framework for studying numerical methods for FIR filter design was given. In this context, we presented the weighted-least-squares (WLS), the Chebyshev (or minimax), and the WLS–Chebyshev schemes. The former minimizes the stopband total energy for a given passband energy level. The second method, which is commonly implemented with the Remez exchange algorithm, is able to minimize the maximum error between the designed and the desired responses. There are two interesting cases where the optimal solution for this problem presents a closed-form solution: when the passband is equiripple and the stopband is monotonically decreasing, or when the passband is monotonically decreasing and the stopband is equiripple. The interested reader may refer to (Saramäki, 1993) to investigate this subject further. The WLS–Chebyshev was introduced as a numerical method able to unite the good performances of the WLS and Chebyshev methods with respect to the total energy level and minimum attenuation level in the stopband, simultaneously.

Finally, the FIR filter design problem was analyzed using the MATLAB software tool. MATLAB was shown to present a series of commands for designing FIR filters. Although the WLS–Chebyshev method is not part of any MATLAB toolbox, we have supplied a description of it in pseudo-code (Subsection 5.6.3), so that it can be easily implemented using any version of MATLAB. It must be emphasized that a very powerful built-in graphic user interface exists in MATLAB that allows one to perform several filter designs with only a few clicks of the mouse.

5.9   Exercises

5.1   Assume that a periodic signal has four sinusoidal components at frequencies $\omega_0, 2\omega_0, 4\omega_0, 6\omega_0$. Design a nonrecursive filter, as simple as possible, that eliminates only the components $2\omega_0, 4\omega_0, 6\omega_0$.

5.2   Given a lowpass FIR filter with transfer function $H(z)$, describe what happens to the filter frequency response when:

(a) z is replaced by $-z$.
(b) z is replaced by z^{-1}.
(c) z is replaced by z^2.

5.3   Complementary filters are such that their frequency responses add to a delay. Given an Mth-order linear-phase FIR filter with transfer function $H(z)$, deduce the conditions on L and M such that the overall filter shown in Figure 5.26 is complementary to $H(z)$.

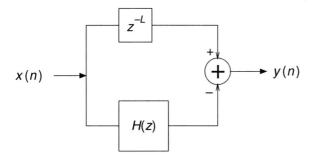

Figure 5.26 Overall filter block diagram.

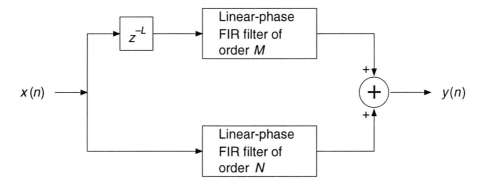

Figure 5.27 Overall filter block diagram.

5.4   Determine the relationship between L, M, and N which means that the overall filter in Figure 5.27 has linear phase.

5.5   Design a highpass filter satisfying the specification below using the frequency sampling method:

$M = 40$
$\Omega_r = 1.0 \text{ rad/s}$
$\Omega_p = 1.5 \text{ rad/s}$
$\Omega_s = 5.0 \text{ rad/s}$

5.6   Plot and compare the characteristics of the Hamming window and the corresponding magnitude response for $M = 5, 10, 15, 20$.

5.7   Plot and compare the rectangular, triangular, Bartlett, Hamming, Hanning, and Blackman window functions and the corresponding magnitude responses for $M = 20$.

5.8   Determine the ideal impulse response associated with the magnitude response shown in Figure 5.28, and compute the corresponding practical filter of orders $M = 10, 20, 30$ using the Hamming window.

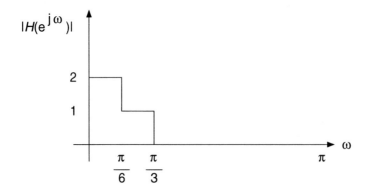

Figure 5.28 Ideal magnitude response.

5.9   Design a bandpass filter satisfying the specification below using the Hamming, Hanning, and Blackman windows:

$M = 10$
$\Omega_{c_1} = 1.125$ rad/s
$\Omega_{c_2} = 2.5$ rad/s
$\Omega_s = 10$ rad/s

5.10   Plot and compare the characteristics of the Kaiser window function and corresponding magnitude response for $M = 20$ and different values of β.

5.11   Design the following filters using the Kaiser window:

(a) $A_p = 1.0$ dB
     $A_r = 40$ dB
     $\Omega_p = 1000$ rad/s
     $\Omega_r = 1200$ rad/s
     $\Omega_s = 5000$ rad/s

(b) $A_p = 1.0$ dB
     $A_r = 40$ dB
     $\Omega_r = 1000$ rad/s
     $\Omega_p = 1200$ rad/s
     $\Omega_s = 5000$ rad/s

    (c) $A_p = 1.0$ dB

       $A_r = 50$ dB

       $\Omega_{r_1} = 800$ rad/s

       $\Omega_{p_1} = 1000$ rad/s

       $\Omega_{p_2} = 1100$ rad/s

       $\Omega_{r_2} = 1400$ rad/s

       $\Omega_s = 10\,000$ rad/s

5.12   Determine a complete procedure for designing differentiators using the Kaiser window.

5.13   Repeat Exercise 5.11(a) using the Dolph–Chebyshev window. Compare the transition bandwidths and the stopband attenuation levels for the two resulting filters.

5.14   Design a maximally flat lowpass filter satisfying the specification below:

$\omega_c = 0.4\pi$ rad/sample

$T_r = 0.2\pi$ rad/sample

where T_r is the transition band as defined in Section 5.5. Find also the direct-form coefficients $h(n)$, for $n = 0, \ldots, M$, and the alternative coefficients $\hat{d}(n)$, for $n = 0, \ldots, K - 1$.

5.15   (a) Show that Type-II linear-phase FIR filters can be put in the form of equation (5.115), by substituting equation (5.114) into equation (5.113).

       (b) Repeat item (a) for Type-III filters, showing that equation (5.119) can be obtained from substituting equation (5.118) into equation (5.117).

       (c) Repeat item (a) for Type-IV filters, showing that equation (5.123) can be obtained from substituting equation (5.122) into equation (5.121).

5.16   Design three narrowband filters, centered on the frequencies 770 Hz, 852 Hz, and 941 Hz, satisfying the specification below, using the minimax approach:

$M = 98$

$\Omega_s = 2\pi \times 5$ kHz

5.17   Using the minimax approach, design a bandpass filter for tone detection with a center frequency of 700 Hz, given that the sampling frequency is 8000 Hz and the order is 95. Use the following band parameters:

- Band 1
  - Edges: 0 and 555.2 Hz

-   – Objective: 0
-   – Weight: 1
-   • Band 2
    -   – Edges: 699.5 Hz and 700.5 Hz
    -   – Objective: 1
    -   – Weight: 1
-   • Band 3
    -   – Edges: 844.8 Hz and 4000 Hz
    -   – Objective: 0
    -   – Weight: 1

5.18   Design a Hilbert transformer of order $M = 98$ using a Type-IV structure.

5.19   The following relationship estimates the order of a lowpass filter designed with the minimax approach (Rabiner et al., 1975). Design a series of lowpass filters and verify the validity of this estimate (Ω_s is the sampling frequency):

$$M \approx \frac{D_\infty(\delta_p, \delta_r) - f(\delta_p, \delta_r)(\Delta F)^2}{\Delta F} + 1$$

where

$$D_\infty(\delta_p, \delta_r) = \{0.005\,309[\log_{10}(\delta_p)]^2 + 0.071\,140[\log_{10}(\delta_p)] - 0.4761\}\log_{10}(\delta_r)$$
$$\qquad - \{0.002\,660[\log_{10}(\delta_p)]^2 + 0.594\,100[\log_{10}(\delta_p)] + 0.4278\}$$
$$\Delta F = \frac{\Omega_r - \Omega_p}{\Omega_s}$$
$$f(\delta_p, \delta_r) = 11.012 + 0.512\,44[\log_{10}(\delta_p) - \log_{10}(\delta_r)]$$

5.20   Repeat Exercise 5.19 with the following order estimate (Kaiser, 1974):

$$M \approx \frac{-20\log_{10}\left(\sqrt{\delta_p\delta_r}\right) - 13}{2.3237\left(\omega_r - \omega_p\right)} + 1$$

where ω_p and ω_r are the digital passband and stopband edges. Which estimate tends to be more accurate?

5.21   Design a bandpass filter satisfying the specification below using the WLS and

Chebyshev methods. Discuss the tradeoff between the stopband minimum attenuation and total stopband energy when using the WLS–Chebyshev scheme.

$M = 50$

$\Omega_{r_1} = 100 \text{ rad/s}$

$\Omega_{p_1} = 150 \text{ rad/s}$

$\Omega_{p_2} = 200 \text{ rad/s}$

$\Omega_{r_2} = 300 \text{ rad/s}$

$\Omega_s = 1000 \text{ rad/s}$

6 IIR filter approximations

6.1 Introduction

This chapter deals with the design methods in which a desired frequency response is approximated by a transfer function consisting of a ratio of polynomials. In general, this type of transfer function yields an impulse response of infinite duration. Therefore, the systems approximated in this chapter are commonly referred to as infinite-duration impulse-response (IIR) filters.

In general, IIR filters are able to approximate a prescribed frequency response with fewer multiplications than FIR filters. For that matter, IIR filters can be more suitable for some practical applications, especially those ones involving real-time signal processing.

In Section 6.2, we study the classical methods of analog filter approximation, namely the Butterworth, Chebyshev, and elliptic approximations. These methods are the most widely used for approximations meeting prescribed magnitude specifications. They originated in the continuous-time domain and their use in the discrete-time domain requires an appropriate transformation.

We then address, in Section 6.3, two approaches that transform a continuous-time transfer function into a discrete-time transfer function, the impulse-invariance and bilinear transformation methods.

Section 6.4 deals with frequency transformation methods in the discrete-time domain. These methods allow the mapping of a given filter type to another, for example the transformation of a given lowpass into a desired bandpass filter.

In applications where magnitude and phase specifications are imposed, we can approximate the desired magnitude specifications by one of the classical transfer functions and design a phase equalizer to meet the phase specifications. As an alternative, we can carry out the design entirely in the digital domain, by using optimization methods to design transfer functions satisfying the magnitude and phase specifications simultaneously. Section 6.5 covers a procedure to approximate a given frequency response iteratively, employing a nonlinear optimization algorithm.

In Section 6.6, we address the situations where an IIR digital filter must present an impulse response similar to a given discrete-time sequence. This problem is commonly known as time-domain approximation.

Finally, in Section 6.7, the role of MATLAB in the approximation of IIR filters is briefly discussed.

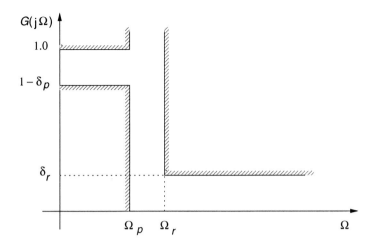

Figure 6.1 Typical gain specifications of a lowpass filter.

6.2   Analog filter approximations

This section covers the classical approximations for normalized-lowpass analog filters.[1] The other types of filters, such as the denormalized-lowpass, highpass, bandstop, and bandpass filters, are obtained from the normalized-lowpass prototype through frequency transformations, which are also addressed in this section.

6.2.1   Analog filter specification

An important step in the design of an analog filter is the definition of the desired magnitude and/or phase specifications that should be satisfied by the filter frequency response. Usually, a classical analog filter is specified through a region of the $\Omega \times H(j\Omega)$ plane[2] where its frequency response must be contained. This is illustrated in Figure 6.1 for a lowpass filter. In this figure, Ω_p and Ω_r denote the passband and stopband edge frequencies, respectively. The frequency region between Ω_p and Ω_r is the so-called transition band where no specification is provided. In addition, the maximum ripples in the passband and the stopband are denoted by δ_p and δ_r, respectively.

Alternatively, the specifications can be given in decibels (dB), as shown in Figure 6.2a, in the case of gain specifications. Figure 6.2b shows the same filter specified in terms of attenuation instead of gain. The relationships between the parameters of these three representations are given in Table 6.1.

---

[1] Normalized filters are derived from standard ones through a simple variable scaling. The original filter is then determined by reversing the frequency transformation previously applied. In this section, to avoid any source of confusion, a normalized analog frequency is always denoted by a primed variable such as Ω'.

[2] Note that once more Ω usually refers to analog frequency and ω to digital frequency.

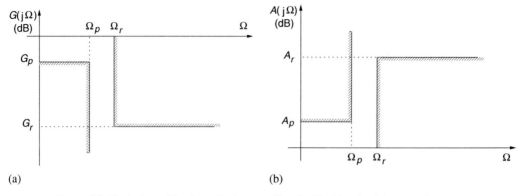

Figure 6.2 Typical specifications of a lowpass filter in dB: (a) gain; (b) attenuation.

Table 6.1. *Relationships among the parameters for the gain, gain in dB, and attenuation in dB specification formats.*

	Ripple	Gain [dB]	Attenuation [dB]
Passband	δ_p	$G_p = 20\log_{10}(1-\delta_p)$	$A_p = -G_p$
Stopband	δ_r	$G_r = 20\log_{10}\delta_r$	$A_r = -G_r$

For historical reasons, in this chapter, we work with the attenuation specifications in dB. Using the relationships given in Table 6.1, readers should be able to transform any other format into the set of parameters that characterize the attenuation in dB.

6.2.2 Butterworth approximation

Usually, the attenuation of an all-pole normalized-lowpass filter is expressed by an equation of the following type:

$$\left|A(j\Omega')\right|^2 = 1 + \left|E(j\Omega')\right|^2 \tag{6.1}$$

where $A(s')$ is the desired attenuation function and $E(s')$ is a polynomial which has low magnitude at low frequencies and large magnitude at high frequencies.

The Butterworth approximation is characterized by a maximally flat magnitude response at $\Omega' = 0$. In order to achieve this property, we choose $E(j\Omega')$ as

$$E(j\Omega') = \epsilon\left(j\Omega'\right)^n \tag{6.2}$$

where ϵ is a constant and n is the filter order. Equation (6.1) then becomes

$$\left|A(j\Omega')\right|^2 = 1 + \epsilon^2\left(\Omega'\right)^{2n} \tag{6.3}$$

resulting in the fact that the first $(2n-1)$ derivatives of the attenuation function at $\Omega' = 0$ are equal to zero, as desired in the Butterworth approximation.

The choice of the parameter ϵ depends on the maximum attenuation A_p allowed in the passband. In that manner, since

$$A_{\mathrm{dB}}(\Omega') = 20 \log_{10} |A(j\Omega')| = 10 \log_{10} \left[1 + \epsilon^2 (\Omega')^{2n} \right] \tag{6.4}$$

at $\Omega' = 1$, we must have that

$$A_p = A_{\mathrm{dB}}(1) = 10 \log_{10} \left(1 + \epsilon^2 \right) \tag{6.5}$$

and then

$$\epsilon = \sqrt{10^{0.1 A_p} - 1} \tag{6.6}$$

To determine the filter order required to meet the attenuation specification, A_r, in the stopband, at $\Omega' = \Omega'_r$ we must have that

$$A_r = A_{\mathrm{dB}}(\Omega'_r) = 10 \log_{10} \left[1 + \epsilon^2 (\Omega'_r)^{2n} \right] \tag{6.7}$$

Therefore, n should be smallest integer such that

$$n \geq \frac{\log_{10} \left(\frac{10^{0.1 A_r} - 1}{\epsilon^2} \right)}{2 \log_{10} \Omega'_r} \tag{6.8}$$

with ϵ as in equation (6.6).

With n and ϵ available, one has to find the transfer function $A(s')$. We can factor $|A(j\Omega')|^2$ in equation (6.3) as

$$|A(j\Omega')|^2 = A(-j\Omega')A(j\Omega') = 1 + \epsilon^2 \Omega'^{2n} = 1 + \epsilon^2 [-(j\Omega')^2]^n \tag{6.9}$$

Using the analytical continuation for complex variables (Churchill, 1975), that is, replacing $j\Omega'$ by s', we have that

$$A(s')A(-s') = 1 + \epsilon^2 (-s'^2)^n \tag{6.10}$$

In order to determine $A(s')$, we must then find the roots of $[1 + \epsilon^2 (-s'^2)^n]$ and then choose which ones belong to $A(s')$ and which ones belong to $A(-s')$. The solutions of

$$1 + \epsilon^2 (-s'^2)^n = 0 \tag{6.11}$$

are

$$s_i = \frac{1}{\epsilon^{\frac{1}{n}}} e^{j\frac{\pi}{2} \left(\frac{2i+n+1}{n} \right)} \tag{6.12}$$

with $i = 1, 2, \ldots, 2n$. These $2n$ roots are located at equally spaced positions on the circumference of radius $\epsilon^{\frac{1}{n}}$ centered at the origin of the s plane. In order to obtain a stable filter, we choose the n roots p_i on the left-hand side of the s plane to belong to the polynomial $A(s')$. As a result, the normalized transfer function is obtained as

$$H'(s') = \frac{H'_0}{A(s')} = \frac{H'_0}{\prod\limits_{i=1}^{n}(s' - p_i)} \tag{6.13}$$

where H'_0 is chosen so that $|H'(j0)| = 1$, and thus

$$H'_0 = \prod_{i=1}^{n}(-p_i) \tag{6.14}$$

An important characteristic of the Butterworth approximation is that its attenuation increases monotonically with frequency. In addition, it increases very slowly in the passband and quickly in the stopband. In the Butterworth approximation, if one wants to increase the attenuation one has to increase the filter order. However, if one sacrifices the monotonicity of the attenuation, a higher attenuation in the stopband can be obtained for the same filter order. A classic example of one such approximation is the Chebyshev approximation.

6.2.3 Chebyshev approximation

The attenuation function of a normalized-lowpass Chebyshev filter is characterized by

$$\left|A(j\Omega')\right|^2 = 1 + \epsilon^2 C_n^2(\Omega') \tag{6.15}$$

where $C_n(\Omega')$ is a Chebyshev function of order n, which can be written in its trigonometric form as

$$C_n(\Omega') = \begin{cases} \cos(n\cos^{-1}\Omega'), & 0 \leq \Omega' \leq 1 \\ \cosh(n\cosh^{-1}\Omega'), & \Omega' > 1 \end{cases} \tag{6.16}$$

These functions $C_n(\Omega')$ have the following properties

$$\begin{cases} 0 \leq C_n^2(\Omega') \leq 1, & 0 \leq \Omega' \leq 1 \\ C_n^2(\Omega') > 1, & \Omega' > 1 \end{cases} \tag{6.17}$$

As a consequence, for the attenuation function defined in equation (6.15), the passband is placed in the frequency range $0 \leq \Omega' \leq 1$, the rejection band is in the range $\Omega' > 1$, as desired, and the parameter ϵ once again determines the maximum passband ripple.

The Chebyshev functions defined above can also be expressed in polynomial form as

$$\begin{aligned} C_{n+1}(\Omega') + C_{n-1}(\Omega') &= \cos[(n+1)\cos^{-1}\Omega'] + \cos[(n-1)\cos^{-1}\Omega'] \\ &= 2\cos(\cos^{-1}\Omega')\cos(n\cos^{-1}\Omega') \\ &= 2\Omega' C_n(\Omega') \end{aligned} \tag{6.18}$$

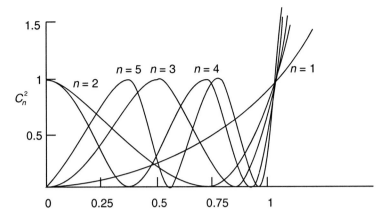

Figure 6.3 Chebyshev functions for $n = 1, 2, \ldots, 5$.

with $C_0(\Omega') = 1$ and $C_1(\Omega') = \Omega'$. We can then generate higher order Chebyshev polynomials through the recursive relation above, that is

$$C_2(\Omega') \quad = 2\Omega'^2 - 1$$
$$C_3(\Omega') \quad = 4\Omega'^3 - 3\Omega'$$
$$\vdots$$
$$C_{n+1}(\Omega') = 2\Omega' C_n(\Omega') - C_{n-1}(\Omega') \tag{6.19}$$

Figure 6.3 depicts the Chebyshev functions for several values of n.

Since $C_n(\Omega') = 1$ at $\Omega' = 1$, we have that

$$A_p = A_{\mathrm{dB}}(1) = 10\log_{10}(1 + \epsilon^2) \tag{6.20}$$

and then

$$\epsilon = \sqrt{10^{0.1 A_p} - 1} \tag{6.21}$$

From equations (6.15) and (6.16), when $\Omega' = \Omega'_r$, we find

$$A_r = A_{\mathrm{dB}}(\Omega'_r) = 10\log_{10}\left[1 + \epsilon^2 \cosh^2\left(n\cosh^{-1}\Omega'_r\right)\right] \tag{6.22}$$

and thus the order of the normalized-lowpass Chebyshev filter that satisfies the required stopband attenuation is the smallest integer number that satisfies

$$n \geq \frac{\cosh^{-1}\sqrt{\dfrac{10^{0.1 A_r} - 1}{\epsilon^2}}}{\cosh^{-1}\Omega'_r} \tag{6.23}$$

Similarly to the Butterworth case (see equation (6.9)), we can now continue the approximation process by evaluating the zeros of $A(s')A(-s')$, with $s' = j\Omega'$. Since zero attenuation can never occur in the stopband, these zeros are in the passband region $0 \leq \Omega' \leq 1$, and thus, from equation (6.16), we have

$$\cos\left(n\cos^{-1}\frac{s'}{j}\right) = \pm\frac{j}{\epsilon} \tag{6.24}$$

The above equation can be solved for s' by defining a complex variable p as

$$p = x_1 + jx_2 = \cos^{-1}\left(\frac{s'}{j}\right) \tag{6.25}$$

Replacing this value of p in equation (6.24), we arrive at

$$\cos n(x_1 + jx_2) = (\cos nx_1 \cosh nx_2) - j(\sin nx_1 \sinh nx_2) = \pm\frac{j}{\epsilon} \tag{6.26}$$

By equating the real parts of both sides of the above equation, we can deduce that

$$\cos nx_1 \cosh nx_2 = 0 \tag{6.27}$$

and considering that

$$\cosh nx_2 \geq 1, \qquad \forall n, x_2 \tag{6.28}$$

we then have

$$\cos nx_1 = 0 \tag{6.29}$$

which yields the following $2n$ solutions:

$$x_{1i} = \frac{2i + 1}{2n}\pi \tag{6.30}$$

for $i = 0, 1, \ldots, (2n - 1)$. Now, by equating the imaginary parts of both sides of equation (6.26) and using the values of x_{1i} obtained in equation (6.30), it follows that

$$\sin nx_{1i} = \pm 1 \tag{6.31}$$

$$x_2 = \frac{1}{n}\sinh^{-1}\left(\frac{1}{\epsilon}\right) \tag{6.32}$$

Since, from equations (6.24) and (6.25), the zeros of $A(s')A(-s')$ are given by

$$s'_i = \sigma'_i \pm j\Omega'_i = j\cos(x_{1i} + jx_2) = \sin x_{1i} \sinh x_2 + j\cos x_{1i}\cosh x_2 \tag{6.33}$$

for $i = 0, 1, \ldots, (2n - 1)$, we have, from equations (6.25) and (6.30), that

$$\sigma_i = \pm\sin\frac{\pi}{2}\left(\frac{2i + 1}{n}\right)\sinh\left(\frac{1}{n}\sinh^{-1}\frac{1}{\epsilon}\right) \tag{6.34}$$

$$\Omega_i = \cos\frac{\pi}{2}\left(\frac{2i + 1}{n}\right)\cosh\left(\frac{1}{n}\sinh^{-1}\frac{1}{\epsilon}\right) \tag{6.35}$$

The calculated zeros belong to $A(s')A(-s')$. Analogously to the Butterworth case, we associate the n zeros, p_i, with negative real part to $A(s')$, in order to guarantee the filter stability.

The above equations indicate that the zeros of a Chebyshev approximation are placed on an ellipse in the s plane, since equation (6.34) implies the following relation:

$$\left[\frac{\sigma_i}{\sinh(\frac{1}{n}\sinh^{-1}\frac{1}{\epsilon})}\right]^2 + \left[\frac{\Omega_i}{\cosh(\frac{1}{n}\sinh^{-1}\frac{1}{\epsilon})}\right]^2 = 1 \tag{6.36}$$

The transfer function of the Chebyshev filter is then given by

$$H'(s') = \frac{H'_0}{A(s')} = \frac{H'_0}{\prod\limits_{i=1}^{n}(s' - p_i)} \tag{6.37}$$

where H'_0 is chosen so that $A(s')$ satisfies equation (6.15), that is

$$H'_0 = \begin{cases} \prod\limits_{i=1}^{n}(-p_i), & \text{for } n \text{ odd} \\ 10^{-0.05A_P}\prod\limits_{i=1}^{n}(-p_i), & \text{for } n \text{ even} \end{cases} \tag{6.38}$$

It is interesting to note that in the Butterworth case the frequency response is monotone in both passband and stopband, and is maximally flat at $\Omega = 0$. In the case of the Chebyshev filters, the smooth passband characteristics are exchanged for steeper transition bands for the same filter orders. In fact, for a given prescribed specification, Chebyshev filters usually require lower-order transfer functions than Butterworth filters, owing to their equiripple behavior in the passband.

6.2.4 Elliptic approximation

The two approximations discussed so far, namely the lowpass Butterworth and Chebyshev approximations, lead to transfer functions whose numerator is a constant and the denominator is a polynomial in s. These are called all-pole filters, because all their zeros are located at infinity. When going from the Butterworth to the Chebyshev filters, we have traded-off monotonicity and maximal flatness in the passband for higher attenuation in the stopband. At this point, it is natural to wonder whether we could also exchange the monotonicity in the stopband possessed by the Butterworth and Chebyshev filters for an even steeper transition band without increasing the filter order. This is indeed the case, as approximations with finite-frequency zeros can have transition bands with very steep slopes.

In practice, there are transfer function approximations with finite zeros which have equiripple characteristics in the passband and in the stopband, with the advantage that their coefficients can be computed using closed formulas. These filters are usually called elliptic filters, as their closed-form equations are derived based on elliptic functions, but they are also known as Cauer or Zolotarev filters (Daniels, 1974).

This section covers the lowpass elliptic filter approximation. (The derivations are not detailed here, as they are beyond the scope of this book.) In the following, we describe an algorithm to calculate the coefficients of elliptic filters which is based on the procedure described in the benchmark book (Antoniou, 1993).

Consider the following lowpass filter transfer function:

$$|H(j\Omega')| = \frac{1}{\sqrt{1 + R_n^2(\Omega')}} \tag{6.39}$$

where

$$R_n(\Omega') = \begin{cases} k_p \dfrac{\displaystyle\prod_{i=1}^{\frac{n}{2}}(\Omega_i^2\Omega_r - \Omega'^2)}{\displaystyle\prod_{i=1}^{\frac{n}{2}}(\Omega_r - \Omega_i'^2)}, & \text{for } n \text{ even} \\[3em] k_i \Omega' \dfrac{\displaystyle\prod_{i=1}^{\frac{n-1}{2}}(\Omega_i^2\Omega_r - \Omega'^2)}{\displaystyle\prod_{i=1}^{\frac{n-1}{2}}(\Omega_r - \Omega_i^2\Omega'^2)}, & \text{for } n \text{ odd} \end{cases} \tag{6.40}$$

The computation of $R_n(\Omega')$ requires the use of some elliptic functions. The normalization procedure for the elliptic approximation is rather distinct from the one for the Butterworth and Chebyshev filters. Here, the frequency normalization factor is given by

$$\Omega_c = \sqrt{\Omega_p\Omega_r} \tag{6.41}$$

In that manner, we have that

$$\Omega'_p = \sqrt{\frac{\Omega_p}{\Omega_r}} \tag{6.42}$$

$$\Omega'_r = \sqrt{\frac{\Omega_r}{\Omega_p}} \tag{6.43}$$

Defining

$$k = \frac{\Omega_p'}{\Omega_r'} \tag{6.44}$$

$$q_0 = \frac{1}{2} \left[\frac{1 - (1 - k^2)^{\frac{1}{4}}}{1 + (1 - k^2)^{\frac{1}{4}}} \right] \tag{6.45}$$

$$q = q_0 + 2q_0^5 + 15q_0^9 + 150q_0^{13} \tag{6.46}$$

$$\epsilon = \sqrt{\frac{10^{0.1A_p} - 1}{10^{0.1A_r} - 1}} \tag{6.47}$$

the specifications are satisfied if the filter order n is chosen through the following relation:

$$n \geq \frac{\log_{10} \dfrac{16}{\epsilon^2}}{\log_{10} \dfrac{1}{q}} \tag{6.48}$$

Having the filter order n, we can then determine the following parameters before proceeding with the computation of the filter coefficients:

$$\Theta = \frac{1}{2n} \ln \frac{10^{0.05A_p} + 1}{10^{0.05A_p} - 1} \tag{6.49}$$

$$\sigma = \left| \frac{2q^{\frac{1}{4}} \displaystyle\sum_{j=0}^{\infty} (-1)^j q^{j(j+1)} \sinh[(2j+1)\Theta]}{1 + 2 \displaystyle\sum_{j=1}^{\infty} (-1)^j q^{j^2} \cosh(2j\Theta)} \right| \tag{6.50}$$

$$W = \sqrt{(1 + k\sigma^2)\left(1 + \frac{\sigma^2}{k}\right)} \tag{6.51}$$

Also, for $i = 1, 2, \ldots, l$, we compute

$$\rho_i = \frac{2q^{\frac{1}{4}} \displaystyle\sum_{j=0}^{\infty} (-1)^j q^{j(j+1)} \sin \dfrac{(2j+1)\pi u}{n}}{1 + 2 \displaystyle\sum_{j=1}^{\infty} (-1)^j q^{j^2} \cos \dfrac{2j\pi u}{n}} \tag{6.52}$$

$$V_i = \sqrt{(1 - k\rho_i^2)\left(1 - \frac{\rho_i^2}{k}\right)} \tag{6.53}$$

where

$$
\left.\begin{array}{ll}
u = i, & \text{for } n \text{ odd} \\
u = i - \frac{1}{2}, & \text{for } n \text{ even}
\end{array}\right\}
\tag{6.54}
$$

The infinite summations in equations (6.50) and (6.52) converge extremely quickly, and only two or three terms are sufficient to reach a very accurate result.

The transfer function of a normalized-lowpass elliptic filter can be written as

$$
H'(s') = \frac{H_0'}{(s' + \sigma)^m} \prod_{i=1}^{l} \frac{s'^2 + b_{2i}}{s'^2 + a_{1i}s' + a_{2i}}
\tag{6.55}
$$

where

$$
\left.\begin{array}{ll}
m = 0 \text{ and } l = \frac{n}{2}, & \text{for } n \text{ even} \\
m = 1 \text{ and } l = \frac{n-1}{2}, & \text{for } n \text{ odd}
\end{array}\right\}
\tag{6.56}
$$

The coefficients of the above transfer function are calculated based on the parameters obtained from equations (6.44)–(6.53) as

$$
b_{2i} = \frac{1}{\rho_i^2}
\tag{6.57}
$$

$$
a_{2i} = \frac{(\sigma V_i)^2 + (\rho_i W)^2}{(1 + \sigma^2 \rho_i^2)^2}
\tag{6.58}
$$

$$
a_{1i} = \frac{2\sigma V_i}{1 + \sigma^2 \rho_i^2}
\tag{6.59}
$$

$$
H_0' = \begin{cases}
\sigma \displaystyle\prod_{i=1}^{l} \frac{a_{2i}}{b_{2i}}, & \text{for } n \text{ odd} \\[2em]
10^{-0.05 A_p} \sigma \displaystyle\prod_{i=1}^{l} \frac{a_{2i}}{b_{2i}}, & \text{for } n \text{ even}
\end{cases}
\tag{6.60}
$$

Following this procedure, the resulting minimum stopband attenuation is slightly better than the specified value, being precisely given by

$$
A_r = 10 \log_{10} \left(\frac{10^{0.1 A_p} - 1}{16 q^n} + 1 \right)
\tag{6.61}
$$

6.2.5   Frequency transformations

The approximation methods presented so far are meant for designing normalized-lowpass filters. In this subsection, we address the issue of how a transfer function of a general lowpass, highpass, symmetric bandpass, or symmetric bandstop filter can be transformed into a normalized-lowpass transfer function, and vice versa. The procedure used here, the so-called frequency transformation technique, consists of replacing the variable s' in the normalized-lowpass filter by an appropriate function of s. In the following, we make a detailed analysis of the normalized-lowpass ↔ bandpass transformation. The analyses of the other transformations are similar, and their expressions are summarized in Table 6.2.

A normalized-lowpass transfer function $H'(s')$ can be transformed into a symmetric bandpass transfer function by applying the following substitution of variables

$$s' \leftrightarrow \frac{1}{a} \frac{s^2 + \Omega_0^2}{Bs} \tag{6.62}$$

where Ω_0 is the central frequency of the bandpass filter, B is the filter passband width, and a is a normalization parameter that depends upon the filter type, as follows:

$$\Omega_0 = \sqrt{\Omega_{p_1} \Omega_{p_2}} \tag{6.63}$$

$$B = \Omega_{p_2} - \Omega_{p_1} \tag{6.64}$$

$$a = \begin{cases} 1, & \text{for any Butterworth or Chebyshev filter} \\[2ex] \sqrt{\dfrac{\Omega_r}{\Omega_p}}, & \text{for a lowpass elliptic filter} \\[2ex] \sqrt{\dfrac{\Omega_p}{\Omega_r}}, & \text{for a highpass elliptic filter} \\[2ex] \sqrt{\dfrac{\Omega_{r_2} - \Omega_{r_1}}{\Omega_{p_2} - \Omega_{p_1}}}, & \text{for a bandpass elliptic filter} \\[2ex] \sqrt{\dfrac{\Omega_{p_2} - \Omega_{p_1}}{\Omega_{r_2} - \Omega_{r_1}}}, & \text{for a bandstop elliptic filter} \end{cases} \tag{6.65}$$

The value of a is different from unity for the elliptic filters because in this case the normalization is not $\Omega'_p = 1$, but $\sqrt{\Omega'_p \Omega'_r} = 1$ (see equations (6.41)–(6.43)).

The frequency transformation in equation (6.62) has the following properties:

- The frequency $s' = j0$ is transformed into $s = \pm j\Omega_0$.
- Any complex frequency $s' = -j\Omega'$, corresponding to an attenuation of A_{dB} in the normalized-lowpass filter, is transformed into the two distinct frequencies

$$\Omega_1 = -\frac{1}{2}aB\Omega' + \sqrt{\frac{1}{4}a^2 B^2 \Omega'^2 + \Omega_0^2} \tag{6.66}$$

$$\overline{\Omega}_1 = -\frac{1}{2}aB\Omega' - \sqrt{\frac{1}{4}a^2 B^2 \Omega'^2 + \Omega_0^2} \tag{6.67}$$

where Ω_1 is a positive frequency and $\overline{\Omega}_1$ is a negative frequency, both corresponding to the attenuation A_{dB}.

In addition, a complex frequency $s' = j\Omega'$, which also corresponds to an attenuation of A_{dB}, is transformed into two frequencies with the same attenuation level, that is,

$$\Omega_2 = \frac{1}{2}aB\Omega' + \sqrt{\frac{1}{4}a^2B^2\Omega'^2 + \Omega_0^2} \tag{6.68}$$

$$\overline{\Omega}_2 = \frac{1}{2}aB\Omega' - \sqrt{\frac{1}{4}a^2B^2\Omega'^2 + \Omega_0^2} \tag{6.69}$$

and it can be seen that $\overline{\Omega}_1 = -\Omega_2$ and $\overline{\Omega}_2 = -\Omega_1$.

The positive frequencies Ω_1 and Ω_2 are the ones we are interested in analyzing. They can be expressed in a single equation as follows:

$$\Omega_{1,2} = \mp\frac{1}{2}aB\Omega' + \sqrt{\frac{1}{4}a^2B^2\Omega'^2 + \Omega_0^2} \tag{6.70}$$

from which we get

$$\Omega_2 - \Omega_1 = aB\Omega' \tag{6.71}$$

$$\Omega_1\Omega_2 = \Omega_0^2 \tag{6.72}$$

These relationships indicate that, in this kind of transformation, for each frequency with attenuation A_{dB} there is another frequency geometrically symmetric with respect to the central frequency Ω_0 with the same attenuation.

- From the above the cutoff frequency of the normalized-lowpass filter Ω'_p is mapped into the frequencies

$$\Omega_{p1,2} = \mp\frac{1}{2}aB\Omega'_p + \sqrt{\frac{1}{4}a^2B^2\Omega'^2_p + \Omega_0^2} \tag{6.73}$$

such that

$$\Omega_{p2} - \Omega_{p1} = aB\Omega'_p \tag{6.74}$$

$$\Omega_{p1}\Omega_{p2} = \Omega_0^2 \tag{6.75}$$

- Similarly, the stopband edge frequency Ω'_r of the normalized-lowpass prototype is transformed into the frequencies

$$\Omega_{r1,2} = \mp\frac{1}{2}aB\Omega'_r + \sqrt{\frac{1}{4}a^2B^2\Omega'^2_r + \Omega_0^2} \tag{6.76}$$

such that

$$\Omega_{r2} - \Omega_{r1} = aB\Omega'_r \tag{6.77}$$

$$\Omega_{r1}\Omega_{r2} = \Omega_0^2 \tag{6.78}$$

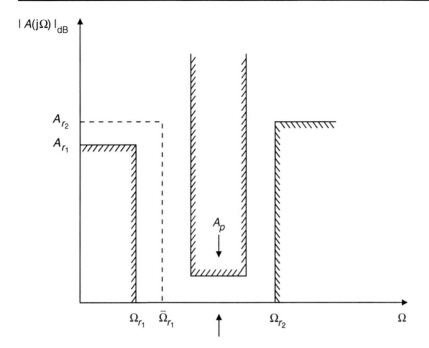

Figure 6.4 Nonsymmetric bandpass filter specifications.

The above analysis leads to the conclusion that this normalized-lowpass ↔ bandpass transformation works for bandpass filters which are geometrically symmetric with respect to the central frequency. However, bandpass filter specifications are not usually geometrically symmetric in practice. Fortunately, we can generate geometrically symmetric bandpass specifications satisfying the minimum stopband attenuation requirements by the following procedure (see Figure 6.4):

(i) Compute $\Omega_0^2 = \Omega_{p_1}\Omega_{p_2}$.

(ii) Compute $\overline{\Omega}_{r_1} = \frac{\Omega_0^2}{\Omega_{r_2}}$, and if $\overline{\Omega}_{r_1} > \Omega_{r_1}$, replace Ω_{r_1} with $\overline{\Omega}_{r_1}$, as illustrated in Figure 6.4.

(iii) If $\overline{\Omega}_{r_1} \leq \Omega_{r_1}$, then compute $\overline{\Omega}_{r_2} = \frac{\Omega_0^2}{\Omega_{r_1}}$, and replace Ω_{r_2} with $\overline{\Omega}_{r_2}$.

(iv) If $A_{r_1} \neq A_{r_2}$, choose $A_r = \max\{A_{r_1}, A_{r_2}\}$.

Once the geometrically symmetric bandpass filter specifications are available, we need to determine the normalized frequencies Ω_p' and Ω_r', in order to have the corresponding normalized-lowpass filter completely specified. According to equations (6.74) and (6.77), they can be computed as follows:

$$\Omega_p' = \frac{1}{a} \tag{6.79}$$

$$\Omega_r' = \frac{1}{a}\frac{\Omega_{r_2} - \Omega_{r_1}}{\Omega_{p_2} - \Omega_{p_1}} \tag{6.80}$$

Table 6.2. *Analog frequency transformations.*

Transformation	Normalization	Denormalization
lowpass(Ω) $\leftrightarrow$ lowpass(Ω')	$\Omega'_p = \dfrac{1}{a}$ $\Omega'_r = \dfrac{1}{a}\dfrac{\Omega_r}{\Omega_p}$	$s' \leftrightarrow \dfrac{1}{a}\dfrac{s}{\Omega_p}$
highpass(Ω) $\leftrightarrow$ lowpass(Ω')	$\Omega'_p = \dfrac{1}{a}$ $\Omega'_r = \dfrac{1}{a}\dfrac{\Omega_p}{\Omega_r}$	$s' \leftrightarrow \dfrac{1}{a}\dfrac{\Omega_p}{s}$
bandpass(Ω) $\leftrightarrow$ lowpass(Ω')	$\Omega'_p = \dfrac{1}{a}$ $\Omega'_r = \dfrac{1}{a}\dfrac{\Omega_{r_2}-\Omega_{r_1}}{\Omega_{p_2}-\Omega_{p_1}}$	$s' \leftrightarrow \dfrac{1}{a}\dfrac{s^2+\Omega_0^2}{Bs}$
bandstop(Ω) $\leftrightarrow$ lowpass(Ω')	$\Omega'_p = \dfrac{1}{a}$ $\Omega'_r = \dfrac{1}{a}\dfrac{\Omega_{p_2}-\Omega_{p_1}}{\Omega_{r_2}-\Omega_{r_1}}$	$s' \leftrightarrow \dfrac{1}{a}\dfrac{Bs}{s^2+\Omega_0^2}$

It is worth noting that bandstop filter specifications must also be geometrically symmetric. In this case, however, in order to satisfy the minimum stopband attenuation requirements, the stopband edges must be preserved, while the passband edges should be modified analogously to the procedure described above.

A summary of all types of transformations, including the respective correspondence among the lowpass prototype frequencies and desired filter specifications, are shown in Table 6.2.

The general procedure to approximate a standard analog filter using frequency transformations can be summarized as follows:

(i)   Determine the specifications for the lowpass, highpass, bandpass, or bandstop analog filter.

(ii)   When designing a bandpass or bandstop filter, make sure the specifications are geometrically symmetric following the proper procedure described earlier in this subsection.

(iii) Determine the normalized-lowpass specifications equivalent to the desired filter, following the relationships seen in Table 6.2.

(iv) Perform the filter approximation using the Butterworth, Chebyshev, or elliptic methods.

(v)   Denormalize the prototype using the frequency transformations given on the right-hand side of Table 6.2.

Sometimes the approximation of analog filters can present poor numerical conditioning, especially when the desired filter has a narrow transition and/or passband. In this case, design techniques employing transformed variables are available (Daniels, 1974; Sedra & Brackett, 1978), which can improve the numerical conditioning by separating the roots of the polynomials involved.

EXAMPLE 6.1

Design a bandpass filter satisfying the specification below using the Butterworth, Chebyshev, and elliptic approximation methods:

$$
\left.\begin{aligned}
A_p &= 1.0 \text{ dB} \\
A_r &= 40 \text{ dB} \\
\Omega_{r_1} &= 1394\pi \text{ rad/s} \\
\Omega_{p_1} &= 1510\pi \text{ rad/s} \\
\Omega_{p_2} &= 1570\pi \text{ rad/s} \\
\Omega_{r_2} &= 1704\pi \text{ rad/s}
\end{aligned}\right\} \tag{6.81}
$$

SOLUTION

Since $\Omega_{p_1}\Omega_{p_2} \neq \Omega_{r_1}\Omega_{r_2}$, the first step in the design is to determine the geometrically symmetric bandpass filter following the procedure described earlier in this subsection. In that manner, we get

$$
\Omega_{r_2} = \bar{\Omega}_{r_2} = \frac{\Omega_0^2}{\Omega_{r_1}} = 1700.6456\pi \text{ rad/s} \tag{6.82}
$$

Finding the corresponding lowpass specifications based on the transformations in Table 6.2, we have

$$
\Omega'_p = \frac{1}{a} \tag{6.83}
$$

$$
\Omega'_r = \frac{1}{a}\frac{\bar{\Omega}_{r_2} - \Omega_{r_1}}{\Omega_{p_2} - \Omega_{p_1}} = \frac{1}{a}5.1108 \tag{6.84}
$$

where

$$
a = \begin{cases} 1, & \text{for the Butterworth and Chebyshev filters} \\ 2.2607, & \text{for the elliptic filter} \end{cases} \tag{6.85}
$$

- Butterworth approximation: from the specifications above, we can compute ϵ from equation (6.6), and, having ϵ, the minimum filter order required to satisfy the specifications from equation (6.8):

$$
\epsilon = 0.5088 \tag{6.86}
$$

$$
n = 4 \tag{6.87}
$$

Table 6.3. *Characteristics of the Butterworth bandpass filter.*
Constant gain: $H_0 = 2.480\,945\,10 \times 10^9$.

Denominator coefficients	Filter poles
$a_0 = 2.997\,128\,62 \times 10^{29}$	$p_1 = -41.793\,563\,49 + j4734.949\,273\,36$
$a_1 = 7.470\,396\,68 \times 10^{24}$	$p_2 = -41.793\,563\,49 - j4734.949\,273\,36$
$a_2 = 5.133\,072\,97 \times 10^{22}$	$p_3 = -102.185\,156\,89 + j4793.520\,899\,05$
$a_3 = 9.585\,099\,76 \times 10^{17}$	$p_4 = -102.185\,156\,89 - j4793.520\,899\,05$
$a_4 = 3.292\,722\,76 \times 10^{15}$	$p_5 = -104.005\,809\,72 + j4878.927\,993\,89$
$a_5 = 4.096\,569\,17 \times 10^{10}$	$p_6 = -104.005\,809\,72 - j4878.927\,993\,89$
$a_6 = 9.376\,154\,35 \times 10^{7}$	$p_7 = -43.613\,531\,30 + j4941.140\,239\,95$
$a_7 = 5.831\,961\,23 \times 10^{2}$	$p_8 = -43.613\,531\,30 - j4941.140\,239\,95$
$a_8 = 1.0$	

From equation (6.12), the zeros of the normalized Butterworth polynomial when $n = 4$ are given by

$$
\left.
\begin{aligned}
s'_{1,2} &= -1.0939 \pm j0.4531 \\
s'_{3,4} &= -0.4531 \pm j1.0939 \\
s'_{5,6} &= 1.0939 \pm j0.4531 \\
s'_{7,8} &= 0.4531 \pm j1.0939
\end{aligned}
\right\}
\tag{6.88}
$$

Selecting the ones with negative real part to be the poles of $H'(s')$, this normalized transfer function becomes

$$
H'(s') = 1.9652 \frac{1}{s'^4 + 3.0940s'^3 + 4.7863s'^2 + 4.3373s' + 1.9652}
\tag{6.89}
$$

The design is completed by applying the lowpass to bandpass transformation in Table 6.2. The resulting bandpass transfer function is then given by

$$
H(s) = H_0 \frac{s^4}{a_8 s^8 + a_7 s^7 + a_6 s^6 + a_5 s^5 + a_4 s^4 + a_3 s^3 + a_2 s^2 + a_1 s + a_0}
\tag{6.90}
$$

where the filter coefficients and poles are listed in Table 6.3.

Figure 6.5 depicts the frequency response of the designed Butterworth bandpass filter.

- Chebyshev approximation: from the normalized specifications in equations (6.81) and (6.82), one can compute ϵ and n based on equations (6.21) and (6.23), respectively, resulting in

$$
\epsilon = 0.5088
\tag{6.91}
$$

$$
n = 3
\tag{6.92}
$$

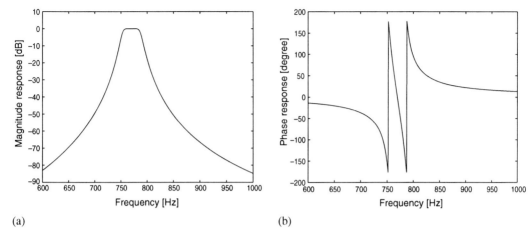

Figure 6.5 Bandpass Butterworth filter: (a) magnitude response; (b) phase response.

Then, from equations (6.24)–(6.35), we have that the poles of the normalized transfer function are:

$$\left.\begin{array}{l} s'_{1,2} = -0.2471 \pm j0.9660 \\ s'_3 = -0.4942 - j0.0 \end{array}\right\} \tag{6.93}$$

This implies that the normalized-lowpass filter has the following transfer function:

$$H'(s') = 0.4913 \frac{1}{s'^3 + 0.9883s'^2 + 1.2384s' + 0.4913} \tag{6.94}$$

The denormalized design is obtained by applying the lowpass to bandpass transformation. The resulting transfer function is of the form

$$H(s) = H_0 \frac{s^3}{a_6 s^6 + a_5 s^5 + a_4 s^4 + a_3 s^3 + a_2 s^2 + a_1 s + a_0} \tag{6.95}$$

where all filter coefficients and poles are listed in Table 6.4.

Figure 6.6 depicts the frequency response of the resulting Chebyshev bandpass filter.

• Elliptic approximation: from equation (6.85), for this elliptic approximation, we have that $a = 2.2607$, and then the normalized specifications are

$$\Omega'_p = 0.4423 \tag{6.96}$$

$$\Omega'_r = 2.2607 \tag{6.97}$$

From equation (6.48), the minimum order required for the elliptic approximation to satisfy the specifications is $n = 3$. Therefore, from equations (6.55)–(6.60), the normalized-lowpass filter has the transfer function

$$H'(s') = 6.3627 \times 10^{-3} \frac{s'^2 + 6.7814}{s'^3 + 0.4362s'^2 + 0.2426s' + 0.0431} \tag{6.98}$$

Table 6.4. *Characteristics of the Chebyshev bandpass filter.*
Constant gain: $H_0 = 3.290\,455\,64 \times 10^6$.

Denominator coefficients	Filter poles
$a_0 = 1.280\,940\,73 \times 10^{22}$	$p_1 = -22.849\,007\,56 + j4746.892\,064\,91$
$a_1 = 1.019\,907\,34 \times 10^{17}$	$p_2 = -22.849\,007\,56 - j4746.892\,064\,91$
$a_2 = 1.643\,410\,66 \times 10^{15}$	$p_3 = -46.574\,482\,26 + j4836.910\,374\,53$
$a_3 = 8.721\,240\,34 \times 10^{9}$	$p_4 = -46.574\,482\,26 - j4836.910\,374\,53$
$a_4 = 7.023\,761\,49 \times 10^{7}$	$p_5 = -23.725\,474\,70 + j4928.978\,525\,35$
$a_5 = 1.862\,979\,29 \times 10^{2}$	$p_6 = -23.725\,474\,70 - j4928.978\,525\,35$
$a_6 = 1.0$	

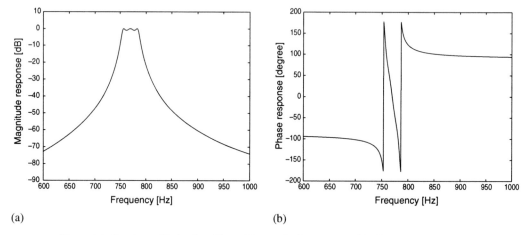

(a)   (b)

Figure 6.6 Bandpass Chebyshev filter: (a) magnitude response; (b) phase response.

The denormalized-bandpass design is then obtained by applying the lowpass to bandpass transformation given in Table 6.2, with $a = 2.2607$. The resulting bandpass transfer function is given by

$$H(s) = H_0 \frac{b_5 s^5 + b_3 s^3 + b_1 s}{a_6 s^6 + a_5 s^5 + a_4 s^4 + a_3 s^3 + a_2 s^2 + a_1 s + a_0} \tag{6.99}$$

where all filter coefficients, zeros, and poles are listed in Table 6.2.5.

Figure 6.7 depicts the frequency response of the resulting elliptic bandpass filter.

$\triangle$

Table 6.5. *Characteristics of the elliptic bandpass filter. Constant gain:* $H_0 = 2.711\,344\,84 \times 10^6$.

Numerator coefficients	Denominator coefficients	Filter zeros	Filter poles
$b_0 = 0.0$	$a_0 = 1.280\,940\,73 \times 10^{22}$		
$b_1 = 5.474\,603\,75 \times 10^{14}$	$a_1 = 1.017\,508\,00 \times 10^{17}$	$z_1 = +j4314.006\,091\,36$	$p_1 = -22.461\,679\,38 + j4746.679\,115\,03$
$b_2 = 0.0$	$a_2 = 1.643\,411\,70 \times 10^{15}$	$z_2 = -j4314.006\,091\,36$	$p_2 = -22.461\,679\,38 - j4746.679\,115\,03$
$b_3 = 4.802\,716\,10 \times 10^7$	$a_3 = 8.700\,779\,63 \times 10^9$	$z_3 = +j5423.699\,146\,03$	$p_3 = -47.142\,751\,63 + j4836.904\,869\,30$
$b_4 = 0.0$	$a_4 = 7.023\,765\,92 \times 10^7$	$z_4 = -j5423.699\,146\,03$	$p_4 = -47.142\,751\,63 - j4836.904\,869\,30$
$b_5 = 1.0$	$a_5 = 1.858\,596\,62 \times 10^2$	$z_5 = 0.0$	$p_5 = -23.325\,399\,83 + j4929.203\,482\,90$
$b_6 = 0.0$	$a_6 = 1.0$		$p_6 = -23.325\,399\,83 - j4929.203\,482\,90$

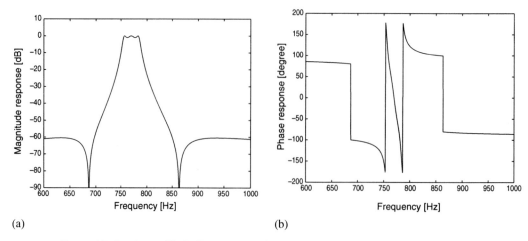

(a)   (b)

Figure 6.7 Bandpass elliptic filter: (a) magnitude response; (b) phase response.

6.3   Continuous-time to discrete-time transformations

As mentioned at the beginning of this chapter, a classical procedure for designing IIR digital filters is to design an analog prototype first and then transform it into a digital filter. In this section, we study two methods of carrying out this transformation, namely, the impulse-invariance method and the bilinear transformation method.

6.3.1   Impulse-invariance method

The intuitive way to implement a digital filtering operation, having an analog prototype as starting point, is the straightforward digitalization of the convolution operation, as follows. The output $y_a(t)$ of an analog filter having impulse response $h_a(t)$ when excited by a signal $x_a(t)$ is

$$y_a(t) = \int_{-\infty}^{\infty} x_a(\tau) h_a(t - \tau) d\tau \qquad (6.100)$$

One possible way to implement this operation in the discrete-time domain is to divide the time axis into slices of size T, replacing the integral by a summation of the areas of rectangles of width T and height $x_a(mT) h_a(t - mT)$, for all integers m. Equation (6.100) then becomes

$$y_a(t) = \sum_{m=-\infty}^{\infty} x_a(mT) h_a(t - mT) T \qquad (6.101)$$

The sampled version of $y_a(t)$ is obtained by substituting t by nT, yielding

$$y_a(nT) = \sum_{m=-\infty}^{\infty} x_a(mT) h_a(nT - mT) T \qquad (6.102)$$

This is clearly equivalent to obtaining the samples $y_a(nT)$ of $y_a(t)$ by filtering the samples $x_a(nT)$ with a digital filter having impulse response $h(n) = h_a(nT)$. That is, the impulse response of the equivalent digital filter would be a sampled version of the impulse response of the analog filter, using the same sampling rate for the input and output signals.

Roughly speaking, if the Nyquist criterion is met by the filter impulse response during the sampling operation, the discrete-time prototype has the same frequency response as the continuous-time one. In addition, a sampled version of a stable analog impulse response is clearly also stable. These are the main properties of this method of generating IIR filters, called the impulse-invariance method. In what follows, we analyze the above properties more precisely, in order to get a better understanding of the main strengths and limitations of this method.

We can begin by investigating the properties of the digital filter with impulse response $h(n) = h_a(nT)$ in the frequency domain. From equation (6.102), the discrete-time Fourier transform of $h(n)$ is

$$H(e^{j\Omega T}) = \frac{1}{T} \sum_{l=-\infty}^{\infty} H_a(j\Omega + j\Omega_s l) \tag{6.103}$$

where $H_a(s)$, $s = \sigma + j\Omega$, is the analog transfer function, and $\Omega_s = \frac{2\pi}{T}$ is the sampling frequency. That is, the digital frequency response is equal to the analog one replicated at intervals $l\Omega_s$. One important consequence of this fact is that if $H_a(j\Omega)$ has much energy for $\Omega > \Omega_s/2$, there will be aliasing, and therefore the digital frequency response will be a severely distorted version of the analog one.

Another way of seeing this is that the digital frequency response is obtained by folding the analog frequency response, for $-\infty < \Omega < \infty$, on the unit circle of the $z = e^{sT}$ plane, with each interval $[\sigma + j(l - \frac{1}{2})\Omega_s, \sigma + j(l + \frac{1}{2})\Omega_s]$, for all integers l, corresponding to one full turn over the unit circle of the z plane. This limits the usefulness of the impulse-invariance method in the design of transfer functions whose magnitude responses decrease monotonically at high frequencies. For example, its use in the direct design of highpass, bandstop, or even elliptic lowpass and bandpass filters is strictly forbidden, and other methods should be considered for designing such filters.

Stability of the digital filter can be inferred from the stability of the analog prototype by analyzing equation (6.103). In fact, based on that equation, we can interpret the impulse-invariance method as a mapping from the s domain to the z domain such that each slice of the s plane given by the interval $[\sigma + j(l - \frac{1}{2})\Omega_s, \sigma + j(l + \frac{1}{2})\Omega_s]$, for all integers l, where $\sigma = \text{Re}\{s\}$, is mapped into the same region of the z plane. Also, the left side of the s plane, that is, where $\sigma < 0$, is mapped into the interior of the unit circle, implying that if the analog transfer function is stable (all poles on the left side of the s plane), then the digital transfer function is also stable (all poles inside the unit circle of the z plane).

In practice, the impulse-invariance transformation is not implemented through equation (6.103), as a simpler procedure can be deduced by expanding an Nth-order $H_a(s)$ as follows:

$$H_a(s) = \sum_{l=1}^{N} \frac{r_l}{s - p_l} \tag{6.104}$$

where it is assumed that $H_a(s)$ does not have multiple poles. The corresponding impulse response is given by

$$h_a(t) = \sum_{l=1}^{N} r_l e^{p_l t} u(t) \tag{6.105}$$

where $u(t)$ is the unitary step function. If we now sample that impulse response, the resulting sequence is

$$h_d(n) = h_a(nT) = \sum_{l=1}^{N} r_l e^{p_l n T} u(nT) \tag{6.106}$$

and the corresponding discrete-time transfer function is given by

$$H_d(z) = \sum_{l=1}^{N} \frac{r_l z}{z - e^{p_l T}} \tag{6.107}$$

This equation shows that a pole $s = p_l$ of the continuous-time filter corresponds to a pole of the discrete-time filter at $z = e^{p_l T}$. In that way, if p_l has negative real part, then $e^{p_l T}$ is inside the unit circle, generating a stable digital filter when we use the impulse-invariance method.

In order to obtain the same passband gain for the continuous- and discrete-time filters, for any value of the sampling period T, we should use the following expression for $H_d(z)$:

$$H_d(z) = \sum_{l=1}^{N} T \frac{r_l z}{z - e^{p_l T}} \tag{6.108}$$

which corresponds to

$$h_d(n) = T h_a(nT) \tag{6.109}$$

Thus, the overall impulse-invariance method consists of writing the analog transfer function $H_a(s)$ in the form of equation (6.104), determining the poles p_l and corresponding residues r_l, and generating $H_d(z)$ according to equation (6.108).

EXAMPLE 6.2

Transform the continuous-time lowpass transfer function given by

$$H(s) = \frac{1}{s^2 + s + 1} \tag{6.110}$$

into a discrete-time transfer function using the impulse-invariance transformation method with $\Omega_s = 10$ rad/s. Plot the corresponding analog and digital magnitude responses.

SOLUTION
A second-order lowpass transfer function can be written as

$$H(s) = \frac{\Omega_0^2}{s^2 + \frac{\Omega_0}{Q}s + \Omega_0^2}$$

$$= \frac{\Omega_0^2}{\sqrt{\frac{\Omega_0^2}{Q^2} - 4\Omega_0^2}} \left(\frac{1}{s + \frac{\Omega_0}{2Q} - \sqrt{\frac{\Omega_0^2}{4Q^2} - \Omega_0^2}} - \frac{1}{s + \frac{\Omega_0}{2Q} + \sqrt{\frac{\Omega_0^2}{4Q^2} - \Omega_0^2}} \right) \quad (6.111)$$

Its poles are located at

$$p_1 = p_2^* = -\frac{\Omega_0}{2Q} + j\sqrt{\Omega_0^2 - \frac{\Omega_0^2}{4Q^2}} \quad (6.112)$$

and the corresponding residues are given by

$$r_1 = r_2^* = \frac{-j\Omega_0^2}{\sqrt{4\Omega_0^2 - \frac{\Omega_0^2}{Q^2}}} \quad (6.113)$$

Applying the impulse-invariance method with $T = 2\pi/10$, the resulting discrete-time transfer function is given by

$$H(z) = \frac{2jTr_1\sin(\mathrm{Im}\{p_1\}T)e^{\mathrm{Re}\{p_1\}T}}{z^2 - 2\cos(\mathrm{Im}\{p_1\}T)e^{\mathrm{Re}\{p_1\}T}z + e^{2\mathrm{Re}\{p_1\}T}}$$

$$= \frac{0.274\,331\,03}{z^2 - 1.249\,825\,52z + 0.533\,488\,09} \quad (6.114)$$

The magnitude responses corresponding to the analog and digital transfer functions are depicted in Figure 6.8. As can be seen, the frequency responses are similar except for the limited stopband attenuation of the discrete-time filter which is due to the aliasing effect.

$\triangle$

We should again emphasize that the impulse-invariance method is suitable only for continuous-time prototypes with frequency responses that decrease monotonically at high frequencies, which limits its applicability a great deal. In the next section, we analyze the bilinear transformation method, which overcomes some of the limitations of the impulse-invariance method.

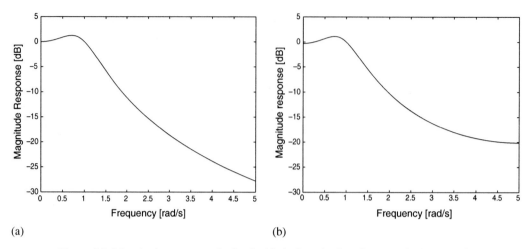

(a)   (b)

Figure 6.8 Magnitude responses obtained with the impulse-invariance method: (a) continuous-time filter; (b) discrete-time filter.

6.3.2   Bilinear transformation method

The bilinear transformation method, like the impulse-invariance method, basically consists of mapping the left-hand side of the s plane into the interior of the unit circle of the z plane. The main difference between them is that in the bilinear transformation method the whole analog frequency range $-\infty < \Omega < \infty$ is squeezed into the unit circle $-\pi \leq \omega \leq \pi$, while in the impulse-invariance method the analog frequency response is folded around the unit circle indefinitely. The main advantage of the bilinear transformation method is that aliasing is avoided, thereby keeping the magnitude response characteristics of the continuous-time transfer function when generating the discrete-time transfer function.

The bilinear mapping is derived by first considering the key points of the s plane and analyzing their corresponding points in the z plane after the transformation. Hence, the left-hand side of the s plane should be uniquely mapped into the interior of the unit circle of the z plane, and so on, as given in Table 6.6.

In order to satisfy the second and third requirements of Table 6.6, the bilinear transformation must have the following form

$$s \to k\frac{f_1(z) - 1}{f_2(z) + 1} \tag{6.115}$$

where $f_1(1) = 1$ and $f_2(-1) = -1$.

Sufficient conditions for the last three mapping requirements to be satisfied can be determined as follows:

$$s = \sigma + j\Omega = k\frac{(\text{Re}\{f_1(z)\} - 1) + j\text{Im}\{f_1(z)\}}{(\text{Re}\{f_2(z)\} + 1) + j\text{Im}\{f_2(z)\}} \tag{6.116}$$

Table 6.6. *Correspondence of key points of the s and z planes using the bilinear transformation method.*

s plane	→	z plane
$\sigma \pm j\Omega$	→	$re^{\pm j\omega}$
$j0$	→	1
$j\infty$	→	-1
$\sigma > 0$	→	$r > 1$
$\sigma = 0$	→	$r = 1$
$\sigma < 0$	→	$r < 1$
$j\Omega$	→	$e^{j\omega}$
$-\infty < \Omega < \infty$	→	$-\pi < \omega < \pi$

Equating the real parts of both sides of the above equation, we have that

$$\sigma = k\frac{(\text{Re}\{f_1(z)\} - 1)(\text{Re}\{f_2(z)\} + 1) + \text{Im}\{f_1(z)\}\text{Im}\{f_2(z)\}}{(\text{Re}\{f_2(z)\} + 1)^2 + (\text{Im}\{f_2(z)\})^2} \tag{6.117}$$

and since $\sigma = 0$ implies that $r = 1$, the following relation is valid:

$$\frac{\text{Re}\{f_1(e^{j\omega})\} - 1}{\text{Im}\{f_1(e^{j\omega})\}} = -\frac{\text{Im}\{f_2(e^{j\omega})\}}{\text{Re}\{f_2(e^{j\omega})\} + 1} \tag{6.118}$$

The condition $\sigma < 0$ is equivalent to

$$\frac{\text{Re}\{f_1(re^{j\omega})\} - 1}{\text{Im}\{f_1(re^{j\omega})\}} < -\frac{\text{Im}\{f_2(re^{j\omega})\}}{\text{Re}\{f_2(re^{j\omega})\} + 1}, \quad r < 1 \tag{6.119}$$

The last two lines of Table 6.6 show the correspondence between the analog frequency and the unit circle of the z plane.

If we want the orders of the discrete-time and continuous-time systems to remain the same after the transformation, then $f_1(z)$ and $f_2(z)$ must be first-order polynomials. In addition, if we wish to satisfy the conditions imposed by equation (6.115), we must choose $f_1(z) = f_2(z) = z$. It is straightforward to verify that equation (6.118), as well as the inequality (6.119), are automatically satisfied with this choice for $f_1(z)$ and $f_2(z)$.

The bilinear transformation is then given by

$$s \to k\frac{z - 1}{z + 1} \tag{6.120}$$

which, for $s = j\Omega$ and $z = e^{j\omega}$, is equivalent to

$$j\Omega \to k\frac{e^{j\omega} - 1}{e^{j\omega} + 1} = k\frac{e^{\frac{j\omega}{2}} - e^{\frac{-j\omega}{2}}}{e^{\frac{j\omega}{2}} + e^{\frac{-j\omega}{2}}} = jk\frac{\sin\frac{\omega}{2}}{\cos\frac{\omega}{2}} = jk\tan\frac{\omega}{2} \tag{6.121}$$

that is

$$\Omega \rightarrow k \tan \frac{\omega}{2} \tag{6.122}$$

For small frequencies, $\tan \frac{\omega}{2} \approx \frac{\omega}{2}$. Hence, to keep the magnitude response of the digital filter approximately the same as the prototype analog filter at low frequencies, we should have for small frequencies $\Omega = \omega \frac{\Omega_s}{2\pi}$, and therefore we should choose $k = \frac{\Omega_s}{\pi} = \frac{2}{T}$. In conclusion, the bilinear transformation of a continuous-time transfer function into a discrete-time transfer function is implemented through the following mapping:

$$H(z) = H_a(s)|_{s=\frac{2}{T}\frac{z-1}{z+1}} \tag{6.123}$$

and therefore, the bilinear transformation maps analog frequencies into digital frequencies as follows:

$$\Omega \rightarrow \frac{2}{T} \tan \frac{\omega}{2} \tag{6.124}$$

For high frequencies, this relationship is highly nonlinear, as seen in Figure 6.9a, corresponding to a large distortion in the frequency response of the digital filter when compared to the analog prototype. The distortion in the magnitude response, also known as the warping effect, can be visualized as in Figure 6.9b.

The warping effect caused by the bilinear transformation can be compensated for by prewarping the frequencies given at the specifications before the analog filter is actually designed. For example, suppose we wish to design a lowpass digital filter with cutoff frequency ω_p and stopband edge ω_r. The prewarped specifications Ω_{a_p} and Ω_{a_r} of the lowpass analog prototype are then given by

$$\Omega_{a_p} = \frac{2}{T} \tan \frac{\omega_p}{2} \tag{6.125}$$

$$\Omega_{a_r} = \frac{2}{T} \tan \frac{\omega_r}{2} \tag{6.126}$$

Following the same line of thought, we may apply prewarping to as many frequencies of interest as specified for the digital filter. If these frequencies are given by ω_i, for $i = 1, 2, \ldots, n$, then the frequencies to be included in the analog filter specifications are

$$\Omega_{a_i} = \frac{2}{T} \tan \frac{\omega_i}{2} \tag{6.127}$$

for $i = 1, 2, \ldots, n$.

Hence, the design procedure using the bilinear transformation method can be summarized as follows:

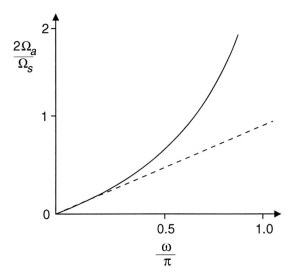

(a)

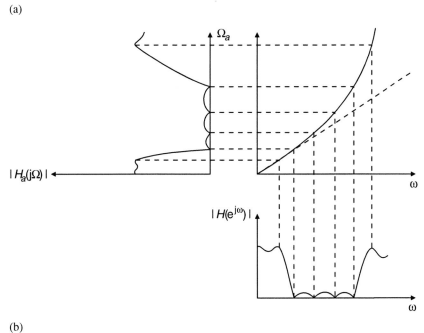

(b)

Figure 6.9 Bilinear transformation method. (a) Relation between the analog and digital frequencies. (b) Warping effect in the magnitude response of a bandstop filter.

(i) Prewarp all the prescribed frequency specifications ω_i, obtaining Ω_{a_i}, for $i = 1, 2, \ldots, n$.

(ii) Generate $H_a(s)$, following the procedure given in Subsection 6.2.5, satisfying the specifications for the frequencies Ω_{a_i}.

(iii) Obtain $H_d(z)$, by replacing s with $\frac{2}{T}\frac{z-1}{z+1}$ in $H_a(s)$.

With the bilinear transformation, we can design Butterworth, Chebyshev, and elliptic digital filters starting with a corresponding analog prototype. The bilinear transformation method always generates stable digital filters as long as the prototype analog filter is stable. Using the prewarping procedure, the method keeps the magnitude characteristics of the prototype but introduces distortions to the phase response.

EXAMPLE 6.3

Design a digital elliptic bandpass filter satisfying the following specifications:

$$
\left.
\begin{aligned}
A_p &= 0.5 \text{ dB} \\
A_r &= 65 \text{ dB} \\
\Omega_{r_1} &= 850 \text{ rad/s} \\
\Omega_{p_1} &= 980 \text{ rad/s} \\
\Omega_{p_2} &= 1020 \text{ rad/s} \\
\Omega_{r_2} &= 1150 \text{ rad/s} \\
\Omega_s &= 10\,000 \text{ rad/s}
\end{aligned}
\right\}
\tag{6.128}
$$

SOLUTION

First, we have to normalize the frequencies above to the range of digital frequencies using the expression $\omega = \Omega \frac{2\pi}{\Omega_s}$. Since $\Omega_s = 10\,000$ rad/s, we have that

$$
\left.
\begin{aligned}
\omega_{r_1} &= 0.5341 \text{ rad/sample} \\
\omega_{p_1} &= 0.6158 \text{ rad/sample} \\
\omega_{p_2} &= 0.6409 \text{ rad/sample} \\
\omega_{r_2} &= 0.7226 \text{ rad/sample}
\end{aligned}
\right\}
\tag{6.129}
$$

Then, by applying equation (6.127), the prewarped frequencies become

$$
\left.
\begin{aligned}
\Omega_{a_{r_1}} &= 870.7973 \text{ rad/s} \\
\Omega_{a_{p_1}} &= 1012.1848 \text{ rad/s} \\
\Omega_{a_{p_2}} &= 1056.4085 \text{ rad/s} \\
\Omega_{a_{r_2}} &= 1202.7928 \text{ rad/s}
\end{aligned}
\right\}
\tag{6.130}
$$

By making $\Omega_{a_{r_1}} = 888.9982$ rad/s to obtain a geometrically symmetric filter, from Table 6.2, we have that

$$\Omega_0 = 1034.0603 \text{ rad/s} \tag{6.131}$$

$$B = 44.2237 \text{ rad/s} \tag{6.132}$$

$$a = 2.6638 \tag{6.133}$$

$$\Omega_p' = 0.3754 \tag{6.134}$$

$$\Omega_r' = 2.6638 \tag{6.135}$$

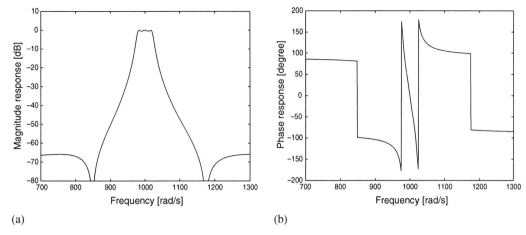

Figure 6.10 Digital elliptic bandpass filter: (a) magnitude response; (b) phase response.

From the filter specifications, the required order for the analog elliptic normalized-lowpass filter is $n = 3$, and the resulting normalized transfer function is

$$H'(s') = 4.0426 \times 10^{-3} \frac{s'^2 + 9.4372}{s'^3 + 0.4696s'^2 + 0.2162s' + 0.0382} \tag{6.136}$$

The denormalized design is then obtained by applying the lowpass to bandpass transformation given in Table 6.2, with $a = 2.6638$. After applying the bilinear transformation the resulting digital bandpass transfer function becomes

$$H(z) = H_0 \frac{b_6 z^6 + b_5 z^5 + b_4 z^4 + b_3 z^3 + b_2 z^2 + b_1 z + b_0}{a_6 z^6 + a_5 z^5 + a_4 z^4 + a_3 z^3 + a_2 z^2 + a_1 z + a_0} \tag{6.137}$$

where all filter coefficients, zeros, and poles are listed in Table 6.3.2.

Figure 6.10 depicts the frequency response of the resulting digital elliptic bandpass filter.

$$\triangle$$

From the previous discussion, we can observe that, as we perform the mapping $s \to z$, either opting for the impulse-invariance method or the bilinear transformation method, we are essentially folding the continuous-time frequency axis around the z-domain unit circle. For that matter, any digital frequency response is periodic, and the interval $-\frac{\Omega_s}{2} \le \Omega \le \frac{\Omega_s}{2}$, or equivalently, $-\pi \le \omega \le \pi$, is the so-called fundamental period. We should bear in mind that in the expressions developed in this subsection the unfolded analog frequencies Ω are related to the digital frequencies ω, which are restricted to the interval $-\pi \le \omega \le \pi$. Therefore, all digital filter specifications should be normalized to the interval $[-\pi, \pi]$ using the expression

$$\omega = \Omega \frac{2\pi}{\Omega_s} \tag{6.138}$$

Table 6.7. *Characteristics of the digital elliptic bandpass filter. Constant gain:* $H_0 = 1.346\,099\,3 \times 10^{-4}$.

Numerator coefficients	Denominator coefficients	Filter zeros	Filter poles
$b_0 = -1.0$	$a_0 = 0.969\,050\,45$		
$b_1 = 3.202\,456\,01$	$a_1 = -4.728\,514\,52$	$z_1 = 0.739\,900\,69 + j0.672\,716\,12$	$p_1 = 0.798\,2249 + j0.595\,7678$
$b_2 = -3.549\,186\,70$	$a_2 = 10.628\,545\,61$	$z_2 = 0.739\,900\,69 - j0.672\,716\,12$	$p_2 = 0.798\,2249 - j0.595\,7678$
$b_3 = 0.0$	$a_3 = -13.726\,132\,28$	$z_3 = 0.861\,327\,32 + j0.508\,050\,36$	$p_3 = 0.813\,3921 + j0.575\,1252$
$b_4 = 3.549\,186\,70$	$a_4 = 10.740\,516\,30$	$z_4 = 0.861\,327\,32 - j0.508\,050\,36$	$p_4 = 0.813\,3921 - j0.575\,1252$
$b_5 = -3.202\,456\,01$	$a_5 = -4.828\,668\,17$	$z_5 = 1.0 + j0.0$	$p_5 = 0.802\,7170 + j0.583\,0217$
$b_6 = 1.0$	$a_6 = 1.0$	$z_6 = -1.0 + j0.0$	$p_6 = 0.802\,7170 - j0.583\,0217$

In the discussion that follows, we assume that $\Omega_s = 2\pi$ rad/s, unless otherwise specified.

6.4   Frequency transformation in the discrete-time domain

Usually, in the approximation of a continuous-time filter, we begin by designing a normalized-lowpass filter and then, through a frequency transformation, the filter with the specified magnitude response is obtained. In the design of digital filters, we can also start by designing a digital lowpass filter and then apply a frequency transformation in the discrete-time domain.

The procedure consists of replacing the variable z by an appropriate function $g(z)$ to generate the desired magnitude response. The function $g(z)$ needs to meet some constraints to be a valid transformation (Constantinides, 1970), namely:

- The function $g(z)$ must be a ratio of polynomials, since the digital filter transfer function must remain a ratio of polynomials after the transformation.
- The mapping $z \rightarrow g(z)$ must be such that the filter stability is maintained, that is, stable filters generate stable transformed filters and unstable filters generate unstable transformed filters. This is equivalent to saying that the transformation maps the interior of the unit circle onto the interior of the unit circle and the exterior of the unit circle onto the exterior of the unit circle.

It can be shown that a function $g(z)$ satisfying the above conditions is of the form

$$g(z) = \pm \left[\prod_{i=1}^{n} \frac{(z - \alpha_i)}{(1 - z\alpha_i^*)} \frac{(z - \alpha_i^*)}{(1 - z\alpha_i)} \right] \left(\prod_{i=n+1}^{m} \frac{z - \alpha_i}{1 - z\alpha_i} \right) \tag{6.139}$$

where α_i^* is complex conjugate of α_i, and α_i is real for $n + 1 \leq i \leq m$.

In the following subsections, we analyze special cases of $g(z)$ that generate lowpass to lowpass, lowpass to highpass, lowpass to bandpass, and lowpass to bandstop transformations.

6.4.1   Lowpass to lowpass transformation

One necessary condition for a lowpass to lowpass transformation is that a magnitude response must keep its original values at $\omega = 0$ and $\omega = \pi$ after the transformation. Therefore, we must have

$$g(1) = 1 \tag{6.140}$$

$$g(-1) = -1 \tag{6.141}$$

Another necessary condition is that the frequency response should only be warped between $\omega = 0$ and $\omega = \pi$, that is, a full turn around the unit circle in z must correspond to a full turn around the unit circle in $g(z)$.

One possible $g(z)$ in the form of equation (6.139) that satisfies these conditions is

$$g(z) = \frac{z - \alpha}{1 - \alpha z} \tag{6.142}$$

where α is real such that $|\alpha| < 1$.

Assuming that the passband edge frequency of the original lowpass filter is given by ω_p, and that we wish to transform the original filter into a lowpass filter with cutoff frequency at ω_{p_1}, the following relation must be valid:

$$e^{j\omega_p} = \frac{e^{j\omega_{p_1}} - \alpha}{1 - \alpha e^{j\omega_{p_1}}} \tag{6.143}$$

and then

$$\alpha = \frac{e^{-j\left(\frac{\omega_p - \omega_{p_1}}{2}\right)} - e^{j\left(\frac{\omega_p - \omega_{p_1}}{2}\right)}}{e^{-j\left(\frac{\omega_p + \omega_{p_1}}{2}\right)} - e^{j\left(\frac{\omega_p + \omega_{p_1}}{2}\right)}} = \frac{\sin\left(\frac{\omega_p - \omega_{p_1}}{2}\right)}{\sin\left(\frac{\omega_p + \omega_{p_1}}{2}\right)} \tag{6.144}$$

The desired transformation is then implemented by replacing z by $g(z)$ given in equation (6.142), with α calculated as indicated in equation (6.144).

6.4.2 Lowpass to highpass transformation

If ω_{p_1} is the highpass filter band edge and ω_p is the lowpass filter cutoff frequency, the lowpass to highpass transformation function is given by

$$g(z) = -\frac{z + \alpha}{\alpha z + 1} \tag{6.145}$$

where

$$\alpha = -\frac{\cos\left(\frac{\omega_p + \omega_{p_1}}{2}\right)}{\cos\left(\frac{\omega_p - \omega_{p_1}}{2}\right)} \tag{6.146}$$

6.4.3 Lowpass to bandpass transformation

The lowpass to bandpass transformation is accomplished if the following mappings occur

$$g(1) = -1 \tag{6.147}$$

$$g(e^{j\omega_{p_1}}) = e^{j\omega_p} \tag{6.148}$$

$$g(e^{-j\omega_{p_2}}) = e^{j\omega_p} \tag{6.149}$$

$$g(-1) = -1 \tag{6.150}$$

where ω_{p_1} and ω_{p_2} are the band edges of the bandpass filter, and ω_p is the band edge of the lowpass filter. Since the bandpass filter has two passband edges, we need a second-order function $g(z)$ to accomplish the lowpass to bandpass transformation. After some manipulation, it can be inferred that the required transformation and its parameters are given by (Constantinides, 1970)

$$g(z) = -\frac{z^2 + \alpha_1 z + \alpha_2}{\alpha_2 z^2 + \alpha_1 z + 1} \tag{6.151}$$

with

$$\alpha_1 = -\frac{2\alpha k}{k+1} \tag{6.152}$$

$$\alpha_2 = \frac{k-1}{k+1} \tag{6.153}$$

where

$$\alpha = \frac{\cos\left(\frac{\omega_{p_2}+\omega_{p_1}}{2}\right)}{\cos\left(\frac{\omega_{p_2}-\omega_{p_1}}{2}\right)} \tag{6.154}$$

$$k = \tan\left(\frac{\omega_{p_2}-\omega_{p_1}}{2}\right)\tan\left(\frac{\omega_p}{2}\right) \tag{6.155}$$

6.4.4   Lowpass to bandstop transformation

The lowpass to bandstop transformation function $g(z)$ is given by

$$g(z) = -\frac{z^2 + \alpha_1 z + \alpha_2}{\alpha_2 z^2 + \alpha_1 z + 1} \tag{6.156}$$

with

$$\alpha_1 = -\frac{2\alpha}{k+1} \tag{6.157}$$

$$\alpha_2 = \frac{1-k}{1+k} \tag{6.158}$$

where

$$\alpha = \frac{\cos\left(\frac{\omega_{p_2}+\omega_{p_1}}{2}\right)}{\cos\left(\frac{\omega_{p_2}-\omega_{p_1}}{2}\right)} \tag{6.159}$$

$$k = \tan\left(\frac{\omega_{p_2}-\omega_{p_1}}{2}\right)\tan\left(\frac{\omega_p}{2}\right) \tag{6.160}$$

6.4.5   Variable cutoff filter design

An interesting application for the frequency transformations, first proposed in Constantinides (1970), is to design highpass and lowpass filters with variable cutoff frequency with the cutoff frequency being directly controlled by a single parameter α. This method can be best understood through an example, as given below.

EXAMPLE 6.4

Consider the lowpass notch filter

$$H(z) = 0.004 \frac{z^2 - \sqrt{2}z + 1}{z^2 - 1.8z + 0.96} \tag{6.161}$$

whose zeros are located at $z = \frac{\sqrt{2}}{2}(1 \pm j)$. Transform this filter into a highpass notch with a zero at frequency $\omega_{p_1} = \frac{\pi}{6}$ rad/sample. Plot the magnitude responses before and after the frequency transformation.

SOLUTION

Using the lowpass to highpass transformation given in equation (6.145), the highpass transfer function is of the form

$$H(z) = H_0 \frac{(\alpha^2 + \sqrt{2}\alpha + 1)(z^2 + 1)(\sqrt{2}\alpha^2 + 4\alpha + \sqrt{2})z}{(0.96\alpha^2 + 1.8\alpha + 1)z^2 + (1.8\alpha^2 + 3.92\alpha + 1.8)z + (\alpha^2 + 1.8\alpha + 0.96)} \tag{6.162}$$

with $H_0 = 0.004$. The parameter α can control the position of the zeros of the highpass notch filter. For instance, in this example, as the original zero is at $\omega_p = \frac{\pi}{4}$ rad/sample and the desired zero is at $\omega_{p_1} = \frac{\pi}{6}$ rad/sample, the parameter α should be, as given in equation (6.146), equal to

$$\alpha = -\frac{\cos\left(\frac{\frac{\pi}{4} + \frac{\pi}{6}}{2}\right)}{\cos\left(\frac{\frac{\pi}{4} - \frac{\pi}{6}}{2}\right)} = -0.8002 \tag{6.163}$$

The magnitude responses corresponding to the lowpass and highpass transfer functions are seen in Figure 6.11. Notice how the new transfer function has indeed a zero at the desired position.

$\triangle$

6.5   Magnitude and phase approximation

In this section, we discuss the approximation of IIR digital filters using optimization techniques aimed at the simultaneous approximation of the magnitude and phase

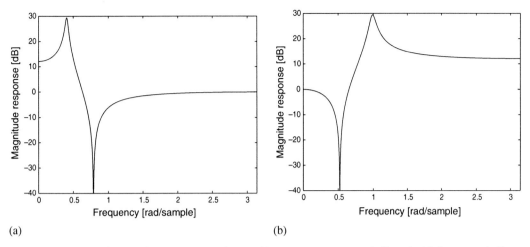

Figure 6.11 Magnitude responses of notch filters. (a) Lowpass notch filter. (b) Highpass notch filter.

responses. The same approach is useful in designing continuous-time filters and FIR filters. However, in the case of FIR filters more efficient approaches exist, as we have seen in Subsections 5.6.2 and 5.6.3.

6.5.1 Basic principles

Assume that $H(z)$ is the transfer function of an IIR digital filter. Then $H(e^{j\omega})$ is a function of the filter coefficients, which are usually grouped into a single vector γ, and of the independent variable $\theta = \omega$.

The frequency response of a digital filter can be expressed as a function of the filter parameters γ and θ, that is $F(\gamma, \theta)$, and a desired frequency response is usually referred to as $f(\theta)$.

The complete specification of an optimization problem involves: definition of an objective function (also known as a cost function), determination of the form of the transfer function $H(z)$ and its coefficients γ, and the solution methods for the optimization problem. These three items are further discussed below.

- Choosing the objective function: a widely used type of objective function in filter design is the weighted L_p norm, defined as (Deczky, 1972)

$$\|L(\gamma)\|_p = \left(\int_0^\pi W(\theta) |F(\gamma, \theta) - f(\theta)|^p d\theta \right)^{\frac{1}{p}} \tag{6.164}$$

where $W(\theta) > 0$ is the so-called weight function.

Problems based on the L_p-norm minimization criteria with different values of p lead, in general, to different solutions. An appropriate choice for the value of p depends on the type of error which is acceptable for the given application. For example, when we wish to minimize the mean-square value of the error between the desired and the designed responses, we should choose $p = 2$. Another problem is the minimization of the maximum deviation between the desired specification and the designed filter by searching the space of parameters. This case, which is known as the Chebyshev or minimax criterion, corresponds to $p \to \infty$. This important result derived from the optimization theory can be stated more formally as (Deczky, 1972):

THEOREM 6.1

For a given coefficient space P and a given angle space X_θ, there is a unique optimal minimax approximation $F(\gamma_\infty^*, \theta)$ for $f(\theta)$. In addition, if the best L_p approximation for the function $f(\theta)$ is denoted by $F(\gamma_p^*, \theta)$, then it can be demonstrated that

$$\lim_{p \to \infty} \gamma_p^* = \gamma_\infty^* \tag{6.165}$$

$\diamond$

This result shows that we can use any minimization program based on the L_p norm to find a minimax (or approximately minimax) solution, by progressively calculating the L_p optimal solution with, for instance, $p = 2, 4, 6$, and so on, indefinitely.

Specifically, the minimax criterion for a continuous frequency function is best defined as

$$\left\| L(\gamma^*) \right\|_\infty = \min_{\gamma \in P} \{ \max_{\theta \in X_\theta} \{ W(\theta) | F(\gamma, \theta) - f(\theta) | \} \} \tag{6.166}$$

In practice, due to several computational aspects, it is more convenient to use a simplified objective function given by

$$L_{2p}(\gamma) = \sum_{k=1}^{K} W(\theta_k) \left(F(\gamma, \theta_k) - f(\theta_k) \right)^{2p} \tag{6.167}$$

where, by minimizing $L_{2p}(\gamma)$, we also minimize $\|L(\gamma)\|_{2p}$. In this case, the minimax solution is obtained by minimizing $L_{2p}(\gamma)$, for $p = 1, 2, 3$, and so on, indefinitely.

The points θ_k are the angles chosen to sample the desired and the prototype frequency responses. These points, lying on the unit circle do not need to be equally spaced. In fact, we usually choose θ_k such that there are denser grids in the regions where the error function has more variations.

The sort of filter designs described here can be applied to a large class of problems, in particular the design of filters with arbitrary magnitude response, phase equalizers, and filters with simultaneous specifications of magnitude and phase responses. The last two classes are illustrated below.

– Phase equalizer: the transfer function of a phase equalizer is

$$H_l(z) = \prod_{i=1}^{M} \frac{a_{2i} z^2 + a_{1i} z + 1}{z^2 + a_{1i} z + a_{2i}} \qquad (6.168)$$

Since its magnitude is unitary, the objective function becomes

$$L_{2p}\tau(\boldsymbol{\gamma}, \tau_0) = \sum_{k=1}^{K} W(\theta_k) \, (\tau_l(\boldsymbol{\gamma}, \theta_k) - \tau_s(\theta_k) + \tau_0)^{2p} \qquad (6.169)$$

where τ_s is the group delay of the original digital filter, $\tau_l(\boldsymbol{\gamma}, \theta_k)$ is the equalizer group delay and τ_0 is a constant delay, whose value minimizes $\sum_{k=1}^{K}(\tau_s(\theta_k) - \tau_0)^{2p}$.

– Simultaneous approximation of magnitude and group-delay responses: for this type of approximation, the objective function can be given by

$$L_{2p,2q} M, \tau(\boldsymbol{\gamma}, \tau_0) = \delta \sum_{k=1}^{K} W_M(\theta_k) \, (M(\boldsymbol{\gamma}, \theta_k) - f(\theta_k))^{2p}$$

$$+ (1 - \delta) \sum_{r=1}^{R} W_\tau(\theta_k) \, (\tau(\boldsymbol{\gamma}, \theta_r) + \tau(\theta_r) - \tau_0)^{2p} \qquad (6.170)$$

where $0 \le \delta \le 1$ and $\tau(\theta_r)$ is the group delay which we wish to equalize.

Usually, in the simultaneous approximation of magnitude and group-delay responses, the numerator of $H(z)$ is forced to have zeros on the unit circle or in reciprocal pairs, such that the group delay is a function of the poles of $H(z)$ only. The task of the zeros would be to shape the magnitude response.

• Choosing the form of the transfer function: one of the most convenient ways to describe an IIR $H(z)$ is the cascade form decomposition, because the filter stability can be easily tested and controlled. In this case, the coefficient vector $\boldsymbol{\gamma}$, according to equation (4.32), is of the form

$$\boldsymbol{\gamma} = (\gamma'_{11}, \gamma'_{21}, m_{11}, m_{21}, \ldots, \gamma'_{1i}, \gamma'_{2i}, m_{1i}, m_{2i}, \ldots, H_0) \qquad (6.171)$$

Unfortunately, the expressions for the magnitude and group delay of $H(z)$, as a function of the coefficients of the second-order sections, are very complicated. The same comment holds for the expressions of the partial derivatives of $H(z)$ with respect to the coefficients, which are also required in the optimization algorithm.

An alternative solution is to use the poles and zeros of the second-order sections represented in polar coordinates as parameters. In this case, the coefficient vector $\boldsymbol{\gamma}$ becomes

$$\boldsymbol{\gamma} = (r_{z1}, \phi_{z1}, r_{p1}, \phi_{p1}, \ldots, r_{zi}, \phi_{zi}, r_{pi}, \phi_{pi}, \ldots, k_0) \qquad (6.172)$$

and the magnitude and group-delay responses are expressed as

$$
M(\gamma, \omega) = k_0 \prod_{i=1}^{m} \left\{ \frac{\left[1 - 2r_{zi}\cos(\omega - \phi_{zi}) + r_{zi}^2\right]^{\frac{1}{2}}}{\left[1 - 2r_{pi}\cos(\omega - \phi_{pi}) + r_{pi}^2\right]^{\frac{1}{2}}} \right\}
$$

$$
\times \left\{ \frac{\left[1 - 2r_{zi}\cos(\omega + \phi_{zi}) + r_{zi}^2\right]^{\frac{1}{2}}}{\left[1 - 2r_{pi}\cos(\omega + \phi_{pi}) + r_{pi}^2\right]^{\frac{1}{2}}} \right\}
\tag{6.173}
$$

$$
\tau(\gamma, \omega) = \sum_{i=1}^{N} \left[\frac{1 - r_{pi}\cos(\omega - \phi_{pi})}{1 - 2r_{pi}\cos(\omega - \phi_{pi}) + r_{pi}^2} + \frac{1 - r_{pi}\cos(\omega + \phi_{pi})}{1 - 2r_{pi}\cos(\omega + \phi_{pi}) + r_{pi}^2} \right.
$$

$$
\left. - \frac{1 - r_{zi}\cos(\omega - \phi_{zi})}{1 - 2r_{zi}\cos(\omega - \phi_{zi}) + r_{zi}^2} - \frac{1 - r_{zi}\cos(\omega + \phi_{zi})}{1 - 2r_{zi}\cos(\omega + \phi_{zi}) + r_{zi}^2} \right]
\tag{6.174}
$$

respectively.

In an optimization problem such as this, the first- and second-order derivatives should be determined using closed-form formulas to speed up the convergence process. In fact, the use of numerical approximation to calculate such derivatives would make the optimization procedure too complex to be implemented even in the fastest personal computers. The partial derivatives of the magnitude and group delay with respect to the radii and angles of the poles and zeros, that are required in the optimization processes, are given below:

$$
\frac{\partial M}{\partial r_{zi}} = M(\gamma, \omega) \left[\frac{r_{zi} - \cos(\omega - \phi_{zi})}{1 - 2r_{zi}\cos(\omega - \phi_{zi}) + r_{zi}^2} + \frac{r_{zi} - \cos(\omega + \phi_{zi})}{1 - 2r_{zi}\cos(\omega + \phi_{zi}) + r_{zi}^2} \right]
\tag{6.175}
$$

$$
\frac{\partial M}{\partial \phi_{zi}} = -M(\gamma, \omega) \left[\frac{r_{zi}\sin(\omega - \phi_{zi})}{1 - 2r_{zi}\cos(\omega + \phi_{zi}) + r_{zi}^2} - \frac{r_{zi}\sin(\omega + \phi_{zi})}{1 - 2r_{zi}\cos(\omega + \phi_{zi}) + r_{zi}^2} \right]
\tag{6.176}
$$

$$
\frac{\partial \tau}{\partial r_{pi}} = \left\{ \frac{(1 + r_{pi}^2)\cos(\omega - \phi_{pi}) - 2r_{pi}}{\left[1 - 2r_{pi}\cos(\omega - \phi_{pi}) + r_{pi}^2\right]^2} + \frac{(1 + r_{pi}^2)\cos(\omega + \phi_{pi}) - 2r_{pi}}{\left[1 - 2r_{pi}\cos(\omega + \phi_{pi}) + r_{pi}^2\right]^2} \right\}
\tag{6.177}
$$

$$
\frac{\partial \tau}{\partial \phi_{pi}} = \left\{ \frac{r_{pi}(1 - r_{pi}^2)\sin(\omega - \phi_{pi})}{\left[1 - 2r_{pi}\cos(\omega - \phi_{pi}) + r_{pi}^2\right]^2} - \frac{r_{pi}(1 - r_{pi}^2)\sin(\omega + \phi_{pi})}{\left[1 - 2r_{pi}\cos(\omega + \phi_{pi}) + r_{pi}^2\right]^2} \right\}
\tag{6.178}
$$

We also need $\frac{\partial M}{\partial r_{pi}}$, $\frac{\partial M}{\partial \phi_{pi}}$, $\frac{\partial \tau}{\partial r_{zi}}$, and $\frac{\partial \tau}{\partial \phi_{zi}}$, which are similar to the expressions above. These derivatives are part of the expressions of the partial derivatives of the objective function with respect to the filter poles and zeros which are the derivatives used in the optimization algorithms employed. These derivatives are:

$$\frac{\partial L_{2p} M(\gamma)}{\partial r_{zi}} = \sum_{k=1}^{K} 2p W_M(\theta_k) \frac{\partial M}{\partial r_{zi}} (M(\gamma, \theta_k) - f(\theta_k))^{2p-1} \tag{6.179}$$

$$\frac{\partial L_{2p} \tau(\gamma)}{\partial r_{zi}} = \sum_{k=1}^{K} 2p W_\tau(\theta_k) \frac{\partial \tau}{\partial r_{zi}} (\tau(\gamma, \theta_k) - f(\theta_k))^{2p-1} \tag{6.180}$$

Analogously, we need the expressions for $\frac{\partial L_{2p} M(\gamma)}{\partial \phi_{zi}}$, $\frac{\partial L_{2p} M(\gamma)}{\partial r_{pi}}$, $\frac{\partial L_{2p} M(\gamma)}{\partial \phi_{pi}}$, $\frac{\partial L_{2p} \tau(\gamma)}{\partial \phi_{zi}}$, $\frac{\partial L_{2p} \tau(\gamma)}{\partial r_{pi}}$, and $\frac{\partial L_{2p} \tau(\gamma)}{\partial \phi_{pi}}$, which are similar to the expressions given above.

It is important to note that we are interested only in generating stable filters. Since we are performing a search for the minimum of the error function on the parameter space P, the region in which the optimal parameter should be searched is a restricted subspace $P_s = \{\gamma \mid r_{pi} < 1, \forall i\}$.

- Choosing the optimization procedure: there are several optimization methods suitable for solving the problem of filter approximation. Choosing the best method depends heavily on the designer's experience in dealing with this problem and on the available computer resources. The optimization algorithms used are such that they will converge only if the error function has a local minimum in the interior of the subspace P_s and not on the boundaries of P_s. In the present case, this is not a cause for concern because the magnitude and group delay of digital filters become large when a pole approaches the unit circle (Deczky, 1972) and, as a consequence, there is no local minimum corresponding to poles on the unit circle. In this manner, if we start the search from the interior of P_s, that is, with all the poles strictly inside the unit circle, and constrain our search to the subspace P_s, a local minimum not located at the boundary of P_s will be reached for certain.

Due to the importance of this step for setting up a procedure for designing IIR digital filters, for the sake of completeness and clarity of presentation, its discussion is left to the next subsection, which is devoted exclusively to it.

6.5.2   Multi-variable function minimization method

An n-variable function $F(\mathbf{x})$ can be approximated by a quadratic function in a small region around a given operating point. For instance, in a region close to a point $\mathbf{x}_k$, we can write that

$$F(\mathbf{x}_k + \boldsymbol{\delta}_k) \approx F(\mathbf{x}_k) + \mathbf{g}^T(\mathbf{x}_k)\boldsymbol{\delta}_k + \frac{1}{2}\boldsymbol{\delta}_k^T \mathbf{H}(\mathbf{x}_k)\boldsymbol{\delta}_k \tag{6.181}$$

where

$$\mathbf{g}^T(\mathbf{x}_k) = \left[\frac{\partial F}{\partial x_1}, \frac{\partial F}{\partial x_2}, \dots, \frac{\partial F}{\partial x_n} \right] \tag{6.182}$$

is the gradient vector of $F(\mathbf{x})$ at the operating point $\mathbf{x}_k$ and $\mathbf{H}(\mathbf{x}_k)$ is the Hessian matrix of $F(\mathbf{x})$ defined as

$$
\mathbf{H}(\mathbf{x}_k) = \begin{bmatrix}
\dfrac{\partial^2 F}{\partial x_1^2} & \dfrac{\partial^2 F}{\partial x_1 \partial x_2} & \cdots & \dfrac{\partial^2 F}{\partial x_1 \partial x_n} \\[2mm]
\dfrac{\partial^2 F}{\partial x_2 \partial x_1} & \dfrac{\partial^2 F}{\partial x_2^2} & \cdots & \dfrac{\partial^2 F}{\partial x_2 \partial x_n} \\[2mm]
\vdots & \vdots & \ddots & \vdots \\[2mm]
\dfrac{\partial^2 F}{\partial x_n \partial x_1} & \dfrac{\partial^2 F}{\partial x_n \partial x_2} & \cdots & \dfrac{\partial^2 F}{\partial x_n^2}
\end{bmatrix}
\tag{6.183}
$$

Clearly, if $F(\mathbf{x})$ is a quadratic function, the right-hand side of equation (6.181) is minimized when

$$
\boldsymbol{\delta}_k = -\mathbf{H}^{-1}(\mathbf{x}_k)\mathbf{g}(\mathbf{x}_k)
\tag{6.184}
$$

If, however, the function $F(\mathbf{x})$ is not quadratic and the operating point is far away from a local minimum, we can devise an algorithm which iteratively searches the minimum in the direction of $\boldsymbol{\delta}_k$ as

$$
\mathbf{x}_{k+1} = \mathbf{x}_k + \boldsymbol{\delta}_k = \mathbf{x}_k - \alpha_k \mathbf{H}^{-1}(\mathbf{x}_k)\mathbf{g}(\mathbf{x}_k)
\tag{6.185}
$$

where the convergence factor, α_k, is a scalar that minimizes, in the kth iteration, $F(\mathbf{x}_k + \boldsymbol{\delta}_k)$ in the direction of $\boldsymbol{\delta}_k$. There are several procedures for determining the value of α_k (Fletcher, 1980; Luenberger, 1984), which can be considered as two classes: exact and inexact line searches. As a general rule of thumb, an inexact line search should be used when the operating point is far from a local minimum, because in these conditions it is appropriate to trade accuracy for faster results. However, when the parameters approach a minimum, accuracy becomes an important issue, and an exact line search is the best choice.

The minimization procedure described above is widely known as the Newton method. The main drawbacks related to this method are the need for computation of the second-order derivatives of the objective function $F(\mathbf{x})$ with respect to the parameters in $\mathbf{x}$ and the necessity of inverting the Hessian matrix.

Due to these two reasons, the most widely used methods for the solution of the simultaneous approximation of magnitude and phase are the so-called quasi-Newton methods (Fletcher, 1980; Luenberger, 1984). These methods are characterized by an attempt to build the inverse of the Hessian matrix, or an approximation of it, using the data obtained during the optimization process. The updated approximation of the Hessian inverse is used in each step of the algorithm in order to define the next direction in which to search for the minimum of the objective function.

A general structure of an optimization algorithm suitable for designing digital filters is given below.

(i)   Algorithm initialization:
      Set the iteration counter as $k = 0$.
      Choose the initial vector $\mathbf{x}_0$ corresponding to a stable filter.
      Use the identity as the first estimate of the Hessian inverse, that is, $\mathbf{P}_k = \mathbf{I}$.
      Compute $F_0 = F(\mathbf{x}_0)$.
(ii) Convergence check:
      Check if convergence was achieved by using an appropriate criterion. For
      example, a criterion would be to verify if $F_k < \epsilon$ where ϵ is a pre-defined error
      threshold. An alternative criterion is to verify that $\| \mathbf{x}_k - \mathbf{x}_{k-1} \|^2 < \epsilon'$.
      If the algorithm converged, go to step (iv), otherwise go on to step (iii).
(iii) Algorithm iteration:
      Compute $\mathbf{g}_k = \mathbf{g}(\mathbf{x}_k)$.
      Set $\mathbf{s}_k = -\mathbf{P}_k\mathbf{g}_k$.
      Compute α_k that minimizes $F(\mathbf{x})$ in the direction of $\mathbf{s}_k$.
      Set $\boldsymbol{\delta}_k = \alpha_k\mathbf{s}_k$.
      Upgrade the coefficient vector, $\mathbf{x}_{k+1} = \mathbf{x}_k + \boldsymbol{\delta}_k$.
      Compute $F_{k+1} = F(\mathbf{x}_{k+1})$.
      Update $\mathbf{P}_k$, generating $\mathbf{P}_{k+1}$ (see discussion below).
      Increment k and return to step (ii).
(iv) Data output:
      Display $\mathbf{x}^* = \mathbf{x}_k$ and $F^* = F(\mathbf{x}^*)$.

We should note that the way that the matrix $\mathbf{P}_k$, an estimate of the Hessian inverse,
is updated was omitted from the above algorithm. In fact, what distinguishes the
different quasi-Newton methods is solely the way that $\mathbf{P}_k$ is updated. The most widely
known quasi-Newton method is the Davidson–Fletcher–Powell method (Fletcher,
1980; Luenberger, 1984), used in Deczky (1972). Such an algorithm updates $\mathbf{P}_k$ in
the form

$$\mathbf{P}_{k+1} = \mathbf{P}_k + \frac{\boldsymbol{\delta}_k\boldsymbol{\delta}_k^{\mathrm{T}}}{\boldsymbol{\delta}_k^{\mathrm{T}}\boldsymbol{\gamma}_k} - \frac{\mathbf{P}_k\boldsymbol{\gamma}_k\boldsymbol{\gamma}_k^{\mathrm{T}}\mathbf{P}_k}{\boldsymbol{\gamma}_k^{\mathrm{T}}\mathbf{P}_k\boldsymbol{\gamma}_k} \tag{6.186}$$

where $\boldsymbol{\gamma}_k = \mathbf{g}_k - \mathbf{g}_{k-1}$.

However, our experience has shown that the Broyden–Fletcher–Goldfarb–Shannon
(BFGS) method (Fletcher, 1980) is more efficient. This algorithm updates $\mathbf{P}_k$ in the
form

$$\mathbf{P}_{k+1} = \mathbf{P}_k + \left(1 + \frac{\boldsymbol{\gamma}_k^{\mathrm{T}}\mathbf{P}_k\boldsymbol{\gamma}_k}{\boldsymbol{\gamma}_k^{\mathrm{T}}\boldsymbol{\delta}_k}\right)\frac{\boldsymbol{\delta}_k\boldsymbol{\delta}_k^{\mathrm{T}}}{\boldsymbol{\gamma}_k^{\mathrm{T}}\boldsymbol{\delta}_k} - \frac{\boldsymbol{\delta}_k\boldsymbol{\gamma}_k^{\mathrm{T}}\mathbf{P}_k + \mathbf{P}_k\boldsymbol{\gamma}_k\boldsymbol{\delta}_k^{\mathrm{T}}}{\boldsymbol{\gamma}_k^{\mathrm{T}}\boldsymbol{\delta}_k} \tag{6.187}$$

with $\boldsymbol{\gamma}_k$ as before.

It is important to notice that, in general, filter designers do not need to implement an optimization routine as they can employ optimization routines already available in a number of computer packages. What the designers are required to do is to express the objective function and optimization problem in a way that can be input to the chosen optimization routine.

6.5.3 Alternative methods

Some methods in addition to phase equalization have been proposed for the simultaneous approximation of magnitude and group delay (Charalambous & Antoniou, 1980; Saramäki & Neüvo, 1984; Cortelazzo & Lightner, 1984).

The work of Charalambous & Antoniou (1980) presents a fast algorithm for phase equalization, satisfying a minimax criterion.

The work in Saramäki & Neüvo (1984) emphasizes the design of digital filters with a reduced number of multiplications. It employs an all-pole IIR filter in cascade with a linear-phase FIR filter. The IIR filter section is designed with the constraint of keeping the group delay equiripple whereas the FIR filter is designed so that the overall cascade meets the desired magnitude response specifications. The work does not address any issue related to the speed of convergence of the algorithms involved. The reduction in the number of multipliers results from proper trading between the orders of the IIR and FIR filters. The drawback to this approach seems to be the large dynamic range of the resulting IIR and FIR filter internal signals, which may require very accurate coefficients in their implementation.

The work of Cortelazzo & Lightner (1984) presents a systematic procedure for the simultaneous approximation of magnitude and group delay based on the concept of multi-criterion optimization. The minimization procedure searches for a solution satisfying the minimax criterion. In this work, the methodology is discussed, but no reference to the efficiency of the optimization algorithms is given. In fact, only low-order examples are presented due to the long computational time required to find a solution, indicating that this methodology will only be useful when a more efficient algorithm is proposed.

In recent years there has been great effort in this field to find specific methods tailored to design IIR digital filters which satisfy magnitude and phase specifications simultaneously. In Holford & Agathoklis (1996), for example, IIR filters with almost linear phase in the passband are designed by applying a technique called model reduction to the state-space form of an FIR filter prototype. This type of design leads to filters with high selectivity in the magnitude response whereas the phase is kept almost linear. Another approach generalizes the traditional designs with prescribed magnitude and phase specifications based on L_p norms or minimax objective functions to filters with equiripple passbands and peak constrained stopbands (Sullivan & Adams, 1998; Lu, 1999).

Table 6.8. *Characteristics of the initial bandpass filter.*
Constant gain: $H_0 = 0.0588$.

Filter zeros (r_{z_i}; ϕ_{z_i} [rad])	Filter poles (r_{p_i}; ϕ_{p_i} [rad])
$r_{z_1} = 1.0$; $\phi_{z_1} = 0.1740$	$r_{p_1} = 0.8182$; $\phi_{p_1} = 0.3030$
$r_{z_2} = 1.0$; $\phi_{z_2} = -0.1740$	$r_{p_2} = 0.8182$; $\phi_{p_2} = -0.3030$
$r_{z_3} = 0.7927$; $\phi_{z_3} = 0.5622$	$r_{p_3} = 0.8391$; $\phi_{p_3} = 0.4837$
$r_{z_4} = 0.7927$; $\phi_{z_4} = -0.5622$	$r_{p_4} = 0.8391$; $\phi_{p_4} = -0.4837$
$r_{z_5} = 1.0$; $\phi_{z_5} = 0.9022$	$r_{p_5} = 0.8346$; $\phi_{p_5} = 0.6398$
$r_{z_6} = 1.0$; $\phi_{z_6} = -0.9022$	$r_{p_6} = 0.8346$; $\phi_{p_6} = -0.6398$
$r_{z_7} = 1.0$; $\phi_{z_7} = 2.6605$	$r_{p_7} = 0.8176$; $\phi_{p_7} = 0.8053$
$r_{z_8} = 1.0$; $\phi_{z_8} = -2.6605$	$r_{p_8} = 0.8176$; $\phi_{p_8} = -0.8053$

EXAMPLE 6.5

Design a bandpass filter satisfying the specifications below:

$$\left.\begin{aligned}
M(\omega) &= 1, && \text{for } 0.2\pi < \omega < 0.5\pi \\
M(\omega) &= 0, && \text{for } 0 < \omega < 0.1\pi \text{ and } 0.6\pi < \omega < \pi \\
\tau(\omega) &= L, && \text{for } 0.2\pi < \omega < 0.5\pi
\end{aligned}\right\} \tag{6.188}$$

where L is constant.

SOLUTION

Since it is a simultaneous magnitude and phase approximation, the objective function is given by equation (6.170), with the expressions for magnitude and group delay, and their derivatives given in equations (6.173)–(6.180). We can start the design with an eighth-order transfer function with the characteristics given in Table 6.8.

This initial filter is designed with the objective of approximating the desired magnitude specifications, and its average delay in the passband is used to estimate an initial value for L. In order to solve this optimization problem, we used a quasi-Newton program based on the BFGS method. Keeping the order of the starting filter at $n = 8$, we ran 100 iterations without obtaining noticeable improvements. We then increased the numerator and denominator orders by two, that is, we made $n = 10$, and after a few iterations the solution described in Table 6.9 was achieved.

Figure 6.12 illustrates the resulting frequency response. The attenuation at the first stopband edges are 18.09 dB and 18.71 dB, respectively. The attenuation at the second stopband edges are 18.06 dB and 19.12 dB, respectively. The passband attenuation at the edges are 0.69 dB and 0.71 dB, the two passband peaks have gains of 0.50 dB and 0.41 dB, whereas the attenuation at the passband minimum point is 0.14 dB. The group-delay values at the beginning of the passband, at the passband minimum, and at

Table 6.9. *Characteristics of the resulting bandpass filter. Constant gain:* $H_0 = 0.058\,772\,50$.

Filter zeros (r_{z_i}; ϕ_{z_i} [rad])		Filter poles (r_{p_i}; ϕ_{p_i} [rad])	
$r_{z_1} = 1.0$;	$\phi_{z_1} = 0.123\,202\,79$	$r_{p_1} = 0.0$;	$\phi_{p_1} = 0.0$
$r_{z_2} = 1.0$;	$\phi_{z_2} = -0.123\,202\,79$	$r_{p_2} = 0.0$;	$\phi_{p_2} = 0.0$
$r_{z_3} = 0.774\,7882$;	$\phi_{z_3} = 0.554\,526\,01$	$r_{p_3} = 0.907\,1557$;	$\phi_{p_3} = 0.244\,276\,28$
$r_{z_4} = 0.774\,7882$;	$\phi_{z_4} = -0.554\,526\,01$	$r_{p_4} = 0.907\,1557$;	$\phi_{p_4} = -0.244\,276\,28$
$r_{z_5} = 1.290\,6753$;	$\phi_{z_5} = 0.554\,526\,01$	$r_{p_5} = 0.865\,3806$;	$\phi_{p_5} = 0.433\,452\,52$
$r_{z_6} = 1.290\,6753$;	$\phi_{z_6} = -0.554\,526\,01$	$r_{p_6} = 0.865\,3806$;	$\phi_{p_6} = -0.433\,452\,52$
$r_{z_7} = 1.0$;	$\phi_{z_7} = 1.000\,632\,17$	$r_{p_7} = 0.874\,0485$;	$\phi_{p_7} = 0.658\,338\,19$
$r_{z_8} = 1.0$;	$\phi_{z_8} = -1.000\,632\,17$	$r_{p_8} = 0.874\,0485$;	$\phi_{p_8} = -0.658\,338\,19$
$r_{z_9} = 1.0$;	$\phi_{z_9} = 2.092\,004\,00$	$r_{p_9} = 0.915\,1922$;	$\phi_{p_9} = 0.860\,447\,32$
$r_{z_{10}} = 1.0$;	$\phi_{z_{10}} = -2.092\,004\,00$	$r_{p_{10}} = 0.915\,1922$;	$\phi_{p_{10}} = -0.860\,447\,32$

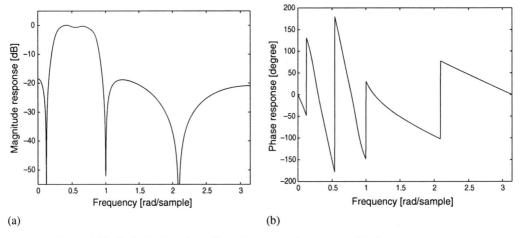

(a)   (b)

Figure 6.12 Optimized bandpass filter: (a) magnitude response; (b) phase response.

the end of the passband are 14.02 s, 12.09 s, and 14.38 s, respectively.

△

6.6 Time-domain approximation

In some applications, time-domain specifications are given to the filter designer. In these cases the objective is to design a transfer function $H(z)$ such that the corresponding impulse response h_n is as close as possible to a given sequence g_n, for $n = 0, 1, \ldots, K - 1$, where

$$H(z) = \frac{b_0 + b_1 z^{-1} + \cdots + b_M z^{-M}}{1 + a_1 z^{-1} + \cdots + a_N z^{-N}} = h_0 + h_1 z^{-1} + h_2 z^{-2} + \cdots \tag{6.189}$$

Since $H(z)$ has $(M + N + 1)$ coefficients, if $K = (M + N + 1)$ there is at least one transfer function available which satisfies the specifications. This solution can be obtained through optimization, as follows.

By equating

$$H(z) = g_0 + g_1 z^{-1} + \cdots + g_{M+N} z^{-(M+N)} + \cdots \tag{6.190}$$

and considering the z-transform products as convolutions in the time domain, we can write, from equations (6.189) and (6.190), that

$$\sum_{n=0}^{N} a_n g_{i-n} = \begin{cases} b_i, & \text{for } i = 0, 1, \ldots, M \\ 0, & \text{for } i > M \end{cases} \tag{6.191}$$

Now, assuming that $g_n = 0$, for all $n < 0$, this equation can be rewritten in matrix form as

$$\begin{bmatrix} b_0 \\ b_1 \\ b_2 \\ \vdots \\ b_M \\ 0 \\ \vdots \\ 0 \end{bmatrix} = \begin{bmatrix} g_0 & 0 & 0 & \cdots & 0 \\ g_1 & g_0 & 0 & \cdots & 0 \\ g_2 & g_1 & g_0 & \cdots & 0 \\ \vdots & \vdots & \vdots & \ddots & \vdots \\ g_M & g_{M-1} & g_{M-2} & \cdots & g_{M-N} \\ \vdots & \vdots & \vdots & \ddots & \vdots \\ \vdots & \vdots & \vdots & \ddots & \vdots \\ g_{M+N} & g_{M+N-1} & g_{M+N-2} & \cdots & g_M \end{bmatrix} \begin{bmatrix} 1 \\ a_1 \\ a_2 \\ \vdots \\ \vdots \\ \vdots \\ \vdots \\ a_N \end{bmatrix} \tag{6.192}$$

which can be partitioned as

$$\begin{bmatrix} b_0 \\ b_1 \\ b_2 \\ \vdots \\ b_M \end{bmatrix} = \begin{bmatrix} g_0 & 0 & 0 & \cdots & 0 \\ g_1 & g_0 & 0 & \cdots & 0 \\ g_2 & g_1 & g_0 & \cdots & 0 \\ \vdots & \vdots & \vdots & \ddots & \vdots \\ g_M & g_{M-1} & g_{M-2} & \cdots & g_{M-N} \end{bmatrix} \begin{bmatrix} 1 \\ a_1 \\ a_2 \\ \vdots \\ a_N \end{bmatrix} \tag{6.193}$$

$$\begin{bmatrix} 0 \\ 0 \\ \vdots \\ 0 \end{bmatrix} = \begin{bmatrix} g_{M+1} & \cdots & g_{M-N+1} \\ \vdots & \ddots & \vdots \\ g_{K-1} & \cdots & g_{K-N-1} \end{bmatrix} \begin{bmatrix} 1 \\ a_1 \\ a_2 \\ \vdots \\ a_N \end{bmatrix} \tag{6.194}$$

or

$$
\begin{bmatrix} \mathbf{b} \\ \mathbf{0} \end{bmatrix} = \begin{bmatrix} \mathbf{G}_1 \\ \mathbf{g}_2 & \mathbf{G}_3 \end{bmatrix} \begin{bmatrix} 1 \\ \mathbf{a} \end{bmatrix}
\tag{6.195}
$$

where $\mathbf{g}_2$ is a column vector and $\mathbf{G}_3$ is an $N \times N$ matrix. If $\mathbf{G}_3$ is nonsingular, the coefficients $\mathbf{a}$ are given by

$$
\mathbf{a} = -\mathbf{G}_3^{-1}\mathbf{g}_2
\tag{6.196}
$$

If $\mathbf{G}_3$ is singular of rank $R < N$, there are infinite solutions, one of which is obtained by forcing the first $(N - R)$ entries of $\mathbf{a}$ to be null.

With $\mathbf{a}$ available, $\mathbf{b}$ can be computed as

$$
\mathbf{b} = \mathbf{G}_1 \begin{bmatrix} 1 \\ \mathbf{a} \end{bmatrix}
\tag{6.197}
$$

The main differences between the filters designed with different values of M and N, while keeping $K = (M + N + 1)$ constant, are the values of h_k, for $k > K$.

Approximate approach

A solution that is in general satisfactory is obtained by replacing the null vector in equation (6.195) by a vector $\hat{\boldsymbol{\epsilon}}$ whose magnitude should be minimized. In that manner, equation (6.195) becomes

$$
\begin{bmatrix} \mathbf{b} \\ \hat{\boldsymbol{\epsilon}} \end{bmatrix} = \begin{bmatrix} \mathbf{G}_1 \\ \mathbf{g}_2 & \mathbf{G}_3 \end{bmatrix} \begin{bmatrix} 1 \\ \mathbf{a} \end{bmatrix}
\tag{6.198}
$$

Given the prescribed g_n and the values of N and M, we then have to find a vector $\mathbf{a}$ such that $(\hat{\boldsymbol{\epsilon}}^{\mathrm{T}}\hat{\boldsymbol{\epsilon}})$ is minimized, with

$$
\hat{\boldsymbol{\epsilon}} = \mathbf{g}_2 + \mathbf{G}_3\mathbf{a}
\tag{6.199}
$$

The value of $\mathbf{a}$ which minimizes $(\hat{\boldsymbol{\epsilon}}^{\mathrm{T}}\hat{\boldsymbol{\epsilon}})$ is the normal equation solution (Evans & Fischel, 1973),

$$
\mathbf{G}_3^{\mathrm{T}}\mathbf{G}_3\mathbf{a} = -\mathbf{G}_3^{\mathrm{T}}\mathbf{g}_2
\tag{6.200}
$$

If the rank of $\mathbf{G}_3$ is N, then the rank of $\mathbf{G}_3^{\mathrm{T}}\mathbf{G}_3$ is also N, and, therefore, the solution is unique, being given by

$$
\mathbf{a} = -[\mathbf{G}_3^{\mathrm{T}}\mathbf{G}_3]^{-1}\mathbf{G}_3^{\mathrm{T}}\mathbf{g}_2
\tag{6.201}
$$

On the other hand, if the rank of $\mathbf{G}_3$ is $R < N$, we should force $a_i = 0$, for $i = 0, 1, \ldots, (R - 1)$, as before, and redefine the problem as described in Burrus & Parks (1970).

It is important to point out that the procedure described above does not lead to a minimum squared error in the specified samples. In fact, the squared error is given by

$$\mathbf{e}^{\mathsf{T}}\mathbf{e} = \sum_{n=0}^{K} (g_n - h_n)^2 \tag{6.202}$$

where g_n and h_n are the desired and obtained impulse responses, respectively.

In order to obtain $\mathbf{b}$ and $\mathbf{a}$ which minimize $\mathbf{e}^{\mathsf{T}}\mathbf{e}$, we need an iterative process, such as the one proposed in Evans & Fischel (1973). The time-domain approximation can also be formulated as a system identification problem, as addressed in Jackson (1996).

EXAMPLE 6.6

Design a digital filter characterized by $M = 3$ and $N = 4$ such that its impulse response approximates the following sequence:

$$g_n = \frac{1}{3}\left[\frac{1}{4^{n+1}} + e^{-n-1} + \frac{1}{(n+2)}\right]u(n) \tag{6.203}$$

for $n = 0, 1, \ldots, 7$.

SOLUTION

If we choose $M = 3$ and $N = 4$, the designed transfer function is given by

$$H(z) = \frac{0.372\,626\,48z^3 - 0.644\,602\,01z^2 + 0.331\,214\,24z - 0.046\,560\,21}{z^4 - 2.205\,045\,15z^3 + 1.654\,455\,18z^2 - 0.487\,708\,53z + 0.047\,275\,63} \tag{6.204}$$

which has the exact desired impulse response for $n = 0, 1, \ldots, 7$.

The impulse response corresponding to the transfer function above is depicted in Figure 6.13, together with the prescribed impulse response. As can be seen, the responses are the same in the first few iterations, and they become distinct for $n > 7$, as expected, because we have only eight coefficients to adjust.

$\triangle$

6.7 IIR filter approximation with MATLAB

- `butter`: Designs Butterworth analog and digital filters.
  Input parameters:

  - The filter order n;
  - The filter normalized (between 0 and 1) cutoff frequency, `wp`, which is the frequency where the magnitude response assumes the value -3 dB. If such parameter is a two-element vector `[w1,w2]`, then the command returns an order $2n$ digital bandpass filter with passband `w1` $\leq \omega \leq$ `w2`;

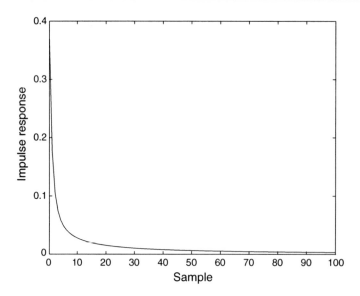

Figure 6.13 Impulse responses: desired (solid line) and obtained (dotted line).

- The filter type specified by a string. The options are 'high' for a highpass filter, and 'stop' for a 2nth-order stopband filter with stopband w1 $\leq \omega \leq$ w2;
- For analog filters, add the string 's'.

There are three possibilities for output parameters. The choice is automatic depending on the number of parameters being requested:

- Direct-form parameters [b,a], where b is the vector of numerator coefficients and a is the vector of denominator coefficients;
- Zero-pole parameters [z,p,k], where z is the set of filter zeros, p is the set of filter poles, and k is the constant gain;
- State-space parameters [A,B,C,D].

Example (direct-form highpass filter):

```
n=11; wp=0.2;
[b,a]=butter(n,wp,'high');
```

- buttord: Order selection for Butterworth filters.
  Input parameters:

  - Passband edge frequency ω_p;
  - Stopband edge frequency ω_r;
  - Passband ripple in dB A_p;
  - Stopband attenuation in dB A_r;
  - For analog filters, add the string 's'.

Output parameters:

- The filter order n;
- The corresponding cutoff frequency for lowpass and highpass filters, and a pair of frequencies for bandpass and bandstop filters corresponding to the passband and stopband edges, respectively.

Example:
```
wp=0.1; wr=0.15; Ap=1; Ar=20;
[n,wn]=buttord(wp,wr,Ap,Ar);
```

- `buttap`: Determines the analog prototype of a Butterworth lowpass filter.
  Input parameter: The filter order n.
  Output parameters: The pole–zero parameters `[z,p,k]`, where z is the set of filter zeros, p is the set of filter poles, and k is the constant gain.
  Example:
```
n=9; [z,p,k]=buttap(n);
```

- `cheby1`: Designs Chebyshev analog and digital filters.
  For input parameters, output parameters, and a similar example, see the command `butter`.

- `cheb1ord`: Order selection for Chebyshev filters.
  For input parameters, output parameters, and a similar example, see the command `buttord`.

- `cheb1ap`: Determines the analog prototype of a Chebyshev lowpass filter.
  For input and output parameters, see the command `buttap`. For Chebyshev filters, however, the passband ripple level should also be provided.
  Example:
```
n=9; Ap=1.5;
[z,p,k]=cheb1ap(n,Ap);
```

- `cheby2`: Designs inverse Chebyshev analog and digital filters (see Exercise 6.7).
  For input parameters, output parameters, and a similar example, see the command `butter`.

- `cheb2ord`: Order selection for inverse Chebyshev filters.
  For input parameters, output parameters, and a similar example, see the command `buttord`.

- `cheb2ap`: Determines the analog prototype of an inverse Chebyshev lowpass filter.
  For input and output parameters, see the command `buttap`. For inverse Chebyshev filters, however, the stopband attenuation level should also be provided.
  Example:

```
n=11; Ar=30;
[z,p,k]=cheb2ap(n,Ar);
```

- `ellip`: Designs elliptic analog and digital filters.
  Input parameters:

  – The filter order n;
  – The desired passband ripple `Ap` and minimum stopband attenuation `Ar` in dB.
  – The filter normalized cutoff frequency, $0 < wp < 1$, which is the frequency value where the magnitude response assumes the value `Ap` dB. If such a parameter is a two-element vector `[w1, w2]`, then the command returns an $2n$th-order digital bandpass filter with passband $w1 \leq \omega \leq w2$;
  – For analog filters, add the string `'s'`.

  Output parameters: as for `butter` command.
  Example:
  ```
  n=5; Ap=1; Ar=25; wp=0.2;
  [b,a]=ellip(n,Ap,Ar,wp);
  ```

- `ellipord`: Order selection for elliptic filters.
  For input parameters, output parameters, and a similar example, see the command `buttord`.

- `ellipap`: Determines the analog prototype of an elliptic lowpass filter.
  For input and output parameters, see the command `buttap`. For elliptic filters, however, the passband ripple and the stopband attenuation levels should also be provided.
  Example:
  ```
  n=9; Ap=1.5; Ar=30;
  [z,p,k]=ellipap(n,Ap,Ar);
  ```

- `lp2lp`: Transforms a normalized analog lowpass filter prototype with cutoff frequency 1 rad/s into a lowpass filter with a given cutoff frequency.
  Input parameters:

  – Direct-form coefficients `b,a`, where `b` is the vector of numerator coefficients and `a` is the vector of denominator coefficients; or state-space parameters `[A,B,C,D]`;
  – Desired normalized cutoff frequency in rad/s.

  Output parameters: The direct-form coefficients `bt,at`; or the state-space parameters `[At,Bt,Ct,Dt]`.
  Example (direct-form filter):
  ```
  n=9; [z,p,k]=buttap(n);
  ```

```
b=poly(z)*k; a=poly(p); wp=0.3;
[bt,at]=lp2lp(b,a,wp);
```

- lp2hp: Transforms a normalized analog lowpass filter prototype with cutoff frequency 1 rad/s into a highpass filter with a given cutoff frequency.
  For input and output parameters, see the command lp2lp.
  Example:

```
n=9; [z,p,k]=buttap(n);
b=poly(z)*k; a=poly(p); wp=1.3;
[bt,at]=lp2hp(b,a,wp);
```

- lp2bp: Transforms a normalized analog lowpass filter prototype with cutoff frequency 1 rad/s into a bandpass filter with given center frequency and passband width.
  For input and output parameters, see the command lp2lp. For bandpass filters, however, one must specify the center frequency and the passband width, instead of the desired cutoff frequency.
  Example:

```
n=9; [z,p,k]=buttap(n);
b=poly(z)*k; a=poly(p); wc=1; Bw=0.2;
[bt,at]=lp2bp(b,a,wc,Bw);
```

- lp2bs: Transforms a normalized analog lowpass filter prototype with cutoff frequency 1 rad/s into a bandstop filter with given center frequency and stopband width.
  For input parameters, output parameters, and a similar example, see the command lp2bp. For bandstop filters, however, one must specify the center frequency and the stopband width, instead of the passband width.

- impinvar: Maps the analog s plane into the digital z plane using the impulse-invariance method. It then transforms analog filters into their discrete-time equivalents, whose impulse response is equal to the analog impulse response scaled by a factor $1/F_s$.
  Input parameters:

  - Direct-form coefficients b,a, where b is the vector of numerator coefficients and a is the vector of denominator coefficients;
  - The scaling factor F_s, whose default value, if none is provided, is 1 Hz.

  Output parameters: The direct-form coefficients bt and at.
  Example:

```
n=9; [z,p,k]=buttap(n); b=poly(z)*k; a=poly(p); Fs=5;
[bt,at]=impinvar(b,a,Fs);
```

- `bilinear`: Maps the analog s plane into the digital z plane using the bilinear transformation. It then transforms analog filters into their discrete-time equivalents. This command requires the denominator order to be greater or equal to the numerator order.
  Input parameters:

  - Direct-form coefficients `b`, `a`, where `b` is the vector of numerator coefficients and `a` is the vector of denominator coefficients; or zero–pole parameters `[z,p,k]`, where `z` is the set of filter zeros, `p` is the set of filter poles, and `k` is the constant gain; or state-space parameters `[A,B,C,D]`;
  - The sampling frequency in Hz;
  - A frequency, in Hz, for which the frequency response remains unchanged before and after the transformation.

  Output parameters: The direct-form coefficients `bt,at`; the zero–pole parameters `[z,p,k]`; or the state-space parameters `[At,Bt,Ct,Dt]`.
  Example (direct-form filter with prewarping):
  ```
  n=9; [z,p,k]=buttap(n);
  b=poly(z)*k; a=poly(p); Fs=5; Fp=2;
  [bt,at]=bilinear(b,a,Fs,Fp);
  ```

- `invfreqz`: Performs digital filter design using a given frequency response. By default, this command minimizes a quadratic (equation-error) functional that may lead to unstable solutions. The minimization of an output-error functional, similar to that given in Subsection 6.5.1, is performed, using the equation-error solution as an initial guess, when the parameters' maximum number of iterations or tolerance are provided by the user.
  Input parameters:

  - A vector with the desired complex frequency response and the corresponding frequency vector;
  - The desired numerator and denominator orders;
  - A weight vector for each frequency specified;
  - A maximum number of iterations for the numerical algorithm to achieve convergence. This parameter forces the minimization of an output-error functional;
  - A tolerance for convergence checking. This parameter forces the minimization of an output-error functional.

  Output parameter: The direct-form coefficients `b,a`.
  Example (Butterworth digital filter with $n = 2$ and $\omega_p = 0.2$):
  ```
  n=2; b=[0.0675 0.1349 0.0675];
  a=[1.0000 -1.1430 0.4128];
  ```

```
[h,w]=freqz(b,a,64);
[bt,at]=invfreqz(h,w,n,n);
```

- invfreqs: Performs analog filter design using a given frequency response. For input parameters, output parameters, and a similar example, see the command invfreqz.

- As mentioned in Chapter 5, MATLAB provides the command sptool that integrates most of the commands above into a unique interface that greatly simplifies the design of standard IIR digital filters, using several methods discussed previously.

- The interested reader may also refer to the MATLAB literature for the commands lpc, maxflat, prony, stmcb, and yulewalk, for more specific design procedures of IIR digital filters.

6.8   Summary

In this chapter, we have covered the classical approximation methods for analog filters as well as two methods of transforming a continuous-time transfer function into a discrete-time transfer function. The transformation methods addressed were impulse-invariance and bilinear transformations. Although other transformation methods exist, the ones presented in this chapter are the most widely used in digital signal processing.

Transformation methods in the discrete-time domain were also addressed. It was shown that some of these transformations are useful in the design of variable-cutoff filters.

The simultaneous approximation of magnitude and phase responses was studied, and an optimization procedure was described.

Finally, the time-domain approximation problem was presented, and some methods aimed at the minimization of the mean-squared error between the prescribed impulse response and the resulting one were briefly discussed.

6.9   Exercises

6.1   Given the analog transfer function $H(s)$ below, design transfer functions corresponding to discrete-time filters using both the impulse-invariance method and the bilinear transformation. Choose $\Omega_s = 12$ rad/s.

$$H(s) = \frac{1}{(s^2 + 0.767\,22s + 1.338\,63)(s + 0.767\,22)}$$

6.2   Design a bandstop elliptic filter satisfying the following specifications:

$A_p = 0.5$ dB
$A_r = 60$ dB
$\Omega_{p_1} = 40$ Hz
$\Omega_{r_1} = 50$ Hz
$\Omega_{r_2} = 70$ Hz
$\Omega_{p_2} = 80$ Hz
$\Omega_s = 240$ Hz

6.3   Design highpass Butterworth, Chebyshev, and elliptic filters that satisfy the following specifications:

$A_p - 1.0$ dB
$A_r = 40$ dB
$\Omega_r = 5912.5$ rad/s
$\Omega_p = 7539.8$ rad/s
$\Omega_s = 50\,265.5$ rad/s

6.4   Design three bandpass digital filters with central frequencies at 770 Hz, 852 Hz, and 941 Hz, respectively. For the first filter, the stopband edges are at frequencies 697 Hz and 852 Hz. For the second filter, the stopband edges are 770 Hz and 941 Hz. For the third filter, the edges are at 852 Hz and 1209 Hz. In all three filters, the minimum stopband attenuation is 40 dB, and use $\Omega_s = 8$ kHz.

6.5   Design an analog elliptic filter satisfying the following specifications:

$A_p = 1.0$ dB
$A_r = 40$ dB
$\Omega_p = 1000$ Hz
$\Omega_r = 1209$ Hz

6.6   Design a digital filter corresponding to the filter in Exercise 6.5, with $\Omega_s = 8$ kHz. Then transform the designed filter into a highpass filter satisfying the specifications of Exercise 6.3, using the frequency transformation of Section 6.4.

6.7   This exercise describes the inverse Chebyshev approximation. The attenuation function of a lowpass inverse Chebyshev filter is characterized as

$$|A(j\Omega')|^2 = 1 + E(j\Omega')E(-j\Omega')$$

$$E(s')E(-s') = \frac{\epsilon'^2 (\frac{s'}{j})^{2n}}{(\frac{s'}{j})^{2n} C_n^2(\frac{j}{s'})}$$

where $C_n(\Omega')$ is a Chebyshev function of order n and

$$\epsilon' = \sqrt{10^{0.1A_r} - 1}$$

The inverse Chebyshev approximation is maximally flat at $\Omega' = 0$, and has a number of transmission zeros at the stopband placed at the inverse of the roots of the corresponding Chebyshev polynomial. Using the equations above, the stopband edge is placed at $\Omega'_r = 1$ rad/s, and this property should be considered when applying denormalization.

(a) Develop expressions for the passband edge, the transmission zeros, and the poles of a normalized filter of order n.

(b) Design the filter of Exercise 6.3 using the inverse Chebyshev approximation.

6.8   Show that the lowpass to bandpass and lowpass to bandstop transformations proposed in Section 6.4 are valid.

6.9   Given the transfer function below, describe a frequency transformation to a bandpass filter with zeros at $\frac{\pi}{6}$ and $\frac{2\pi}{3}$

$$H(z) = 0.06 \frac{z^2 + \sqrt{2}z + 1}{z^2 - 1.18z + 0.94}$$

6.10   Design a phase equalizer for the elliptic filter of Exercise 6.3, with the same order as the filter.

6.11   Design a lowpass filter satisfying the following specifications:

$$M(\Omega T) = 1.0, \quad \text{for } 0.0\Omega_s < \Omega < 0.1\Omega_s$$
$$M(\Omega T) = 0.5, \quad \text{for } 0.2\Omega_s < \Omega < 0.5\Omega_s$$
$$\tau(\Omega T) = 4.0, \quad \text{for } 0.0\Omega_s < \Omega < 0.1\Omega_s$$

6.12   The desired impulse response for a filter is given by $g(n) = \frac{1}{2^n}$. Design a recursive filter such that its impulse response $h(n)$ equals $g(n)$ for $n = 0, 1, \ldots, 5$.

6.13   Design a filter with 10 coefficients such that its impulse response approximates the following sequence:

$$g_n = \left[\frac{1}{6^n} + 10^{-n} + \frac{0.05}{(n+2)} \right] u(n)$$

Choose a few key values for M and N, and discuss which choice leads to the smallest mean-squared error after the 10th sample.

7 Finite-precision effects

7.1 Introduction

In practice, a digital signal processing system is implemented by software on a digital computer, either using a general-purpose digital signal processor, or using dedicated hardware for the given application. In either case, quantization errors are inherent due to the finite-precision arithmetic. These errors are of the following types:

- Errors due to the quantization of the input signals into a set of discrete levels, such as the ones introduced by the analog-to-digital converter.
- Errors in the frequency response of filters, or in transform coefficients, due to the finite-wordlength representation of multiplier constants.
- Errors made when internal data, like outputs of multipliers, are quantized before or after subsequent additions.

All these error forms depend on the type of arithmetic utilized in the implementation. If a digital signal processing routine is implemented on a general-purpose computer, since floating-point arithmetic is in general available, this type of arithmetic becomes the most natural choice. On the other hand, if the building block is implemented on special-purpose hardware, or a fixed-point digital signal processor, fixed-point arithmetic may be the best choice, because it is less costly in terms of hardware and simpler to design. A fixed-point implementation usually implies a lot of savings in terms of chip area as well.

For a given application, the quantization effects are key factors to be considered when assessing the performance of a digital signal processing algorithm. In this chapter, the various quantization effects are introduced along with the most widely used formulations for their analyses.

7.2 Binary number representation

7.2.1 Fixed-point representations

In most of the cases when digital signal processing systems are implemented using fixed-point arithmetic, the numbers are represented in one of the following forms: sign-magnitude, one's-complement, and two's-complement formats. These representations are described in this subsection, where we implicitly assume that all numbers have been previously scaled to the interval $x \in (-1, 1)$.

In order to clarify the definitions of the three types of number representation given below, we associate, for every positive number x such that $x < 2$, a function that returns its representation in base 2, $\mathcal{B}(x)$, which is defined by the equations below:

$$\mathcal{B}(x) = x_0.x_1x_2 \cdots x_n \tag{7.1}$$

and

$$x = \mathcal{B}^{-1}(x_0.x_1x_2 \cdots x_n) = x_0 + x_12^{-1} + x_22^{-2} + \cdots + x_n2^{-n} \tag{7.2}$$

Sign-magnitude representation

The sign-magnitude representation of a given number consists of a sign bit followed by a binary number representing its magnitude. That is

$$[x]_M = s_x.x_1x_2x_3 \cdots x_n \tag{7.3}$$

where s_x is the sign bit, and $x_1x_2x_3 \cdots x_n$ represents the magnitude of the number in base 2, that is, in binary format. Here, we use $s_x = 0$ for positive numbers, and $s_x = 1$ for negative numbers. This means that the number x is given by

$$x = \begin{cases} \mathcal{B}^{-1}(0.x_1x_2 \cdots x_n) = x_12^{-1} + x_22^{-2} + \cdots + x_n2^{-n}, & \text{for } s_x = 0 \\ -\mathcal{B}^{-1}(0.x_1x_2 \cdots x_n) = -(x_12^{-1} + x_22^{-2} + \cdots + x_n2^{-n}), & \text{for } s_x = 1 \end{cases} \tag{7.4}$$

One's-complement representation

The one's-complement representation of a number is given by

$$[x]_{1c} = \begin{cases} \mathcal{B}(x), & \text{if } x \geq 0 \\ \mathcal{B}(2 - 2^{-n} - |x|), & \text{if } x < 0 \end{cases} \tag{7.5}$$

where $\mathcal{B}$ is defined by equations (7.1) and (7.2). Notice that in the case of positive numbers, the one's-complement and the sign-magnitude representations are identical. However, for negative numbers, the one's complement is generated by changing all the 0s to 1s and all the 1s to 0s in the sign-magnitude representation of its absolute value. As before $s_x = 0$ for positive numbers and $s_x = 1$ for negative numbers.

Two's-complement representation

The two's-complement representation of a number is given by

$$[x]_{2c} = \begin{cases} \mathcal{B}(x), & \text{if } x \geq 0 \\ \mathcal{B}(2 - |x|), & \text{if } x < 0 \end{cases} \tag{7.6}$$

where $\mathcal{B}$ is defined by equations (7.1) and (7.2). Again, for positive numbers, the two's-complement representation is identical to the sign-magnitude representation.

The representation of a negative number in two's complement can be obtained by adding 1 to the least significant bit of the number represented in one's complement. Then, as in the one's-complement case, $s_x = 0$ for positive numbers and $s_x = 1$ for negative numbers.

If the two's-complement representation of a number x is given by

$$[x]_{2c} = s_x.x_1 x_2 \cdots x_n \tag{7.7}$$

then, from equation (7.6), we have that

- For $s_x = 0$, then

$$x = \mathcal{B}^{-1}([x]_{2c}) = \mathcal{B}^{-1}(0.x_1 x_2 \cdots x_n) = x_1 2^{-1} + x_2 2^{-2} + \cdots + x_n 2^{-n} \tag{7.8}$$

- For $s_x = 1$, then

$$2 - |x| = 2 + x = \mathcal{B}^{-1}([x]_{2c}) = \mathcal{B}^{-1}(s_x.x_1 x_2 \cdots x_n) \tag{7.9}$$

and then

$$
\begin{aligned}
x &= -2 + \mathcal{B}^{-1}(1.x_1 x_2 \cdots x_n) \\
&= -2 + 1 + x_1 2^{-1} + x_2 2^{-2} + \cdots + x_n 2^{-n} \\
&= -1 + x_1 2^{-1} + x_2 2^{-2} + \cdots + x_n 2^{-n}
\end{aligned} \tag{7.10}
$$

From equations (7.8) and (7.10), we have that x can be generally expressed as

$$x = -s_x + x_1 2^{-1} + x_2 2^{-2} + \cdots + x_n 2^{-n} \tag{7.11}$$

This short notation is very useful for introducing the multiplication operation of binary numbers represented in two's-complement format, as given in Section 12.2.1.

The one's-complement and two's-complement representations are more efficient for addition implementations, whereas the sign-magnitude representation is more efficient for the implementation of multiplications (Hwang, 1979). Overall, the most widely used binary code is the two's-complement representation.

7.2.2 Floating-point representation

Using floating-point representation, a number is represented as

$$x = x_m 2^c \tag{7.12}$$

where x_m is the mantissa and c is the number exponent, with $\frac{1}{2} \le |x_m| < 1$.

When compared to fixed-point representations, the main advantage of the floating-point representation is its large dynamic range. Its main disadvantage is that its implementation is more complex. For example, in floating-point arithmetic, the mantissa must be quantized after both multiplications and additions, whereas in fixed-point arithmetic quantization is required only after multiplications. In this text, we deal with both representation systems. However, we put the focus more on the fixed-point arithmetic, since it is more prone to errors, thus requiring more attention.

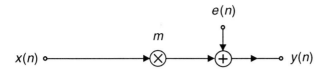

Figure 7.1 Noise model for multiplier.

7.3   Product quantization

A finite-wordlength multiplier can be modeled in terms of an ideal multiplier followed by a single additive noise source $e(n)$, as shown in Figure 7.1. Three distinct approximation schemes can be employed after a multiplication, namely: rounding, truncation, and magnitude truncation. We analyze their effects in numbers represented in two's complement.

Product quantization by rounding leads to a result in finite precision whose value is the closest possible to the actual value. If we assume that the dynamic range throughout the digital filter is much larger than the value of the least significant bit $q = 2^{-b}$ (b corresponds to n in equations (7.1)–(7.11)), the probability density function of the quantization error is depicted in Figure 7.2a. The mean or expected value of the quantization error due to rounding is zero, that is

$$E\{e(n)\} = 0 \tag{7.13}$$

Also, it is easy to show that the variance of the noise $e(n)$ is given by

$$\sigma_e^2 = \frac{q^2}{12} = \frac{2^{-2b}}{12} \tag{7.14}$$

In the case of truncation of a number, where the result is always less than the original value, the probability density function is as shown in Figure 7.2b, the expected value for the quantization error is

$$E\{e(n)\} = -\frac{q}{2} \tag{7.15}$$

and the error variance amounts to

$$\sigma_e^2 = \frac{2^{-2b}}{12} \tag{7.16}$$

This type of quantization is not adequate, in general, because the errors associated with a nonzero mean value, although small, tend to propagate through the filter, and their effect can be sensed at the filter output.

If we apply magnitude truncation, which necessarily implies reducing the magnitude of the number, the probability density function has the form depicted in Figure 7.2c. In this case, the mean value of the quantization error is

$$E\{e(n)\} = 0 \tag{7.17}$$

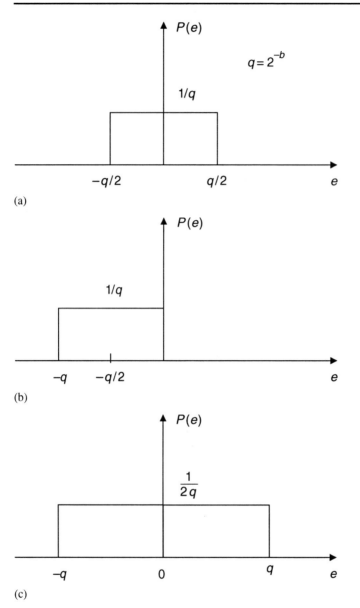

Figure 7.2 Probability density functions for the product-quantization error: (a) rounding; (b) truncation; (c) magnitude truncation.

and the variance is

$$\sigma_e^2 = \frac{2^{-2b}}{3} \tag{7.18}$$

Due to these results, it is easy to understand why rounding is the most attractive form of quantization, since it generates the smallest noise variance while maintaining the mean value of the quantization error equal to zero. Magnitude truncation, besides having a noise variance four times greater than rounding, leads to a higher correlation between the signal and quantization noise (for example, when the signal is positive/negative, the quantization noise is also positive/negative), which is a strong disadvantage, as will soon be clear. However, the importance of magnitude truncation cannot be overlooked, since it can eliminate granular limit cycles in recursive digital filters, as will be shown later in this chapter.

At this point, it is interesting to study how the signal quantization affects the output signal. In order to simplify the analysis of roundoff-noise effects, the following assumptions are made regarding the filter internal signals (Jackson, 1969, 1970a,b):

- Their amplitude is much larger than the quantization error.
- Their amplitude is small enough that overflow never occurs.
- They have a wide spectral content.

These assumptions imply that:

(i) The quantization errors at different instants are uncorrelated, that is, $e_i(n)$ is uncorrelated with $e_i(n + l)$, for $l \neq 0$.
(ii) Errors at different nodes are uncorrelated, that is, $e_i(n)$ is uncorrelated with $e_j(n)$, for $i \neq j$.

From the above considerations, the contributions of different noise sources can be accounted for separately and added to determine the overall roundoff error at the filter output. However, one should note that (i) and (ii) do not hold when magnitude truncation is used, due the inherent correlation between the signal and the quantization error. Therefore, one should bear in mind that, for the magnitude truncation scheme, the analysis that follows does not lead to accurate results.

Denoting the error variance due to each internal signal quantization by σ_e^2, and assuming that the quantization error is the outcome of a white noise process, the power spectral density (PSD) of a given noise source is (Papoulis, 1965)

$$P_e(e^{j\omega}) = \sigma_e^2 \tag{7.19}$$

It is known that, in a linear system having transfer function $H(z)$, if a stationary signal with PSD $P_x(z)$ is input, then the PSD of the output is $P_y(z) = H(z)H(z^{-1})P_x(z)$. Therefore, in a fixed-point digital-filter implementation, the PSD of the output noise is given by

$$P_y(z) = \sigma_e^2 \sum_{i=1}^{K} G_i(z)G_i(z^{-1}) \tag{7.20}$$

where K is the number of multipliers of the filter, and $G_i(z)$ is the transfer function from each multiplier output, $g_i(n)$, to the output of the filter, as indicated in Figure 7.3.

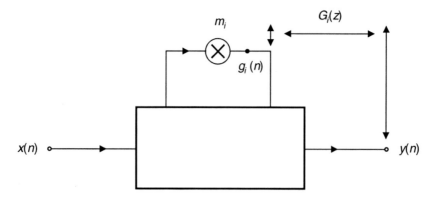

Figure 7.3 Noise transfer function for a digital filter.

A common figure of merit for evaluating the performance of digital filters is the relative power spectral density (RPSD) of the output noise, defined in dB as

$$\text{RPSD} = 10 \log \frac{P_y(e^{j\omega})}{P_e(e^{j\omega})} \tag{7.21}$$

This figure eliminates the dependence of the output noise on the filter wordlength, thus representing a true measure of the extent to which the output noise depends on the internal structure of the filter.

A simpler but useful performance criterion to evaluate the roundoff noise generated in digital filters is the noise gain or the relative noise variance, defined by

$$\frac{\sigma_y^2}{\sigma_e^2} = \frac{1}{\pi} \int_0^\pi \sum_{i=1}^K \left(G_i(z) G_i(z^{-1}) \right)_{z=e^{j\omega}}^2 d\omega$$

$$= \frac{1}{\pi} \int_0^\pi \sum_{i=1}^K |G_i(e^{j\omega})|^2 d\omega$$

$$= \frac{1}{\pi} \sum_{i=1}^K \int_0^\pi |G_i(e^{j\omega})|^2 d\omega$$

$$= \sum_{i=1}^K \| G_i(e^{j\omega}) \|_2^2 \tag{7.22}$$

Another noise source that must be accounted for in digital filters is the roundoff noise generated when the input-signal amplitude is quantized in the process of analog-to-digital conversion. Since the input-signal quantization is similar to product quantization, it can be represented by including a noise source at the input of the digital filter structure.

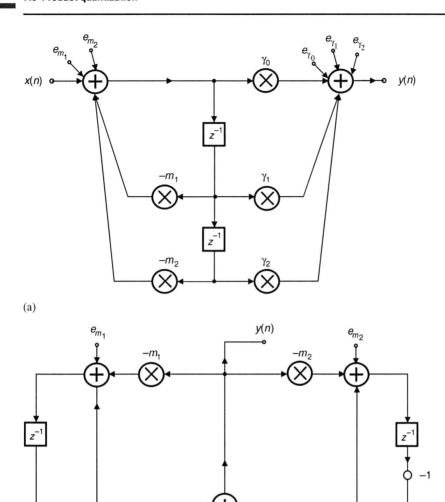

(a)

(b)

Figure 7.4 Second-order networks.

EXAMPLE 7.1

Compute the quantization noise PSD at the output of the networks shown in Figure 7.4, assuming that fixed-point arithmetic is employed.

SOLUTION

- For the structure of Figure 7.4a, we have that

$$G_{m_1}(z) = G_{m_2}(z) = H(z) \tag{7.23}$$

$$G_{\gamma_0}(z) = G_{\gamma_1}(z) = G_{\gamma_2}(z) = 1 \tag{7.24}$$

where

$$H(z) = \frac{\gamma_0 z^2 + \gamma_1 z + \gamma_2}{z^2 + m_1 z + m_2} \tag{7.25}$$

The output PSD is then

$$P_y(z) = \sigma_e^2 \left(2H(z)H(z^{-1}) + 3 \right) \tag{7.26}$$

• In the case of the structure shown in Figure 7.4b

$$G_{m_1}(z) = \frac{z + 1}{D(z)} \tag{7.27}$$

$$G_{m_2}(z) = \frac{z - 1}{D(z)} \tag{7.28}$$

where

$$D(z) = z^2 + (m_1 - m_2)z + (m_1 + m_2 - 1) \tag{7.29}$$

The PSD at the output is then given by

$$P_y(z) = \sigma_e^2 \left(G_{m_1}(z)G_{m_1}(z^{-1}) + G_{m_2}(z)G_{m_2}(z^{-1}) \right) \tag{7.30}$$

$\triangle$

In floating-point arithmetic, quantization errors are introduced not only in products but also in additions. In fact, the sum and product of two numbers x_1 and x_2, in finite precision, have the following characteristics

$$Fl\{x_1 + x_2\} = x_1 + x_2 - (x_1 + x_2)n_a \tag{7.31}$$

$$Fl\{x_1 x_2\} = x_1 x_2 - (x_1 x_2)n_p \tag{7.32}$$

where n_a and n_p are random variables with zero mean, which are independent of any other internal signals and errors in the filter. Their variances are approximately given by (Smith et al., 1992)

$$\sigma_{n_a} \approx 0.165 \times 2^{-2b} \tag{7.33}$$

$$\sigma_{n_p} \approx 0.180 \times 2^{-2b} \tag{7.34}$$

respectively, where b is the number of bits in the mantissa representation, not including the sign bit. Using the above expressions, the roundoff-noise analysis for floating-point arithmetic can be done in the same way as for the fixed-point case (Bomar et al., 1997).

A variant of floating-point arithmetic is the so-called block-floating-point arithmetic, which consists of representing several numbers with a shared exponent term. Block floating point is a compromise between the high complexity hardware of floating-point arithmetic and the short dynamic range inherent to the fixed-point representation. A roundoff-errors analysis of block-floating-point systems is available in Kalliojärvi & Astola. (1996); Ralev & Bauer (1999).

7.4   Signal scaling

It is possible for overflow to occur in any internal node of a digital filter structure. In general, input scaling is often required to reduce the probability of overflow occurring to an acceptable level. Particularly in fixed-point implementations, signal scaling should be applied to make the probability of overflow the same at all internal nodes of a digital filter. In such cases, the signal-to-noise ratio is maximized.

In a practical digital filter, however, a few internal signals are critical, and they should not exceed the allowed dynamic range for more than some very short periods of time. For these signals, if overflow frequently occurs, serious signal distortions will be observed at the filter output.

If either the one's- or the two's-complement representation is used, an important fact greatly simplifies the scaling in order to avoid overflow distortions: if the sum of two or more numbers is within the available range of representable numbers, the complete sum will always be correct irrespective of the order in which they are added, even if overflow occurs in a partial operation (Antoniou, 1993). This implies that the overflow distortions which are due to signal additions can be recovered by subsequent additions. As a consequence, when using either one's- or two's-complement arithmetic, one only needs to avoid overflow at the multiplier inputs. Therefore, they are the only points that require scaling.

In this case, then, in order to avoid frequent overflows, we should calculate the upper limit for the magnitude of each signal $x_i(n)$, for all possible types of filter inputs $u(n)$. This is shown by the analysis of the decimation-in-time FFT algorithm in the example below.

EXAMPLE 7.2
Perform a roundoff-noise analysis of FFT algorithms.

SOLUTION
Each distinct FFT algorithm requires a specific analysis of the corresponding quantization effects. In this example, we perform a roundoff-noise analysis on the radix-2 FFT algorithm with decimation in time.

The basic cell of the radix-2 FFT algorithm based on the decimation-in-time method is shown in Figure 3.10 and repeated for convenience as Figure 7.5.

From Figure 7.5, we have that

$$|X_i(k)| \leq 2 \max\{|X_{ie}(k)|, |X_{io}(k)|\} \tag{7.35}$$

$$\left| X_i \left(k + \frac{L}{2} \right) \right| \leq 2 \max\{|X_{ie}(k)|, |X_{io}(k)|\} \tag{7.36}$$

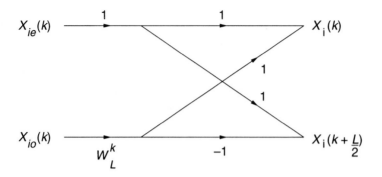

Figure 7.5 Basic cell of the radix-2 FFT algorithm using decimation in time.

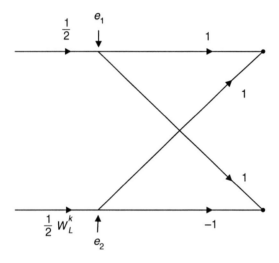

Figure 7.6 Basic cell with input scaling.

Therefore, to avoid overflow in such structure, using fixed-point arithmetic, a factor of $\frac{1}{2}$ on each cell input should be employed, as seen in Figure 7.6. There, one can clearly discern the two noise sources that result from both signal scaling and multiplication by W_L^k.

In the common case of rounding, the noise variance, as given in equation (7.14), is equal to

$$\sigma_e^2 = \frac{2^{-2b}}{12} \tag{7.37}$$

where b is the number of fixed-point bits. In Figure 7.6, the noise source e_1 models the scaling noise on a cell input which is a complex number. By considering the real and imaginary noise contributions to be uncorrelated, we have that

$$\sigma_{e_1}^2 = 2\sigma_e^2 = \frac{2^{-2b}}{6} \tag{7.38}$$

Meanwhile, the noise source, e_2, models the scaling noise due to the multiplication of two complex numbers, which involves four different terms. By considering these four terms to be uncorrelated, we have that

$$\sigma_{e_2}^2 = 4\sigma_e^2 = \frac{2^{-2b}}{3} \tag{7.39}$$

This completes the noise analysis in a single basic cell.

To extend these results to the overall FFT algorithm, let us assume that all noise sources in the algorithm are uncorrelated. Therefore, the output-noise variance can be determined by the addition of all individual noise variances, from all the basic cells involved.

Naturally, the noise generated in the first stage appears at the output scaled by $(\frac{1}{2})^{(l-1)}$, where l is the overall number of stages, and the noise generated at stage k is multiplied by $(\frac{1}{2})^{(l-k)}$. To determine the total number of cells, we note that each FFT output is directly connected to one basic cell in the last stage, two in the next-to-last stage, and so on, up to $2^{(l-1)}$ cells in the first stage, with each cell presenting two noise sources similar to e_1 and e_2, as discussed above.

Each stage then has $2^{(l-k)}$ cells, and their output-noise contribution is given by

$$2^{(l-k)} \left(\frac{1}{2}\right)^{(2l-2k)} \left(\sigma_{e_1}^2 + \sigma_{e_2}^2\right) \tag{7.40}$$

Consequently, the overall noise variance in each FFT output sample is given by

$$\sigma_o^2 = \left(\sigma_{e_1}^2 + \sigma_{e_2}^2\right) \sum_{k=1}^{l} 2^{l-k} \left(\frac{1}{2}\right)^{2l-2k} = 6\sigma_e^2 \sum_{k=1}^{l} \left(\frac{1}{2}\right)^{l-k} = 6\sigma_e^2 \left(2 - \frac{1}{2^{l-1}}\right) \tag{7.41}$$

A similar analysis for other FFT algorithms can be determined in an analogous way. For a roundoff analysis on the DCT, readers should refer to (Yun & Lee, 1995), for instance.

$\triangle$

The general case of scaling analysis is illustrated in Figure 7.7, where $F_i(z)$ represents the transfer function before scaling from the input node to the input of the multiplier m_i. Assuming zero initial conditions, we have that

$$x_i(n) = \sum_{k=0}^{\infty} f_i(k)u(n-k) \tag{7.42}$$

If $u(n)$ is bounded in magnitude by u_m, for all n, the above equation implies that

$$|x_i(n)| \leq u_m \sum_{k=0}^{\infty} |f_i(k)| \tag{7.43}$$

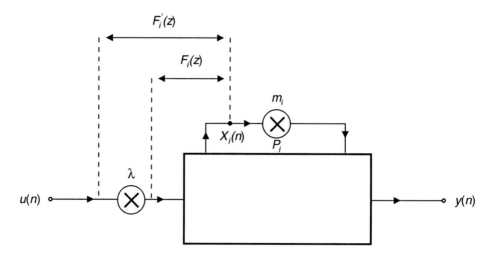

Figure 7.7 Signal scaling.

If we want the magnitude of the signal $x_i(n)$ to also be upper bounded by u_m, for all types of input sequences, then the associated scaling must ensure that

$$\sum_{k=0}^{\infty} |f_i'(k)| \leq 1 \tag{7.44}$$

This is a necessary and sufficient condition to avoid overflow for any input signal. However, the condition in equation (7.44) is not useful in practice, as it cannot be easily implemented. In addition, for a large class of input signals, it leads to a very stringent scaling. A more practical scaling strategy, aimed at more specific classes of input signals, is presented in the following. Since

$$U(z) = \sum_{n=-\infty}^{\infty} u(n)z^{-n} \tag{7.45}$$

and

$$X_i(z) = F_i(z)U(z) \tag{7.46}$$

then, in the time domain, $x_i(n)$ is given by

$$x_i(n) = \frac{1}{2\pi j} \oint_C X_i(z)z^{n-1}dz \tag{7.47}$$

where C is in the convergence region common to $F_i(z)$ and $U(z)$. Accordingly, in the frequency domain, $x_i(n)$ is given by

$$x_i(n) = \frac{1}{2\pi} \int_0^{2\pi} F_i(e^{j\omega})U(e^{j\omega})e^{j\omega n}d\omega \tag{7.48}$$

Let, now, the L_p norm be defined for any periodic function $F(e^{j\omega})$ as follows:

$$\|F(e^{j\omega})\|_p = \left(\frac{1}{2\pi}\int_0^{2\pi}|F(e^{j\omega})|^p d\omega\right)^{\frac{1}{p}} \tag{7.49}$$

for all $p \geq 1$, such that $\int_0^{2\pi}|F(e^{j\omega})|^p d\omega \leq \infty$.

If $F(e^{j\omega})$ is continuous, the limit when $p \to \infty$ of equation (7.49) exists, and is given by

$$\|F(e^{j\omega})\|_\infty = \max_{0\leq\omega\leq 2\pi}\left\{|F(e^{j\omega})|\right\} \tag{7.50}$$

Assuming that $|U(e^{j\omega})|$ is upper bounded by U_m, that is, $\|U(e^{j\omega})\|_\infty \leq U_m$, it clearly follows from equation (7.48) that

$$|x_i(n)| \leq \frac{U_m}{2\pi}\int_0^{2\pi}|F_i(e^{j\omega})|d\omega \tag{7.51}$$

that is

$$|x_i(n)| \leq \|F_i(e^{j\omega})\|_1\|U(e^{j\omega})\|_\infty \tag{7.52}$$

Following a similar reasoning, we have that

$$|x_i(n)| \leq \|F_i(e^{j\omega})\|_\infty\|U(e^{j\omega})\|_1 \tag{7.53}$$

Also, from the Schwartz inequality (Jackson, 1969)

$$|x_i(n)| \leq \|F_i(e^{j\omega})\|_2\|U(e^{j\omega})\|_2 \tag{7.54}$$

Equations (7.52)–(7.54) are special cases of a more general relation, known as the Hölder inequality (Jackson, 1969; Hwang, 1979). It states that, if $\frac{1}{p} + \frac{1}{q} = 1$, then

$$|x_i(n)| \leq \|F_i(e^{j\omega})\|_p\|U(e^{j\omega})\|_q \tag{7.55}$$

If $|u(n)| \leq u_m$, for all n, and if its Fourier transform exists, then there is U_m such that $\|U(e^{j\omega})\|_q \leq U_m$, for any $q \geq 1$. If we then want $|x_i(n)|$ to be upper limited by U_m, for all n, equation (7.55) indicates that a proper scaling factor should be calculated such that, in Figure 7.7, we have that

$$\|F_i'(e^{j\omega})\|_p \leq 1 \tag{7.56}$$

which leads to

$$\lambda \leq \frac{1}{\|F_i(e^{j\omega})\|_p} \tag{7.57}$$

In practice, when the input signal is deterministic, the most common procedures for determining λ are:

- When $U(e^{j\omega})$ is bounded, and therefore $\|U(e^{j\omega})\|_{\infty}$ can be precisely determined, one may use λ as in equation (7.57), with $p = 1$.
- When the input signal has finite energy, that is,

$$E = \sum_{n=-\infty}^{\infty} u^2(n) = \|U(e^{j\omega})\|_2^2 < \infty \tag{7.58}$$

then λ can be obtained from equation (7.57) with $p = 2$.

- If the input signal has a dominant frequency component, such as a sinusoidal signal, this means that it has an impulse in the frequency domain. In this case, neither $\|U(e^{j\omega})\|_{\infty}$ nor $\|U(e^{j\omega})\|_2$ are defined, and thus only the L_1 norm can be used. Then the scaling factor comes from equation (7.57) with $p = \infty$, which is the most strict case for λ.

For the random-input case, the above analysis does not apply directly, since the z transform of $u(n)$ is not defined. In this case, if $u(n)$ is stationary, we use the autocorrelation function defined by

$$a_u(n) = E\{u(m)u(n + m)\} \tag{7.59}$$

The z transform of $a_u(n)$, denoted by $A_u(z)$, can then be defined. Assuming that $E\{u(n)\} = 0$, the expression for the variance, or the average power of $u(n)$ becomes

$$a_u(0) = E\{u^2(m)\} = \sigma_u^2 \tag{7.60}$$

Applying the inverse z transform, it follows that

$$\sigma_u^2 = \frac{1}{2\pi j} \oint_C A_u(z)z^{-1}dz \tag{7.61}$$

and then, for $z = e^{j\omega}$,

$$\sigma_u^2 = \frac{1}{2\pi} \int_0^{2\pi} A_u(e^{j\omega})d\omega \tag{7.62}$$

The z transform of the autocorrelation function of an internal signal $x_i(n)$ is given by

$$A_{x_i}(z) = F_i(z)F_i(z^{-1})A_u(z) \tag{7.63}$$

Therefore, in the frequency domain, we get

$$A_{x_i}(e^{j\omega}) = |F_i(e^{j\omega})|^2 A_u(e^{j\omega}) \tag{7.64}$$

Hence, the variance of the internal signal $x_i(n)$ is given by

$$\sigma_{x_i}^2 = \frac{1}{2\pi} \int_0^{2\pi} |F_i(e^{j\omega})|^2 A_u(e^{j\omega})d\omega \tag{7.65}$$

Applying the Hölder inequality (equation (7.55)) to the above equation, we have that, if $\frac{1}{p} + \frac{1}{q} = 1$, then

$$\sigma_{x_i}^2 \leq \|F_i^2(e^{j\omega})\|_p \|A_u(e^{j\omega})\|_q \tag{7.66}$$

or, alternatively,

$$\sigma_{x_i}^2 \leq \|F_i(e^{j\omega})\|_{2p}^2 \|A_u(e^{j\omega})\|_q \tag{7.67}$$

For stochastic processes, the most interesting cases in practice are:

- If we consider $q = 1$, then $p = \infty$, and noting that $\sigma_u^2 = \|A_u(e^{j\omega})\|_1$, we have, from equation (7.67), that

$$\sigma_{x_i}^2 \leq \|F_i(e^{j\omega})\|_\infty^2 \sigma_u^2 \tag{7.68}$$

and a λ, such that $\sigma_{x_i}^2 \leq \sigma_u^2$, is

$$\lambda = \frac{1}{\|F_i(e^{j\omega})\|_\infty} \tag{7.69}$$

- If the input signal is a white noise, $A_u(e^{j\omega}) = \sigma_u^2$, for all ω, and then, from equation (7.64),

$$\sigma_{x_i}^2 = \|F_i(e^{j\omega})\|_2^2 \sigma_u^2 \tag{7.70}$$

and an appropriate λ is

$$\lambda = \frac{1}{\|F_i(e^{j\omega})\|_2} \tag{7.71}$$

In practical implementations, the use of powers of two to represent the scaling multiplier coefficients is a common procedure, as long as these coefficients satisfy the constraints to control overflow. In this manner, the scaling multipliers can be implemented using simple shift operations.

In the general case when we have m multipliers, the following single scaling can be used at the input:

$$\lambda = \frac{1}{\max\left\{\|F_1(e^{j\omega})\|_p, \ldots, \|F_m(e^{j\omega})\|_p\right\}} \tag{7.72}$$

For cascade and parallel realizations, a scaling multiplier is employed at the input of each section. For some types of second-order sections, used in cascade realizations, the scaling factor of a given section can be incorporated into the output multipliers of the previous section. In general, this procedure leads to a reduction in the output quantization noise. In the case of second-order sections, it is possible to calculate the L_2 and L_∞ norms of the internal transfer functions in closed form (Diniz & Antoniou, 1986; Bomar & Joseph, 1987; Laakso, 1992), as given in Chapter 11.

7.5 Coefficient quantization

During the approximation step, the coefficients of a digital filter are calculated with the high accuracy inherent to the computer employed in the design. When these coefficients are quantized for practical implementations, commonly using rounding, the time and frequency responses of the realized digital filter deviate from the ideal response. In fact, the quantized filter may even fail to meet the prescribed specifications. The sensitivity of the filter response to errors in the coefficients is highly dependent on the type of structure. This fact is one of the motivations for considering alternative realizations having low sensitivity, such as those presented in Chapter 11.

Among the several sensitivity criteria that evaluate the effect of the fixed-point coefficient quantization on the digital filter transfer function, the most widely used are

$$
{}_{I}S_{m_i}^{H(z)}(z) = \frac{\partial H(z)}{\partial m_i} \tag{7.73}
$$

$$
{}_{II}S_{m_i}^{H(z)}(z) = \frac{1}{H(z)} \frac{\partial H(z)}{\partial m_i} \tag{7.74}
$$

For the floating-point representation, the sensitivity criterion must take into account the relative variation of $H(z)$ due to a relative variation in a multiplier coefficient. We then must use

$$
{}_{III}S_{m_i}^{H(z)}(z) = \frac{m_i}{H(z)} \frac{\partial H(z)}{\partial m_i} \tag{7.75}
$$

With such a formulation, it is possible to use the value of the multiplier coefficient to determine ${}_{III}S_{m}^{H(z)}(z)$. A simple example illustrating the importance of this fact is given by the quantization of the system

$$
y(n) = (1 + m)x(n); \quad \text{for } |m| \ll 1 \tag{7.76}
$$

Using equation (7.73), ${}_{I}S_{m_i}^{H(z)}(z) = 1$, regardless of the value of m, while using equation (7.75), ${}_{III}S_{m_i}^{H(z)}(z) = m/(m + 1)$, indicating that a smaller value for the magnitude of m leads to a smaller sensitivity of $H(z)$ with respect to m. This is true for the floating-point representation, as long as the number of bits in the exponent is enough to represent the exponent of m.

7.5.1 Deterministic sensitivity criterion

In practice, one is often interested in the variation of the magnitude of the transfer function, $|H(e^{j\omega})|$, with coefficient quantization. Taking into account the contributions of all multipliers, a useful figure of merit related to this variation would be

$$
S(e^{j\omega}) = \sum_{i=1}^{K} \left| S_{m_i}^{|H(e^{j\omega})|}(e^{j\omega}) \right| \tag{7.77}
$$

where K is the total number of multipliers in the structure, and $S_{m_i}^{\left|H(e^{j\omega})\right|}(e^{j\omega})$, is computed according to one of the equations (7.73)–(7.75), depending on the case.

However, in general, the sensitivities of $H(e^{j\omega})$ to coefficient quantization are much easier to derive than the ones of $\left|H(e^{j\omega})\right|$, and thus it would be convenient if the first one could be used as an estimate of the second. In order to investigate this possibility, we write the frequency response in terms of its magnitude and phase as

$$H(e^{j\omega}) = \left|H(e^{j\omega})\right| e^{j\Theta(\omega)} \tag{7.78}$$

Then the sensitivity measures, defined in equations (7.73)–(7.75), can be written as

$$\left|{}_{I}S_{m_i}^{H(e^{j\omega})}(e^{j\omega})\right| = \sqrt{\left({}_{I}S_{m_i}^{\left|H(e^{j\omega})\right|}(e^{j\omega})\right)^2 + \left|H(e^{j\omega})\right|^2 \left(\frac{\partial\Theta(\omega)}{\partial m_i}\right)^2} \tag{7.79}$$

$$\left|{}_{II}S_{m_i}^{H(e^{j\omega})}(e^{j\omega})\right| = \sqrt{\left({}_{II}S_{m_i}^{\left|H(e^{j\omega})\right|}(e^{j\omega})\right)^2 + \left(\frac{\partial\Theta(\omega)}{\partial m_i}\right)^2} \tag{7.80}$$

$$\left|{}_{III}S_{m_i}^{H(e^{j\omega})}(e^{j\omega})\right| = \sqrt{\left({}_{III}S_{m_i}^{\left|H(e^{j\omega})\right|}(e^{j\omega})\right)^2 + \left|m_i\right|^2 \left(\frac{\partial\Theta(\omega)}{\partial m_i}\right)^2} \tag{7.81}$$

From equations (7.79)–(7.81), one can see that $|S_{m_i}^{H(e^{j\omega})}(e^{j\omega})| \geq |S_{m_i}^{\left|H(e^{j\omega})\right|}(e^{j\omega})|$. Thus, $|S_{m_i}^{H(e^{j\omega})}(e^{j\omega})|$ can be used as a conservative estimate of $|S_{m_i}^{\left|H(e^{j\omega})\right|}(e^{j\omega})|$, in the sense that it guarantees that the transfer function variation will be below a specified tolerance.

Moreover, it is known that low sensitivity is more critical when implementing filters with poles close to the unit circle, and, in such cases, for ω close to a pole frequency ω_0, we can show that $|S_{m_i}^{H(e^{j\omega})}(e^{j\omega})| \approx |S_{m_i}^{\left|H(e^{j\omega})\right|}(e^{j\omega})|$. Therefore, we can rewrite equation (7.77), yielding the following practical sensitivity figure of merit:

$$S(e^{j\omega}) = \sum_{i=1}^{K} \left|S_{m_i}^{H(e^{j\omega})}(e^{j\omega})\right| \tag{7.82}$$

where, depending on the case, $S_{m_i}^{H(e^{j\omega})}(e^{j\omega})$ is given by one of the equations (7.73)–(7.75).

EXAMPLE 7.3

Design a lowpass elliptic filter with the following specifications:

$$\left.\begin{array}{l} A_p = 1.0 \text{ dB} \\ A_r = 40 \text{ dB} \\ \omega_p = 0.3\pi \text{ rad/sample} \\ \omega_r = 0.4\pi \text{ rad/sample} \end{array}\right\} \tag{7.83}$$

Table 7.1. *Filter coefficients for the specifications (7.83).*

Numerator coefficients	Denominator coefficients
$b_0 = 0.028\,207\,76$	$a_0 = 1.000\,000\,00$
$b_1 = -0.001\,494\,75$	$a_1 = -3.028\,484\,73$
$b_2 = 0.031\,747\,58$	$a_2 = 4.567\,772\,20$
$b_3 = 0.031\,747\,58$	$a_3 = -3.900\,153\,49$
$b_4 = -0.001\,494\,75$	$a_4 = 1.896\,641\,38$
$b_5 = 0.028\,207\,76$	$a_5 = -0.418\,854\,19$

Perform the fixed-point sensitivity analysis for the direct-order structure, determining the variation on the ideal magnitude response for a 11-bit quantization of the fractional part, including the sign bit.

SOLUTION

The coefficients of the lowpass elliptic filter are given in Table 7.1.

For the general direct-form structure described by

$$H(z) = \frac{B(z)}{A(z)} = \frac{b_0 + b_1 z^{-1} + \cdots + b_N z^{-N}}{1 + a_1 z^{-1} + \cdots + a_N z^{-N}} \tag{7.84}$$

it is easy to find that the sensitivities as defined in equation (7.73) with respect to the numerator and denominator coefficients are given by

$$_I S_{b_i}^{H(z)}(z) = \frac{z^{-i}}{A(z)} \tag{7.85}$$

$$_I S_{a_i}^{H(z)}(z) = -\frac{z^{-i} H(z)}{A(z)} \tag{7.86}$$

respectively. The magnitude of these functions for the designed fifth-order elliptic filter are seen in Figure 7.8.

The figure of merit $S(e^{j\omega})$, as given in equation (7.82), for the general direct-form realization can be written as

$$S(e^{j\omega}) = \frac{(N+1) + N|H(z)|}{|A(z)|} \tag{7.87}$$

For this example, the function is depicted in Figure 7.9a.

We can then estimate the variation of the ideal magnitude response using the approximation

$$\Delta|H(e^{j\omega})| \approx \Delta m_i S(e^{j\omega}) \tag{7.88}$$

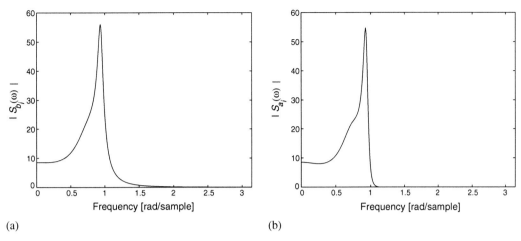

(a)   (b)

Figure 7.8 Magnitudes of the sensitivity functions of $H(z)$ with respect to: (a) numerator coefficients b_i; (b) denominator coefficients a_i.

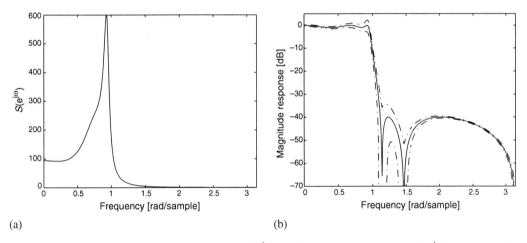

(a)   (b)

Figure 7.9 (a) Sensitivity measurement $S(e^{j\omega})$. (b) Worst-case variation of $|H(e^{j\omega})|$ with an 11-bit fixed-point quantization.

For a fixed-point 11-bit rounding quantization, including the sign bit, $\max\{\Delta m_i\} = 2^{-11}$. In this case, Figure 7.9b depicts the ideal magnitude response of a fifth-order elliptic filter, satisfying the specifications in equation (7.83), along with the corresponding worst-case margins due to the coefficient quantization.

$\triangle$

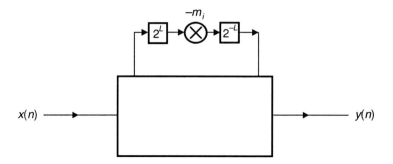

Figure 7.10 Implementation of a multiplication in a pseudo-floating-point representation.

It is worth noting that the sensitivity measurement given by equation (7.82) is also useful as a figure of merit if one uses the so-called pseudo-floating-point representation, that consists of implementing multiplication between a signal and a coefficient with small magnitude in the following form, as depicted in Figure 7.10

$$[x \times m_i]_Q = [(x \times m_i \times 2^L) \times 2^{-L}]_Q \tag{7.89}$$

where L is the exponent of m_i when represented in floating point. Note that in the pseudo-floating-point scheme, all operations are actually performed using in fixed-point arithmetic.

7.5.2 Statistical forecast of the wordlength

Suppose we have designed a digital filter with frequency response $H(e^{j\omega})$, and that the ideal magnitude response is $H_d(e^{j\omega})$, with a tolerance given by $\rho(\omega)$. When the filter coefficients are quantized, then

$$\left| H_Q(e^{j\omega}) \right| = \left| H(e^{j\omega}) \right| + \Delta \left| H(e^{j\omega}) \right| \tag{7.90}$$

Obviously, for a meaningful design, $\left| H_Q(e^{j\omega}) \right|$ must not deviate from $H_d(e^{j\omega})$ by more than $\rho(\omega)$, that is

$$\left| \left(\left| H_Q(e^{j\omega}) \right| - H_d(e^{j\omega}) \right) \right| = \left| \left| H(e^{j\omega}) \right| + \Delta \left| H(e^{j\omega}) \right| - H_d(e^{j\omega}) \right| \le \rho(\omega) \tag{7.91}$$

or, more strictly,

$$\left| \Delta \left| H(e^{j\omega}) \right| \right| \le \rho(\omega) - \left| \left| H(e^{j\omega}) \right| - H_d(e^{j\omega}) \right| \tag{7.92}$$

The variation in the magnitude response of the digital filter due to the variations in the multiplier coefficients m_i can be approximated by

$$\Delta \left| H(e^{j\omega}) \right| \approx \sum_{i=1}^{K} \frac{\partial \left| H(e^{j\omega}) \right|}{\partial m_i} \Delta m_i \tag{7.93}$$

If we consider that:

- the multiplier coefficients are rounded;
- the quantization errors are statistically independent;
- all Δm_i are uniformly distributed;

then the variance of the error in each coefficient, based on equation (7.14), is given by

$$\sigma^2_{\Delta m_i} = \frac{2^{-2b}}{12} \tag{7.94}$$

where b is the number of bits not including the sign bit.

With the assumptions above, the variance of $\Delta |H(e^{j\omega})|$ is given by

$$\sigma^2_{\Delta |H(e^{j\omega})|} \approx \sigma^2_{\Delta m_i} \sum_{i=1}^{K} \left(\frac{\partial |H(e^{j\omega})|}{\partial m_i} \right)^2 = \sigma^2_{\Delta m_i} S^2(e^{j\omega}) \tag{7.95}$$

where $S^2(e^{j\omega})$ is given by equations (7.73) and (7.82).

If we further assume that $\Delta |H(e^{j\omega})|$ is Gaussian (Avenhaus, 1972), we can estimate the probability of $\Delta |H(e^{j\omega})|$ being less than or equal to $x\sigma_{\Delta |H(e^{j\omega})|}$ by

$$\Pr\left\{ \Delta |H(e^{j\omega})| \leq x\sigma_{\Delta |H(e^{j\omega})|} \right\} = \frac{2}{\sqrt{\pi}} \int_0^x e^{-x'^2} dx' \tag{7.96}$$

To guarantee that equation (7.92) holds with a probability less than or equal to the one given in equation (7.96), it suffices that

$$x\sigma_{\Delta m_i} S(e^{j\omega}) \leq \rho(\omega) - \left| |H(e^{j\omega})| - H_d(e^{j\omega}) \right| \tag{7.97}$$

Now, assume that the wordlength, including the sign bit, is given by

$$B = I + F + 1 \tag{7.98}$$

where I and F are the numbers of bits in the integer and fractional parts, respectively. The value of I depends on the required order of magnitude of the coefficient, and F can be estimated from equation (7.97) to guarantee that equation (7.92) holds with probability as given in equation (7.96). To satisfy the inequality in (7.97), the value of 2^{-b}, from equation (7.94), should be given by

$$2^{-b} = \sqrt{12} \min_{\omega \in C} \left\{ \left| \frac{\rho(\omega) - \left| |H(e^{j\omega})| - H_d(e^{j\omega}) \right|}{x S(e^{j\omega})} \right| \right\} \tag{7.99}$$

where C is the set of frequencies not belonging to the filter transition bands. Then, an estimate for F is

$$F \approx -\log_2 \left(\sqrt{12} \min_{\omega \in C} \left\{ \left| \frac{\rho(\omega) - \left| |H(e^{j\omega})| - H_d(e^{j\omega}) \right|}{x S(e^{j\omega})} \right| \right\} \right) \tag{7.100}$$

This method for estimating the wordlength is also useful in iterative procedures to design filters with minimum wordlength (Avenhaus, 1972; Crochiere & Oppenheim, 1975).

An alternative procedure that is widely used in practice to evaluate the design of digital filters with finite-coefficient wordlength, is to design the filters with tighter specifications than required, quantize the coefficients, and check if the prescribed specifications are still met. Obviously, in this case, the success of the design is highly dependent on the designer's experience.

EXAMPLE 7.4

Determine the total number of bits required for the filter designed in Example 7.3 to satisfy the following specifications, after coefficient quantization:

$$\left.\begin{array}{l} A_p = 1.2 \text{ dB} \\ A_r = 39 \text{ dB} \\ \omega_p = 0.3\pi \text{ rad/sample} \\ \omega_r = 0.4\pi \text{ rad/sample} \end{array}\right\} \tag{7.101}$$

SOLUTION

Using the specifications in equation (7.101), we determine

$$\delta_p = 1-10^{-A_p/20} = 0.1482 \tag{7.102}$$

$$\delta_r = 10^{-A_r/20} = 0.0112 \tag{7.103}$$

and define

$$\rho(\omega) = \begin{cases} \delta_p, & \text{for } 0 \le \omega \le 0.3\pi \\ \delta_r, & \text{for } 0.4\pi \le \omega \le \pi \end{cases} \tag{7.104}$$

$$H_d(e^{j\omega}) = \begin{cases} 1, & \text{for } 0 \le \omega \le 0.3\pi \\ 0, & \text{for } 0.4\pi \le \omega \le \pi \end{cases} \tag{7.105}$$

A reasonable certainty margin is about 90%, yielding, from equation (7.96), $x = 1.1631$. We use the filter designed in Example 7.3 as $H(e^{j\omega})$, with the corresponding sensitivity function $S(e^{j\omega})$, as given in equation (7.87) and depicted in Figure 7.9a.

Based on these values, we can compute the number of bits F for the fractional part, using equation (7.100), resulting in $F \approx 11.5993$, which we round up to $F = 12$ bits. From Table 7.1, we observe that $I = 3$ bits are necessary to represent the integer part of the filter coefficients which fall in the range -4 to $+4$. Therefore, the total number of bits required, including the sign bit, is

$$B = I + F + 1 = 16 \tag{7.106}$$

Table 7.2. *Quantized filter coefficients for specifications (7.101).*

Numerator coefficients	Denominator coefficients
$b_0 = 0.028\,320\,31$	$a_0 = 1.000\,000\,00$
$b_1 = -0.001\,464\,84$	$a_1 = -3.028\,564\,45$
$b_2 = 0.031\,738\,28$	$a_2 = 4.567\,871\,09$
$b_3 = 0.031\,738\,28$	$a_3 = -3.900\,146\,48$
$b_4 = -0.001\,464\,84$	$a_4 = 1.896\,728\,52$
$b_5 = 0.028\,320\,31$	$a_5 = -0.418\,945\,31$

Table 7.3. *Filter characteristics as a function of the number of fractional bits F.*

F	A_p [dB]	A_r [dB]
15	1.0100	40.0012
14	1.0188	40.0106
13	1.0174	40.0107
12	**1.1625**	**39.7525**
11	1.1689	39.7581
10	1.2996	39.7650
9	1.2015	40.0280
8	2.3785	40.2212

Table 7.2 shows the filter coefficients after quantization.

Table 7.3 includes the resulting passband ripple and stopband attenuation for several values of F, from which we can clearly see that using the predicted $F = 12$, the specifications in equation (7.101) are satisfied, even after quantization of the filter coefficients.

$\triangle$

7.6   Limit cycles

A serious practical problem that affects the implementation of recursive digital filters is the possible occurrence of parasitic oscillations. These oscillations can be classified, according to their origin, as either granular or overflow limit cycles, as presented below.

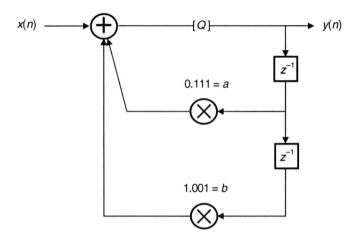

Figure 7.11 Second-order section with a quantizer.

7.6.1 Granular limit cycles

Any stable digital filter, if implemented with idealized infinite-precision arithmetic, should have an asymptotically decreasing response when the input signal becomes zero after a given instant of time $n_0 T$. However, if the filter is implemented with finite-precision arithmetic, the noise signals generated at the quantizers become highly correlated from sample to sample and from source to source. This correlation can cause autonomous oscillations, referred to as granular limit cycles, originating from quantization performed in the least significant signal bits, as indicated in the example that follows.

EXAMPLE 7.5

Suppose the filter of Figure 7.11 has the following input signal:

$$x(n) = \begin{cases} 0.111, & \text{for } n = 1 \\ 0.000, & \text{for } n \neq 1 \end{cases} \tag{7.107}$$

where the numbers are represented in two's complement. Determine the output signal in the case when the quantizer performs rounding, for $n = 1, \ldots, 40$.

SOLUTION

The output in the time domain, assuming that the quantizer rounds the signal, is given in Table 7.4, where one can easily see that an oscillation is sustained at the output, even after the input becomes zero.

$\triangle$

In many practical applications, where the signal levels in a digital filter can be constant or very low, even for short periods of time, limit cycles are highly

Table 7.4. *Output signal of network shown in Figure 7.11:*
$y(n) = Q[ay(n-1) + by(n-2) + x(n)]$.

n	$y(n)$
1	0.111
2	$Q(0.110\,001 + 0.000\,000) = 0.110$
3	$Q(0.101\,010 + 1.001\,111) = 1.111$
4	$Q(1.111\,001 + 1.010\,110) = 1.010$
5	$Q(1.010\,110 + 0.000\,111) = 1.100$
6	$Q(1.100\,100 + 0.101\,010) = 0.010$
7	$Q(0.001\,110 + 0.011\,100) = 0.101$
8	$Q(0.100\,011 + 1.110\,010) = 0.011$
9	$Q(0.010\,101 + 1.011\,101) = 1.110$
10	$Q(1.110\,010 + 1.101\,011) = 1.100$
11	$Q(1.100\,100 + 0.001\,110) = 1.110$
12	$Q(1.110\,010 + 0.011\,100) = 0.010$
13	$Q(0.001\,110 + 0.001\,110) = 0.100$
14	$Q(0.011\,100 + 1.110\,010) = 0.010$
15	$Q(0.001\,110 + 1.100\,100) = 1.110$
16	$Q(1.110\,010 + 1.110\,010) = 1.100$
17	$Q(1.100\,100 + 0.001\,110) = 1.110$
18	$Q(1.110\,010 + 0.011\,100) = 0.010$
19	$Q(0.001\,110 + 0.001\,110) = 0.100$
20	$Q(0.011\,100 + 1.110\,010) = 0.010$
21	$Q(0.001\,110 + 1.100\,100) = 1.110$
22	$Q(1.110\,010 + 1.110\,010) = 1.100$
⋮	⋮

undesirable, and should be eliminated or at least have their amplitude bounds strictly limited.

7.6.2   Overflow limit cycles

Overflow limit cycles can occur when the magnitudes of the internal signals exceed the available register range. In order to avoid the increase of the signal wordlength in recursive digital filters, overflow nonlinearities must be applied to the signal. Such nonlinearities influence the most significant bits of the signal, possibly causing severe distortion. An overflow can give rise to self-sustained, high-amplitude oscillations, widely known as overflow limit cycles. Overflow can occur in any structure in the presence of an input signal, and input-signal scaling is crucial to reduce the probability of overflow to an acceptable level.

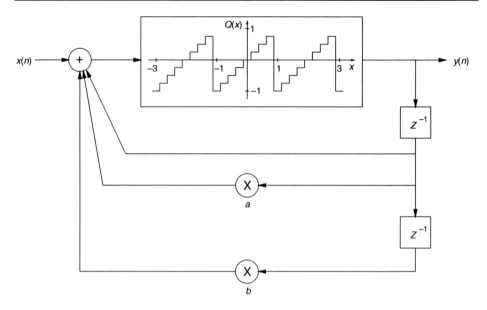

Figure 7.12 Second-order section with an overflow quantizer.

EXAMPLE 7.6

Consider the filter of Figure 7.12 with $a = 0.9606$ and $b = 0.9849$, where the overflow nonlinearity employed is the two's complement with 3-bit quantization (see Figure 7.12). Its analytic expression is given by

$$Q(x) = \frac{1}{4}[(\lceil 4x - 0.5\rceil + 4) \mod 8] - 1 \tag{7.108}$$

where $\lceil x \rceil$ means the smallest interger larger than or equal to x.

Determine the output signal of such a filter for zero input, given the initial conditions $y(-2) = 0.50$ and $y(-1) = -1.00$.

SOLUTION

With $a = 0.9606$, $b = -0.9849$, $y(-2) = 0.50$ and $y(-1) = -1.00$, we have that

$$
\begin{aligned}
y(0) &= Q[1.9606(-1.00) - 0.9849(0.50)] &&= Q[-2.4530] = -0.50 \\
y(1) &= Q[1.9606(-0.50) - 0.9849(-1.00)] &&= Q[0.0046] &&= 0.00 \\
y(2) &= Q[1.9606(0.00) - 0.9849(-0.50)] &&= Q[0.4924] &&= 0.50 \\
y(3) &= Q[1.9606(0.50) - 0.9849(0.00)] &&= Q[0.9803] &&= -1.00 \\
&\quad\vdots
\end{aligned}
\tag{7.109}
$$

Since $y(2) = y(-2)$ and $y(3) = y(-1)$, we have that, although there is no excitation, the output signal in nonzero and periodic with a period of 4, thus indicating the existence of overflow limit cycles.

$\triangle$

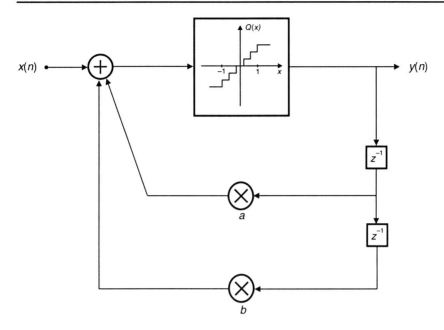

Figure 7.13 Second-order section with a rounding and saturating quantizer.

A digital filter structure is considered free from overflow limit cycles if the error introduced in the filter after an overflow decreases with time in such a way that the output of the nonlinear filter (including the quantizers) converges to the output of the ideal linear filter (Claasen et al., 1975).

In practice a quantizer incorporates nonlinearities corresponding to both granular quantization and overflow. Figure 7.13 illustrates a digital filter using a quantizer that implements rounding as the granular quantization and saturation arithmetic as the overflow nonlinearity.

7.6.3 Elimination of zero-input limit cycles

A general IIR filter can be depicted as in Figure 7.14a, where the linear N-port network consists of interconnections of multipliers and adders. In a recursive filter implemented with fixed-point arithmetic, each internal loop contains a quantizer. Assuming that the quantizers are placed at the delay inputs (the state variables), as shown in Figure 7.14b, we can describe the digital filter, including the quantizers, using the following state-space formulation:

$$\begin{aligned} \mathbf{x}(n+1) &= [\mathbf{A}\mathbf{x}(n) + \mathbf{b}u(n)]_Q \\ y(n) &= \mathbf{c}^T\mathbf{x}(n) + du(n) \end{aligned} \tag{7.110}$$

where $[x]_Q$ indicates the quantized value of x, $\mathbf{A}$ is the state matrix, $\mathbf{b}$ is the input vector, $\mathbf{c}$ is the output vector, and d represents the direct connection between the input and output of the filter.

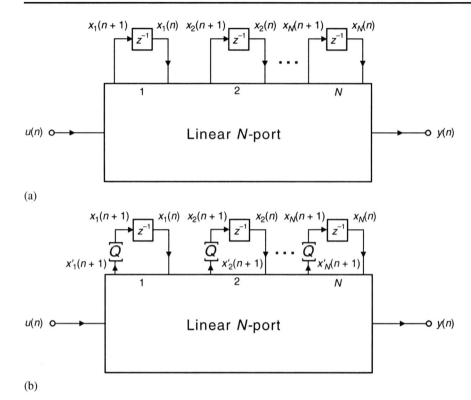

Figure 7.14 Digital filter networks: (a) ideal; (b) with quantizers at the state variables.

In order to analyze zero-input limit cycles, it is sufficient to consider the recursive part of the state equation given by

$$\mathbf{x}(k+1) = [\mathbf{A}\mathbf{x}(k)]_Q = [\mathbf{x}'(k+1)]_Q \qquad (7.111)$$

where the quantization operations $[\cdot]_Q$ are nonlinear operations such as truncation, rounding, or overflow.

The basis for the elimination of nonlinear oscillations is given by Theorem 7.1 below.

THEOREM 7.1
If a stable digital filter has a state matrix $\mathbf{A}$ and, for any $N \times 1$ vector $\hat{\mathbf{x}}$ there exists a diagonal positive-definite matrix $\mathbf{G}$, such that

$$\hat{\mathbf{x}}^{\mathrm{T}}(\mathbf{G} - \mathbf{A}^{\mathrm{T}}\mathbf{G}\mathbf{A})\hat{\mathbf{x}} \geq 0 \qquad (7.112)$$

then the granular zero-input limit cycles can be eliminated if the quantization is performed through magnitude truncation.

◇

PROOF

Consider a non-negative pseudo-energy Lyapunov function given by (Willems, 1970)

$$p(\mathbf{x}(n)) = \mathbf{x}^T(n)\mathbf{G}\mathbf{x}(n) \tag{7.113}$$

The energy variation in a single iteration can be defined as

$$\begin{aligned}
\Delta p(n+1) &= p(\mathbf{x}(n+1)) - p(\mathbf{x}(n)) \\
&= \mathbf{x}^T(n+1)\mathbf{G}\mathbf{x}(n+1) - \mathbf{x}^T(n)\mathbf{G}\mathbf{x}(n) \\
&= [\mathbf{x}'^T(n+1)]_Q\mathbf{G}[\mathbf{x}'(n+1)]_Q - \mathbf{x}^T(n)\mathbf{G}\mathbf{x}(n) \\
&= [\mathbf{A}\mathbf{x}(n)]_Q^T\mathbf{G}[\mathbf{A}\mathbf{x}(n)]_Q - \mathbf{x}^T(n)\mathbf{G}\mathbf{x}(n) \\
&= [\mathbf{A}\mathbf{x}(n)]^T\mathbf{G}[\mathbf{A}\mathbf{x}(n)] - \mathbf{x}^T(n)\mathbf{G}\mathbf{x}(n) - \sum_{i=1}^{N}(x_i'^2(n+1) - x_i^2(n+1))g_i \\
&= \mathbf{x}^T(n)[\mathbf{A}^T\mathbf{G}\mathbf{A} - \mathbf{G}]\mathbf{x}(n) - \sum_{i=1}^{N}(x_i'^2(n+1) - x_i^2(n+1))g_i \tag{7.114}
\end{aligned}$$

where g_i are the diagonal elements of $\mathbf{G}$.

If quantization is performed through magnitude truncation, then the errors due to granular quantization and overflow are such that

$$|x_i(n+1)| \le |x_i'(n+1)| \tag{7.115}$$

for all i and n. Therefore, if equation (7.112) holds, from equation (7.114), we have that

$$\Delta p(n+1) \le 0 \tag{7.116}$$

If a digital filter is implemented with finite-precision arithmetic, within a finite number of samples after the input signal becomes zero, the output signal will become either a periodic oscillation or zero. Periodic oscillations, with nonzero amplitude, cannot be sustained if $\Delta p(n+1) \le 0$, as shown above. Therefore, equations (7.112) and (7.115) are sufficient conditions to guarantee the elimination of granular zero-input limit cycles on a recursive digital filter.

$\square$

Note that the condition given in equation (7.112) is equivalent to requiring that $\mathbf{F}$ be positive semidefinite, where

$$\mathbf{F} = (\mathbf{G} - \mathbf{A}^T\mathbf{G}\mathbf{A}) \tag{7.117}$$

It is worth observing that for any stable state matrix $\mathbf{A}$, its eigenvalues are inside the unit circle, and there will always be a positive-definite and symmetric matrix $\mathbf{G}$

such that $\mathbf{F}$ is symmetric and positive semidefinite. However, if $\mathbf{G}$ is not diagonal, the quantization process required to eliminate zero-input limit cycles is extremely complicated (Meerkötter, 1976), as the quantization operation in each quantizer is coupled to the others. On the other hand, if there is a matrix $\mathbf{G}$ which is diagonal and positive definite such that $\mathbf{F}$ is positive semidefinite, then zero-input limit cycles can be eliminated by simple magnitude truncation.

In the following theorem, we will state more specific conditions regarding the elimination of zero-input limit cycles in second-order systems.

THEOREM 7.2
Given a 2×2 stable state matrix $\mathbf{A}$, there is a diagonal positive-definite matrix $\mathbf{G}$, such that $\mathbf{F}$ is positive semidefinite, if and only if (Mills et al., 1978)

$$a_{12}a_{21} \geq 0 \tag{7.118}$$

or

$$\left. \begin{array}{l} a_{12}a_{21} < 0 \\ |a_{11} - a_{22}| + \det(\mathbf{A}) \leq 1 \end{array} \right\} \tag{7.119}$$

$\diamond$

PROOF
Let $\mathbf{G} = (\mathbf{T}^{-1})^2$ be a diagonal positive-definite matrix, such that $\mathbf{T}$ is a diagonal nonsingular matrix. Therefore, we can write $\mathbf{F}$ as

$$\mathbf{F} = \mathbf{T}^{-1}\mathbf{T}^{-1} - \mathbf{A}^{\mathrm{T}}\mathbf{T}^{-1}\mathbf{T}^{-1}\mathbf{A} \tag{7.120}$$

and then

$$\begin{aligned} \mathbf{T}^{\mathrm{T}}\mathbf{F}\mathbf{T} &= \mathbf{T}^{\mathrm{T}}\mathbf{T}^{-1}\mathbf{T}^{-1}\mathbf{T} - \mathbf{T}^{\mathrm{T}}\mathbf{A}^{\mathrm{T}}\mathbf{T}^{-1}\mathbf{T}^{-1}\mathbf{A}\mathbf{T} \\ &= \mathbf{I} - (\mathbf{T}^{-1}\mathbf{A}\mathbf{T})^{\mathrm{T}}(\mathbf{T}^{-1}\mathbf{A}\mathbf{T}) \\ &= \mathbf{I} - \mathbf{M} \end{aligned} \tag{7.121}$$

with $\mathbf{M} = (\mathbf{T}^{-1}\mathbf{A}\mathbf{T})^{\mathrm{T}}(\mathbf{T}^{-1}\mathbf{A}\mathbf{T})$, as $\mathbf{T}^{\mathrm{T}} = \mathbf{T}$.

Since the matrix $(\mathbf{I} - \mathbf{M})$ is symmetric and real, its eigenvalues are real. This matrix is then positive semidefinite if and only if its eigenvalues are non-negative (Strang, 1980), or, equivalently, if and only if its trace and determinant are non-negative. We then have that

$$\det\{\mathbf{I} - \mathbf{M}\} = 1 + \det\{\mathbf{M}\} - \mathrm{tr}\{\mathbf{M}\} = 1 + (\det\{\mathbf{A}\})^2 - \mathrm{tr}\{\mathbf{M}\} \tag{7.122}$$

$$\mathrm{tr}\{\mathbf{I} - \mathbf{M}\} = 2 - \mathrm{tr}\{\mathbf{M}\} \tag{7.123}$$

For a stable digital filter, it is easy to verify that $\det\{\mathbf{A}\} < 1$, and then

$$\mathrm{tr}\{\mathbf{I} - \mathbf{M}\} > \det\{\mathbf{I} - \mathbf{M}\} \tag{7.124}$$

Hence, the condition $\det\{\mathbf{I} - \mathbf{M}\} \geq 0$ is necessary and sufficient to guarantee that $(\mathbf{I} - \mathbf{M})$ is positive semidefinite.

From the definition of $\mathbf{M}$, and using $\alpha = t_{22}/t_{11}$, then

$$\det\{\mathbf{I} - \mathbf{M}\} = 1 + (\det\{\mathbf{A}\})^2 - \left(a_{11}^2 + \alpha^2 a_{12}^2 + \frac{a_{21}^2}{\alpha^2} + a_{22}^2\right) \tag{7.125}$$

By calculating the maximum of the equation above with respect to α, we get

$$\alpha^{*2} = |a_{21}/a_{12}| \tag{7.126}$$

and then

$$\det{}^*\{\mathbf{I} - \mathbf{M}\} = 1 + (\det\{\mathbf{A}\})^2 - (a_{11}^2 + 2|a_{12}a_{21}| + a_{22}^2)$$
$$= (1 + \det\{\mathbf{A}\})^2 - (\mathrm{tr}\{\mathbf{A}\})^2 + 2(a_{12}a_{21} - |a_{12}a_{21}|) \tag{7.127}$$

where the $*$ indicates the maximum value of the respective determinant. We now analyze two separate cases to guarantee that $\det{}^*\{\mathbf{I} - \mathbf{M}\} \geq 0$.

- If

$$a_{12}a_{21} \geq 0 \tag{7.128}$$

  then

$$\det{}^*\{\mathbf{I} - \mathbf{M}\} = (1 + \det\{\mathbf{A}\})^2 - (\mathrm{tr}\{\mathbf{A}\})^2$$
$$= (1 + \alpha_2)^2 - (-\alpha_1)^2$$
$$= (1 + \alpha_1 + \alpha_2)(1 - \alpha_1 + \alpha_2) \tag{7.129}$$

  where $\alpha_1 = -\mathrm{tr}\{\mathbf{A}\}$ and $\alpha_2 = \det\{\mathbf{A}\}$ are the filter denominator coefficients. It can be verified that, for a stable filter, $(1 + \alpha_1 + \alpha_2)(1 - \alpha_1 + \alpha_2) > 0$, and then equation (7.128) implies that $(\mathbf{I} - \mathbf{M})$ is positive definite.

- If

$$a_{12}a_{21} < 0 \tag{7.130}$$

  then

$$\det{}^*\{\mathbf{I} - \mathbf{M}\} = 1 + (\det\{\mathbf{A}\})^2 - (a_{11}^2 - 2a_{12}a_{21} + a_{22}^2)$$
$$= (1 - \det\{\mathbf{A}\})^2 - (a_{11} - a_{22})^2 \tag{7.131}$$

  This equation is greater than or equal to zero, if and only if

$$|a_{11} - a_{22}| + \det\{\mathbf{A}\} \leq 1 \tag{7.132}$$

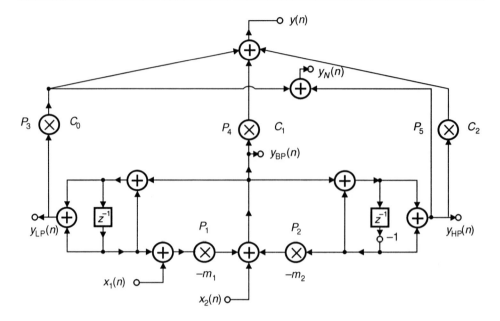

Figure 7.15 General-purpose network.

Therefore, either equation (7.128) or equations (7.130) and (7.132) are the necessary and sufficient conditions for the existence of a diagonal matrix

$$\mathbf{T} = \text{diag}\{t_{11}\ t_{22}\} \tag{7.133}$$

with

$$\frac{t_{22}}{t_{11}} = \sqrt{\left|\frac{a_{21}}{a_{12}}\right|} \tag{7.134}$$

such that $\mathbf{F}$ is positive semidefinite.

$\square$

It is worth observing that the above theorem gives the conditions for the matrix $\mathbf{F}$ to be positive semidefinite for second-order sections. In the example below, we illustrate the limit-cycle elimination process by showing, without resorting to Theorem 7.2, that a given second-order structure is free from zero-input limit cycles. The reader is encouraged to apply the theorem to show the same result.

EXAMPLE 7.7
Examine the possibility of eliminating limit cycles in the network of Figure 7.15 (Diniz & Antoniou, 1988).

SOLUTION
The structure in Figure 7.15 realizes lowpass, bandpass, and highpass transfer

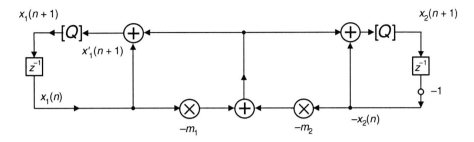

Figure 7.16 Recursive part of the network in Figure 7.15.

functions simultaneously (with subscripts LP, BP, and HP, respectively). The structure also realizes a transfer function with zeros on the unit circle, using the minimum number of multipliers. The characteristic polynomial of the structure is given by

$$D(z) = z^2 + (m_1 - m_2)z + m_1 + m_2 - 1 \qquad (7.135)$$

In order to guarantee stability, the multiplier coefficients m_1 and m_2 should fall in the range

$$\left.\begin{array}{l} m_1 > 0 \\ m_2 > 0 \\ m_1 + m_2 < 2 \end{array}\right\} \qquad (7.136)$$

Figure 7.16 depicts the recursive part of the structure in Figure 7.15, including the quantizers.

The zero-input state-space equation for the structure in Figure 7.16 is

$$\mathbf{x}'(n+1) = \left[\begin{array}{c} x_1'(n+1) \\ x_2'(n+1) \end{array}\right] = \mathbf{A} \left[\begin{array}{c} x_1(n) \\ x_2(n) \end{array}\right] \qquad (7.137)$$

with

$$\mathbf{A} = \left[\begin{array}{cc} (1-m_1) & m_2 \\ -m_1 & (m_2-1) \end{array}\right] \qquad (7.138)$$

By applying quantization to $\mathbf{x}'(n+1)$, we find

$$\mathbf{x}(n+1) = [\mathbf{x}'(n+1)]_Q = [\mathbf{A}\mathbf{x}(n)]_Q \qquad (7.139)$$

A quadratic positive-definite function can be defined as

$$p(\mathbf{x}(n)) = \mathbf{x}^{\mathrm{T}}(n)\mathbf{G}\mathbf{x}(n) = \frac{x_1^2}{m_2} + \frac{x_2^2}{m_1} \qquad (7.140)$$

with

$$
\mathbf{G} = \begin{bmatrix} \frac{1}{m_2} & 0 \\ 0 & \frac{1}{m_1} \end{bmatrix}
\tag{7.141}
$$

which is positive definite, since from equation (7.136), $m_1 > 0$ and $m_2 > 0$.

An auxiliary energy increment is then

$$
\begin{aligned}
\Delta p_0(n+1) &= p(\mathbf{x}'(n+1)) - p(\mathbf{x}(n)) \\
&= \mathbf{x}'^{\mathrm{T}}(n+1)\mathbf{G}\mathbf{x}'(n+1) - \mathbf{x}^{\mathrm{T}}(n)\mathbf{G}\mathbf{x}(n) \\
&= \mathbf{x}^{\mathrm{T}}(n)[\mathbf{A}^{\mathrm{T}}\mathbf{G}\mathbf{A} - \mathbf{G}]\mathbf{x}(n) \\
&= (m_1 + m_2 - 2)\left(x_1(n)\sqrt{\frac{m_1}{m_2}} - x_2(n)\sqrt{\frac{m_2}{m_1}} \right)^2
\end{aligned}
\tag{7.142}
$$

Since from equation (7.136), $m_1 + m_2 < 2$, then

$$
\left.
\begin{aligned}
\Delta p_0(n+1) &= 0, \quad \text{for } x_1(n) = x_2(n)\frac{m_2}{m_1} \\
\Delta p_0(n+1) &< 0, \quad \text{for } x_1(n) \neq x_2(n)\frac{m_2}{m_1}
\end{aligned}
\right\}
\tag{7.143}
$$

Now, if magnitude truncation is applied to quantize the state variables, then $p(\mathbf{x}(n)) \leq p(\mathbf{x}'(n))$, which implies that

$$
\Delta p(\mathbf{x}(n)) = p(\mathbf{x}(n+1)) - p(\mathbf{x}(n)) \leq 0
\tag{7.144}
$$

and then $p(\mathbf{x}(n))$ is a Lyapunov function.

Overall, when no quantization is applied in the structure of Figure 7.15, no self-sustained oscillations occur if the stability conditions of equation (7.136) are satisfied. If, however, quantization is applied to the structure, as shown in Figure 7.16, oscillations may occur. Using magnitude truncation, then $|x_i(n)| \leq |x_i'(n)|$, and under these circumstances, $p(\mathbf{x}(n))$ decreases during the following iterations, and eventually the oscillations disappear, with

$$
\mathbf{x}(n) = \begin{bmatrix} 0 \\ 0 \end{bmatrix}
\tag{7.145}
$$

being the only possible equilibrium point.

$\triangle$

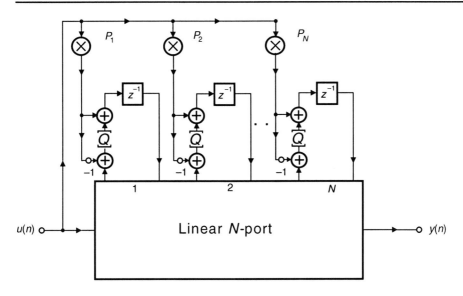

Figure 7.17 Modified Nth-order network for the elimination of constant-input limit cycles.

7.6.4   Elimination of constant-input limit cycles

As seen above, the sufficient conditions for the elimination of zero-input limit cycles are well established. However, if the system input is a nonzero constant, limit cycles may still occur. It is worth observing that the response of a stable linear system to a constant input signal should also be a constant signal. In Diniz & Antoniou (1986), a theorem is presented that establishes how constant-input limit cycles can also be eliminated in digital filters in which zero-input limit cycles are eliminated. This theorem is as follows.

THEOREM 7.3

Assume that the general digital filter in Figure 7.14b does not sustain zero-input limit cycles and that

$$\left.\begin{array}{l} \mathbf{x}(n+1) = [\mathbf{Ax}(n) + \mathbf{B}u(n)]_Q \\ y(n) = \mathbf{C}^T\mathbf{x}(n) + du(n) \end{array}\right\} \tag{7.146}$$

Constant-input limit cycles can also be eliminated, by modifying the structure in Figure 7.14b, as shown in Figure 7.17, where

$$\mathbf{p} = [p_1 \ p_2 \cdots p_n]^T = (\mathbf{I} - \mathbf{A})^{-1}\mathbf{B} \tag{7.147}$$

and $\mathbf{p}u_0$ must be representable in the machine wordlength, where u_0 is a constant input signal.

$\diamond$

PROOF

Since the structure of Figure 7.14b is free from zero-input limit cycles, the autonomous system

$$\mathbf{x}(n+1) = [\mathbf{A}\mathbf{x}(n)]_Q \tag{7.148}$$

is such that

$$\lim_{n\to\infty} \mathbf{x}(n) = [0\ 0\ldots 0]^T \tag{7.149}$$

If $\mathbf{p}$ is as given in equation (7.147), the modified structure of Figure 7.17 is described by

$$
\begin{aligned}
\mathbf{x}(n+1) &= [\mathbf{A}\mathbf{x}(n) - \mathbf{p}u_0 + \mathbf{B}u_0]_Q + \mathbf{p}u_0 \\
&= [\mathbf{A}\mathbf{x}(n) - \mathbf{I}(\mathbf{I}-\mathbf{A})^{-1}\mathbf{B}u_0 + (\mathbf{I}-\mathbf{A})(\mathbf{I}-\mathbf{A})^{-1}\mathbf{B}u_0]_Q + \mathbf{p}u_0 \\
&= [\mathbf{A}(\mathbf{x}(n) - \mathbf{p}u_0)]_Q + \mathbf{p}u_0
\end{aligned}
\tag{7.150}
$$

Defining

$$\hat{\mathbf{x}}(n) = \mathbf{x}(n) - \mathbf{p}u_0 \tag{7.151}$$

then, from equation (7.150), we can write that

$$\hat{\mathbf{x}}(n+1) = [\mathbf{A}\hat{\mathbf{x}}(n)]_Q \tag{7.152}$$

Which is the same as equation (7.148), except for the transformation in the state variable. Hence, as $\mathbf{p}u_0$ is machine representable (that is, $\mathbf{p}u_0$ can be exactly calculated with the available wordlength), equation (7.152) also represents a stable system free from constant-input limit cycles.

□

If the quantization of the structure depicted in Figure 7.14b is performed using magnitude truncation, the application of the strategy of Theorem 7.3 leads to the so-called controlled rounding proposed in Butterweck (1975).

The constraints imposed by requiring that $\mathbf{p}u_0$ be machine representable reduce the number of structures in which the technique described by Theorem 7.3 applies. However, there are a large number of second-order sections and wave digital filter structures in which these requirements are automatically met. In fact, a large number of research papers have been published proposing new structures which are free from zero-input limit cycles (Meerkötter & Wegener, 1975; Fettweis & Meerkötter, 1975b; Diniz & Antoniou, 1988), and free from constant-input limit cycles (Verkroost & Butterweck, 1976; Verkroost, 1977; Liu & Turner, 1983; Diniz, 1988; Sarcinelli & Diniz, 1990). However, the analysis procedures for the generation of these structures are not unified and here we have aimed to provide a unified framework leading to a general procedure to generate structures which are free from granular limit cycles.

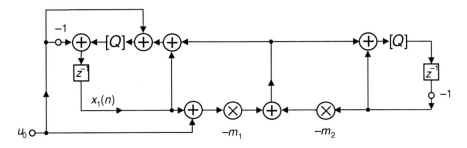

Figure 7.18 Elimination of constant-input limit cycles in the structure of Figure 7.15.

EXAMPLE 7.8

Show that by placing the input signal at the point denoted by $x_1(n)$, the structure in Figure 7.15 is free from constant-input limit cycles.

SOLUTION

The second-order section of Figure 7.15, with a constant input $x_1(n) = u_0$, can be described by

$$\mathbf{x}(n+1) = \mathbf{A}\mathbf{x}(n) + \begin{bmatrix} -m_1 \\ -m_1 \end{bmatrix} u_0 \tag{7.153}$$

with $\mathbf{p}$ such that

$$\mathbf{p} = \begin{bmatrix} m_1 & -m_2 \\ m_1 & 2-m_2 \end{bmatrix}^{-1} \begin{bmatrix} -m_1 \\ -m_1 \end{bmatrix} = \begin{bmatrix} -1 \\ 0 \end{bmatrix} \tag{7.154}$$

Therefore, $\mathbf{p}u_0$ is clearly machine representable, for any u_0, and the constant-input limit cycles can be eliminated, as depicted in Figure 7.18.

$\triangle$

7.6.5   Forced-response stability of digital filters with nonlinearities due to overflow

The stability analysis of the forced response of digital filters that include nonlinearities to control overflow must be performed considering input signals for which in the ideal linear system the overflow level is never reached after a given instant n_0. In this way, we can verify whether the real system output will recover after an overflow has occurred before instant n_0. Although the input signals considered are in a particular class of signals, it can be shown that if the real system recovers for these signals, it will also recover after each overflow, for any input signal, if the recovery period is shorter than the time between two consecutive overflows (Claasen et al., 1975).

Consider the ideal linear system as depicted in Figure 7.19a and the real nonlinear system as depicted in Figure 7.19b.

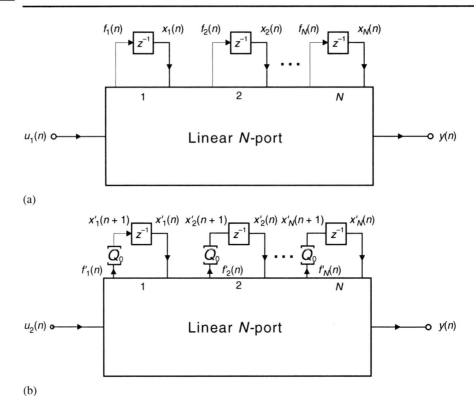

Figure 7.19 General digital filter networks: (a) ideal; (b) with quantizers at the state variables.

The linear system illustrated in Figure 7.19a is described by the equations

$$\mathbf{f}(n) = \mathbf{A}\mathbf{x}(n) + \mathbf{B}u_1(n) \tag{7.155}$$

$$\mathbf{x}(n) = \mathbf{f}(n-1) \tag{7.156}$$

and the nonlinear system illustrated in Figure 7.19b is described by the equations

$$\mathbf{f}'(n) = \mathbf{A}\mathbf{x}'(n) + \mathbf{B}u_2(n) \tag{7.157}$$

$$\mathbf{x}'(n) = [\mathbf{f}'(n-1)]_{Q_0} \tag{7.158}$$

where $[u]_{Q_0}$ denotes quantization of u, in the case where an overflow occurs.

We assume that the output signal of the nonlinear system is properly scaled so that no oscillation due to overflow occurs if it does not occur at the state variables.

The response of the nonlinear system of Figure 7.19b is stable if, when $u_1(n) = u_2(n)$, the difference between the outputs of the linear N-port system of Figure 7.19a, $\mathbf{f}(n)$, and the outputs of the linear N-port system of Figure 7.19b, $\mathbf{f}'(n)$, tends to zero as $n \to \infty$. In other words, if we define an error signal $\mathbf{e}(n) = \mathbf{f}'(n) - \mathbf{f}(n)$, then

$$\lim_{n \to \infty} \mathbf{e}(n) = [0\,0\ldots0]^{\mathrm{T}} \tag{7.159}$$

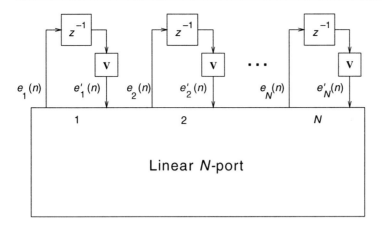

Figure 7.20 Nonlinear system relating the signals $\mathbf{e}'(n)$ and $\mathbf{e}(n)$.

If the difference between the output signals of the two linear N-port systems converges to zero, this implies that the difference between the state variables of both systems will also tend to zero. This can be deduced from equations (7.155) and (7.157), which yield

$$\mathbf{e}(n) = \mathbf{f}'(n) - \mathbf{f}(n) = \mathbf{A}[\mathbf{x}'(n) - \mathbf{x}(n)] = \mathbf{A}\mathbf{e}'(n) \tag{7.160}$$

where $\mathbf{e}'(n) = \mathbf{x}'(n) - \mathbf{x}(n)$ is the difference between the state variables of both systems.

Equation (7.160) is equivalent to saying that $\mathbf{e}(n)$ and $\mathbf{e}'(n)$ are the output and input signals of a linear N-port system described by matrix $\mathbf{A}$, which is the transition matrix of the original system. Then, from equation (7.159), the forced-response stability of the system in Figure 7.19b is equivalent to the zero-input response of the same system, regardless of the quantization characteristics $[\cdot]_{Q_0}$.

Substituting equations (7.156) and (7.158) in equation (7.160), we have that

$$\mathbf{e}'(n) = [\mathbf{f}'(n-1)]_{Q_0} - \mathbf{f}(n-1) = [\mathbf{e}(n-1) + \mathbf{f}(n-1)]_{Q_0} - \mathbf{f}(n-1) \tag{7.161}$$

By defining the time-varying vector $\mathbf{v}(\mathbf{e}(n), n)$ as

$$\mathbf{v}(\mathbf{e}(n), n) = [\mathbf{e}(n) + \mathbf{f}(n)]_{Q_0} - \mathbf{f}(n) \tag{7.162}$$

equation (7.161) can be rewritten as

$$\mathbf{e}'(n) = \mathbf{v}(\mathbf{e}(n-1), (n-1)) \tag{7.163}$$

The nonlinear system described by equations (7.160)–(7.163) is depicted in Figure 7.20.

As we saw in Subsection 7.6.3, a system such as the one in Figure 7.20 is free

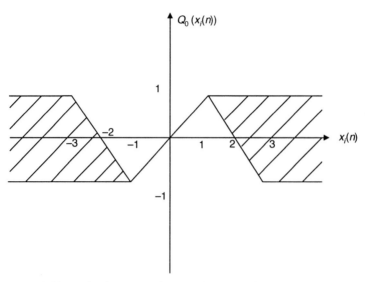

Figure 7.21 Region for the overflow nonlinearity which guarantees forced-response stability in networks which satisfy Theorem 7.1.

from zero-input nonlinear oscillations if the nonlinearity $\mathbf{v}(\cdot, n)$ is equivalent to the magnitude truncation, that is

$$|\mathbf{v}(e_i(n), n)| < |e_i(n)|, \quad \text{for } i = 1, \ldots, N \tag{7.164}$$

If we assume that the internal signals are such that $|f_i(n)| \leq 1$, for $n > n_0$, then it can be shown that equation (7.164) remains valid whenever the quantizer Q_0 has overflow characteristics within the shaded region of Figure 7.21 (see exercise 7.16).

Figure 7.21 can be interpreted as follows:

- If $-1 \leq x_i(n) \leq 1$, then there should be no overflow.
- If $1 \leq x_i(n) \leq 3$, then the overflow nonlinearity should be such that $2 - x_i(n) \leq Q_0(x_i(n)) \leq 1$.
- If $-3 \leq x_i(n) \leq -1$, then the overflow nonlinearity should be such that $-1 \leq Q_0(x_i(n)) \leq -2 - x_i(n)$.
- If $x_i(n) \geq 3$ or $x_i(n) \leq -3$, then $-1 \leq Q_0(x_i(n)) \leq 1$.

It is important to note that the overflow nonlinearity of the saturation type (equation (7.108)) satisfies the requirements in Figure 7.21.

Summarizing the above reasoning, one can state that a digital filter which is free of zero-input limit cycles, according to the condition of equation (7.112), is also forced-input stable, provided that the overflow nonlinearities are in the shaded regions of Figure 7.21.

7.7   Summary

This chapter has presented an introduction to the finite-wordlength effects on the performance of digital signal processing systems, taking into consideration that all internal signals and parameters within these systems are quantized to a finite set of discrete values.

Initially, the basic concepts of binary number representation were briefly reviewed. Then the effects of quantizing the internal signals of digital filters were analyzed in some detail. A procedure to control the internal dynamic range, so that overflows are avoided, was given.

Section 7.5 introduced some tools to analyze the quantization effects on the system parameters, such as digital filter coefficients. Several forms of sensitivity measures were introduced and their merits briefly discussed. In addition, we presented a statistical method to predict the required coefficient wordlength for a filter to meet prescribed specifications.

The chapter concluded by studying granular and overflow limit cycles that arise in recursive digital filters. The emphasis was placed on procedures to eliminate these nonlinear oscillations in filters implemented with fixed-point arithmetic, which in many cases is crucial from the designer's point of view. Due to space limitations, we could not address many useful techniques to eliminate or control the amplitude of nonlinear oscillations. An interesting example is the so-called error spectrum shaping technique (Laakso et al., 1992), which is seen in detail in Subsection 11.2.3. This technique has also been successful in reducing roundoff noise (Diniz & Antoniou, 1985) and in eliminating limit cycles in digital filters implemented using floating-point arithmetic (Laakso et al., 1994). There are also a number of research papers dealing with the analysis of the various types of limit cycles, including techniques to determine their frequencies and amplitude bounds (Munson et al., 1984; Bauer & Wang, 1993).

7.8   Exercises

7.1   Deduce the probability density function of the quantization error for the sign-magnitude and one's complement representations. Consider the cases of rounding, truncation, and magnitude truncation.

7.2   Calculate the scaling factor for the digital filter in Figure 7.22, using the L_2 and L_∞ norms.

7.3   Show that, for the Type 1 canonic direct-form realization for IIR filters (Figure 4.12), the scaling coefficient is given by $\lambda = \|D(z)\|_p$.

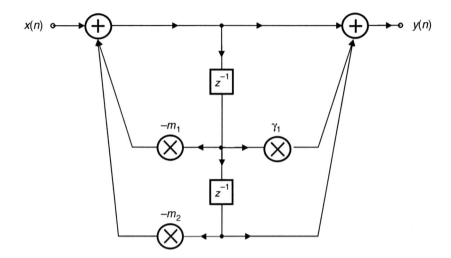

Figure 7.22   Second-order digital filter: $m_1 = -1.933\,683$, $m_2 = 0.951\,89$, $\gamma_1 = -1.889\,37$.

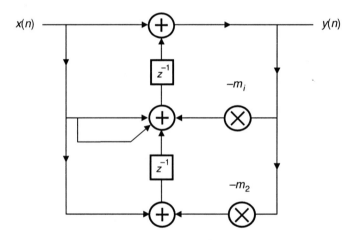

Figure 7.23   Second-order digital filter.

7.4   Show that, for the Type 2 canonic direct-form realization for IIR filters (Figure 4.13), the scaling coefficient is given by

$$\lambda = \frac{1}{\max\{1, \|H(z)\|_p\}}$$

7.5   Calculate the relative output-noise variance in dB for the filter of Figure 7.22 using fixed-point and floating-point arithmetics.

7.6   Calculate the output-noise RPSD for the filter in Figure 7.23 using fixed-point and floating-point arithmetics.

7.7   Derive equations (7.79)–(7.81).

7.8   Plot the magnitude response of the filters designed in Exercise 4.1 when the coefficients are quantized to 7 bits. Compare the results with the ones for the original filters.

7.9   Calculate the maximum expected value of the transfer function deviation, given by the maximum value of the tolerance function, with a confidence factor of 95%, for the digital filter in Figure 7.22 implemented with 6 bits. Use $H(e^{j\omega}) = H_d(e^{j\omega})$ in equation (7.97).

7.10   Determine the minimum number of bits a multiplier should have to keep a signal-to-noise ratio above 80 dB at its output. Consider that the type of quantization is rounding.

7.11   Discuss whether granular limit cycles can be eliminated in the filter of Figure 7.23 by using magnitude truncation quantizers at the state variables.

7.12   Verify whether it is possible to eliminate granular and overflow limit cycles in the structure of Figure 7.4b.

7.13   Verify whether it is possible to eliminate granular and overflow limit cycles in the structure of Exercise 4.2.

7.14   Plot the overflow characteristics of simple one's-complement arithmetic.

7.15   Suppose there is a structure that is free from zero-input limit cycles, when employing magnitude truncation, and that it is also forced-input stable when using saturation arithmetic. Discuss if overflow oscillations occur when the overflow quantization is removed, with the numbers represented in two's-complement arithmetic.

7.16   Show that equation (7.164) is satisfied whenever the quantizer has overflow characteristics as in Figure 7.21, following the steps:

    (i)   Express inequality (7.164) as a function of $f_i'(n)$, $[f_i'(n)]_{Q_0}$, and $f_i(n)$.

    (ii)   Determine the regions of the plane $f_i'(n) \times [f_i'(n)]_{Q_0}$, as a function of $f_i(n)$, where inequality (7.164) applies.

    (iii)   Use the fact that inequality (7.164) must be valid for all $|f_i(n)| < 1$.

    (iv)   Suppose that the output of an overflow quantizer is limited to $[-1, 1]$.

8 Multirate systems

8.1 Introduction

In many applications of digital signal processing, it is necessary for different sampling rates to coexist within a given system. One common example is when two subsystems working at different sampling rates have to communicate and the sampling rates must be made compatible. Another case is when a wideband digital signal is decomposed into several non-overlapping narrowband channels in order to be transmitted. In such a case, each narrowband channel may have its sampling rate decreased until its Nyquist limit is reached, thereby saving transmission bandwidth.

Here, we describe such systems which are generally referred to as multirate systems. Multirate systems are used in several applications, ranging from digital filter design to signal coding and compression, and have been increasingly present in modern digital systems.

First, we study the basic operations of decimation and interpolation, and show how arbitrary rational sampling-rate changes can be implemented with them. The design of decimation and interpolation filters is also addressed. Then, we deal with filter design techniques which use decimation and interpolation in order to achieve a prescribed set of filter specifications. Finally, MATLAB functions which aid in the design and implementation of multirate systems are briefly described.

8.2 Basic principles

Intuitively, any sampling-rate change can be effected by recovering the band-limited analog signal $x_a(t)$ from its samples $x(m)$, and then resampling it with a different sampling rate, thus generating a different discrete version of the signal, $x'(n)$. Of course the intermediate analog signal $x_a(t)$ must be filtered so that it can be re-sampled without aliasing. One possible way to do so is described here.

Suppose we have a digital signal $x(m)$ that was generated from an analog signal $x_a(t)$ with sampling period T_1, that is, $x(m) = x_a(mT_1)$, for $m \in \mathbb{Z}$. In order to avoid aliasing in the process, it is assumed that $x_a(t)$ is band-limited to $[-\frac{\pi}{T_1}, \frac{\pi}{T_1}]$. Therefore, replacing each sample of the signal by an impulse proportional to it, we have that the equivalent analog signal is

$$x_i(t) = \sum_{m=-\infty}^{\infty} x(m)\delta(t - mT_1) \tag{8.1}$$

whose spectrum is periodic with period $\frac{2\pi}{T_1}$. In order to recover the original analog signal $x_a(t)$ from $x_i(t)$, the repetitions of the spectrum must be discarded. Therefore, as seen in Section 1.5, $x_i(t)$ must be filtered with a filter $h(t)$ whose ideal frequency response $H(j\omega)$ is:

$$H(j\omega) = \begin{cases} 1, & \omega \in [-\frac{\pi}{T_1}, \frac{\pi}{T_1}] \\ 0, & \text{otherwise} \end{cases} \tag{8.2}$$

and also

$$x_a(t) = x_i(t) * h(t) = \frac{1}{T_1} \sum_{m=-\infty}^{\infty} x(m)\text{sinc}\left[\frac{\pi}{T_1}(t - mT_1)\right] \tag{8.3}$$

Then, resampling $y(t)$ with period T_2 to generate the digital signal $x'(n) = x_a(nT_2)$, for $n \in \mathbb{Z}$, we have that

$$x'(n) = \frac{1}{T_1} \sum_{m=-\infty}^{\infty} x(m)\text{sinc}\left[\frac{\pi}{T_1}(nT_2 - mT_1)\right] \tag{8.4}$$

This is the general equation governing sampling-rate changes. Observe that there is no restriction on the values of T_1 and T_2. Of course, if $T_2 > T_1$ and aliasing is to be avoided, the filter in equation (8.2) must have a frequency response equal to zero for $\omega \notin [-\frac{\pi}{T_2}, \frac{\pi}{T_2}]$. As seen in Section 1.5, since equation (8.4) consists of infinite summations involving the sinc function, it is not of practical use. In general, for rational sampling-rate changes, which covers most cases of interest, one can derive expressions working solely in the discrete-time domain. This is covered in the next sections, where three special cases are considered: decimation by an integer factor M, interpolation by an integer factor L, and sampling-rate change by a rational factor L/M.

8.3   Decimation

To decimate or subsample a digital signal $x(m)$ by a factor of M is to reduce its sampling rate M times. This is equivalent to keeping only every Mth sample of the signal. This operation is represented as in Figure 8.1 and exemplified in Figure 8.2 for the case $M = 2$.

The relation between the decimated signal and the original one is therefore very straightforward, that is

$$x_d(n) = x(nM) \tag{8.5}$$

Figure 8.1 Block diagram representing the decimation by a factor of M.

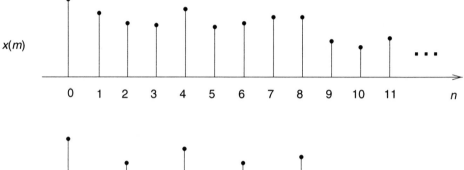

Figure 8.2 Decimation by 2. $x(m) = \ldots x(0)\, x(1)\, x(2)\, x(3)\, x(4) \ldots; x_d(n) = \ldots x(0)\, x(2)\, x(4)$ $x(6)\, x(8) \ldots$

In the frequency domain, if the spectrum of $x(m)$ is $X(\mathrm{e}^{\mathrm{j}\omega})$, the spectrum of the decimated signal, $X_d(\mathrm{e}^{\mathrm{j}\omega})$, becomes

$$X_d(\mathrm{e}^{\mathrm{j}\omega}) = \frac{1}{M}\sum_{k=0}^{M-1} X(\mathrm{e}^{\mathrm{j}\frac{\omega-2\pi k}{M}}) \tag{8.6}$$

Such a result is reached by first defining $x'(m)$ as

$$x'(m) = \begin{cases} x(m), & m = nM,\, n \in \mathbb{Z} \\ 0, & \text{otherwise} \end{cases} \tag{8.7}$$

which can also be written as

$$x'(m) = x(m) \sum_{n=-\infty}^{\infty} \delta(m - nM) \tag{8.8}$$

The Fourier transform $X_d(e^{j\omega})$ is then given by

$$
\begin{aligned}
X_d(e^{j\omega}) &= \sum_{n=-\infty}^{\infty} x_d(n)e^{-j\omega n} \\
&= \sum_{n=-\infty}^{\infty} x(nM)e^{-j\omega n} \\
&= \sum_{n=-\infty}^{\infty} x'(nM)e^{-j\omega n} \\
&= \sum_{\theta=-\infty}^{\infty} x'(\theta)e^{-j\frac{\omega}{M}\theta} \\
&= X'(e^{j\frac{\omega}{M}})
\end{aligned}
\tag{8.9}
$$

But, from equation (8.8) (see also Exercise 2.22),

$$
\begin{aligned}
X'(e^{j\omega}) &= \frac{1}{2\pi} X(e^{j\omega}) * \mathcal{F}\left\{ \sum_{n=-\infty}^{\infty} \delta(m-nM) \right\} \\
&= \frac{1}{2\pi} X(e^{j\omega}) * \frac{2\pi}{M} \sum_{k=0}^{M-1} \delta\left(\omega - \frac{2\pi k}{M}\right) \\
&= \frac{1}{M} \sum_{k=0}^{M-1} X(e^{j(\omega - \frac{2\pi k}{M})})
\end{aligned}
\tag{8.10}
$$

Then, from equation (8.9),

$$
X_d(e^{j\omega}) = X'(e^{j\frac{\omega}{M}}) = \frac{1}{M} \sum_{k=0}^{M-1} X(e^{j\frac{\omega-2\pi k}{M}})
\tag{8.11}
$$

which is the same as equation (8.6).

As illustrated in Figure 8.3 for $M = 2$, equation (8.6) means that the spectrum of $x_d(n)$ is composed of copies of the spectrum of $x(m)$ expanded by M and repeated with period 2π (which is equivalent to copies of the spectrum of $x(m)$ repeated with period $\frac{2\pi}{M}$ and then expanded by M). This implies that, in order to avoid aliasing after decimation, the bandwidth of the signal $x(m)$ must be limited to the interval $[-\frac{\pi}{M}, \frac{\pi}{M}]$. Therefore, the decimation operation is generally preceded by a lowpass filter (see Figure 8.4), which approximates the following frequency response:

$$
H_d(e^{j\omega}) = \begin{cases} 1, & \omega \in [-\frac{\pi}{M}, \frac{\pi}{M}] \\ 0, & \text{otherwise} \end{cases}
\tag{8.12}
$$

If we then include the filtering operation, the decimated signal is obtained by obtaining every Mth sample of the convolution of the signal $x(m)$ with the filter

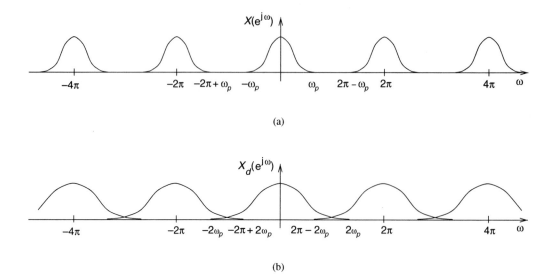

Figure 8.3 (a) Spectrum of the original digital signal. (b) Spectrum of the signal decimated by a factor of 2.

impulse response $h_d(m)$, that is

$$x_d(n) = \sum_{m=-\infty}^{\infty} x(m)h_d(nM - m) \qquad (8.13)$$

Some important facts must be noted about the decimation operation (Crochiere & Rabiner, 1983):

- It is time varying, that is, if the input signal $x(m)$ is shifted, the output signal will not in general be a shifted version of the previous output. More precisely, let $\mathcal{D}_M$ be the decimation-by-M operator. If $x_d(n) = \mathcal{D}_M\{x(m)\}$, then in general $\mathcal{D}_M\{x(m-k)\} \neq x_d(n - l)$, unless $k = rM$, when $\mathcal{D}_M\{x(m - k)\} = x_d(n - r)$). Because of this property, the decimation is referred to as a periodically time-invariant operation.

- Referring to equation (8.13), one can see that, if the filter $H_d(z)$ is FIR, its outputs need only be computed every M samples, which implies that its implementation complexity is M times smaller than that of a usual filtering operation (Peled & Liu, 1985). This is not valid in general for IIR filters, because in such cases one needs all past outputs to compute the present output, unless the transfer function is of the type $H(z) = \frac{N(z)}{D(z^M)}$ (Martinez & Parks, 1979; Ansari & Liu, 1983).

- If the frequency range of interest for the signal $x(m)$ is $[-\omega_p, \omega_p]$, with $\omega_p < \frac{\pi}{M}$, one can afford aliasing outside this range. Therefore, the constraints upon the filter

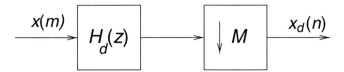

Figure 8.4   General decimation operation.

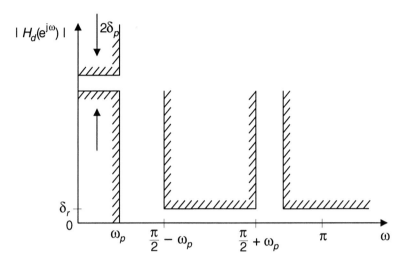

Figure 8.5   Specifications of a decimation filter for $M = 4$.

can be relaxed, yielding the following specifications for $H_d(z)$

$$H_d(e^{j\omega}) = \begin{cases} 1, & |\omega| \in [0, \omega_p] \\ 0, & |\omega| \in [\frac{2\pi k}{M} - \omega_p, \frac{2\pi k}{M} + \omega_p], \quad k = 1, 2, \ldots, M - 1 \end{cases} \qquad (8.14)$$

The decimation filter can be efficiently designed using the optimum FIR approximation methods described in Chapter 5. In order to do so, one has to define the following parameters:

$$\left.\begin{array}{l} \delta_p : \text{ passband ripple} \\ \delta_r : \text{ stopband attenuation} \\ \omega_p : \text{ passband cutoff frequency} \\ \omega_{r_1} = (\frac{2\pi}{M} - \omega_p) : \text{ first stopband edge} \end{array}\right\} \qquad (8.15)$$

However, in general, it is more efficient to design a multiband filter according to equation (8.14), as illustrated in Figure 8.5.

EXAMPLE 8.1
A signal that carries useful information only in the range $0 \leq \omega \leq 0.1\omega_s$ must be decimated by a factor of $M = 4$. Design a linear-phase decimation filter satisfying the

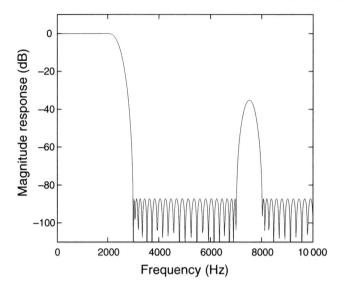

Figure 8.6 Magnitude response of the decimation filter for $M = 4$.

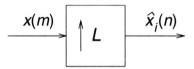

Figure 8.7 Interpolation by a factor of L.

following specifications:

$$\left.\begin{array}{l} \delta_p = 0.001 \\ \delta_r = 5 \times 10^{-5} \\ \Omega_s = 20\,000 \text{ Hz} \end{array}\right\} \qquad (8.16)$$

SOLUTION

In order to satisfy the specifications, the minimum required order is 85. The resulting magnitude response is shown in Figure 8.6. Table 8.1 shows the filter coefficients, where, since the filter is linear-phase, only half of the coefficients are included.

$\triangle$

8.4   Interpolation

To interpolate or upsample a digital signal $x(m)$ by a factor of L is to include $L - 1$ zeros between its samples. This operation is represented as Figure 8.7.

Table 8.1. *Impulse response of the decimation filter.*

$h(0) = 3.820\,781\,32\text{E}{-}06$	$h(15) = 2.517\,008\,36\text{E}{-}03$	$h(29) = -1.321\,676\,37\text{E}{-}02$
$h(1) = -1.507\,805\,29\text{E}{-}04$	$h(16) = 3.701\,624\,44\text{E}{-}03$	$h(30) = -2.990\,579\,33\text{E}{-}03$
$h(2) = -2.448\,756\,47\text{E}{-}04$	$h(17) = 2.045\,639\,63\text{E}{-}03$	$h(31) = 1.279\,759\,72\text{E}{-}02$
$h(3) = -3.435\,647\,29\text{E}{-}04$	$h(18) = -5.802\,170\,25\text{E}{-}04$	$h(32) = 2.557\,482\,46\text{E}{-}02$
$h(4) = -3.788\,346\,54\text{E}{-}04$	$h(19) = -4.016\,433\,37\text{E}{-}03$	$h(33) = 2.356\,110\,89\text{E}{-}02$
$h(5) = 1.485\,714\,57\text{E}{-}06$	$h(20) = -6.309\,219\,85\text{E}{-}03$	$h(34) = 7.655\,099\,06\text{E}{-}03$
$h(6) = 3.509\,150\,06\text{E}{-}04$	$h(21) = -4.100\,211\,14\text{E}{-}03$	$h(35) = -1.970\,338\,33\text{E}{-}02$
$h(7) = 7.404\,404\,53\text{E}{-}04$	$h(22) = 1.834\,005\,55\text{E}{-}04$	$h(36) = -4.486\,666\,60\text{E}{-}02$
$h(8) = 9.475\,605\,22\text{E}{-}04$	$h(23) = 6.051\,101\,25\text{E}{-}03$	$h(37) = -4.765\,930\,50\text{E}{-}02$
$h(9) = 2.536\,401\,58\text{E}{-}04$	$h(24) = 1.018\,897\,23\text{E}{-}02$	$h(38) = -2.078\,454\,67\text{E}{-}02$
$h(10) = -5.233\,496\,25\text{E}{-}04$	$h(25) = 7.514\,458\,17\text{E}{-}03$	$h(39) = 3.942\,389\,34\text{E}{-}02$
$h(11) = -1.450\,931\,72\text{E}{-}03$	$h(26) = 8.311\,957\,34\text{E}{-}04$	$h(40) = 1.194\,177\,20\text{E}{-}01$
$h(12) = -1.996\,616\,53\text{E}{-}03$	$h(27) = -8.812\,788\,68\text{E}{-}03$	$h(41) = 1.921\,554\,78\text{E}{-}01$
$h(13) = -8.658\,689\,55\text{E}{-}04$	$h(28) = -1.603\,665\,23\text{E}{-}02$	$h(42) = 2.380\,381\,17\text{E}{-}01$
$h(14) = 6.363\,479\,43\text{E}{-}04$		

The interpolated signal is then given by

$$\hat{x}_i(n) = \begin{cases} x(n/L), & n = kL,\ k \in \mathbb{Z} \\ 0, & \text{otherwise} \end{cases} \tag{8.17}$$

The interpolation operation is exemplified in Figure 8.8 for the case $L - 2$.

In the frequency domain, if the spectrum of $x(m)$ is $X(\text{e}^{\text{j}\omega})$, it is straightforward to see that the spectrum of the interpolated signal, $\hat{X}_i(\text{e}^{\text{j}\omega})$, becomes (Crochiere & Rabiner, 1983)

$$\hat{X}_i(\text{e}^{\text{j}\omega}) = X(\text{e}^{\text{j}\omega L}) \tag{8.18}$$

Figure 8.9 shows the spectra of the signals $x(m)$ and $\hat{x}_i(n)$ for an interpolation factor of L.

Since the spectrum of the original digital signal is periodic with period 2π, the spectrum of the interpolated signal has period $\frac{2\pi}{L}$. Therefore, in order to obtain a smooth interpolated version of $x(m)$, the spectrum of the interpolated signal must have the same shape as the spectrum of $x(m)$. This can be obtained by filtering out the repetitions of the spectrum of $\hat{x}_i(n)$ outside $[-\frac{\pi}{L}, \frac{\pi}{L}]$. Thus, the interpolation operation is generally followed by a lowpass filter (see Figure 8.10) which approximates the following frequency response:

$$H_i(\text{e}^{\text{j}\omega}) = \begin{cases} L, & \omega \in [-\frac{\pi}{L}, \frac{\pi}{L}] \\ 0, & \text{otherwise} \end{cases} \tag{8.19}$$

The interpolation operation is thus equivalent to the convolution of the interpolation filter impulse response, $h_i(n)$, with the signal $\hat{x}_i(n)$ defined in equation (8.17).

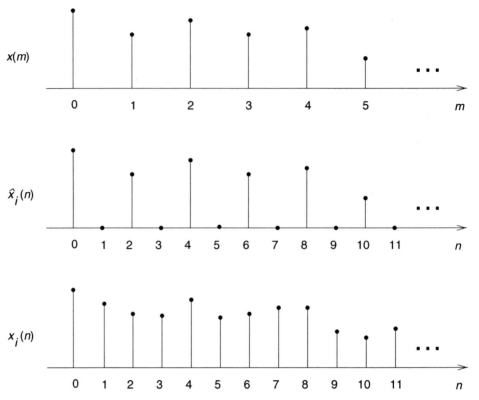

Figure 8.8 Interpolation by 2. $x(m)$: original signal; $\hat{x}_i(n)$: signal with zeros inserted between samples; $x_i(n)$: interpolated signal after filtering by $H_i(z)$. ($x(m) = \dots x(0)\, x(1)\, x(2)\, x(3)\, x(4)$ $x(5)\, x(6) \dots; \hat{x}_i(n) = \dots x(0)\, 0\, x(1)\, 0\, x(2)\, 0\, x(3) \dots; x_i(n) = \dots x_i(0)\, x_i(1)\, x_i(2)\, x_i(3)\, x_i(4)$ $x_i(5)\, x_i(6) \dots$)

Considering that the only nonzero samples of $\hat{x}_i(n)$ are the ones having a multiple of L index, equation (8.17) can be rewritten as

$$\hat{x}_i(kL) = \begin{cases} x(k), & k \in \mathbb{Z} \\ 0, & \text{otherwise} \end{cases} \tag{8.20}$$

With the help of the above equation, it is easy to see that, in the time domain, the filtered interpolated signal becomes

$$x_i(n) = \sum_{m=-\infty}^{\infty} \hat{x}_i(m)h(n-m) = \sum_{k=-\infty}^{\infty} x(k)h(n-kL) \tag{8.21}$$

Some important facts must be noted about the interpolation operation (Crochiere & Rabiner, 1983):

- As opposed to the decimation operation, the interpolation is time invariant. More precisely, if $\mathcal{I}_L$ is the interpolation-by-L operator, $x_i(n) = \mathcal{I}_L\{x(m)\}$ implies that $\mathcal{I}_L\{x(m-k)\} = x_i(n-kL)$.

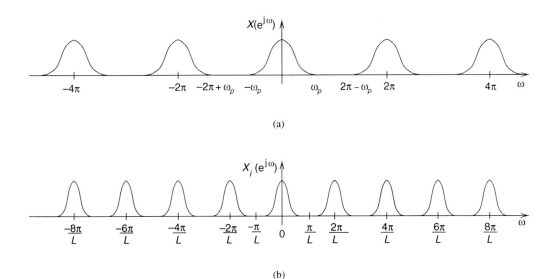

Figure 8.9 (a) Spectrum of the original digital signal. (b) Spectrum of the signal interpolated by a factor of L.

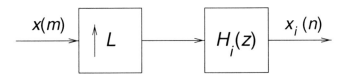

Figure 8.10 General interpolation operation.

- Referring to equation (8.21), one can see that the computation of the output of the filter $H_i(z)$ uses only one out of every L samples of the input signal, because the remaining samples are zero. This means that its implementation complexity can be made L times simpler than that of a usual filtering operation.
- If the signal $x(m)$ is band-limited to $[-\omega_p, \omega_p]$, the repetitions of the spectrum will only appear in a neighborhood of radius $\frac{\omega_p}{L}$ around the frequencies $\frac{2\pi k}{L}$, $k = 1, 2, \ldots, L - 1$. Therefore, the constraints upon the filter can be relaxed as in the decimation case, yielding

$$H_i(e^{j\omega}) = \begin{cases} L, & |\omega| \in [0, \frac{\omega_p}{L}] \\ 0, & |\omega| \in [\frac{2\pi k - \omega_p}{L}, \frac{2\pi k + \omega_p}{L}], \quad k = 1, 2, \ldots, L - 1 \end{cases} \quad (8.22)$$

The gain factor L in equations (8.19) and (8.22) can be understood by noting that since we are maintaining one out of every L samples of the signal, the average energy of the signal decreases by a factor L^2, and therefore the gain of the interpolating filter must be L to compensate for that loss.

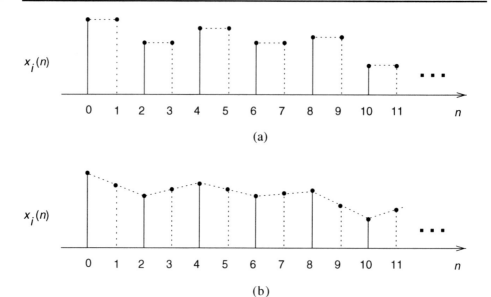

(a)

(b)

Figure 8.11 Examples of interpolators: (a) zero-order hold; (b) linear interpolator.

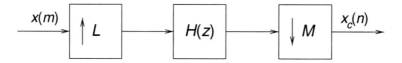

Figure 8.12 Sampling rate change by a factor of $\frac{L}{M}$.

8.4.1   Examples of interpolators

Supposing $L = 2$, two common examples can be devised as shown in Figure 8.11:

- Zero-order hold: $x(2n + 1) = x(2n)$. From equation (8.21), this is equivalent to having $h(0) = h(1) = 1$, that is, $H_i(z) = 1 + z^{-1}$.
- Linear interpolator: $x(2n + 1) = \frac{1}{2}[x(2n) + x(2n + 2)]$. From equation (8.21), this is equivalent to having $h(-1) = \frac{1}{2}$, $h(0) = 1$, and $h(1) = \frac{1}{2}$, that is, $H_i(z) = \frac{1}{2}(z + 2 + z^{-1})$.

8.5   Rational sampling-rate changes

A rational sampling-rate change by a factor $\frac{L}{M}$ can be implemented by cascading an interpolator by a factor of L with a decimator by a factor of M, as represented in Figure 8.12.

Since $H(z)$ is an interpolation filter, its cutoff frequency must be less than $\frac{\pi}{L}$. However, since it is also a decimation filter, its cutoff frequency must be less than

$\frac{\pi}{M}$. Therefore, it must approximate the following frequency response:

$$H(e^{j\omega}) = \begin{cases} L, & |\omega| \leq \min\{\frac{\pi}{L}, \frac{\pi}{M}\} \\ 0, & \text{otherwise} \end{cases} \tag{8.23}$$

Similarly to the cases of decimation and interpolation, the specifications of $H(z)$ can be relaxed if the bandwidth of the signal is smaller than ω_p. The relaxed specifications are the result of cascading the specifications in equation (8.22) and the specifications in equation (8.14) with ω_p replaced by $\frac{\omega_p}{L}$. Since L and M can be assumed, without loss of generality, to be relatively prime, this yields

$$H(e^{j\omega}) = \begin{cases} L, & |\omega| < \min\{\frac{\omega_p}{L}, \frac{\pi}{M}\} \\ 0, & \min\{\frac{2\pi}{L} - \frac{\omega_p}{L}, \frac{2\pi}{M} - \frac{\omega_p}{L}\} \leq |\omega| \leq \pi \end{cases} \tag{8.24}$$

8.6 Inverse operations

At this point, a natural question to ask is: are the decimation-by-M ($\mathcal{D}_M$) and interpolation-by-M ($\mathcal{I}_M$) operators inverses of each other? In other words, does $\mathcal{D}_M \mathcal{I}_M = \mathcal{I}_M \mathcal{D}_M = $ identity?

It is easy to see that $\mathcal{D}_M \mathcal{I}_M = $ identity, because the $(M-1)$ zeros between samples inserted by the interpolation operation are removed by the decimation as long as the two operations are properly aligned, otherwise a null signal will result.

On the other hand, $\mathcal{I}_M \mathcal{D}_M$ is not the identity operator in general, since the decimation operation removes $(M-1)$ out of M samples of the signal and the interpolation operation inserts $(M-1)$ zeros between samples, and then their cascade is equivalent to replacing $(M-1)$ out of M samples of the signal with zeros. However, if the decimation-by-M operation is preceded by a band-limiting filter for the interval $[-\frac{\pi}{M}, \frac{\pi}{M}]$ (see equation (8.12)), and the interpolation operation is followed by the same filter (as illustrated in Figure 8.13), then $\mathcal{I}_M \mathcal{D}_M$ becomes the identity operation. This can be easily confirmed in the frequency domain, as the band-limiting filter avoids aliasing after decimation, and the spectrum of the decimated signal is inside $[-\pi, \pi]$. After interpolation by M, there are images of the spectrum of the signal in the intervals $[\frac{\pi k}{M}, \frac{\pi(k+1)}{M}]$, for $k = -M, -M+1, \ldots, M-1$. However, the second band-limiting filter keeps only the image inside $[-\frac{\pi}{M}, \frac{\pi}{M}]$, which corresponds to the spectrum of the original signal.

8.7 Decimation and interpolation for efficient filter implementation

In Chapter 10, efficient structures for FIR filters will be analyzed in detail. In this section, we show how the concepts of decimation and interpolation can be employed

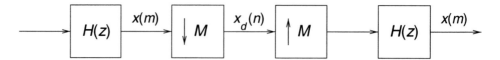

Figure 8.13 Decimation followed by interpolation.

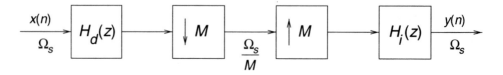

Figure 8.14 Filter using decimation/interpolation.

to generate efficient FIR filter implementations with respect to the number of multiplications per output sample. First, we deal with the case of efficiently implementing a narrowband FIR filter. Then, we give a brief introduction to the frequency masking approach, that will be dealt with in detail in Subsection 10.6.3.

8.7.1 Narrowband FIR filters

Consider the system in Figure 8.14, consisting of the cascade of a decimator and an interpolator by M.

From equations (8.6) and (8.18), one can easily infer the relation between the Fourier transforms of $y(n)$ and $x(n)$, which is

$$Y(e^{j\omega}) = \frac{H_i(e^{j\omega})}{M} \left\{ \sum_{k=0}^{M-1} \left[X(e^{j(\omega-2\pi\frac{k}{M})}) H_d(e^{j(\omega-2\pi\frac{k}{M})}) \right] \right\} \tag{8.25}$$

Supposing that both the decimation filter, $H_d(z)$, and the interpolation filter, $H_i(z)$, have been properly designed, the spectrum repetitions in the above equation are canceled, yielding the following relation:

$$\frac{Y(e^{j\omega})}{X(e^{j\omega})} = \frac{H_d(e^{j\omega})H_i(e^{j\omega})}{M} = H(e^{j\omega}) \tag{8.26}$$

This result shows that the cascading of the decimation and interpolation operations of the same order M is equivalent to just cascading the decimation and interpolation filters, provided that both bandwidths are smaller than $\frac{\pi}{M}$. At a first glance, this structure is entirely equivalent to cascading two filters, and it presents no special advantage. However, one must bear in mind that in the implementation of the decimation operation, there is a reduction by M in the number of multiplications, and the same is true for the interpolation operation. Therefore, this structure can provide a dramatic reduction in the overall number of multiplications. Actually, this reduction increases with the value of M, and we should choose M as large as possible, such that the bandwidth of the desired filter becomes smaller than $\frac{\pi}{M}$.

If we wish to design a filter with passband ripple δ_p and stopband ripple δ_r, it is enough to design interpolation and decimation filters, each having passband ripple $\frac{\delta_p}{2}$ and stopband ripple δ_r.

EXAMPLE 8.2

Using the concepts of decimation and interpolation, design a lowpass filter satisfying the following specifications:

$$\left.\begin{array}{l} \delta_p = 0.001 \\ \delta_r = 5 \times 10^{-4} \\ \Omega_p = 0.025\Omega_s \\ \Omega_r = 0.045\Omega_s \\ \Omega_s = 2\pi \text{ rad/s} \end{array}\right\} \tag{8.27}$$

SOLUTION

With the given set of specifications, the maximum possible value of M is 11. Using the minimax method, $H_d(z)$ and $H_i(z)$ can be made identical, and they must be at least of order 177 each. The magnitude response for these filters is shown in Figure 8.15, and the corresponding coefficients are given in Table 8.2. Once again, only half of the coefficients are listed as the filters have linear phase. It is left to the reader to confirm that the cascading of these two filters satisfy the problem specifications.

With the conventional approach, also using the minimax method, the total number of multiplications per sample would be 94 (as a linear-phase filter of order 187 is required). Using decimation and interpolation, the total number of multiplications per output sample is only 178/11 (89/11 for the decimation and 89/11 for the interpolation), a significant reduction in the overall complexity.

In fact, even greater reductions in complexity can be achieved if the decimators and interpolators in Figure 8.14 are composed of several decimation stages followed by several interpolation stages.

$\triangle$

Although we have considered only lowpass design, the procedure of Figure 8.14 can also be used to design narrowband bandpass filters. All we have to do is to choose M such that the desired filter passband and transition bands are contained in an interval of the form $[\frac{i\pi}{M}, \frac{(i+1)\pi}{M}]$, for only one value of i (Crochiere & Rabiner, 1983). In such cases, the interpolation and decimation filters are bandpass (see Subsection 9.2.1). Highpass and bandstop filters can be implemented based on lowpass and bandpass designs.

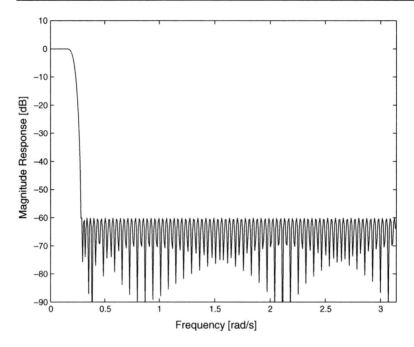

Figure 8.15 Magnitude response of interpolation and decimation filters.

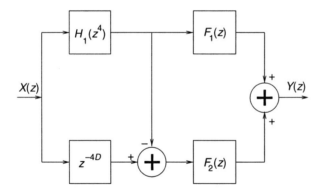

Figure 8.16 Filter design with the frequency masking approach using interpolation.

8.7.2 Wideband FIR filters with narrow transition bands

Another interesting application of interpolation is in the design of sharp cutoff filters with low computational complexity using the so-called frequency response masking approach (Lim, 1986), which uses the fact that an interpolated filter has a transition band L times smaller than the prototype filter. The complete process is sketched in Figure 8.16, for an interpolation ratio of $L = 4$, and exemplified as follows.

Suppose, for instance, that we want to design a normalized-lowpass filter having $\omega_p = 5\pi/8$ and $\omega_r = 11\pi/16$, such that the transition width is $\pi/16$. Using the

Table 8.2. *Interpolator and decimator filter coefficients.*

$h(0) = 5.344\,784\,38\text{E}{-}04$	$h(30) = 8.005\,450\,17\text{E}{-}04$	$h(60) = 5.132\,083\,73\text{E}{-}04$
$h(1) = 8.197\,057\,28\text{E}{-}05$	$h(31) = 4.624\,537\,83\text{E}{-}04$	$h(61) = -1.496\,343\,11\text{E}{-}03$
$h(2) = 6.992\,537\,51\text{E}{-}05$	$h(32) = 5.496\,821\,94\text{E}{-}05$	$h(62) = -3.651\,504\,06\text{E}{-}03$
$h(3) = 4.412\,695\,90\text{E}{-}05$	$h(33) = -4.066\,280\,87\text{E}{-}04$	$h(63) = -5.854\,680\,00\text{E}{-}03$
$h(4) = 4.605\,328\,56\text{E}{-}06$	$h(34) = -9.015\,885\,78\text{E}{-}04$	$h(64) = -7.995\,369\,40\text{E}{-}03$
$h(5) = -4.796\,283\,593\,9\text{E}{-}05$	$h(35) = -1.405\,320\,21\text{E}{-}03$	$h(65) = -9.954\,017\,99\text{E}{-}03$
$h(6) = -1.114\,582\,71\text{E}{-}04$	$h(36) = -1.889\,395\,93\text{E}{-}03$	$h(66) = -1.160\,684\,74\text{E}{-}02$
$h(7) = -1.828\,893\,86\text{E}{-}04$	$h(37) = -2.323\,890\,77\text{E}{-}03$	$h(67) = -1.283\,094\,61\text{E}{-}02$
$h(8) = -2.579\,580\,69\text{E}{-}04$	$h(38) = -2.677\,738\,68\text{E}{-}03$	$h(68) = -1.350\,981\,29\text{E}{-}02$
$h(9) = -3.317\,010\,13\text{E}{-}04$	$h(39) = -2.921\,846\,62\text{E}{-}03$	$h(69) = -1.353\,882\,61\text{E}{-}02$
$h(10) = -3.981\,814\,42\text{E}{-}04$	$h(40) = -3.029\,469\,38\text{E}{-}03$	$h(70) = -1.283\,059\,42\text{E}{-}02$
$h(11) = -4.512\,841\,32\text{E}{-}04$	$h(41) = -2.979\,827\,53\text{E}{-}03$	$h(71) = -1.131\,979\,89\text{E}{-}02$
$h(12) = -4.846\,378\,62\text{E}{-}04$	$h(42) = -2.757\,493\,08\text{E}{-}03$	$h(72) = -8.967\,305\,33\text{E}{-}03$
$h(13) = -4.925\,515\,41\text{E}{-}04$	$h(43) = -2.357\,195\,16\text{E}{-}03$	$h(73) = -5.763\,393\,82\text{E}{-}03$
$h(14) = -4.699\,729\,10\text{E}{-}04$	$h(44) = -1.780\,105\,12\text{E}{-}03$	$h(74) = -1.729\,786\,40\text{E}{-}03$
$h(15) = -4.134\,301\,81\text{E}{-}04$	$h(45) = -1.041\,517\,02\text{E}{-}03$	$h(75) = 3.079\,435\,77\text{E}{-}03$
$h(16) = -3.209\,825\,39\text{E}{-}04$	$h(46) = -1.645\,513\,78\text{E}{-}04$	$h(76) = 8.578\,453\,01\text{E}{-}03$
$h(17) = -1.931\,135\,81\text{E}{-}04$	$h(47) = 8.178\,995\,53\text{E}{-}04$	$h(77) = 1.465\,143\,65\text{E}{-}02$
$h(18) = -3.249\,472\,99\text{E}{-}05$	$h(48) = 1.861\,319\,28\text{E}{-}03$	$h(78) = 2.115\,588\,07\text{E}{-}02$
$h(19) = 1.553\,780\,29\text{E}{-}04$	$h(49) = 2.914\,637\,10\text{E}{-}03$	$h(79) = 2.792\,679\,99\text{E}{-}02$
$h(20) = 3.627\,346\,21\text{E}{-}04$	$h(50) = 3.920\,531\,77\text{E}{-}03$	$h(80) = 3.478\,246\,31\text{E}{-}02$
$h(21) = 5.791\,743\,44\text{E}{-}04$	$h(51) = 4.819\,109\,49\text{E}{-}03$	$h(81) = 4.153\,052\,09\text{E}{-}02$
$h(22) = 7.925\,630\,33\text{E}{-}04$	$h(52) = 5.549\,528\,98\text{E}{-}03$	$h(82) = 4.797\,530\,12\text{E}{-}02$
$h(23) = 9.890\,163\,10\text{E}{-}04$	$h(53) = 6.054\,167\,62\text{E}{-}03$	$h(83) = 5.392\,463\,32\text{E}{-}02$
$h(24) = 1.154\,177\,37\text{E}{-}03$	$h(54) = 6.281\,318\,79\text{E}{-}03$	$h(84) = 5.919\,747\,67\text{E}{-}02$
$h(25) = 1.273\,462\,99\text{E}{-}03$	$h(55) = 6.189\,301\,39\text{E}{-}03$	$h(85) = 6.363\,031\,84\text{E}{-}02$
$h(26) = 5.749\,036\,49\text{E}{-}03$	$h(56) = 1.333\,586\,07\text{E}{-}03$	$h(86) = 6.708\,432\,82\text{E}{-}02$
$h(27) = 1.322\,900\,58\text{E}{-}03$	$h(57) = 4.947\,265\,16\text{E}{-}03$	$h(87) = 6.945\,000\,48\text{E}{-}02$
$h(28) = 1.232\,999\,94\text{E}{-}03$	$h(58) = 3.788\,283\,89\text{E}{-}03$	$h(88) = 7.065\,226\,28\text{E}{-}02$
$h(29) = 1.058\,917\,06\text{E}{-}03$	$h(59) = 2.295\,793\,64\text{E}{-}03$	

frequency masking approach, we design such a filter starting from a prototype half-band lowpass filter having $\omega_p = \pi/2$ and a transition bandwidth four times larger than the one needed, in this case, $\pi/4$. Therefore, the implementation complexity of this prototype filter is much smaller than the original one. From this prototype, the complementary filter $H_2(z)$ is generated by a simple delay and subtraction as

$$H_2(z) = z^{-D} - H_1(z) \tag{8.28}$$

On Figure 8.16, $H_2(z)$ would represent the transfer function between the output of the summation following the delay z^{-D} and $X(z)$.

The magnitude responses of $H_1(z)$ and $H_2(z)$ are illustrated in Figure 8.17a. After interpolation, their responses are related as shown in Figure 8.17b. The filters $F_1(z)$

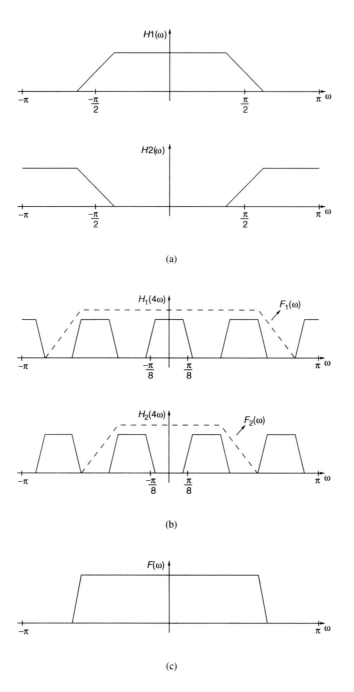

Figure 8.17 (a) Prototype half-band filter $H_1(z)$ and its complementary filter $H_2(z)$. (b) Frequency responses of $H_1(z)$ and $H_2(z)$ after interpolation by a factor of $L = 4$. (c) Frequency response of the equivalent filter $F(z)$.

and $F_2(z)$ are then used to select the parts of the interpolated spectra of $H_1(z)$ and $H_2(z)$ that will be used in composing the desired filter response $F(z) = \frac{Y(z)}{X(z)}$. It is interesting to note that $F_1(z)$ and $F_2(z)$, besides being interpolation filters, are allowed to have large transition bandwidths, and therefore have very low implementation complexity. As can be seen from Figure 8.17c, one can then generate large bandwidth sharp cutoff filters with low implementation complexity.

8.8   Multirate systems with MATLAB

The functions described below are part of the MATLAB Signal Processing toolbox.

- `upfirdn`: Upsamples, applies a specified filter and then downsamples a vector. Input parameters:

  - The vector x containing the input signal;
  - The filter h to be applied after interpolation;
  - The interpolation factor p and the decimation factor q.

  Output parameter: The vector y holding the filtered signal.
  Example 1 (downsample a signal by a factor of 3):
  ```
  x=rand(100,1); h=[1]; p=1; q=3;
  y=upfirdn(x,h,p,q);
  ```
  Example 2 (sampling-rate change by a factor of $\frac{5}{4}$ using a filter h):
  ```
  x=rand(100,1); h=[1 2 3 4 5 4 3 2 1]/5; p=5; q=4;
  y=upfirdn(x,h,p,q);
  ```

- `decimate`: Downsamples after lowpass filtering. Input parameters:

  - The vector x containing the input signal;
  - The decimation factor r;
  - The order n of the lowpass filter;
  - The type of the lowpass filter. The default is a Chebyshev lowpass filter with cutoff $0.8\frac{f_s}{2r}$. 'FIR' specifies FIR filtering.

  Output parameter: The vector y holding the downsampled signal.
  Example 1 (downsample a signal by a factor of 3 using a 10th-order Chebyshev lowpass filter):
  ```
  x=rand(100,1); r=3; n=10
  y=decimate(x,r,10);
  ```
  Example 2 (downsample a signal by a factor of 5 using a 50th-order FIR lowpass filter):

```
x=rand(1000,1); r=5; n=50
y=decimate(x,r,50,'FIR');
```

- `interp`: Interpolates a signal.
  Input parameters:

  - The vector x containing the input signal;
  - The interpolation factor r;
  - The number of original sample values l used to compute each interpolated sample;
  - The bandwidth alpha of the original signal.

  Output parameters:

  - The vector y holding the downsampled signal;
  - The vector b holding the coefficients of the interpolating filter.

  Example 1 (interpolate a signal by a factor of 3 using an FIR filter of order 12):
  ```
  x=rand(100,1); r=3; l=4;
  y=interp(x,r,l);
  ```
  Example 2 (interpolate a signal band-limited to a quarter of the sampling rate by a factor of 4 using an FIR filter of order 12):
  ```
  x=rand(100,1); r=4; l=3, alpha=0.5;
  y=interp(x,r,l,alpha);
  ```

- `resample`: Changes the sampling rate of a signal.
  Input parameters:

  - The vector x containing the input signal;
  - The interpolation factor p;
  - The decimation factor q;
  - The number n, which controls the number of original sample values used to compute each output sample. This number is equal to `2*n*max(1,q/p)`;
  - The filter b used to filter the input signal, or, alternatively, the parameter beta of the Kaiser window used to design the filter;
  - The bandwidth alpha of the original signal.

  Output parameters:

  - The vector y holding the downsampled signal;
  - The vector b holding the coefficients of the interpolating filter.

  Example 1 (change the sampling rate of a signal by a factor of $\frac{5}{4}$):
  ```
  x=rand(100,1); p=5; q=4;
  y=resample(x,p,q);
  ```

Example 2 (change the sampling rate of a signal by a factor of $\frac{2}{3}$ using, for each output sample, 12 original samples of the input signal, and an FIR filter designed with a Kaiser window with `beta=5`):

```
x=rand(500,1); p=2; q=3; n=4; beta=5;
[y,b]=resample(x,p,q,n,beta);
```

- `intfilt`: Interpolation and decimation FIR filter design.
  Input parameters:

  - The interpolation or decimation factor `r`;
  - The factor `l`, equal to `(n+2)/r`, where `n` is the filter order;
  - The fraction `alpha` of the sampling frequency corresponding to the filter bandwidth;
  - In cases where one wants to do Lagrangian interpolation, the filter order n and the parameter `'lagrange'` are provided instead of `l` and `alpha`, respectively.

  Output parameter: The vector b containing the filter coefficients.
  Example 1:
  ```
  r=2; l=3; alpha=0.4;
  b=intfilt(r,l,alpha);
  ```
  Example 2 (Lagrangian interpolator of fifth order for a sequence with two zeros between nonzero samples):
  ```
  r=3; n=5;
  b=intfilt(r,n,'lagrange');
  ```

8.9   Summary

In this chapter, we have studied the concepts of decimation and interpolation from the digital signal processing point of view. Suitable models for representing decimation and interpolation operations have been presented.

The specifications of the filters necessary in the decimation and interpolation processes have been studied and several design alternatives for those filters have been mentioned. We have also presented the use of interpolators and decimators in digital filter design.

Finally, MATLAB functions which aid in the design and implementation of multi-rate systems were presented.

Although this subject goes far beyond that which we have discussed here, it is expected that the material presented is sufficient for the solution of many practical problems. For more specific problems, the reader is advised to resort to one of the many excellent books on the subject, for instance Crochiere & Rabiner (1983).

8.10   Exercises

8.1   Show that decimation is a time-varying, but periodically time-invariant, operation and that interpolation is a time-invariant operation.

8.2   Design two interpolation filters, one lowpass and the other multiband (equation (8.22)). The specifications are:

$$\delta_p = 0.0002$$
$$\delta_r = 0.0001$$
$$\Omega_s = 10\,000 \text{ rad/s}$$
$$L = 10$$

One is interested in the information within $0 \leq \Omega \leq 0.2\Omega_s$. Compare the computational efficiency of the resulting filters.

8.3   Show one efficient FIR structure for decimation and another for interpolation, based on the considerations given on pages 358 and 363.

8.4   Repeat Exercise 8.3 for the case of an IIR filter.

8.5   Show, through an example, that it is more efficient to implement a decimator-by-50 through several decimation stages than using just one. Use the formula for estimating the order of a lowpass FIR filter presented in Exercise 5.19.

8.6   Design a lowpass filter using one decimation/interpolation stage satisfying the following specifications:

$$\delta_p = 0.001$$
$$\delta_r \leq 0.0001$$
$$\Omega_p = 0.01\Omega_s$$
$$\Omega_r = 0.02\Omega_s$$
$$\Omega_s = 1 \text{ rad/s}$$

Repeat the problem using two decimation/interpolation stages and compare the computational complexity of the results.

8.7   Design the bandpass filter for tone detection presented in Exercise 5.17, with center frequency of 700 Hz, using the concept of decimation/interpolation. Compare your results with the ones obtained in Exercise 5.17.

8.8   Design a filter satisfying the specifications in Subsection 8.7.2 using the frequency masking approach.

9 Filter banks and wavelets

9.1 Introduction

In the previous chapter, we dealt with multirate systems in general, that is, systems in which more than one sampling rate coexist. Operations of decimation, interpolation, and sampling-rate changes were studied, as well as some filter design techniques using multirate concepts.

In a number of applications, it is necessary to split a digital signal into several frequency bands. After such decomposition, the signal is represented by more samples than in the original stage. However, we can attempt to decimate each band, ending up with a digital signal decomposed into several frequency bands without increasing the overall number of samples. The question is whether it is possible to exactly recover the original signal from the decimated bands. Systems which achieve this are generally called filter banks.

In this chapter, we first deal with filter banks, showing several ways in which a signal can be decomposed into critically decimated frequency bands, and recovered from them with minimum error. Later, wavelet transforms are considered. They are a relatively recent development of functional analysis that is generating great interest in the signal processing community, because of their ability to represent and analyze signals with varying time and frequency resolutions. Their digital implementation can be regarded as a special case of critically decimated filter banks. Finally, we provide a brief description of functions from the MATLAB Wavelet toolbox which are useful for wavelets and filter banks implementation.

9.2 Filter banks

In some applications, such as signal analysis, signal transmission, and signal coding, a digital signal $x(n)$ is decomposed into several frequency bands, as depicted in Figure 9.1.

In such cases, the signal in each of the bands $x_k(n)$ for $k = 0, \ldots, M - 1$, has at least the same number of samples as the original signal $x(n)$. This implies that after the M-band decomposition, the signal is represented with at least M times more samples than the original one. However, there are many cases in which this expansion of the

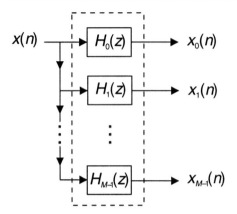

Figure 9.1 Decomposition of a digital signal into M frequency bands.

number of samples is highly undesirable. One such case is signal transmission (Vetterli & Kovačević, 1995), where more samples mean more bandwidth and consequently increased transmission costs.

In the common case where the signal is uniformly split in the frequency domain, that is, each of the frequency bands $x_k(n)$ has the same bandwidth, a natural question to ask is: since each band has bandwidth M times smaller than the one of the original signal, could the bands $x_k(n)$ be decimated by a factor of M (critically decimated) without destroying the original information? If this were possible, then one would have a digital signal decomposed into several frequency bands with the same overall number of samples as its original version.

In the next subsection, we analyze the general problem of decimating a bandpass signal and performing its inverse operation.

9.2.1   Decimation of a bandpass signal

As was seen in Section 8.3, equation (8.6), if the input signal $x(m)$ is lowpass and band-limited to $[-\frac{\pi}{M}, \frac{\pi}{M}]$, the aliasing after decimation by a factor of M can be avoided. However, if a signal is split into M uniform real frequency bands using the scheme in Figure 9.2, the kth band will be confined to $[-\frac{\pi(k+1)}{M}, -\frac{\pi k}{M}] \cup [\frac{\pi k}{M}, \frac{\pi(k+1)}{M}]$ (see Subsection 2.4.7).

This implies that band k, for $k \neq 0$, is necessarily not confined to $[-\frac{\pi}{M}, \frac{\pi}{M}]$. However, by examining equation (8.6), one can see that aliasing is still avoided in this case. The only difference is that, after decimation, the spectrum contained in $[-\frac{\pi(k+1)}{M}, -\frac{\pi k}{M}]$ is mapped into $[0, \pi]$, if k is odd, or into $[-\pi, 0]$, if k is even. Similarly, the spectrum contained in the interval $[\frac{\pi k}{M}, \frac{\pi(k+1)}{M}]$ is mapped into $[-\pi, 0]$, if k is odd, or into $[0, \pi]$, if k is even (Crochiere & Rabiner, 1983). Then, the decimated band k of Figure 9.2 will be as shown in Figures 9.3a and 9.3b, for k odd and even, respectively.

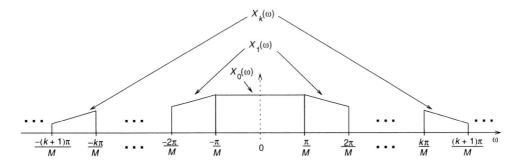

Figure 9.2 Uniform split of a signal into M real bands.

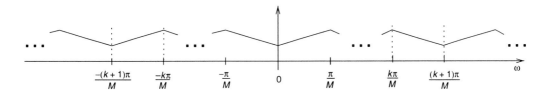

Figure 9.3 Spectrum of band k decimated by a factor of M: (a) k odd; (b) k even.

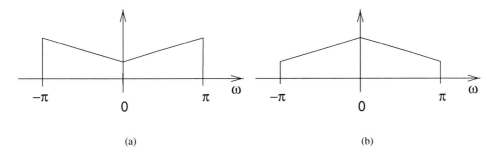

Figure 9.4 Spectrum of band k after decimation and interpolation by a factor of M for k odd.

9.2.2   Inverse decimation of a bandpass signal

We have just seen that a bandpass signal can be decimated by M without aliasing, provided that its spectrum is confined to $[-\frac{\pi(k+1)}{M}, -\frac{\pi k}{M}] \cup [\frac{\pi k}{M}, \frac{\pi(k+1)}{M}]$. The next natural question is: can the original bandpass signal be recovered from its decimated version by an interpolation operation? The case of lowpass signals was examined in Section 8.6. Here we analyze the bandpass case.

The spectrum of a decimated bandpass signal is shown in Figure 9.3. After interpolation by M, the spectrum for k odd will be as given in Figure 9.4.

If we want to recover band k, as in Figure 9.2, it suffices to keep the region of the spectrum in Figure 9.4 within $[-\frac{\pi(k+1)}{M}, -\frac{\pi k}{M}] \cup [\frac{\pi k}{M}, \frac{\pi(k+1)}{M}]$. For k even, the procedure is entirely analogous.

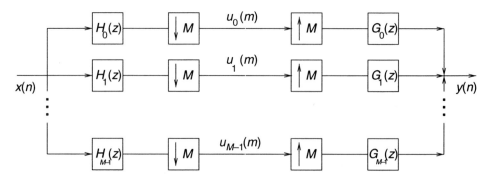

Figure 9.5 Block diagram of an M-band filter bank.

As a general conclusion, the process of decimating and interpolating a bandpass signal is similar to the case of a lowpass signal, seen in Figure 8.13, with the difference that for the bandpass case $H(z)$ must be a bandpass filter with bandwidth $[-\frac{\pi(k+1)}{M}, -\frac{\pi k}{M}] \cup [\frac{\pi k}{M}, \frac{\pi(k+1)}{M}]$.

9.2.3   Critically decimated M-band filter banks

It is clear from the previous discussion that, if a signal $x(m)$ is decomposed into M non-overlapping bandpass channels B_k, with $k = 0, \ldots, M-1$, such that $\bigcup_{k=0}^{M-1} B_k = [-\pi, \pi]$, then it can be recovered by just summing these M channels. However, as conjectured above, exact recovery of the original signal may not be possible if each channel is decimated by M.

In Subsections 9.2.1 and 9.2.2, we examined a way to recover the bandpass signal from its decimated version. In fact, all that is needed are interpolations followed by filters with passband $[-\frac{\pi(k+1)}{M}, -\frac{\pi k}{M}] \cup [\frac{\pi k}{M}, \frac{\pi(k+1)}{M}]$.

This whole process of decomposing a signal and restoring it from the frequency bands is depicted in Figure 9.5. We often refer to it as an M-band filter bank. The signals $u_k(m)$ occupy distinct frequency bands, which are collectively called sub-bands. If the input signal can be recovered exactly from its sub-bands, the structure is called an M-band perfect reconstruction filter bank. Figure 9.6 represents a perfect reconstruction filter bank for the 2-band case.

However, the filters required for the M-band perfect reconstruction filter bank described above are not realizable, that is, at best they can be only approximated (see Chapter 5). Therefore, in a first analysis, the original signal would be only approximately recoverable from its decimated frequency bands.

Figure 9.7 depicts a 2-band filter bank using realizable filters. One can see that since the filters $H_0(z)$ and $H_1(z)$ are not ideal, the sub-bands $s_l(m)$ and $s_h(m)$ have aliasing. In other words, the signals $x_l(n)$ and $x_h(n)$ cannot be correctly recovered from $s_l(m)$ and $s_h(m)$, respectively. Nevertheless, by closely examining Figure 9.7, one can see

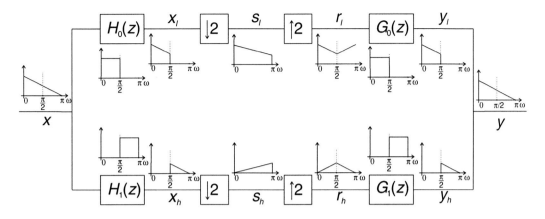

Figure 9.6 A 2-band perfect reconstruction filter bank using ideal filters.

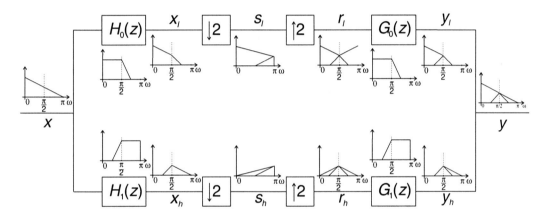

Figure 9.7 Two-band filter bank using realizable filters.

that since $y_l(n)$ and $y_h(n)$ are added in order to obtain $y(n)$, the aliased components of $y_l(n)$ can be combined with the ones of $y_h(n)$. In that manner, in principle, there is no reason why these aliased components could not be made to cancel each other, yielding $y(n)$ equal to $x(n)$. In such a case, the original signal could be recovered from its sub-band components. This setup can be used not only for the 2-band case, but for the general M-band case as well (Vaidyanathan, 1993).[1]

In the remainder of this chapter we will examine methods to design the analysis filters $H_k(z)$ and the synthesis filters $G_k(z)$ so that perfect reconstruction can be achieved, or, at least, arbitrarily approximated.

---

[1] By examining Figure 9.7, one can get the impression that the aliasing cannot be perfectly canceled, because only non-negative quantities are being added. However, only the magnitudes of the signals are shown, and in additions of the spectra the phase should obviously be considered as well as the magnitude.

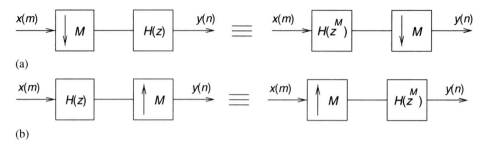

Figure 9.8 Noble identities: (a) decimation; (b) interpolation.

9.3   Perfect reconstruction

9.3.1   Noble identities

The noble identities are depicted in Figure 9.8. They have to do with the commutation of the filtering and decimation or interpolation operations, and are very useful in analyzing multirate systems and filter banks.

The identity in Figure 9.8a means that to decimate a signal by M and then filter it with $H(z)$ is equivalent to filtering the signal with $H(z^M)$ and then decimating the result by M. A filter $H(z^M)$ is one whose impulse response is equal to the impulse response of $H(z)$ with $(M - 1)$ zeros inserted between adjacent samples. Mathematically, it can be stated as

$$\mathcal{D}_M\{X(z)\}H(z) = \mathcal{D}_M\{X(z)H(z^M)\} \tag{9.1}$$

where $\mathcal{D}_M$ is the decimation-by-M operator.

The identity in Figure 9.8b means that to filter a signal with $H(z)$ and then interpolate it by M is equivalent to interpolating it by M and then filtering it with $H(z^M)$. Mathematically, it is stated as

$$\mathcal{I}_M\{X(z)H(z)\} = \mathcal{I}_M\{X(z)\}H(z^M) \tag{9.2}$$

where $\mathcal{I}_M$ is the interpolation-by-M operator.

In order to prove the identity in Figure 9.8a, one begins by rewriting equation (8.6), which gives the Fourier transform of the decimated signal $x_d(n)$ as a function of the input signal $x(m)$, in the z domain, that is

$$X_d(z) = \frac{1}{M} \sum_{k=0}^{M-1} X(z^{\frac{1}{M}} e^{-j\frac{2\pi k}{M}}) \tag{9.3}$$

For the decimator followed by filter $H(z)$, we have that

$$Y(z) = H(z)X_d(z) = \frac{1}{M} H(z) \sum_{k=0}^{M-1} X(z^{\frac{1}{M}} e^{-j\frac{2\pi k}{M}}) \tag{9.4}$$

For the filter $H(z^M)$ followed by the decimator, if $U(z) = X(z)H(z^M)$, we have, from equation (9.3), that

$$
\begin{aligned}
Y(z) &= \frac{1}{M} \sum_{k=0}^{M-1} U(z^{\frac{1}{M}} e^{-j\frac{2\pi k}{M}}) \\
&= \frac{1}{M} \sum_{k=0}^{M-1} X(z^{\frac{1}{M}} e^{-j\frac{2\pi k}{M}}) H(ze^{-j\frac{2\pi Mk}{M}}) \\
&= \frac{1}{M} \sum_{k=0}^{M-1} X(z^{\frac{1}{M}} e^{-j\frac{2\pi k}{M}}) H(z)
\end{aligned}
\tag{9.5}
$$

which is the same as equation (9.4), and the identity is proved.

Proof of the identity in Figure 9.8b is straightforward, as $H(z)$ followed by an interpolator gives $Y(z) = H(z^M)X(z^M)$, which is the same as the expression for an interpolator followed by $H(z^M)$.

9.3.2   Polyphase decompositions

The z transform $H(z)$ of a filter $h(n)$ can be written as

$$
\begin{aligned}
H(z) &= \sum_{k=-\infty}^{+\infty} h(k)z^{-k} \\
&= \sum_{l=-\infty}^{+\infty} h(Ml)z^{-Ml} + \sum_{l=-\infty}^{+\infty} h(Ml+1)z^{-(Ml+1)} + \cdots \\
&\quad + \sum_{l=-\infty}^{+\infty} h(Ml+M-1)z^{-(Ml+M-1)} \\
&= \sum_{l=-\infty}^{+\infty} h(Ml)z^{-Ml} + z^{-1} \sum_{l=-\infty}^{+\infty} h(Ml+1)z^{-Ml} + \cdots \\
&\quad + z^{-M+1} \sum_{l=-\infty}^{+\infty} h(Ml+M-1)z^{-Ml} \\
&= \sum_{j=0}^{M-1} z^{-j} E_j(z^M)
\end{aligned}
\tag{9.6}
$$

Equation (9.6) represents the polyphase decomposition (Vaidyanathan, 1993) of the filter $H(z)$, and

$$
E_j(z) = \sum_{l=-\infty}^{+\infty} h(Ml+j)z^{-l}
\tag{9.7}
$$

are called the polyphase components of $H(z)$. In such a decomposition, the filter $H(z)$ is split into M filters: the first one with every sample of $h(m)$, whose indexes are

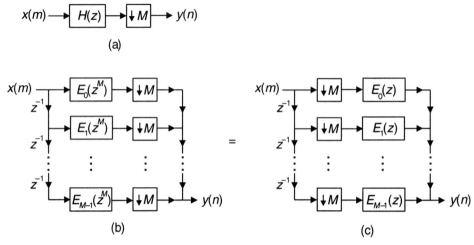

Figure 9.9 (a) Decimation by a factor of M. (b) Decimation using polyphase decompositions. (c) Decimation using polyphase decompositions and the noble identities.

multiples of M, the second one with every sample of $h(m)$, whose indexes are 1 plus a multiple of M, and so on.

Let us now analyze the basic operation of filtering followed by decimation represented in Figure 9.9a. Using the polyphase decomposition, such processing can be visualized as in Figure 9.9b, and applying the noble identity in equation (9.1), we arrive at Figure 9.9c, which provides an interesting and useful interpretation of the operation represented in Figure 9.9a. In fact, Figure 9.9c shows that the whole operation is equivalent to filtering the samples of $x(m)$ whose indexes are equal to an integer k plus a multiple of M, with a filter composed of only the samples of $h(m)$ whose indexes are equal to the same integer k plus a multiple of M, for $k = 0, \ldots, M - 1$.

The polyphase decompositions also provide useful insights into the interpolation operation followed by filtering, as depicted in Figure 9.10 but, in this case, a variation of equation (9.6) is usually employed. Defining $R_j(z) = E_{M-1-j}(z)$, the polyphase decomposition becomes

$$H(z) = \sum_{j=0}^{M-1} z^{-(M-1-j)} R_j(z^M) \tag{9.8}$$

Based on this equation, and applying the noble identity in equation (9.2), the complete operation can be represented as depicted in Figure 9.10b.

9.3.3   Commutator models

The operations described in Figures 9.9c and 9.10b can also be interpreted in terms of rotary switches. These interpretations are referred to as commutator models. In them,

(a)

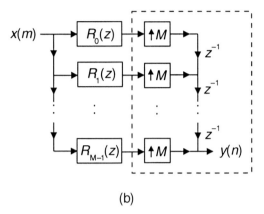

(b)

Figure 9.10 (a) Interpolation by a factor of M. (b) Interpolation using polyphase decompositions and the noble identities.

the decimators and delays are replaced by rotary switches as depicted in Figure 9.11 (Vaidyanathan, 1993).

In Figure 9.11a, the model with decimators and delays is noncausal, having 'advances' instead of delays. However, in real-time systems, one wants to avoid noncausal operations. In these cases, the causal model of Figure 9.12 is usually preferred.

9.3.4 M-band filter banks in terms of polyphase components

In an M-band filter bank as shown in Figure 9.5, the filters $H_k(z)$ and $G_k(z)$ are usually referred to as the analysis and synthesis filters of the filter bank.

As we will see next, polyphase decompositions can give us a valuable insight into the properties of M-channel filter banks. By substituting each of the filters $H_k(z)$ and $G_k(z)$ by their polyphase components, according to equations (9.6) and (9.8), we have that

$$H_k(z) = \sum_{j=0}^{M-1} z^{-j} E_{kj}(z^M)$$

(9.9)

$$G_k(z) = \sum_{j=0}^{M-1} z^{-(M-1-j)} R_{jk}(z^M)$$

(9.10)

where $E_{kj}(z)$ is the jth polyphase component of $H_k(z)$, and $R_{jk}(z)$ is the jth polyphase component of $G_k(z)$. By defining the matrices $\mathbf{E}(z)$ and $\mathbf{R}(z)$ as those having entries $E_{ij}(z)$ and $R_{ij}(z)$, for $i, j = 0, \ldots, M-1$, we have that the polyphase

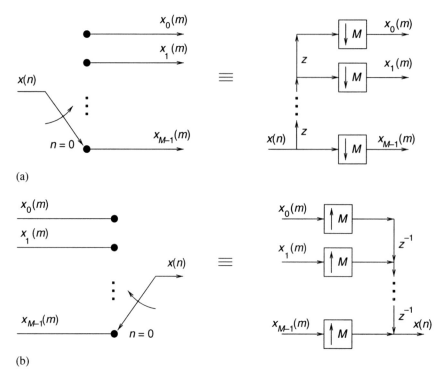

Figure 9.11 Commutator models for: (a) decimation; (b) interpolation.

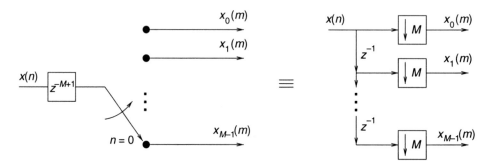

Figure 9.12 Causal commutator model for decimation.

decompositions in equations (9.9) and (9.10) can be expressed as

$$
\begin{bmatrix} H_0(z) \\ H_1(z) \\ \vdots \\ H_{M-1}(z) \end{bmatrix} = \mathbf{E}(z^M) \begin{bmatrix} 1 \\ z^{-1} \\ \vdots \\ z^{-(M-1)} \end{bmatrix}
\tag{9.11}
$$

$$
\begin{bmatrix} G_0(z) \\ G_1(z) \\ \vdots \\ G_{M-1}(z) \end{bmatrix} = \mathbf{R}^T(z^M) \begin{bmatrix} z^{-(M-1)} \\ z^{-(M-2)} \\ \vdots \\ 1 \end{bmatrix}
\tag{9.12}
$$

Therefore, the M-band filter bank in Figure 9.5 can be represented as in Figure 9.13a (Vaidyanathan, 1993), which turns into Figure 9.13b once the noble identities are applied.

9.3.5 Perfect reconstruction M-band filter banks

In Figure 9.13b, if $\mathbf{R}(z)\mathbf{E}(z) = \mathbf{I}$, where $\mathbf{I}$ is the identity matrix, the M-band filter bank becomes that shown in Figure 9.14.

By substituting the decimators and interpolators in Figure 9.14 by the commutator models of Figures 9.12 and 9.11b, respectively, we arrive at the scheme depicted in Figure 9.15, which is clearly equivalent to a pure delay. Therefore, the condition $\mathbf{R}(z)\mathbf{E}(z) = \mathbf{I}$ guarantees perfect reconstruction for the M-band filter bank (Vaidyanathan, 1993). It should be noted that if $\mathbf{R}(z)\mathbf{E}(z)$ is equal to a pure delay perfect reconstruction still holds. Therefore, the weaker condition $\mathbf{R}(z)\mathbf{E}(z) = z^{-\Delta}\mathbf{I}$ is sufficient for perfect reconstruction.

EXAMPLE 9.1
Let $M = 2$, and

$$
\mathbf{E}(z) = \begin{bmatrix} \frac{1}{2} & \frac{1}{2} \\ 1 & -1 \end{bmatrix}
\tag{9.13}
$$

$$
\mathbf{R}(z) = \begin{bmatrix} 1 & \frac{1}{2} \\ 1 & -\frac{1}{2} \end{bmatrix}
\tag{9.14}
$$

Show that these matrices characterize a perfect reconstruction filter bank, and find the analysis and synthesis filters and their corresponding polyphase components.

SOLUTION
Clearly $\mathbf{R}(z)\mathbf{E}(z) = \mathbf{I}$, and the filter yields perfect reconstruction. The polyphase

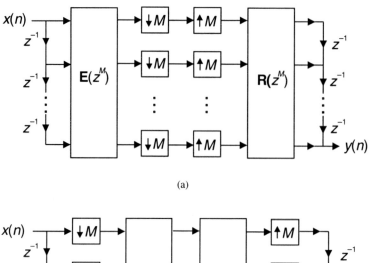

(a)

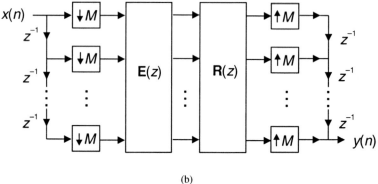

(b)

Figure 9.13 M-band filter bank in terms of the polyphase components: (a) before application of the noble identities; (b) after application of the noble identities.

components $E_{kj}(z)$ of the analysis filters $H_k(z)$, and $R_{jk}(z)$ of the synthesis filters $G_k(z)$ are then

$$E_{00}(z) = \frac{1}{2}, \qquad E_{01}(z) = \frac{1}{2}, \qquad E_{10}(z) = 1, \qquad E_{11}(z) = -1 \qquad (9.15)$$

$$R_{00}(z) = 1, \qquad R_{01}(z) = \frac{1}{2}, \qquad R_{10}(z) = 1, \qquad R_{11}(z) = -\frac{1}{2} \qquad (9.16)$$

From equations (9.15) and (9.6), we can find $H_k(z)$, and, from equations (9.16) and (9.8), we can find $G_k(z)$. They are

$$H_0(z) = \frac{1}{2}(1 + z^{-1}) \qquad (9.17)$$

$$H_1(z) = 1 - z^{-1} \qquad (9.18)$$

$$G_0(z) = 1 + z^{-1} \qquad (9.19)$$

$$G_1(z) = -\frac{1}{2}(1 - z^{-1}) \qquad (9.20)$$

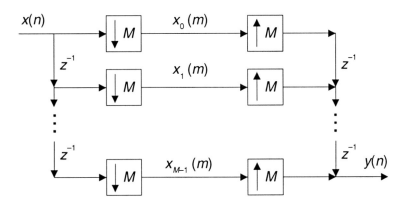

Figure 9.14 M-band filter bank when $\mathbf{R}(z)\mathbf{E}(z) = \mathbf{I}$.

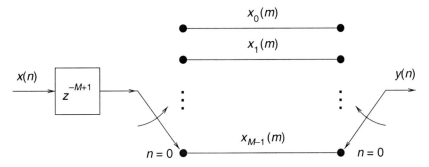

Figure 9.15 The commutator model of an M-band filter bank when $\mathbf{R}(z)\mathbf{E}(z) = \mathbf{I}$ is equivalent to a pure delay.

This is known as the Haar filter bank. The normalized magnitude responses of $H_0(z)$ and $H_1(z)$ are depicted in Figure 9.16. From equations (9.17)–(9.20), we see that the magnitude response of $G_k(z)$ is equal to the one of $H_k(z)$ for $k = 0, 1$. One can see that perfect reconstruction could be achieved with filters that are far from being ideal. In other words, despite the fact that each sub-band is highly aliased, one can still recover the original signal exactly at the output.

$\triangle$

EXAMPLE 9.2
Repeat Example 9.1 for the case when

$$
\mathbf{E}(z) = \begin{bmatrix} -\dfrac{1}{8} + \dfrac{3}{4}z^{-1} - \dfrac{1}{8}z^{-2} & \dfrac{1}{4} + \dfrac{1}{4}z^{-1} \\[2ex] \dfrac{1}{2} + \dfrac{1}{2}z^{-1} & -1 \end{bmatrix}
\tag{9.21}
$$

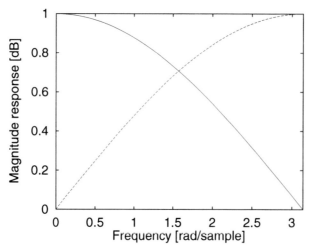

Figure 9.16 Frequency responses of the filters described by equations (9.17) and (9.18). (Solid line – $H_0(z)$; dashed line – $H_1(z)$.)

$$\mathbf{R}(z) = \begin{bmatrix} 1 & \dfrac{1}{4} + \dfrac{1}{4}z^{-1} \\[2ex] \dfrac{1}{2} + \dfrac{1}{2}z^{-1} & \dfrac{1}{8} - \dfrac{3}{4}z^{-1} + \dfrac{1}{8}z^{-2} \end{bmatrix} \tag{9.22}$$

SOLUTION

Since

$$\mathbf{R}(z)\mathbf{E}(z) = \begin{bmatrix} z^{-1} & 0 \\ 0 & z^{-1} \end{bmatrix} = z^{-1}\mathbf{I} \tag{9.23}$$

then the filter bank has perfect reconstruction. From equations (9.21) and (9.22), the polyphase components, $E_{kj}(z)$, of the analysis filters $H_k(z)$, and $R_{jk}(z)$ of the synthesis filters $G_k(z)$ are

$$\left. \begin{aligned} E_{00}(z) &= -\frac{1}{8} + \frac{3}{4}z^{-1} - \frac{1}{8}z^{-2} \\[2ex] E_{01}(z) &= \frac{1}{4} + \frac{1}{4}z^{-1} \\[2ex] E_{10}(z) &= \frac{1}{2} + \frac{1}{2}z^{-1} \\[2ex] E_{11}(z) &= -1 \end{aligned} \right\} \tag{9.24}$$

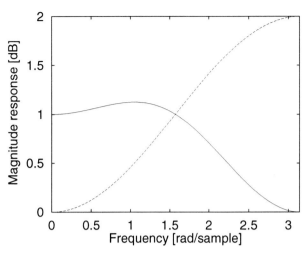

Figure 9.17 Frequency responses of the filters described by equations (9.26) and (9.27). (Solid line – $H_0(z)$; dashed line – $H_1(z)$.)

$$
\left.
\begin{aligned}
R_{00}(z) &= 1 \\[4pt]
R_{01}(z) &= \frac{1}{4} + \frac{1}{4}z^{-1} \\[4pt]
R_{10}(z) &= \frac{1}{2} + \frac{1}{2}z^{-1} \\[4pt]
R_{11}(z) &= \frac{1}{8} - \frac{3}{4}z^{-1} + \frac{1}{8}z^{-2}
\end{aligned}
\right\}
\tag{9.25}
$$

From equations (9.24) and (9.6), we can find $H_k(z)$, and, from equations (9.25) and (9.8), we can find $G_k(z)$. They are

$$
H_0(z) = -\frac{1}{8} + \frac{1}{4}z^{-1} + \frac{3}{4}z^{-2} + \frac{1}{4}z^{-3} - \frac{1}{8}z^{-4}
\tag{9.26}
$$

$$
H_1(z) = \frac{1}{2} - z^{-1} + \frac{1}{2}z^{-2}
\tag{9.27}
$$

$$
G_0(z) = \frac{1}{2} + z^{-1} + \frac{1}{2}z^{-2}
\tag{9.28}
$$

$$
G_1(z) = \frac{1}{8} + \frac{1}{4}z^{-1} - \frac{3}{4}z^{-2} + \frac{1}{4}z^{-3} + \frac{1}{8}z^{-4}
\tag{9.29}
$$

The magnitude responses of the analysis filters are depicted in Figure 9.17.

$\triangle$

9.3.6 Transmultiplexers

If two identical zero-delay M-channel perfect reconstruction filter banks as in Figure 9.5 are cascaded, we have that the signal corresponding to $u_k(m)$ in one filter

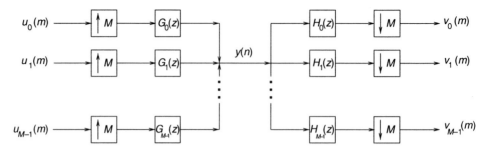

Figure 9.18 *M*-band transmultiplexer.

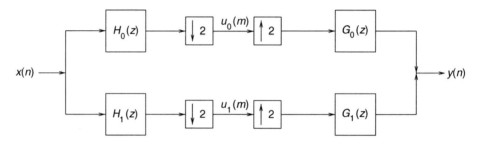

Figure 9.19 Two-band filter bank.

bank is identical to the corresponding signal in the other filter bank, for each $k = 0, \ldots, M - 1$. Therefore, with the same filters as in Figure 9.5, one can build a perfect reconstruction transmultiplexer as in Figure 9.18, which can combine the M signals $u_k(m)$ into one single signal $y(n)$, and then recover the signals $v_k(m)$ that are identical to $u_k(m)$ (Vaidyanathan, 1993). One important application of such transmultiplexers is the multiplexing of M signals so that they can be easily recovered without any cross-interference. What is very interesting in this case is that the filters do not need to be at all selective for this kind of transmultiplexer to work.[2] This is the basis of generalized interpretations of digital multiple access systems, such as TDMA, FDMA, and CDMA (Hettling et al., 1999, Chapter 1).

9.4 General 2-band perfect reconstruction filter banks

The general 2-band case is shown in Figure 9.19.

Representing the filters $H_0(z)$, $H_1(z)$, $G_0(z)$, and $G_1(z)$ in terms of their polyphase

---

[2] If the filter banks are not zero-delay, this kind of transmultiplexer still works. It can be shown that the signals $v_k(m)$ are equal to delayed versions of $u_k(m)$, provided that a suitable delay is introduced between the filter banks.

components, using equations (9.6) and (9.8), we then have that

$$H_0(z) = H_{00}(z^2) + z^{-1} H_{01}(z^2) \tag{9.30}$$

$$H_1(z) = H_{10}(z^2) + z^{-1} H_{11}(z^2) \tag{9.31}$$

$$G_0(z) = z^{-1} G_{00}(z^2) + G_{01}(z^2) \tag{9.32}$$

$$G_1(z) = z^{-1} G_{10}(z^2) + G_{11}(z^2) \tag{9.33}$$

The matrices $\mathbf{E}(z)$ and $\mathbf{R}(z)$ in Figure 9.13b are then

$$\mathbf{E}(z) = \begin{bmatrix} H_{00}(z) & H_{01}(z) \\ H_{10}(z) & H_{11}(z) \end{bmatrix} \tag{9.34}$$

$$\mathbf{R}(z) = \begin{bmatrix} G_{00}(z) & G_{10}(z) \\ G_{01}(z) & G_{11}(z) \end{bmatrix} \tag{9.35}$$

If $\mathbf{R}(z)\mathbf{E}(z) = \mathbf{I}$, we have perfect reconstruction, as represented in Figures 9.14 and 9.15. In fact, from Figure 9.15, we see that the output signal $y(n)$ will be equal to $x(n)$ delayed by $(M - 1)$ samples, which for a 2-band filter bank is equal to just one sample. In the general case, we have $\mathbf{R}(z)\mathbf{E}(z) = \mathbf{I}z^{-\Delta}$, which makes the output signal of the 2-band filter bank, $y(n)$, equal to $x(n)$ delayed by $(2\Delta + 1)$ samples. Therefore, the 2-band filter bank will be equivalent to a delay of $(2\Delta + 1)$ samples if

$$\mathbf{R}(z) = z^{-\Delta} \mathbf{E}^{-1}(z) \tag{9.36}$$

From equations (9.34) and (9.35), this implies that

$$\begin{bmatrix} G_{00}(z) & G_{10}(z) \\ G_{01}(z) & G_{11}(z) \end{bmatrix} = \frac{z^{-\Delta}}{H_{00}(z)H_{11}(z) - H_{01}(z)H_{10}(z)} \begin{bmatrix} H_{11}(z) & -H_{01}(z) \\ -H_{10}(z) & H_{00}(z) \end{bmatrix} \tag{9.37}$$

This result is sufficient for IIR filter bank design, as long as stability constraints are taken into consideration. However, if we want the filters to be FIR, as is often the case, the term in the denominator must be proportional to a pure delay, that is

$$H_{00}(z)H_{11}(z) - H_{01}(z)H_{10}(z) = cz^{-l} \tag{9.38}$$

From equations (9.30)–(9.33), we can express the polyphase components in terms of the filters $H_k(z)$ and $G_k(z)$ as

$$H_{00}(z^2) = \frac{H_0(z) + H_0(-z)}{2}, \qquad H_{01}(z^2) = \frac{H_0(z) - H_0(-z)}{2z^{-1}}$$

$$H_{10}(z^2) = \frac{H_1(z) + H_1(-z)}{2}, \qquad H_{11}(z^2) = \frac{H_1(z) - H_1(-z)}{2z^{-1}} \tag{9.39}$$

$$G_{00}(z^2) = \frac{G_0(z) - G_0(-z)}{2z^{-1}}, \qquad G_{01}(z^2) = \frac{G_0(z) + G_0(-z)}{2}$$

$$G_{10}(z^2) = \frac{G_1(z) - G_1(-z)}{2z^{-1}}, \qquad G_{11}(z^2) = \frac{G_1(z) + G_1(-z)}{2} \tag{9.40}$$

Substituting equation (9.39) into equation (9.38), we have that

$$H_0(-z)H_1(z) - H_0(z)H_1(-z) = 2cz^{-2l-1} \tag{9.41}$$

Now, substituting equation (9.38) into equation (9.37), and computing the $G_k(z)$ from equations (9.32) and (9.33), we arrive at

$$G_0(z) = -\frac{z^{2(l-\Delta)}}{c}H_1(-z) \tag{9.42}$$

$$G_1(z) = \frac{z^{2(l-\Delta)}}{c}H_0(-z) \tag{9.43}$$

Equations (9.41)–(9.43) suggest a possible way to design 2-band perfect reconstruction filter banks. The design procedure is as follows (Vetterli & Kovačević, 1995):

(i)   Find a polynomial $P(z)$ such that $P(-z) - P(z) = 2cz^{-2l-1}$.

(ii)   Factorize $P(z)$ into two factors, $H_0(z)$ and $H_1(-z)$. Care must be taken to guarantee that $H_0(z)$ and $H_1(-z)$ are lowpass filters.

(iii) Find $G_0(z)$ and $G_1(z)$ using equations (9.42) and (9.43).

Some important points should be noted in this case:

- If one wants the filter bank to be composed of linear-phase filters, it suffices to find a linear-phase product filter $P(z)$, and make linear-phase factorizations of it.
- If the delay, Δ, is zero, some of the filters are certainly noncausal: for l negative, either $H_0(z)$ or $H_1(z)$ must be noncausal (see equation (9.41)); for l positive, either $G_0(z)$ or $G_1(z)$ must be noncausal. Therefore, a causal perfect reconstruction filter bank will always have nonzero delay.
- The magnitude responses $|G_0(e^{j\omega})|$ and $|H_1(e^{j\omega})|$ are mirror images of each other around $\omega = \frac{\pi}{2}$ (equation (9.42)). The same happens to $|H_0(e^{j\omega})|$ and $|G_1(e^{j\omega})|$ (see equation (9.43)).

EXAMPLE 9.3

A product filter $P(z)$ satisfying $P(z) - P(-z) = 2z^{-2l-1}$ is

$$P(z) = \frac{1}{16}(-1 + 9z^{-2} + 16z^{-3} + 9z^{-4} - z^{-6}) = \frac{1}{16}(1 + z^{-1})^4(-1 + 4z^{-1} - z^{-2}) \tag{9.44}$$

Find two possible factorizations of $P(z)$ and plot the magnitude responses of their corresponding analysis filters.

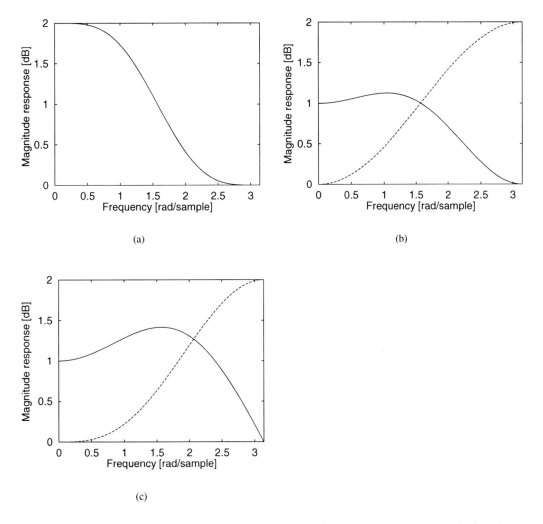

Figure 9.20 Magnitude responses: (a) $P(z)$ from equation (9.44); (b) $H_0(z)$ and $H_1(z)$ from the factorizations in equations (9.45) and (9.46); (c) $H_0(z)$ and $H_1(z)$ from the factorizations in equations (9.49) and (9.50). (Solid line – $H_0(z)$; dashed line – $H_1(z)$.)

SOLUTION

We can see from the magnitude response in Figure 9.20a that $P(z)$ is a lowpass filter.

One possible factorization of $P(z)$ results in the following filter bank, which is the same as the one in Example 9.2. This filter bank is the popular symmetric short-kernel filter bank (Le Gall & Tabatabai, 1988; Vetterli & Kovačević, 1995):

$$H_0(z) = \frac{1}{8}(-1 + 2z^{-1} + 6z^{-2} + 2z^{-3} - z^{-4}) \tag{9.45}$$

$$H_1(z) = \frac{1}{2}(1 - 2z^{-1} + z^{-2}) \tag{9.46}$$

$$G_0(z) = \frac{1}{2}(1 + 2z^{-1} + z^{-2}) \tag{9.47}$$

$$G_1(z) = \frac{1}{8}(1 + 2z^{-1} - 6z^{-2} + 2z^{-3} + z^{-4}) \tag{9.48}$$

The magnitude responses of the analysis filters are depicted in Figure 9.20b.

Another possible factorization is as follows:

$$H_0(z) = \frac{1}{4}(-1 + 3z^{-1} + 3z^{-2} - z^{-3}) \tag{9.49}$$

$$H_1(z) = \frac{1}{4}(1 - 3z^{-1} + 3z^{-2} - z^{-3}) \tag{9.50}$$

$$G_0(z) = \frac{1}{4}(1 + 3z^{-1} + 3z^{-2} + z^{-3}) \tag{9.51}$$

$$G_1(z) = \frac{1}{4}(1 + 3z^{-1} - 3z^{-2} - z^{-3}) \tag{9.52}$$

The corresponding magnitude responses of the analysis filters are depicted in Figure 9.20c.

△

In the following sections, we examine some particular cases of filter bank design which have become popular in the technical literature.

9.5   QMF filter banks

One of the earliest proposed approaches for the design of 2-band filter banks is the so-called quadrature mirror filter (QMF) bank (Croisier et al., 1976), where the analysis highpass filter is designed to alternate the signs of the impulse-response samples of the lowpass filter, that is

$$H_1(z) = H_0(-z) \tag{9.53}$$

Please note that it is assumed the filters have real coefficients. For this choice of the analysis filter bank, the magnitude response of the highpass filter, $|H_1(e^{j\omega})|$, is the mirror image of the lowpass filter magnitude response, $|H_0(e^{j\omega})|$, with respect to the quadrature frequency $\frac{\pi}{2}$. Hence the QMF nomenclature.

The 2-band filter bank illustrated in Figure 9.19 and discussed in Section 9.4 can also be analyzed in the following way. The signals after the analysis filter are described by

$$X_k(z) = H_k(z)X(z) \tag{9.54}$$

for $k = 0, 1$. The decimated signals are

$$U_k(z) = \frac{1}{2}(X_k(z^{1/2}) + X_k(-z^{1/2})) \tag{9.55}$$

for $k = 0, 1$, whereas the signals after the interpolator are

$$U_k(z^2) = \frac{1}{2}(X_k(z) + X_k(-z)) = \frac{1}{2}(H_k(z)X(z) + H_k(-z)X(-z)) \tag{9.56}$$

Then, the reconstructed signal is represented as

$$
\begin{aligned}
Y(z) &= G_0(z)U_0(z^2) + G_1(z)U_1(z^2) \\
&= \frac{1}{2}(H_0(z)G_0(z) + H_1(z)G_1(z))X(z) \\
&\quad + \frac{1}{2}(H_0(-z)G_0(z) + H_1(-z)G_1(z))X(-z) \\
&= \frac{1}{2} \begin{bmatrix} X(z) & X(-z) \end{bmatrix} \begin{bmatrix} H_0(z) & H_1(z) \\ H_0(-z) & H_1(-z) \end{bmatrix} \begin{bmatrix} G_0(z) \\ G_1(z) \end{bmatrix}
\end{aligned} \tag{9.57}
$$

The last equality is the so-called modulation-matrix representation of a 2-band filter bank. The aliasing effect is represented by the terms containing $X(-z)$. A possible solution to avoid aliasing is to choose the synthesis filters such that

$$G_0(z) = H_1(-z) \tag{9.58}$$
$$G_1(z) = -H_0(-z) \tag{9.59}$$

This choice keeps the filters $G_0(z)$ and $G_1(z)$ as lowpass and highpass filters, respectively, as desired. Also, the aliasing is now canceled by the synthesis filters, instead of being totally avoided by the analysis filters, relieving the specifications of the latter filters (see Subsection 9.2.3).

The overall transfer function of the filter bank, after the aliasing component is eliminated, is given by

$$H(z) = \frac{1}{2}(H_0(z)G_0(z) + H_1(z)G_1(z)) = \frac{1}{2}(H_0(z)H_1(-z) - H_1(z)H_0(-z)) \tag{9.60}$$

where in the second equality we employed the aliasing elimination constraint of equations (9.58) and (9.59).

In the original QMF design, the aliasing elimination condition is combined with the alternating-sign choice for the highpass filter of equation (9.53). In such a case, the overall transfer function is given by

$$H(z) = \frac{1}{2}(H_0^2(z) - H_0^2(-z)) \tag{9.61}$$

The expression above can be rewritten in a more convenient form by employing the polyphase decomposition of the lowpass filter $H_0(z) = E_0(z^2) + z^{-1}E_1(z^2)$, as follows:

$$H(z) = \frac{1}{2}(H_0(z) + H_0(-z))(H_0(z) - H_0(-z)) = 2z^{-1}(E_0(z^2)E_1(z^2)) \tag{9.62}$$

Note that the QMF design approach of 2-band filter banks consists of designing the lowpass filter $H_0(z)$. The above equation also shows that perfect reconstruction is achievable only if the polyphase components of the lowpass filter (that is, $E_0(z)$ and $E_1(z)$) are simple delays. This constraint limits the selectivity of the generated filters.

Alternatively, we can adopt an approximate solution by choosing $H_0(z)$ to be an Nth-order FIR linear-phase lowpass filter, eliminating any phase distortion of the overall transfer function $H(z)$, which, in this case, will also have linear phase. The filter bank transfer function can then be written as

$$H(e^{j\omega}) = \frac{e^{-j\omega N}}{2}(|H_0(e^{j\omega})|^2 + |H_0(e^{j(\omega-\pi)})|^2) = \frac{e^{-j\omega N}}{2}(|H_0(e^{j\omega})|^2 + |H_1(e^{j\omega})|^2) \tag{9.63}$$

for N odd. For N even, the $+$ sign becomes a $-$ sign, generating an undesirable zero at $\omega = \frac{\pi}{2}$.

The design procedure goes on to minimize the following objective function using an optimization algorithm:

$$\xi = \delta \int_{\omega_s}^{\pi} \left|H_0(e^{j\omega})\right|^2 d\omega + (1-\delta) \int_0^{\pi} \left|H(e^{j\omega}) - \frac{e^{-j\omega N}}{2}\right|^2 d\omega \tag{9.64}$$

where ω_s is the stopband edge, usually chosen slightly above 0.5π. The parameter $0 < \delta < 1$ provides a tradeoff between the stopband attenuation of the lowpass filter and the amplitude distortion of the filter bank. Although this objective function has local minima, a good starting point for the coefficients of the lowpass filter and an adequate nonlinear optimization algorithm lead to good results, that is, filter banks with low amplitude distortions and good selectivity of the filters. Usually, a simple window-based design provides a good starting point for the lowpass filter. Overall, the simplicity of the QMF design makes it widely used in practice. Figure 9.21a depicts the magnitude responses of the analysis filters of order $N = 15$ of a QMF design, and 9.21b depicts the amplitude response of the complete filter bank. Johnston was among the first to provide QMF coefficients for several designs (Johnston, 1980). Due to his pioneering work, such QMF filters are usually termed Johnston filter banks.

The name QMF filter banks is also used to denote M-band maximally decimated filter banks. For M-band QMF filter banks there are two design approaches that are widely used, namely the perfect reconstruction QMF filter banks and the pseudo-QMF filter banks. The perfect reconstruction QMF designs require the use of sophisticated nonlinear optimization programs because the objective function is a nonlinear function of the filter parameters (Vaidyanathan, 1993). In particular, for a large number of sub-bands, the number of parameters becomes excessively large. Meanwhile, the pseudo-QMF designs consist of designing a prototype filter, with the sub-filters of the analysis bank being obtained by the modulation of the prototype. As a consequence, the pseudo-QMF filter has a very efficient design procedure. The pseudo-QMF filter

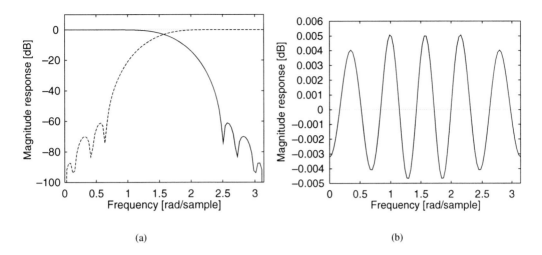

Figure 9.21 QMF design of order $N = 15$: (a) magnitude responses of the analysis filters (solid line – $H_0(z)$; dashed line – $H_1(z)$); (b) overall magnitude response.

banks are also known as cosine-modulated filter banks, since they are designed by applying cosine modulation to a lowpass prototype filter. There are cosine-modulated filter banks which achieve perfect reconstruction (Koilpillai & Vaidyanathan, 1992; Nguyen & Koilpillai, 1996), and the procedure for their design is given in Section 9.8.

9.6   CQF filter banks

In the QMF design designing the highpass filter from the lowpass prototype by alternating the signs of its impulse response is quite simple, but the possibility of getting perfect reconstruction is lost except in a few trivial cases. In Smith & Barnwell (1986), however, it was disclosed that by time-reversing the impulse response and alternating the signs of the lowpass filter, one can design perfect reconstruction filter banks with more selective sub-filters. The resulting filters became known as the conjugate quadrature filter (CQF) banks.

In the CQF design, we have that the analysis highpass filter is given by

$$H_1(z) = -z^{-N} H_0(-z^{-1})$$
(9.65)

By verifying again that the order N of the filters must be odd, the filter bank transfer function is given by

$$H(e^{j\omega}) = \frac{e^{-j\omega N}}{2} \left(H_0(e^{j\omega}) H_0(e^{-j\omega}) + H_0(-e^{j\omega}) H_0(-e^{-j\omega}) \right)$$

$$= \frac{e^{-j\omega N}}{2} \left(P(e^{j\omega}) + P(-e^{j\omega}) \right)$$
(9.66)

From equations (9.42) and (9.43), with $c = -1$ we have that, in order to guarantee perfect reconstruction, the synthesis filters should be given by

$$G_0(z) = z^{-N} H_0(z^{-1}) \tag{9.67}$$
$$G_1(z) = -H_0(-z) \tag{9.68}$$

It is a fact that perfect reconstruction is achieved when the time-domain response of the filter bank equals a delayed impulse, that is

$$h(n) = \delta(n - \hat{N}) \tag{9.69}$$

Now, by examining $H(e^{j\omega})$ in equation (9.66), one can easily infer that perfect reconstruction is equivalent to having the time-domain representation of $P(z)$ satisfying

$$p(n)[1 + (-1)^n] = 2\delta(n - \hat{N} + N) \tag{9.70}$$

Therefore, the design procedure consists of the following steps:

(i) Noting that $p(n) = 0$, for n even, except for $n = (\hat{N} - N)$, we start by designing a half-band filter, namely a filter such that $\frac{\omega_p + \omega_r}{2} = \frac{\pi}{2}$, with order $2N$ and the same ripple δ_{hb} in the passband and stopband. The resulting half-band filter will have null samples on its impulse response for every even n, such that $n \neq (\hat{N} - N)$. This half-band filter can be designed using the standard Chebyshev approach for FIR filters, previously presented in Section 5.6.2. However, since the product filter $P(e^{j\omega})$ has to be positive, we should add $\frac{\delta_{hb}}{2}$ to the entire frequency response of the half-band filter. As a rule of thumb, to simplify the design procedure, the stopband attenuation of the half-band filter should be at least twice the desired stopband attenuation plus 6 dB (Vaidyanathan, 1993).

(ii) The usual approach is to decompose $P(z) = H_0(z)H_0(z^{-1})$ such that $H_0(z)$ has either near linear phase or has minimum phase. In order to obtain near linear phase, one can select the zeros of $H_0(z)$ to be alternately from inside and outside the unit circle as the frequency is increased. Minimum phase is obtained when all zeros are either inside or on the unit circle of the z plane.

If we wanted the filter bank to be composed of linear-phase filters, we would have to find a linear-phase product filter $P(z)$ and make linear-phase factorizations of it, as seen in Section 9.4. However, the only linear-phase 2-band filter banks that satisfy equation 9.65 are the ones composed of trivial linear-phase filters such as the ones described by equations (9.17)–(9.20). Therefore, there is no point in looking for linear-phase factorizations of $P(z)$. This is why in step (ii) above the usual approach is to look for either minimum phase or near linear-phase factorizations of $P(z)$.

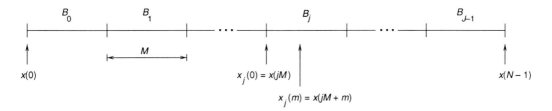

Figure 9.22 Division of a length-N signal into J non-overlapping blocks of length M.

9.7 Block transforms

Perhaps the most popular example of M-band perfect reconstruction filter banks is given by the block transforms (see discrete transforms in Chapter 3). For instance, the discrete cosine transform (DCT) does essentially the same job as a filter bank: it divides a signal into several frequency components. The main difference is that, given a length-N signal, the DCT divides it into N frequency channels, whereas a filter bank divides it into M channels, $M < N$. However, in many applications, one wants to divide a length-N signal into J blocks, each having length M, and separately apply the transform to each one. This is done, for example, in the MPEG2 standard employed in digital video transmission (Le Gall, 1992; Whitaker, 1999). In it, instead of transmitting the individual pixels of a video sequence, each frame is first divided into 8×8 blocks. Then a DCT is applied to each block, and the DCT coefficients are transmitted instead. In this section, we show that such 'block transforms' are equivalent to M-band filter banks.

Consider a signal $x(n)$, $n = 0, \ldots, N - 1$, divided into J blocks B_j, with $j = 0, \ldots, J - 1$, each of size M. Block B_j then consists of the signal $x_j(m)$, given by

$$x_j(m) = x(jM + m) \tag{9.71}$$

for $j = 0, \ldots, J - 1$ and $m = 0, \ldots, M - 1$. This is depicted in Figure 9.22.

Suppose that $y_j(k)$, for $k = 0, \ldots, M - 1$ and $j = 0, \ldots, J - 1$, is the transform of a block $x_j(m)$, where the direct and inverse transforms are described by

$$y_j(k) = \sum_{m=0}^{M-1} c_k(m) x_j(m) \tag{9.72}$$

$$x_j(m) = \sum_{k=0}^{M-1} c_k(m) y_j(k) \tag{9.73}$$

where $c_k(m)$, for $m = 0, \ldots, M - 1$, is the kth basis function of the transform, or, alternatively, the kth row of the transform matrix. We can then regroup the transforms of the blocks by 'sequency', that is, group all the $y_j(0)$, all the $y_j(1)$, and so on. This is equivalent to creating M signals $u_k(j) = y_j(k)$, for $j = 0, \ldots, J - 1$ and $k =$

$0, \ldots, M-1$. Since $x_j(m) = x(Mj+m)$, from equations (9.72) and (9.73), we have that

$$u_k(j) = \sum_{m=0}^{M-1} c_k(m)x(Mj+m) \tag{9.74}$$

$$x(Mj+m) = \sum_{k=0}^{M-1} c_k(m)u_k(j) \tag{9.75}$$

From equation (9.74), we can interpret $u_k(j)$ as the convolution of $x(n)$ with $c_k(-n)$ sampled at the points Mj. This is the same as filtering $x(n)$ with a filter of impulse response $c_k(-n)$, and decimating its output by M.

Likewise, if we define $u_{k_i}(j)$ as the signal $u_k(j)$ interpolated by M, we have that $u_{k_i}(Mj-m) = 0$, for $m = 1, \ldots, M-1$ (see equation (8.17)). This implies that

$$c_k(m)u_k(j) = \sum_{m=0}^{M-1} c_k(m)u_{k_i}(Mj-m) \tag{9.76}$$

Such an expression can be interpreted as $c_k(m)u_k(j)$ being the result of interpolating $u_k(j)$ by M, and filtering it with a filter of impulse response $c_k(m)$.

Substituting equation (9.76) in equation (9.75), we arrive at the following expression:

$$x(Mj+m) = \sum_{k=0}^{M-1} \sum_{m=0}^{M-1} c_k(m)u_{k_i}(Mj-m) \tag{9.77}$$

Therefore, from equations (9.74) and (9.77), the direct and inverse block-transform operations can be interpreted as in Figure 9.23. Comparing it to Figure 9.5, we can see that a block transform is equivalent to a perfect reconstruction filter bank having the impulse responses of band k analysis and synthesis filters equal to $c_k(-m)$ and $c_k(m)$, respectively.

EXAMPLE 9.4
As seen in Chapter 3, the DCT is defined by

$$c_k(m) = \alpha(k)\cos\left[\frac{\pi(2m+1)k}{2M}\right] \tag{9.78}$$

where $\alpha(0) = \sqrt{\frac{1}{M}}$ and $\alpha(k) = \sqrt{\frac{2}{M}}$, for $k = 1, \ldots, M-1$. Plot the impulse response and the magnitude response of the analysis filters corresponding to the length-10 DCT. In addition, determine whether the filter bank has linear phase or not.

SOLUTION
The impulse responses of the analysis filters for the length-10 DCT are depicted

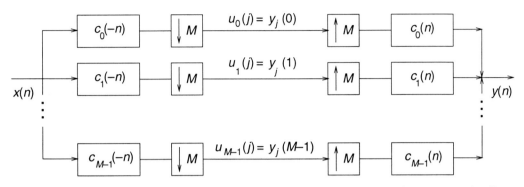

Figure 9.23 Interpretation of direct and inverse block transforms as a perfect reconstruction filter bank.

in Figure 9.24, and their magnitude responses are depicted in Figure 9.25. From equation (9.78), we can see that $c_k(m) = (-1)^k c_k(M - 1 - m)$, and therefore the filter bank has linear phase. This also implies that the magnitude responses of the analysis and synthesis filters are the same.

$\triangle$

9.8   Cosine-modulated filter banks

The cosine-modulated filter banks are an attractive choice for the design and implementation of filter banks with a large number of sub-bands. Their main features are:

- Simple design procedure, consisting of generating a lowpass prototype whose impulse response satisfies some constraints required to achieve perfect reconstruction.
- Low cost of implementation measured in terms of multiplication count, since the resulting analysis and synthesis filter banks rely on a type of DCT, which is amenable to fast implementation and can share the prototype implementation cost with each sub-filter.

In the cosine-modulated filter bank design, we begin by finding a linear-phase prototype lowpass filter $H(z)$ of order N, with passband edge $\omega_p = (\frac{\pi}{2M} - \rho)$ and stopband edge $\omega_r = (\frac{\pi}{2M} + \rho)$, where 2ρ is the width of the transition band. For convenience, we assume that the length $N + 1$ is an even multiple of the number M of sub-bands, that is, $N = (2LM - 1)$. Although the actual length of the prototype can be arbitrary, this assumption greatly simplifies our analysis.[3]

---

[3] It is also worth mentioning that the prototype filter does not need to be linear phase. In fact, we describe here only a particular type of cosine-modulated filter bank, many others exist.

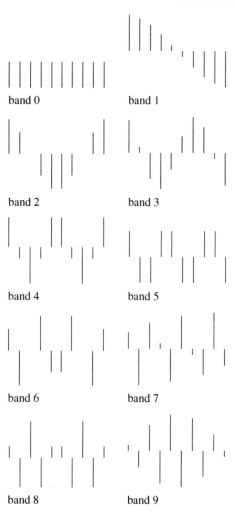

Figure 9.24 Impulse responses of the filters of the 10-band DCT.

Given the prototype filter, we can generate cosine-modulated versions of it in order to obtain the analysis and synthesis filter banks as follows:

$$h_m(n) = 2h(n) \cos\left[(2m+1)\frac{\pi}{2M}\left(n - \frac{N}{2}\right) + (-1)^m\frac{\pi}{4}\right] \tag{9.79}$$

$$g_m(n) = 2h(n) \cos\left[(2m+1)\frac{\pi}{2M}\left(n - \frac{N}{2}\right) - (-1)^m\frac{\pi}{4}\right] \tag{9.80}$$

for $n = 0, \ldots, N$ and $m = 0, \ldots, M-1$, with $N = (2LM - 1)$. We should note that in equation (9.79) the term that multiplies $h(n)$ is related to the (m, n) entry of an $(M \times 2LM)$ DCT-type matrix $\mathbf{C}$ (Sorensen & Burrus, 1993), given by

$$c_{m,n} = 2\cos\left[(2m+1)\frac{\pi}{2M}\left(n - \frac{N}{2}\right) + (-1)^m\frac{\pi}{4}\right] \tag{9.81}$$

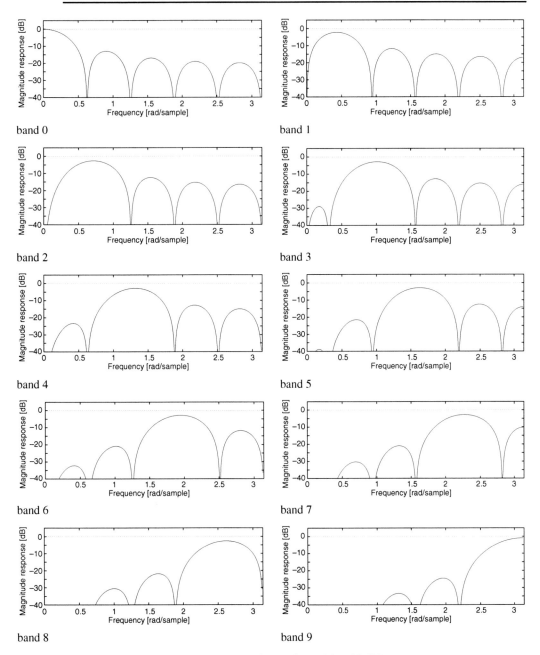

Figure 9.25 Magnitude responses of the filters of the 10-band DCT.

The prototype filter can be decomposed into $2M$ polyphase components as follows

$$H(z) = \sum_{l=0}^{L-1} \sum_{j=0}^{2M-1} h(2lM + j)z^{-(2lM+j)} = \sum_{j=0}^{2M-1} z^{-j} E_j(z^{2M}) \qquad (9.82)$$

where $E_j(z) = \sum_{l=0}^{L-1} h(2lM + j)z^{-l}$ are the polyphase components of the filter $H(z)$. With this formulation, the analysis filter bank can be described as

$$
\begin{aligned}
H_m(z) &= \sum_{n=0}^{N-1} h_m(n)z^{-n} \\
&= \sum_{n=0}^{2LM-1} c_{m,n} h(n)z^{-n} \\
&= \sum_{l=0}^{L-1} \sum_{j=0}^{2M-1} c_{m,2lM+j} h(2lM + j)z^{-(2lM+j)}
\end{aligned}
\tag{9.83}
$$

This expression can be further simplified, if we explore the following property:

$$
\begin{aligned}
&\cos\left\{ (2m + 1)\frac{\pi}{2M}\left[(n + 2kM) - \frac{N}{2} \right] + \phi \right\} \\
&= (-1)^k \cos\left[(2m + 1)\frac{\pi}{2M}\left(n - \frac{N}{2} \right) + \phi \right]
\end{aligned}
\tag{9.84}
$$

which leads to

$$
c_{m,n+2kM} = (-1)^k c_{m,n}
\tag{9.85}
$$

and, therefore, substituting j for n, and l for k, we get

$$
c_{m,j+2lM} = (-1)^l c_{m,j}
\tag{9.86}
$$

With this relation, and after some manipulation, we can rewrite equation (9.83) as

$$
H_m(z) = \sum_{j=0}^{2M-1} c_{m,j}z^{-j} \sum_{l=0}^{L-1}(-1)^l h(2lM + j)z^{-2lM} = \sum_{j=0}^{2M-1} c_{m,j}z^{-j} E_j(-z^{2M})
\tag{9.87}
$$

which can be rewritten in a compact form as

$$
\mathbf{e}(z) =
\begin{bmatrix}
H_0(z) \\
H_1(z) \\
\vdots \\
H_{M-1}(z)
\end{bmatrix}
=
\begin{bmatrix} \mathbf{C}_1 & \mathbf{C}_2 \end{bmatrix}
\begin{bmatrix}
E_0(-z^{2M}) \\
z^{-1}E_1(-z^{2M}) \\
\vdots \\
z^{-(2M-1)}E_{2M-1}(-z^{2M})
\end{bmatrix}
\tag{9.88}
$$

where $\mathbf{C}_1$ and $\mathbf{C}_2$ are $M \times M$ matrices whose (m, j) elements are $c_{m,j}$ and $c_{m,j+M}$, respectively, for $m = 0, \ldots, M - 1$ and $j = 0, \ldots, M - 1$.

Now, defining $\mathbf{d}(z) = [1 \quad z^{-1} \quad \cdots \quad z^{-M+1}]^T$, equation (9.88) can be expressed in a convenient form as

$$
\mathbf{e}(z) = \begin{bmatrix} \mathbf{C}_1 & \mathbf{C}_2 \end{bmatrix} \begin{bmatrix} E_0(-z^{2M}) & & & \mathbf{0} \\ & E_1(-z^{2M}) & & \\ & & \ddots & \\ \mathbf{0} & & & E_{2M-1}(-z^{2M}) \end{bmatrix} \begin{bmatrix} \mathbf{d}(z) \\ z^{-M}\mathbf{d}(z) \end{bmatrix}
$$

$$
= \left\{ \mathbf{C}_1 \begin{bmatrix} E_0(-z^{2M}) & & & \mathbf{0} \\ & E_1(-z^{2M}) & & \\ & & \ddots & \\ \mathbf{0} & & & E_{M-1}(-z^{2M}) \end{bmatrix} \right.
$$

$$
\left. + z^{-M} \mathbf{C}_2 \begin{bmatrix} E_M(-z^{2M}) & & & \mathbf{0} \\ & E_{M+1}(-z^{2M}) & & \\ & & \ddots & \\ \mathbf{0} & & & E_{2M-1}(-z^{2M}) \end{bmatrix} \right\} \mathbf{d}(z)
$$

$$
= \mathbf{E}(z^M)\mathbf{d}(z) \tag{9.89}
$$

where $\mathbf{E}(z)$ is the polyphase matrix as in equation (9.11).

To achieve perfect reconstruction in a filter bank with M bands, we should have $\mathbf{E}(z)\mathbf{R}(z) = \mathbf{R}(z)\mathbf{E}(z) = \mathbf{I}z^{-\Delta}$. However, it is well known, see Vaidyanathan (1993), that the polyphase matrix of the analysis filter bank can be designed to be paraunitary or lossless, that is, $\mathbf{E}^T(z^{-1})\mathbf{E}(z) = \mathbf{I}$, where $\mathbf{I}$ is an identity matrix of dimension M. In this case, the synthesis filters can be easily obtained from the analysis filter bank using either equation (9.80) or

$$
\mathbf{R}(z) = z^{-\Delta}\mathbf{E}^{-1}(z) = z^{-\Delta}\mathbf{E}^T(z^{-1}) \tag{9.90}
$$

It only remains to show how the prototype filter can be constrained so that the polyphase matrix of the analysis filter bank becomes paraunitary. The desired result is the following:

PROPERTY 9.1
The polyphase matrix of the analysis filter bank becomes paraunitary, for a real coefficient prototype filter, if and only if

$$
E_j(z^{-1})E_j(z) + E_{j+M}(z^{-1})E_{j+M}(z) = \frac{1}{2M} \tag{9.91}
$$

for $j = 0, \ldots, M - 1$. If the prototype filter has linear phase, these constraints can be reduced by half, because they are unique only for $j = 0, \ldots, \frac{M-1}{2}$, in which case M is odd, and for $j = 0, \ldots, \frac{M}{2} - 1$, in which case M is even.

OUTLINE

In order to prove the desired result, we need the following properties related to the matrixes $\mathbf{C}_1$ and $\mathbf{C}_2$:

$$\mathbf{C}_1^{\mathsf{T}}\mathbf{C}_1 = 2M[\mathbf{I} + (-1)^{L-1}\mathbf{J}] \tag{9.92}$$

$$\mathbf{C}_2^{\mathsf{T}}\mathbf{C}_2 = 2M[\mathbf{I} - (-1)^{L-1}\mathbf{J}] \tag{9.93}$$

$$\mathbf{C}_1^{\mathsf{T}}\mathbf{C}_2 = \mathbf{C}_2^{\mathsf{T}}\mathbf{C}_1 = \mathbf{0} \tag{9.94}$$

where $\mathbf{I}$ is the identity matrix, $\mathbf{J}$ is the reverse identity matrix, and $\mathbf{0}$ is a matrix with all elements equal to zero. All these matrices are square with order M. These results are widely discussed in the literature, and can be found, for instance, in Vaidyanathan (1993).

With equations (9.92)–(9.94), it is straightforward to show that

$$\mathbf{E}^{\mathsf{T}}(z^{-1})\mathbf{E}(z) =$$

$$\begin{bmatrix} E_0(-z^{-2}) & & & \mathbf{0} \\ & E_1(-z^{-2}) & & \\ & & \ddots & \\ \mathbf{0} & & & E_{M-1}(-z^{-2}) \end{bmatrix} \mathbf{C}_1^{\mathsf{T}}\mathbf{C}_1 \begin{bmatrix} E_0(-z^2) & & & \mathbf{0} \\ & E_1(-z^2) & & \\ & & \ddots & \\ \mathbf{0} & & & E_{M-1}(-z^2) \end{bmatrix}$$

$$+ \begin{bmatrix} E_M(-z^{-2}) & & & \mathbf{0} \\ & E_{M+1}(-z^{-2}) & & \\ & & \ddots & \\ \mathbf{0} & & & E_{2M-1}(-z^{-2}) \end{bmatrix} \mathbf{C}_2^{\mathsf{T}}\mathbf{C}_2 \begin{bmatrix} E_M(-z^2) & & & \mathbf{0} \\ & E_{M+1}(-z^2) & & \\ & & \ddots & \\ \mathbf{0} & & & E_{2M-1}(-z^2) \end{bmatrix}$$

$$\tag{9.95}$$

Since the prototype is a linear-phase filter, it can be shown that after some manipulation

$$\begin{bmatrix} E_0(-z^{-2}) & & & \mathbf{0} \\ & E_1(-z^{-2}) & & \\ & & \ddots & \\ \mathbf{0} & & & E_{M-1}(-z^{-2}) \end{bmatrix} \mathbf{J} \begin{bmatrix} E_0(-z^2) & & & \mathbf{0} \\ & E_1(-z^2) & & \\ & & \ddots & \\ \mathbf{0} & & & E_{M-1}(-z^2) \end{bmatrix}$$

$$= \begin{bmatrix} E_M(-z^{-2}) & & & \mathbf{0} \\ & E_{M+1}(-z^{-2}) & & \\ & & \ddots & \\ \mathbf{0} & & & E_{2M-1}(-z^{-2}) \end{bmatrix} \mathbf{J} \begin{bmatrix} E_M(-z^2) & & & \mathbf{0} \\ & E_{M+1}(-z^2) & & \\ & & \ddots & \\ \mathbf{0} & & & E_{2M-1}(-z^2) \end{bmatrix}$$

$$\tag{9.96}$$

This result allows some simplification in equation (9.95), after we apply the expressions for $\mathbf{C}_1^T\mathbf{C}_1$ and $\mathbf{C}_2^T\mathbf{C}_2$ given in equations (9.92) and (9.93), yielding

$$
\mathbf{E}^T(z^{-1})\mathbf{E}(z) = 2M \left\{ \begin{bmatrix} E_0(-z^{-2}) & & & 0 \\ & E_1(-z^{-2}) & & \\ & & \ddots & \\ 0 & & & E_{M-1}(-z^{-2}) \end{bmatrix} \begin{bmatrix} E_0(-z^2) & & & 0 \\ & E_1(-z^2) & & \\ & & \ddots & \\ 0 & & & E_{M-1}(-z^2) \end{bmatrix} \right.
$$
$$
\left. + \begin{bmatrix} E_M(-z^{-2}) & & & 0 \\ & E_{M+1}(-z^{-2}) & & \\ & & \ddots & \\ 0 & & & E_{2M-1}(-z^{-2}) \end{bmatrix} \begin{bmatrix} E_M(-z^2) & & & 0 \\ & E_{M+1}(-z^2) & & \\ & & \ddots & \\ 0 & & & E_{2M-1}(-z^2) \end{bmatrix} \right\}
$$

$$(9.97)$$

If the matrix above is equal to the identity matrix, we achieve perfect reconstruction. This is equivalent to requiring that polyphase components of the prototype filter are pairwise power complementary, which is exactly the result of equation (9.91). In fact, this last property can be explored to further reduce the computational complexity of these types of filter banks, by implementing the power complementary pairs with lattice realizations, which are structures especially suited for this task, as described in Vaidyanathan (1993).

Equation (9.88) suggests the structure of Figure 9.26 for the implementation of the cosine-modulated filter bank.

This structure can be implemented using the DCT-IV, which is represented by the $\mathbf{C}_M^{IV}$ matrix as described in Section 3.6.2 (Sorensen & Burrus, 1993), by noting that the matrices $\mathbf{C}_1$ and $\mathbf{C}_2$ can be expressed as follows (Vaidyanathan, 1993):

$$\mathbf{C}_1 = \sqrt{M}(-1)^{\frac{L}{2}}\mathbf{C}_M^{IV}(\mathbf{I} - \mathbf{J}) \tag{9.98}$$

$$\mathbf{C}_2 = -\sqrt{M}(-1)^{\frac{L}{2}}\mathbf{C}_M^{IV}(\mathbf{I} + \mathbf{J}) \tag{9.99}$$

for L even, and

$$\mathbf{C}_1 = \sqrt{M}(-1)^{\frac{L-1}{2}}\mathbf{C}_M^{IV}(\mathbf{I} + \mathbf{J}) \tag{9.100}$$

$$\mathbf{C}_2 = \sqrt{M}(-1)^{\frac{L-1}{2}}\mathbf{C}_M^{IV}(\mathbf{I} - \mathbf{J}) \tag{9.101}$$

for L odd, where, from equation (3.145) and Table 3.1

$$\{\mathbf{C}_M^{IV}\}_{m,n} = \sqrt{\frac{2}{M}}\cos\left[(2m+1)\left(n+\frac{1}{2}\right)\frac{\pi}{2M}\right] \tag{9.102}$$

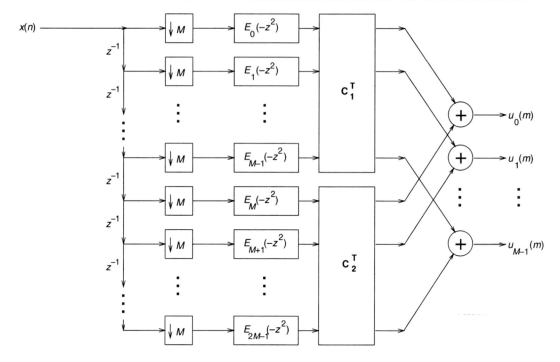

Figure 9.26 Cosine-modulated filter bank.

equations (9.98)–(9.101) can be put in the following form:

$$\mathbf{C}_1 = \sqrt{M}(-1)^{\lfloor \frac{L}{2} \rfloor} \mathbf{C}_M^{IV}[\mathbf{I} - (-1)^L \mathbf{J}] \tag{9.103}$$

$$\mathbf{C}_2 = -\sqrt{M}(-1)^{\lfloor \frac{L}{2} \rfloor} \mathbf{C}_M^{IV}[(-1)^L \mathbf{I} + \mathbf{J}] \tag{9.104}$$

where $\lfloor x \rfloor$ represents the largest integer smaller than or equal to x.

From equation (9.88) and the above equations, the structure in Figure 9.27 immediately follows. One of the main advantages of such a structure is that it can benefit from the fast implementation algorithms for the DCT-IV.

EXAMPLE 9.5
Design a filter bank with $M = 10$ sub-bands using the cosine-modulated method with $L = 3$.

SOLUTION
For the prototype linear-phase design, we employ the least-squares design method described in Chapter 5, using as the objective the minimization of the filter stopband energy. The length of the resulting prototype filter is 60, and the stopband attenuation is $A_r \approx 20$ dB, as shown in Figure 9.28. Its impulse response is depicted in Figure 9.29, and the values of the filter coefficients are listed in Table 9.1. Since the prototype filter has linear phase, only half of its coefficients are shown.

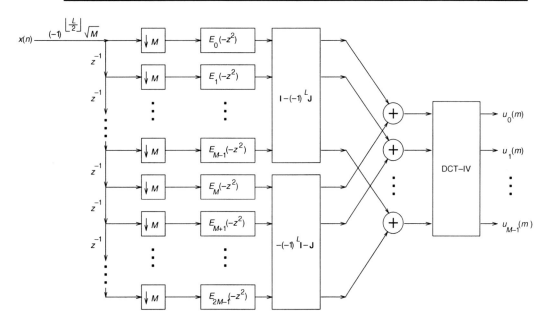

Figure 9.27 Implementation of the cosine-modulated filter bank using the DCT-IV.

Table 9.1. *Prototype filter coefficients.*

$h(0) = -4.982\,2343\text{E}{-}002$	$h(10) = -8.504\,1835\text{E}{-}002$	$h(20) = 4.722\,6358\text{E}{-}001$
$h(1) = -5.044\,4424\text{E}{-}002$	$h(11) = -5.423\,1162\text{E}{-}001$	$h(21) = 5.229\,2769\text{E}{-}001$
$h(2) = -4.717\,6505\text{E}{-}002$	$h(12) = -2.432\,3259\text{E}{-}002$	$h(22) = 5.697\,2514\text{E}{-}001$
$h(3) = -4.056\,7101\text{E}{-}002$	$h(13) = 3.171\,4945\text{E}{-}003$	$h(23) = 6.129\,0705\text{E}{-}001$
$h(4) = -3.208\,6097\text{E}{-}002$	$h(14) = 2.707\,7445\text{E}{-}002$	$h(24) = 6.531\,3114\text{E}{-}001$
$h(5) = -1.520\,4235\text{E}{-}001$	$h(15) = 1.833\,5337\text{E}{-}001$	$h(25) = 7.105\,7841\text{E}{-}001$
$h(6) = -1.521\,6756\text{E}{-}001$	$h(16) = 2.346\,0887\text{E}{-}001$	$h(26) = 7.310\,9022\text{E}{-}001$
$h(7) = -1.445\,9253\text{E}{-}001$	$h(17) = 2.865\,1892\text{E}{-}001$	$h(27) = 7.480\,6145\text{E}{-}001$
$h(8) = -1.309\,1745\text{E}{-}001$	$h(18) = 3.389\,8306\text{E}{-}001$	$h(28) = 7.609\,2844\text{E}{-}001$
$h(9) = -1.117\,9673\text{E}{-}001$	$h(19) = 3.915\,3918\text{E}{-}001$	$h(29) = 7.690\,5871\text{E}{-}001$

By applying cosine modulation to the prototype filter, we obtain the coefficients of the analysis and synthesis filters belonging to the filter bank. The impulse responses of the resulting analysis filters, each of length 60, for the bank with $M = 10$ sub-bands, are shown in Figure 9.30 and their normalized magnitude responses in Figure 9.31.

$\triangle$

The filter banks discussed in this section have the disadvantage of having the nonlinear-phase characteristic of the analysis filters. This problem is solved with the filter banks described in the next section.

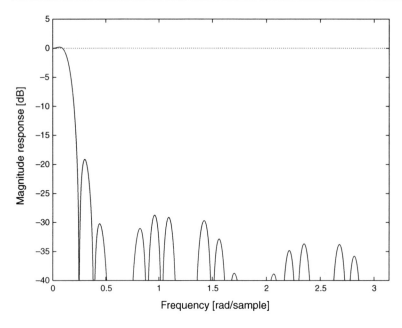

Figure 9.28 Magnitude response of the prototype filter.

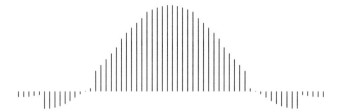

Figure 9.29 Impulse response of the prototype filter.

9.9　Lapped transforms

The lapped orthogonal transforms (LOT) were originally proposed to be block transforms whose basis functions extended beyond the block boundaries, that is, the basis functions from neighboring blocks overlapped with each other. Its main goal was to reduce blocking effects, usually present under quantization of the block transform coefficients, as is the case of the DCT. Blocking effects are discontinuities that appear across the block boundaries. They occur because each block is transformed and quantized independently of the others, and this type of distortion is particularly annoying in images (Malvar, 1992). If the transform basis functions are allowed to overlap, the blocking effects are greatly reduced.

Since block transforms are equivalent to M-band perfect reconstruction filter banks, one idea would be to replace the DCT by a cosine-modulated filter bank. However,

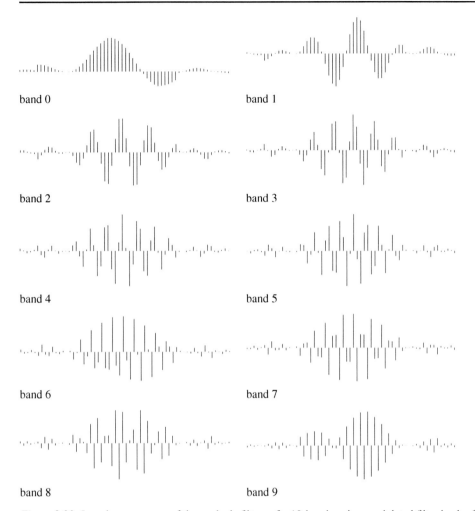

Figure 9.30 Impulse responses of the analysis filters of a 10-band cosine-modulated filter bank of length 60.

the cosine-modulated filter banks discussed in the previous section have the nonlinear phase of the analysis filters, an undesirable feature in applications such as image coding. LOT-based filter banks are very attractive because they lead to linear-phase analysis filters and have fast implementation. The LOT-based filter banks are members of the family of paraunitary FIR perfect reconstruction filter banks with linear phase. Although there are a number of possible designs for linear-phase filter banks with perfect reconstruction, the LOT-based design is simple to derive and to implement. The term LOT applies to the cases where the analysis filters have length $2M$. Generalizations of the LOT to longer analysis and synthesis filters (length LM) are available. They are known as the extended lapped transforms (ELT) proposed in Malvar (1992) and the generalized LOT (GenLOT) proposed in de Queiroz et al.

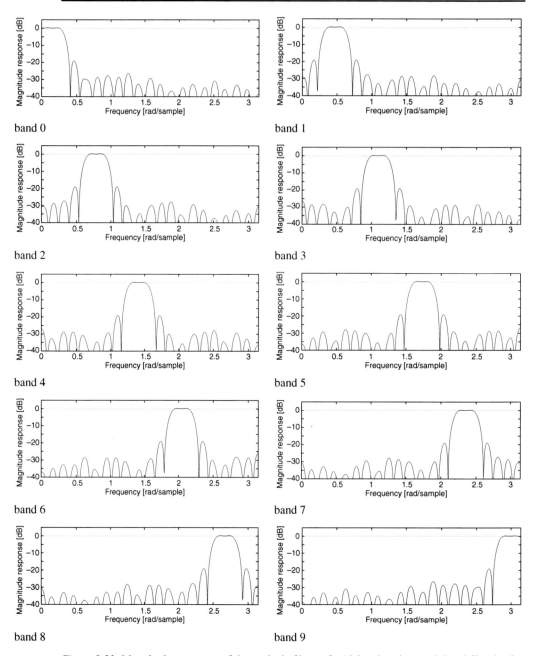

Figure 9.31 Magnitude responses of the analysis filters of a 10-band cosine-modulated filter bank of length 60.

(1996). The ELT is a cosine-modulated filter bank and does not produce linear-phase analysis filters. The GenLOT is a good choice when long analysis filters with high selectivity are required together with linear phase.

In this section we begin by briefly discussing the LOT filter bank, where the analysis and synthesis filter banks have lengths $2M$.

Once again, similarly to the cosine-modulated filter banks, the LOT analysis filters are given by

$$
\mathbf{e}(z) =
\begin{bmatrix}
H_0(z) \\
H_1(z) \\
\vdots \\
H_{M-1}(z)
\end{bmatrix}
=
\begin{bmatrix}
\hat{\mathbf{C}}_1 & \hat{\mathbf{C}}_2
\end{bmatrix}
\begin{bmatrix}
1 \\
z^{-1} \\
\vdots \\
z^{-(2M-1)}
\end{bmatrix}
\tag{9.105}
$$

or, as in equation (9.89),

$$
\mathbf{e}(z) =
\begin{bmatrix}
H_0(z) \\
H_1(z) \\
\vdots \\
H_{M-1}(z)
\end{bmatrix}
= (\hat{\mathbf{C}}_1 + z^{-M}\hat{\mathbf{C}}_2)\mathbf{d}(z) = \mathbf{E}(z^M)\mathbf{d}(z)
\tag{9.106}
$$

where $\hat{\mathbf{C}}_1$ and $\hat{\mathbf{C}}_2$ are $M \times M$ DCT-type matrices, and $\mathbf{E}(z)$ is the polyphase matrix of the analysis filter bank. The perfect reconstruction condition with paraunitary polyphase matrices is generated if

$$
\mathbf{R}(z) = z^{-\Delta}\mathbf{E}^{-1}(z) = z^{-\Delta}\mathbf{E}^T(z^{-1})
\tag{9.107}
$$

We then have the result that follows.

PROPERTY 9.2
The polyphase matrix of the analysis filter bank becomes paraunitary, for a real coefficient prototype filter, if the following conditions are satisfied:

$$
\hat{\mathbf{C}}_1\hat{\mathbf{C}}_1^T + \hat{\mathbf{C}}_2\hat{\mathbf{C}}_2^T = \mathbf{I}
\tag{9.108}
$$

$$
\hat{\mathbf{C}}_1\hat{\mathbf{C}}_2^T = \hat{\mathbf{C}}_2\hat{\mathbf{C}}_1^T = \mathbf{0}
\tag{9.109}
$$

PROOF
From equation (9.106), we have that

$$
\mathbf{E}(z) = \hat{\mathbf{C}}_1 + z^{-1}\hat{\mathbf{C}}_2
\tag{9.110}
$$

Perfect reconstruction requires that $\mathbf{E}(z)\mathbf{E}^T(z^{-1}) = \mathbf{I}$. Therefore,

$$
(\hat{\mathbf{C}}_1 + z^{-1}\hat{\mathbf{C}}_2)(\hat{\mathbf{C}}_1^T + z\hat{\mathbf{C}}_2^T) = \hat{\mathbf{C}}_1\hat{\mathbf{C}}_1^T + \hat{\mathbf{C}}_2\hat{\mathbf{C}}_2^T + z\hat{\mathbf{C}}_1\hat{\mathbf{C}}_2^T + z^{-1}\hat{\mathbf{C}}_2\hat{\mathbf{C}}_1^T = \mathbf{I}
\tag{9.111}
$$

and then equations (9.108) and (9.109) immediately follow.

$\square$

Equation (9.108) implies that the rows of $\hat{\mathbf{C}} = \begin{bmatrix} \hat{\mathbf{C}}_1 & \hat{\mathbf{C}}_2 \end{bmatrix}$ are orthogonal. Meanwhile, equation (9.109) implies that the rows of $\hat{\mathbf{C}}_1$ are orthogonal to the rows of $\hat{\mathbf{C}}_2$, which is the same as saying that the overlapping tails of the basis functions of the LOT are orthogonal.

A simple construction for the matrices above, based on the DCT, and leading to linear-phase filters, results from choosing

$$\hat{\mathbf{C}}_1 = \frac{1}{2} \begin{bmatrix} \mathbf{C}_e - \mathbf{C}_o \\ \mathbf{C}_e - \mathbf{C}_o \end{bmatrix} \tag{9.112}$$

$$\hat{\mathbf{C}}_2 = \frac{1}{2} \begin{bmatrix} (\mathbf{C}_e - \mathbf{C}_o)\mathbf{J} \\ -(\mathbf{C}_e - \mathbf{C}_o)\mathbf{J} \end{bmatrix} = \frac{1}{2} \begin{bmatrix} \mathbf{C}_e + \mathbf{C}_o \\ -\mathbf{C}_e - \mathbf{C}_o \end{bmatrix} \tag{9.113}$$

where $\mathbf{C}_e$ and $\mathbf{C}_o$ are $\frac{M}{2} \times M$ matrices consisting of the even and odd DCT basis of length M, respectively (see equation (9.78)). Observe that $\mathbf{C}_e\mathbf{J} = \mathbf{C}_e$ and $\mathbf{C}_o\mathbf{J} = -\mathbf{C}_o$, because the even basis functions are symmetric and the odd ones are antisymmetric. The reader can easily verify that the above choice satisfies the relations (9.108) and (9.109). With this, we can build an initial LOT whose polyphase matrix, as in equation (9.110), is given by

$$\mathbf{E}(z) = \frac{1}{2} \begin{bmatrix} \mathbf{C}_e - \mathbf{C}_o + z^{-1}(\mathbf{C}_e - \mathbf{C}_o)\mathbf{J} \\ \mathbf{C}_e - \mathbf{C}_o - z^{-1}(\mathbf{C}_e - \mathbf{C}_o)\mathbf{J} \end{bmatrix} = \frac{1}{2} \begin{bmatrix} \mathbf{I} & z^{-1}\mathbf{I} \\ \mathbf{I} & -z^{-1}\mathbf{I} \end{bmatrix} \begin{bmatrix} \mathbf{I} & -\mathbf{I} \\ \mathbf{I} & \mathbf{I} \end{bmatrix} \begin{bmatrix} \mathbf{C}_e \\ \mathbf{C}_o \end{bmatrix} \tag{9.114}$$

This expression suggests the structure of Figure 9.32 for the implementation of the LOT filter bank. This structure consists of the implementation of the polyphase components of the prototype filter using a DCT-based matrix, followed by several orthogonal matrices $\mathbf{T}_0, \mathbf{T}_1, \ldots, \mathbf{T}_{\frac{M}{2}-2}$, which are discussed next. Actually, we can pre-multiply the right-hand term in equation (9.114) by an orthogonal matrix $\mathbf{L}_1$, and still keep the perfect reconstruction conditions.[4] The polyphase matrix is then given by

$$\mathbf{E}(z) = \frac{1}{2}\mathbf{L}_1 \begin{bmatrix} \mathbf{I} & z^{-1}\mathbf{I} \\ \mathbf{I} & -z^{-1}\mathbf{I} \end{bmatrix} \begin{bmatrix} \mathbf{I} & -\mathbf{I} \\ \mathbf{I} & \mathbf{I} \end{bmatrix} \begin{bmatrix} \mathbf{C}_e \\ \mathbf{C}_o \end{bmatrix} \tag{9.115}$$

The basic construction of the LOT presented above is equivalent to the one proposed in Malvar (1992) that utilizes a block transform formulation to generate lapped transforms. Note that in the block transform formulation, equation (9.105) is equivalent to a block transform whose transform matrix $\mathbf{L}_{\text{LOT}}$ is equal to $\begin{bmatrix} \hat{\mathbf{C}}_1 & \hat{\mathbf{C}}_2 \end{bmatrix}$. Referring to Figure 9.22, it is important to point out that since the transform matrix $\mathbf{L}_{\text{LOT}}$ has dimensions $M \times 2M$, in order to compute the transform coefficients $y_j(k)$

---

[4] It can easily be seen that equations (9.108) and (9.109) remain valid, if $\hat{\mathbf{C}}_1$ and $\hat{\mathbf{C}}_2$ are pre-multiplied by an orthogonal matrix. However, $\mathbf{L}_1$ should not destroy the symmetry or the anitsymmetry of the analysis filter's impulse response.

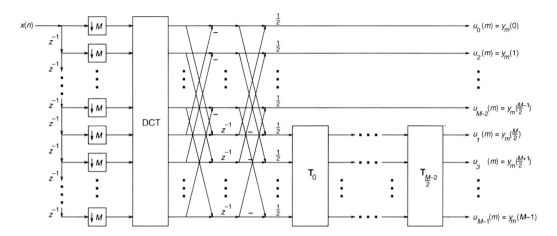

Figure 9.32 Implementation of the lapped orthogonal transform.

of a block B_j, one needs the samples $x_j(m)$ and $x_{j+1}(m)$ of the blocks B_j and B_{j+1}, respectively. The term 'lapped transform' comes from the fact that the block B_{j+1} is needed in the computation of both $y_j(k)$ and $y_{j+1}(k)$, that is, the transforms of the two blocks overlap. This is expressed formally by the following equation:

$$
\begin{bmatrix} y_j(0) \\ \vdots \\ y_j(M-1) \end{bmatrix} = \mathbf{L}_{\text{LOT}} \begin{bmatrix} x_j(0) \\ \vdots \\ x_j(M-1) \\ x_{j+1}(0) \\ \vdots \\ x_{j+1}(M-1) \end{bmatrix}
$$

$$
= \begin{bmatrix} \hat{\mathbf{C}}_1 & \hat{\mathbf{C}}_2 \end{bmatrix} \begin{bmatrix} x_j(0) \\ \vdots \\ x_j(M-1) \\ x_{j+1}(0) \\ \vdots \\ x_{j+1}(M-1) \end{bmatrix}
$$

$$
= \hat{\mathbf{C}}_1 \begin{bmatrix} x_j(0) \\ \vdots \\ x_j(M-1) \end{bmatrix} + \hat{\mathbf{C}}_2 \begin{bmatrix} x_{j+1}(0) \\ \vdots \\ x_{j+1}(M-1) \end{bmatrix} \tag{9.116}
$$

Malvar's formultaion for the LOT differs somewhat from the one in equations (9.105)–(9.115). In fact, Malvar starts with an orthogonal matrix based on the DCT having the following form:

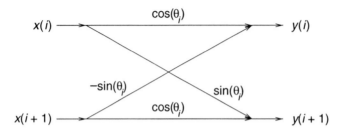

Figure 9.33 Implementation of multiplication by $\mathbf{T}_i$.

$$\mathbf{L}_0 = \frac{1}{2} \begin{bmatrix} \mathbf{C}_e - \mathbf{C}_o & (\mathbf{C}_e - \mathbf{C}_o)\mathbf{J} \\ \mathbf{C}_e - \mathbf{C}_o & -(\mathbf{C}_e - \mathbf{C}_o)\mathbf{J} \end{bmatrix} \tag{9.117}$$

The choice is not at random. First, it satisfies the conditions of equations (9.108) and (9.109). Also, the first half of the basis functions are symmetric, whereas the second half are antisymmetric, thus keeping the phase linear, as desired. The choice of $\mathbf{L}_0$ based on the DCT is the key to generating a fast implementation algorithm. Starting with $\mathbf{L}_0$, we can generate a family of more selective analysis filters in the following form:

$$\mathbf{L}_{\text{LOT}} = \begin{bmatrix} \hat{\mathbf{C}}_1 & \hat{\mathbf{C}}_2 \end{bmatrix} = \mathbf{L}_1\mathbf{L}_0 \tag{9.118}$$

where the matrix $\mathbf{L}_1$ should be orthogonal and should also be amenable to fast implementation. Usually, the matrix $\mathbf{L}_1$ is of the form

$$\mathbf{L}_1 = \begin{bmatrix} \mathbf{I} & \mathbf{0} \\ \mathbf{0} & \mathbf{L}_2 \end{bmatrix} \tag{9.119}$$

where $\mathbf{L}_2$ is a square matrix of dimension $\frac{M}{2}$ consisting of a set of plane rotations. More specifically,

$$\mathbf{L}_2 = \mathbf{T}_{\frac{M}{2}-2} \cdots \mathbf{T}_1\mathbf{T}_0 \tag{9.120}$$

where

$$\mathbf{T}_i = \begin{bmatrix} \mathbf{I}_i & \mathbf{0} & \mathbf{0} \\ \mathbf{0} & \mathbf{Y}(\theta_i) & \mathbf{0} \\ \mathbf{0} & \mathbf{0} & \mathbf{I}_{\frac{M}{2}-2-i} \end{bmatrix} \tag{9.121}$$

and

$$\mathbf{Y}(\theta_i) = \begin{bmatrix} \cos\theta_i & -\sin\theta_i \\ \sin\theta_i & \cos\theta_i \end{bmatrix} \tag{9.122}$$

The rotation angles θ_i are submitted to optimization aimed at maximizing the coding gain, when using the filter bank in sub-band coders, or at improving the selectivity of

Table 9.2. *LOT analysis filter coefficients: band 0.*

$h(0) = 6.293\,7276\text{E}{-}002$	$h(4) = -1.242\,1729\text{E}{-}001$	$h(7) = -3.182\,1409\text{E}{-}001$
$h(1) = 4.221\,4260\text{E}{-}002$	$h(5) = -1.957\,8374\text{E}{-}001$	$h(8) = -3.563\,7073\text{E}{-}001$
$h(2) = 1.412\,0175\text{E}{-}003$	$h(6) = -2.629\,9060\text{E}{-}001$	$h(9) = -3.740\,8240\text{E}{-}001$
$h(3) = -5.596\,1133\text{E}{-}002$		

the analysis and synthesis filters (Malvar, 1992). Since in the multiplication by $\mathbf{T}_i$ only the samples i and $i + 1$ are modified, a flowgraph that implements it is as depicted in Figure 9.33.

A simple fast LOT implementation not allowing any type of optimization consists of, instead of implementing $\mathbf{L}_2$ as a cascade of rotations $\mathbf{T}_i$, implementing $\mathbf{L}_2$ as a cascade of square matrices of dimension $\hat{M} = \frac{M}{2}$ comprised of DCTs Type II and Type IV (see Subsection 3.6.2), the elements of which are, from equation (3.145) and Table 3.1,

$$c_{l,n} = \alpha_{\hat{M}}(l)\sqrt{\frac{2}{\hat{M}}}\cos\left[(2l + 1)\frac{\pi}{2\hat{M}}(n)\right] \tag{9.123}$$

and

$$c_{l,n} = \sqrt{\frac{2}{\hat{M}}}\cos\left[(2l + 1)\frac{\pi}{2\hat{M}}\left(n + \frac{1}{2}\right)\right] \tag{9.124}$$

respectively, where $\alpha_{\hat{M}}(l) = 1/\sqrt{2}$, for $l = 0$, or $l = \hat{M}$, and $\alpha_{\hat{M}}(l) = 1$ otherwise. The implementation is fast because the DCTs of Types II and IV have fast algorithms.

EXAMPLE 9.6
Design a filter bank with $M = 10$ sub-bands using the LOT.

SOLUTION
The length of the analysis and synthesis filters should be 20. For this design, it is assumed that the input signal is an autoregressive process generated by filtering a white noise with a first-order lowpass digital filter with the pole at $z = 0.9$. With this assumption, we optimize matrix $\mathbf{L}_1$ so that the maximum energy compaction on the transform coefficient is obtained, that is, most energy concentrates in the smallest number of transform coefficients. The resulting analysis filters' impulse responses are shown in Figure 9.34. The coefficients of the first analysis filter are listed in Table 9.2. Observe that the LOT has linear phase, and thus only the first half of the coefficients are shown.

Figure 9.35 depicts the normalized magnitude responses of the analysis filters for a LOT filter bank with $M = 10$ sub-bands.

$\triangle$

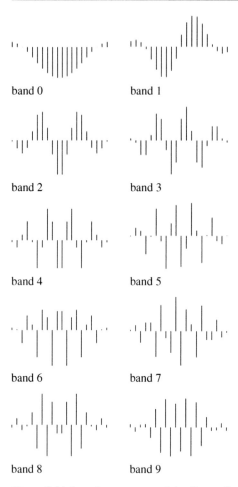

band 0   band 1

band 2   band 3

band 4   band 5

band 6   band 7

band 8   band 9

Figure 9.34 Impulse responses of the filters of a 10-band LOT.

Table 9.3. *Fast LOT analysis filter coefficients: band 0.*

$h(0) = -6.273\,9944\text{E}{-}002$	$h(4) = 1.231\,3407\text{E}{-}001$	$h(7) = 3.162\,2777\text{E}{-}001$
$h(1) = -4.112\,1233\text{E}{-}002$	$h(5) = 1.930\,9369\text{E}{-}001$	$h(8) = 3.573\,4900\text{E}{-}001$
$h(2) = -2.775\,5576\text{E}{-}017$	$h(6) = 2.596\,2924\text{E}{-}001$	$h(9) = 3.789\,6771\text{E}{-}001$
$h(3) = 5.659\,8521\text{E}{-}002$		

EXAMPLE 9.7

Design a filter bank with $M = 10$ sub-bands using the fast LOT.

SOLUTION

For the fast LOT design, we use a factorable $\mathbf{L}_1$ matrix composed of a cascade of a $\mathbf{C}^{\text{II}}$ transform and a transposed $\mathbf{C}^{\text{IV}}$ transform on its lower part, to allow fast implementation. The resulting analysis filter impulse responses are shown in

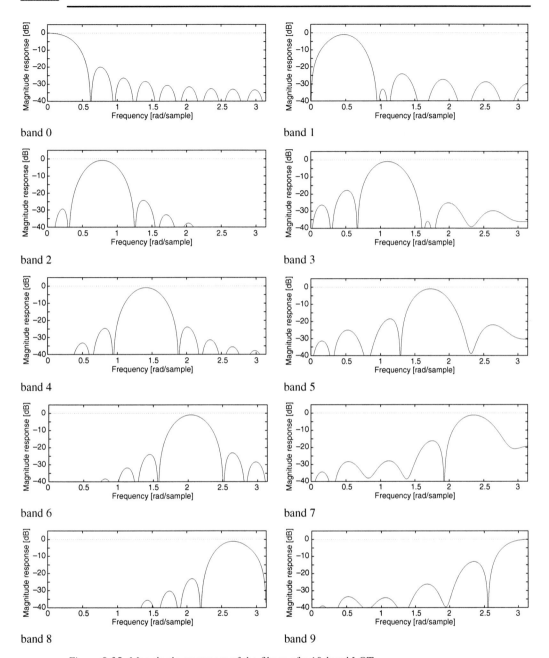

Figure 9.35 Magnitude responses of the filters of a 10-band LOT.

Figure 9.36. The coefficients of the first analysis filter are listed in Table 9.3. Again, because of the linear-phase property, only half of the coefficients are shown.

Figure 9.37 depicts the normalized magnitude responses of the analysis filters, each of length 20, for a fast LOT filter bank with $M = 10$ sub-bands.

$\triangle$

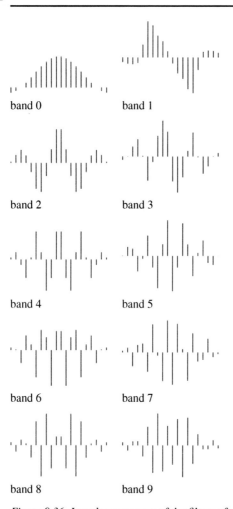

band 0   band 1

band 2   band 3

band 4   band 5

band 6   band 7

band 8   band 9

Figure 9.36 Impulse responses of the filters of a 10-band fast LOT.

9.9.1 Fast algorithms and biorthogonal LOT

We now present a general construction of a fast algorithm for the LOT. We start by defining two matrices as follows:

$$\hat{\mathbf{C}}_3 = \hat{\mathbf{C}}_1 \hat{\mathbf{C}}_1^{\mathrm{T}} \tag{9.125}$$

$$\hat{\mathbf{C}}_4 = \hat{\mathbf{C}}_1 + \hat{\mathbf{C}}_2 \tag{9.126}$$

Using the results in equation (9.109), one can show that

$$\hat{\mathbf{C}}_1 = \hat{\mathbf{C}}_3 \hat{\mathbf{C}}_4 \tag{9.127}$$

$$\hat{\mathbf{C}}_2 = (\mathbf{I} - \hat{\mathbf{C}}_3) \hat{\mathbf{C}}_4 \tag{9.128}$$

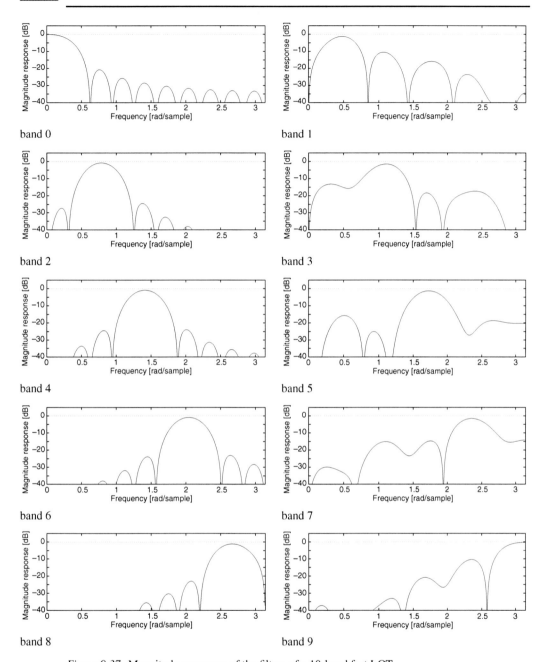

Figure 9.37 Magnitude responses of the filters of a 10-band fast LOT.

With the above relations, it is straightforward to show that the polyphase components of the analysis filter can be written as

$$\mathbf{E}(z) = [\hat{\mathbf{C}}_3 + z^{-1}(\mathbf{I} - \hat{\mathbf{C}}_3)]\hat{\mathbf{C}}_4 \tag{9.129}$$

The initial solution for the LOT matrix discussed previously can be analyzed in

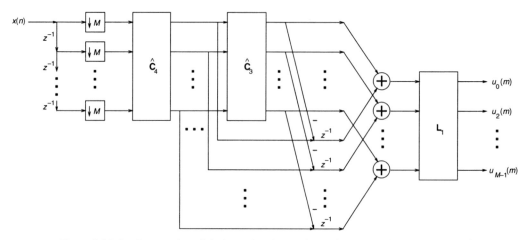

Figure 9.38 Implementation of the lapped orthogonal transform according to the general formulation in equation (9.129).

the light of this general formulation. Equation (9.129) suggests the implementation in Figure 9.38. After a few manipulations, the matrices of the polyphase description above corresponding to the LOT matrix of equation (9.117) are given by

$$\hat{\mathbf{C}}_3 = \frac{1}{2} \begin{bmatrix} \mathbf{I} & \mathbf{I} \\ \mathbf{I} & \mathbf{I} \end{bmatrix} \tag{9.130}$$

$$\hat{\mathbf{C}}_4 = \frac{1}{2} \begin{bmatrix} \mathbf{C}_e - \mathbf{C}_o + (\mathbf{C}_e - \mathbf{C}_o)\mathbf{J} \\ \mathbf{C}_e - \mathbf{C}_o - (\mathbf{C}_e - \mathbf{C}_o)\mathbf{J} \end{bmatrix} = \begin{bmatrix} \mathbf{C}_e \\ -\mathbf{C}_o \end{bmatrix} \tag{9.131}$$

One can see that in the LOT given by equation (9.131), matrices $\hat{\mathbf{C}}_3$ and $\hat{\mathbf{C}}_4$ have fast implementation algorithms, and Figure 9.38 leads to Figure 9.32. In fact, as long there are fast implementation algorithms for $\hat{\mathbf{C}}_3$ and $\hat{\mathbf{C}}_4$, the LOT will have a fast algorithm.

Biorthogonal lapped transforms can be constructed using the formulation of equation (9.129), if $\hat{\mathbf{C}}_3$ is chosen such that $\hat{\mathbf{C}}_3\hat{\mathbf{C}}_3 = \hat{\mathbf{C}}_3$, and the polyphase matrices are chosen such that

$$\mathbf{E}(z) = [\hat{\mathbf{C}}_3 + z^{-1}(\mathbf{I} - \hat{\mathbf{C}}_3)]\hat{\mathbf{C}}_4 \tag{9.132}$$

$$\mathbf{R}(z) = \hat{\mathbf{C}}_4^{-1}[z^{-1}\hat{\mathbf{C}}_3 + (\mathbf{I} - \hat{\mathbf{C}}_3)] \tag{9.133}$$

9.9.2   Generalized LOT

The formulation described in the previous subsection can be extended to obtain generalized lapped transforms in their orthogonal and biorthogonal forms. These transforms allow the overlapping of multiple blocks of length M. The approach described here follows closely the approach of de Queiroz et al. (1996).

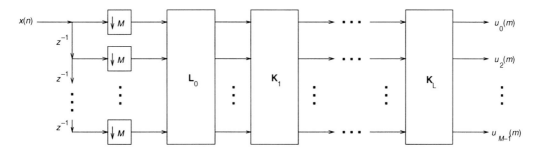

Figure 9.39 Implementation of the GenLOT.

The generalized lapped transforms (GenLOT) can also be constructed if the polyphase matrices are designed as follows

$$
\mathbf{E}(z) = \prod_{j=1}^{L} \left\{ \begin{bmatrix} \mathbf{L}_{3,j} & \mathbf{0} \\ \mathbf{0} & \mathbf{L}_{2,j} \end{bmatrix} [\hat{\mathbf{C}}_{3,j} + z^{-1}(\mathbf{I} - \hat{\mathbf{C}}_{3,j})] \right\} \hat{\mathbf{C}}_4 \tag{9.134}
$$

$$
\mathbf{R}(z) = \hat{\mathbf{C}}_4^{-1} \prod_{j=1}^{L} \left\{ [z^{-1}\hat{\mathbf{C}}_{3,j} + (\mathbf{I} - \hat{\mathbf{C}}_{3,j})] \begin{bmatrix} \mathbf{L}_{3,j}^{-1} & \mathbf{0} \\ \mathbf{0} & \mathbf{L}_{2,j}^{-1} \end{bmatrix} \right\} \tag{9.135}
$$

Biorthogonal GenLOTs are obtained if $\hat{\mathbf{C}}_{3,j}$ is chosen such that $\hat{\mathbf{C}}_{3,j}\hat{\mathbf{C}}_{3,j} = \hat{\mathbf{C}}_{3,j}$. Perfect reconstruction conditions follow directly from equations (9.133) and (9.135).

In order to obtain the GenLOT it is further required that $\hat{\mathbf{C}}_4^{\mathrm{T}}\hat{\mathbf{C}}_4 = \mathbf{I}$, $\mathbf{L}_{3,j}^{\mathrm{T}}\mathbf{L}_{3,j} = \mathbf{I}$, and $\mathbf{L}_{2,j}^{\mathrm{T}}\mathbf{L}_{2,j} = \mathbf{I}$. If we choose $\hat{\mathbf{C}}_3$ and $\hat{\mathbf{C}}_4$ as in equations (9.130) and (9.131), then, we have a valid GenLOT possessing a fast algorithm of the following form:

$$
\begin{aligned}
\mathbf{E}(z) &= \prod_{j=1}^{L} \left(\frac{1}{2} \begin{bmatrix} \mathbf{L}_{3,j} & \mathbf{0} \\ \mathbf{0} & \mathbf{L}_{2,j} \end{bmatrix} \begin{bmatrix} \mathbf{I} + z^{-1}\mathbf{I} & \mathbf{I} - z^{-1}\mathbf{I} \\ \mathbf{I} - z^{-1}\mathbf{I} & \mathbf{I} + z^{-1}\mathbf{I} \end{bmatrix} \right) \hat{\mathbf{C}}_4 \\
&= \prod_{j=1}^{L} \left(\frac{1}{2} \begin{bmatrix} \mathbf{L}_{3,j} & \mathbf{0} \\ \mathbf{0} & \mathbf{L}_{2,j} \end{bmatrix} \begin{bmatrix} \mathbf{I} & \mathbf{I} \\ \mathbf{I} & -\mathbf{I} \end{bmatrix} \begin{bmatrix} \mathbf{I} & \mathbf{0} \\ \mathbf{0} & z^{-1}\mathbf{I} \end{bmatrix} \begin{bmatrix} \mathbf{I} & \mathbf{I} \\ \mathbf{I} & -\mathbf{I} \end{bmatrix} \right) \hat{\mathbf{C}}_4 \\
&= \prod_{j=1}^{L} (\mathbf{K}_j) \, \hat{\mathbf{C}}_4 \tag{9.136}
\end{aligned}
$$

If, in order to have a fast algorithm, we fix $\hat{\mathbf{C}}_4$ as in equation (9.131), the degrees of freedom to design the filter bank are the choices of matrices $\mathbf{L}_{3,j}$ and $\mathbf{L}_{2,j}$, which are constrained to real and orthogonal matrices, to allow a valid GenLOT. The effectiveness in terms of computational complexity is highly dependent on how fast we can compute these matrices.

Equation (9.136) suggests the structure of Figure 9.39 for the implementation of the GenLOT filter bank. This structure consists of the implementation of the matrix $\mathbf{L}_0$

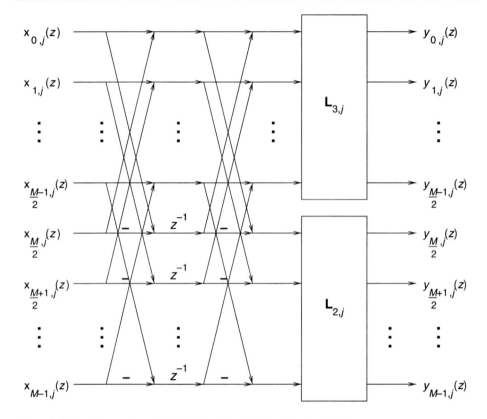

Figure 9.40 Implementation of the building blocks $\mathbf{K}_j$ of the GenLOT.

in cascade with a set of similar building blocks denoted as $\mathbf{K}_j$. The structure for the implementation of each $\mathbf{K}_j$ is depicted in Figure 9.40.

In the formulation presented here, the sub-filter lengths are constrained to be multiples of the number of sub-bands. A more general formulation for the design of sub-filters with arbitrary lengths is proposed in Tran et al. (2000).

EXAMPLE 9.8

Design an orthogonal length-40 GenLOT with $M = 10$ sub-bands.

SOLUTION

The resulting analysis filter impulse responses are shown in Figure 9.41. The co-efficients of the first analysis filter are listed in Table 9.4. Again, because of the linear-phase property, only half of the coefficients are shown.

Figure 9.42 depicts the normalized magnitude responses of the analysis filters for the length-40 GenLOT filter bank with $M = 10$ sub-bands.

$\triangle$

Table 9.4. *GenLOT analysis filter coefficients: band 0.*

$h(0) = -0.000\,734$	$h(7) = 0.006\,070$	$h(14) = -0.096\,310$
$h(1) = -0.001\,258$	$h(8) = 0.004\,306$	$h(15) = -0.191\,556$
$h(2) = 0.000\,765$	$h(9) = 0.003\,363$	$h(16) = -0.268\,528$
$h(3) = 0.000\,017$	$h(10) = 0.053\,932$	$h(17) = -0.318\,912$
$h(4) = 0.000\,543$	$h(11) = 0.030\,540$	$h(18) = -0.354\,847$
$h(5) = 0.005\,950$	$h(12) = -0.013\,838$	$h(19) = -0.383\,546$
$h(6) = 0.000\,566$	$h(13) = -0.055\,293$	

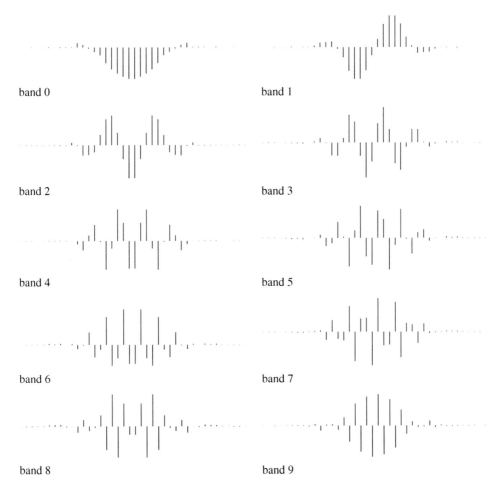

band 0   band 1

band 2   band 3

band 4   band 5

band 6   band 7

band 8   band 9

Figure 9.41 Impulse responses of the analysis filters of the 10-band GenLOT of length 40.

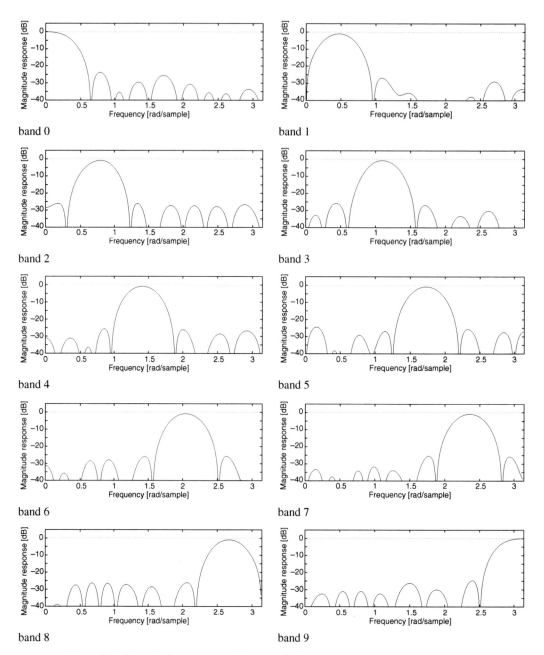

Figure 9.42 Magnitude responses of the analysis filters of the 10-band GenLOT of length 40.

Table 9.5. *Biorthogonal lapped transform analysis and synthesis filter coefficients: band 0.*

	Analysis filter	
$h(0) = 0.000\,541$	$h(6) = -0.005\,038$	$h(11) = 0.104\,544$
$h(1) = -0.000\,688$	$h(7) = -0.023\,957$	$h(12) = 0.198\,344$
$h(2) = -0.001\,211$	$h(8) = -0.026\,804$	$h(13) = 0.293\,486$
$h(3) = -0.000\,078$	$h(9) = -0.005\,171$	$h(14) = 0.369\,851$
$h(4) = -0.000\,030$	$h(10) = 0.032\,177$	$h(15) = 0.415\,475$
$h(5) = -0.003\,864$		

	Synthesis filter	
$h(0) = 0.000\,329$	$h(6) = -0.011\,460$	$h(11) = 0.125\,759$
$h(1) = -0.000\,320$	$h(7) = -0.027\,239$	$h(12) = 0.255\,542$
$h(2) = -0.001\,866$	$h(8) = -0.036\,142$	$h(13) = 0.359\,832$
$h(3) = -0.000\,146$	$h(9) = -0.023\,722$	$h(14) = 0.408\,991$
$h(4) = -0.004\,627$	$h(10) = 0.019\,328$	$h(15) = 0.425\,663$
$h(5) = -0.004\,590$		

EXAMPLE 9.9

Design a length-32 linear-phase lapped biorthogonal transform with $M = 8$ sub-bands.

SOLUTION

The resulting analysis and synthesis filter impulse responses are shown in Figures 9.43 and 9.45, respectively. The coefficients of the first analysis and synthesis filters are listed in Table 9.5.

Figures 9.44 and 9.46 depict the normalized magnitude responses of the analysis and synthesis filters for the length-32 lapped biorthogonal transform filter bank with $M = 8$ sub-bands.

$\triangle$

9.10 Wavelet transforms

Wavelet transforms are a relatively recent development in functional analysis that have attracted a great deal of attention from the signal processing community (Daubechies, 1991). The wavelet transform of a function belonging to $\mathcal{L}^2\{\mathbb{R}\}$, the space of the square integrable functions, is its decomposition in a base formed by expansions, compressions, and translations of a single mother function $\psi(t)$, called a wavelet.

The applications of wavelet transforms range from quantum physics to signal coding. It can be shown that for digital signals the wavelet transform is a special case of critically decimated filter banks (Vetterli & Herley, 1992). In fact, its numerical

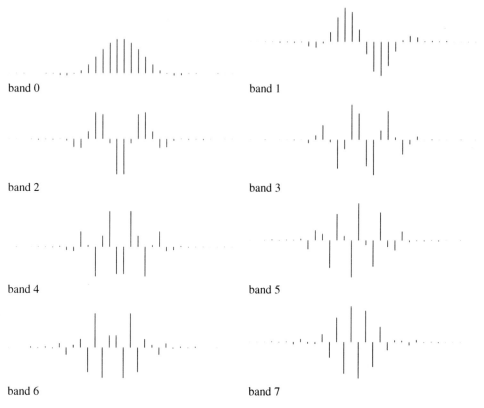

Figure 9.43 Impulse responses of the analysis filters of the 8-band lapped biorthogonal transform filter bank of length 32.

implementation relies heavily on that approach. In what follows, we give a brief introduction to wavelet transforms, emphasizing their relation to filter banks. In the literature there is plenty of good material analyzing wavelet transforms from different points of view. Examples are Daubechies (1991), Vetterli & Kovačević (1995), Strang & Nguyen (1996), and Mallat (1998).

9.10.1 Hierarchical filter banks

The cascading of 2-band filter banks can produce many different kinds of critically decimated decompositions. For example, one can make a 2^k-band uniform decomposition, as depicted in Figure 9.47a, for $k = 3$. Another common type of hierarchical decomposition is the octave-band decomposition, in which only the lowpass band is further decomposed. In Figure 9.47b, one can see a 3-stage octave-band decomposition. In these figures, the synthesis bank is not drawn, because it is entirely analogous to the analysis bank.

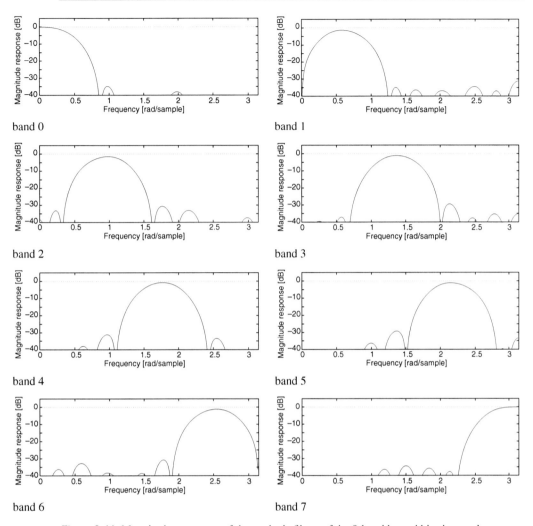

band 0   band 1

band 2   band 3

band 4   band 5

band 6   band 7

Figure 9.44 Magnitude responses of the analysis filters of the 8-band lapped biorthogonal transform filter bank of length 32.

9.10.2 Wavelets

Consider the octave-band analysis and synthesis filter bank in Figure 9.48, where the lowpass bands are recursively decomposed into lowpass and highpass channels. In this framework, the outputs of the lowpass channels after an $(S + 1)$-stage decomposition are $x_{S,n}$, and the outputs of the highpass channels are $c_{S,n}$, with $S \geq 1$.

Applying the noble identities to Figure 9.48, we arrive at Figure 9.49. After $(S + 1)$ stages, and before decimation by a factor of $2^{(S+1)}$, the z transforms of the analysis lowpass and highpass channels, $H_{\text{low}}^{(S)}(z)$ and $H_{\text{high}}^{(S)}(z)$, are

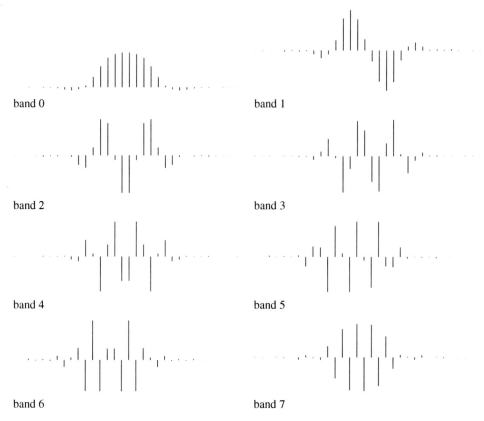

band 0   band 1

band 2   band 3

band 4   band 5

band 6   band 7

Figure 9.45 Impulse responses of the synthesis filters of the 8-band lapped biorthogonal transform filter bank of length 32.

$$H_{\text{low}}^{(S)}(z) = \frac{X_S(z)}{X(z)} = \prod_{k=0}^{S} H_0(z^{2^k}) \tag{9.137}$$

$$H_{\text{high}}^{(S)}(z) = \frac{C_S(z)}{X(z)} = H_1(z^{2^S})H_{\text{low}}^{(S-1)}(z) \tag{9.138}$$

For the synthesis channels, the results are analogous, that is

$$G_{\text{low}}^{(S)}(z) = \prod_{k=0}^{S} G_0(z^{2^k}) \tag{9.139}$$

$$G_{\text{high}}^{(S)}(z) = G_1(z^{2^S})G_{\text{low}}^{(S-1)}(z) \tag{9.140}$$

If $H_0(z)$ has enough zeros at $z = -1$, it can be shown (Vetterli & Kovačević, 1995; Strang & Nguyen, 1996; Mallat, 1998) that the envelope of the impulse response of the filters in equation (9.138) has the same shape for all S. In other words, this envelope

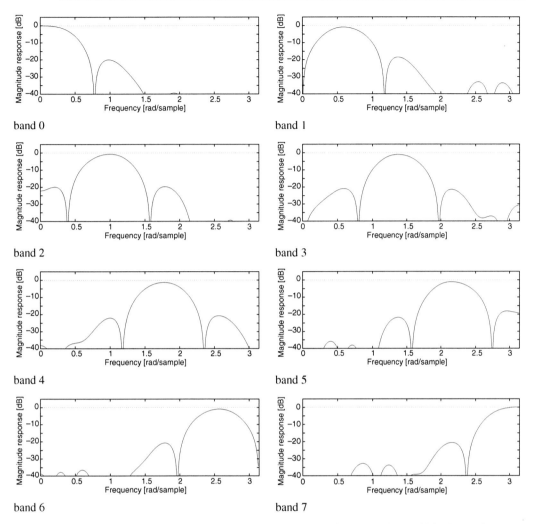

Figure 9.46 Magnitude responses of the synthesis filters of the 8-band lapped biorthogonal transform filter bank of length 32.

can be represented by expansions and contractions of a single function $\psi(t)$, as seen in Figure 9.50 for the analysis filter bank.

In fact, in this setup, the envelopes before and after the decimators are the same. However, it must be noted that, after decimation, we cannot refer to impulse responses in the usual way, because the decimation operation is not time invariant (see Section 8.3).

If Ω_s is the sampling rate at the input of the system in Figure 9.50, we have that this system has the same output as the one in Figure 9.51, where the boxes represent continuous-time filters with impulse responses equal to the envelopes of these signals in Figure 9.50. Note that in this case, sampling with frequency $\frac{\Omega_s}{k}$ is equivalent to decimating by k.

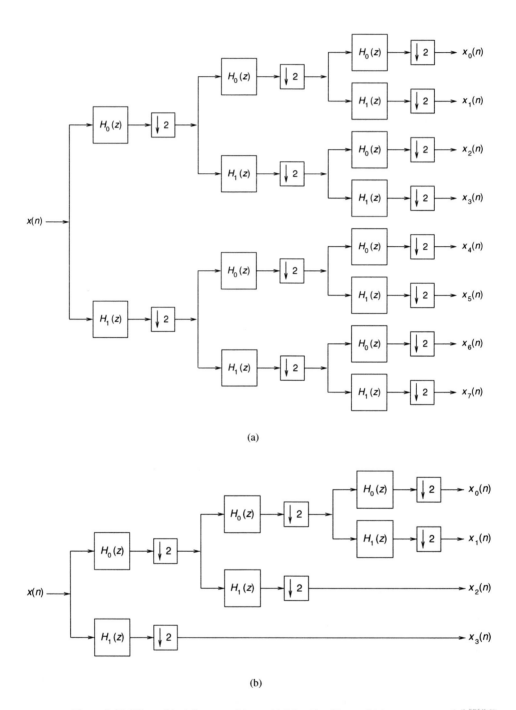

Figure 9.47 Hierarchical decompositions: (a) 8-band uniform; (b) 3-stage octave-band.

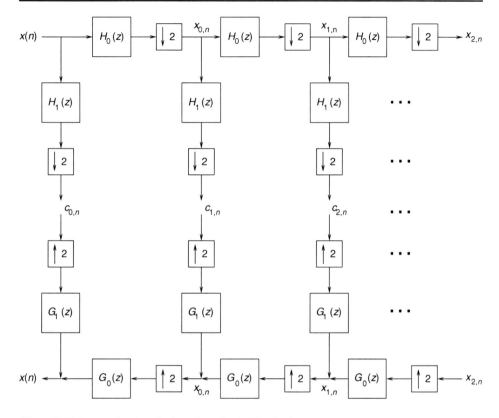

Figure 9.48 Octave-band analysis and synthesis filter bank.

We then observe that the impulse responses of the continuous-time filters of Figure 9.51 are expansions and contractions of a single mother function $\psi(t)$, with the highest sampling rate being $\frac{\Omega_s}{2}$, as stated above. Then, each channel added to the right has an impulse response with double the width, and a sampling rate half of the previous one. There is no impediment in adding channels also to the left of the channel with sampling frequency $\frac{\Omega_s}{2}$. In such cases, each new channel to the left has an impulse response with half the width, and a sampling rate twice the one to the right. If we keep adding channels to both the right and the left indefinitely, we arrive at Figure 9.52, where the input is $x(t)$, and the output is referred to as the wavelet transform of $x(t)$. The mother function $\psi(t)$ is called the wavelet, or, more specifically, the analysis wavelet (Daubechies, 1988; Mallat, 1998).

Assuming, without loss of generality, that $\Omega_s = 2\pi$ (that is, $T_s = 1$), it is straightforward to derive from Figure 9.52 that the wavelet transform of a signal $x(t)$ is given by[5]

---

[5] Actually, in this expression, the impulse response of the filters are expansions and contractions of $\psi(-t)$. In addition, the constant $2^{-\frac{m}{2}}$ is included because, if $\psi(t)$ has unit energy, which can be assumed without loss of generality, $2^{-\frac{m}{2}} \psi(2^{-m}t - n)$ will also have unit energy.

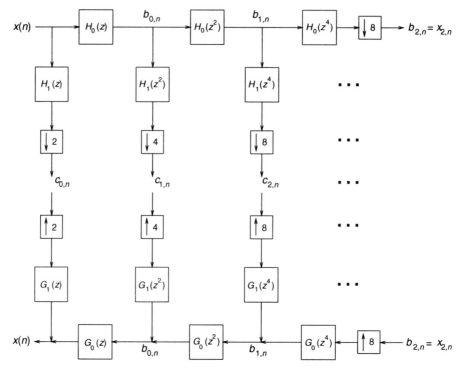

Figure 9.49 Octave-band analysis and synthesis filter bank after the application of the noble identities.

$$c_{m,n} = \int_{-\infty}^{\infty} 2^{-\frac{m}{2}} \psi(2^{-m}t - n)x(t)dt \qquad (9.141)$$

From Figures 9.49–9.52 and equation (9.138), one can see that the wavelet $\psi(t)$ is a bandpass function, because each channel is a cascade of several lowpass filters and a highpass filter with superimposing passbands. Also, when the wavelet is expanded by two, its bandwidth decreases by two, as seen in Figure 9.53. Therefore, the decomposition in Figure 9.52 and equation (9.141) is, in the frequency domain, as shown in Figure 9.54.

In a similar manner, the envelopes of the impulse responses of the equivalent synthesis filters after interpolation (see Figure 9.49 and equation (9.140)) are expansions and contractions of a single mother function $\overline{\psi}(t)$. Using similar reasoning to that leading to Figures 9.50–9.52, one can obtain the continuous-time signal $x(t)$ from the wavelet coefficients $c_{m,n}$ as (Vetterli & Kovačević, 1995; Mallat, 1998)

$$x(t) = \sum_{m=-\infty}^{\infty} \sum_{n=-\infty}^{\infty} c_{m,n} 2^{-\frac{m}{2}} \overline{\psi}(2^{-m}t - n) \qquad (9.142)$$

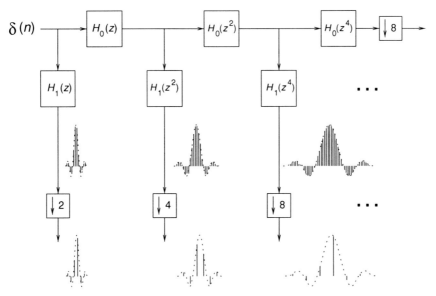

Figure 9.50 The impulse responses of the filters from equation (9.138) have the same shape for every stage.

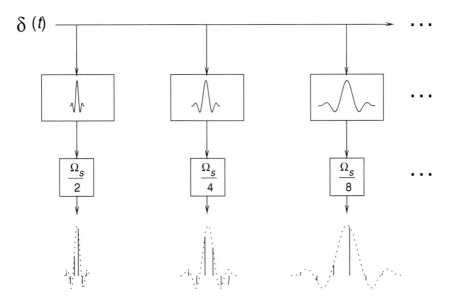

Figure 9.51 Equivalent system to the one in Figure 9.50.

Equations (9.141) and (9.142) represent the direct and inverse wavelet transforms of a continuous-time signal $x(t)$. The wavelet transform of the corresponding discrete-time signal $x(n)$ is merely the octave-band decomposition in Figures 9.48 and 9.49.

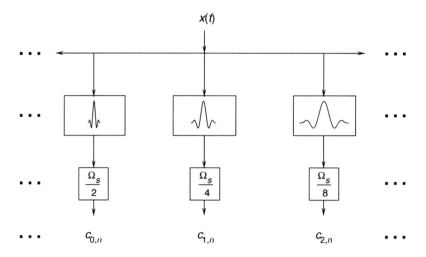

Figure 9.52 Wavelet transform of a continuous signal $x(t)$.

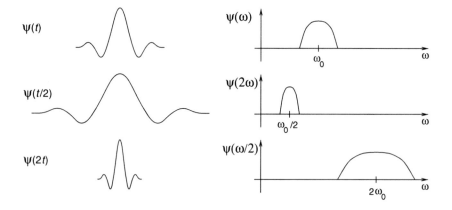

Figure 9.53 Expansions and contractions of the wavelet in the time and frequency domains.

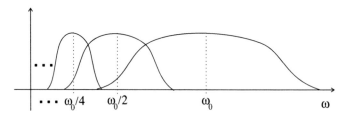

Figure 9.54 Wavelet transform in the frequency domain.

A natural question to ask at this point is: how are the continuous-time signal $x(t)$ and the discrete-time signal $x(n)$ related, if they generate the same wavelet coefficients? In addition, how can the analysis and synthesis wavelets be derived from the filter bank coefficients and vice versa? These questions can be answered using the concept of scaling functions, seen in the next subsection.

9.10.3 Scaling functions

By looking at Figure 9.52 and equations (9.141) and (9.142), we observe that the values of m which are associated with the 'width' of the filters range from $-\infty$ to $+\infty$. Since all signals encountered in practice are somehow band-limited, one can assume, without loss of generality, that the output of the filters with impulse responses $\psi(2^{-m}t)$ are zero, for $m < 0$. Therefore, in practice, m can then vary only from 0 to $+\infty$. Looking at Figures 9.48–9.50, we see that $m \rightarrow +\infty$ means that the lowpass channels will be indefinitely decomposed. However, in practice, the number of stages of decomposition is finite and, after S stages, we have S bandpass channels and one lowpass channel. Therefore, if we restrict the number of decomposing stages in Figures 9.48–9.52, and add a lowpass channel, we can modify equation (9.142) so that m assumes only a finite number of values. This can be done by noting that if $H_0(z)$ has enough zeros at $z = -1$, the envelopes of the analysis lowpass channels given in equation (9.137) will also be expansions and contractions of a single function $\phi(t)$, which is called the analysis scaling function. Likewise, the envelopes of the synthesis lowpass channels are expansions and contractions of the synthesis scaling function $\overline{\phi}(t)$ (Vetterli & Kovačević, 1995; Mallat, 1998). Therefore, if we make an $(S + 1)$-stage decomposition, equation (9.142) becomes

$$x(t) = \sum_{n=-\infty}^{\infty} x_{S,n} 2^{-\frac{S}{2}} \overline{\phi}(2^{-S}t - n) + \sum_{m=0}^{S-1} \sum_{n=-\infty}^{\infty} c_{m,n} 2^{-\frac{m}{2}} \overline{\psi}(2^{-m}t - n) \tag{9.143}$$

where

$$x_{S,n} = \int_{-\infty}^{\infty} 2^{-\frac{S}{2}} \phi(2^{-S}t - n)x(t)\mathrm{d}t \tag{9.144}$$

Hence, the wavelet transform is, in practice, described as in equations (9.141), (9.143), and (9.144). The summations in n will, in general, depend on the supports (that is, the regions where the functions are nonzero) of the signal, wavelets, and scaling functions (Daubechies, 1988; Mallat, 1998).

9.10.4 Relation between $x(t)$ and $x(n)$

Equation (9.144) shows how to compute the coefficients of the lowpass channel after an $(S + 1)$-stage wavelet transform. In Figure 9.48, $x_{S,n}$ are the outputs of a lowpass filter $H_0(z)$ after $(S + 1)$ stages. Since in Figure 9.49 the discrete-time signal $x(n)$ can be regarded as the output of a lowpass filter after 'zero' stages, we can say that

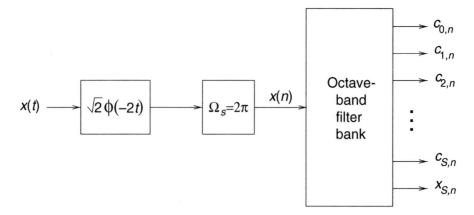

Figure 9.55 Practical way to compute the wavelet transform of a continuous-time signal.

$x(n)$ would be equal to $x_{-1,n}$. In other words, equivalence of the outputs of the octave-band filter bank of Figure 9.48 and the wavelet transform given by equations (9.141) and (9.142) occurs only if the digital signal input to the filter bank of Figure 9.48 is equal to $x_{-1,n}$. From equation (9.144), this means

$$x(n) = \int_{-\infty}^{\infty} \sqrt{2}\phi(2t - n)x(t)\mathrm{d}t \tag{9.145}$$

Such an equation can be interpreted as $x(n)$ being the signal $x(t)$ digitized with a band-limiting filter having as its impulse response $\sqrt{2}\phi(-2t)$. Therefore, a possible way to compute the wavelet transform of a continuous-time signal $x(t)$ is depicted in Figure 9.55, in which $x(t)$ is passed through a filter having as impulse response the scaling function contracted-in-time by 2 and sampled with $T_s = 1$ ($\Omega_s = 2\pi$). The resulting digital signal is then the input to the octave-band filter bank in Figure 9.48 whose filter coefficients will be given in Subsection 9.10.5 by equations (9.146) and (9.148). At this point, it is important to note that, strictly speaking, the wavelet transform is only defined for continuous-time signals. However, it is common practice to refer to the wavelet transform of a discrete-time signal $x(n)$ as the output of the filter bank in Figure 9.48 (Vetterli & Kovačević, 1995). One should also note that, in order for the output signals in Figures 9.50 and 9.51 to be entirely equivalent, the input signal in Figure 9.50 must not be the discrete-time impulse, but the sampled impulse response of the filter $\sqrt{2}\phi(-2t)$. This is nothing less than a sampled version of the scaling function contracted-in-time by 2.

9.10.5 Relation between the wavelets and the filter coefficients

If $h_0(n)$, $h_1(n)$, $g_0(n)$, and $g_1(n)$ are the impulse responses of the analysis lowpass and highpass filters and the synthesis lowpass and highpass filters, respectively, and $\phi(t)$, $\overline{\phi}(t)$, $\psi(t)$, and $\overline{\psi}(t)$ are the analysis and synthesis scaling functions and analysis and synthesis wavelets, respectively, we have that (Vetterli & Herley, 1992)

$$h_0(n) = \int_{-\infty}^{\infty} \phi(t)\sqrt{2}\overline{\phi}(2t+n)dt \tag{9.146}$$

$$g_0(n) = \int_{-\infty}^{\infty} \overline{\phi}(t)\sqrt{2}\phi(2t-n)dt \tag{9.147}$$

$$h_1(n) = \int_{-\infty}^{\infty} \psi(t)\sqrt{2}\overline{\phi}(2t+n)dt \tag{9.148}$$

$$g_1(n) = \int_{-\infty}^{\infty} \overline{\psi}(t)\sqrt{2}\phi(2t-n)dt \tag{9.149}$$

The Fourier transforms of the wavelets, scaling functions, and filters are related by (Vetterli & Kovačević, 1995; Mallat, 1998)

$$\Phi(\Omega) = \prod_{n=1}^{\infty} \frac{1}{\sqrt{2}}H_0(e^{-j\frac{w}{2^n}}) \tag{9.150}$$

$$\overline{\Phi}(\Omega) = \prod_{n=1}^{\infty} \frac{1}{\sqrt{2}}G_0(e^{j\frac{w}{2^n}}) \tag{9.151}$$

$$\Psi(\Omega) = \frac{1}{\sqrt{2}}H_1(e^{-j\frac{w}{2}}) \prod_{n=2}^{\infty} \frac{1}{\sqrt{2}}H_0(e^{-j\frac{w}{2^n}}) \tag{9.152}$$

$$\overline{\Psi}(\Omega) = \frac{1}{\sqrt{2}}G_1(e^{j\frac{w}{2}}) \prod_{n=2}^{\infty} \frac{1}{\sqrt{2}}G_0(e^{j\frac{w}{2^n}}) \tag{9.153}$$

When $\phi(t) = \overline{\phi}(t)$ and $\psi(t) = \overline{\psi}(t)$, the wavelet transform is orthogonal (Daubechies, 1988). Otherwise, it is only biorthogonal (Cohen et al., 1992). It is important to notice that, for the wavelet transform to be defined, the corresponding filter bank must provide perfect reconstruction.

9.10.6 Regularity

From equations (9.150)–(9.153), one can see that the wavelets and scaling functions are derived from the filter bank coefficients by infinite products. Therefore, in order for a wavelet to be defined, these infinite products must converge. In other words, a wavelet transform is not necessarily defined for every 2-band perfect reconstruction filter bank. In fact, there are cases in which the envelope of the impulse responses of the equivalent filters of equations (9.137)–(9.140) is not the same for every S (Vetterli & Kovačević, 1995; Mallat, 1998).

The regularity of a wavelet or scaling function is roughly speaking the number of continuous derivatives that a wavelet has. It gives a measure of the extent of convergence of the products in equations (9.150)–(9.153). In order to define regularity more formally, we first define the following concept (Rioul, 1992; Mallat, 1998).

DEFINITION 9.1

A function $f(t)$ is Lipschitz continuous of order α, with $0 < \alpha \leq 1$, if $\forall x, h \in \mathbb{R}$, we have

$$|f(x + h) - f(x)| \leq ch^{\alpha} \tag{9.154}$$

where c is a constant.

Using this definition, we have the regularity concept.

DEFINITION 9.2

The Hölder regularity of a scaling function $\phi(t)$, such that $\frac{d^N \phi(t)}{dt^N}$ is Lipschitz continuous of order α, is $r = (N + \alpha)$, where N is integer and $0 < \alpha \leq 1$ (Rioul, 1992; Mallat, 1998).

It can be shown that, in order for a scaling function $\phi(t)$ to be regular, $H_0(z)$ must have enough zeros at $z = -1$. In addition, supposing that $\phi(t)$ generated by $H_0(z)$ (equation (9.150)) has regularity r, if we take

$$H_0'(z) = \left(\frac{1 + z^{-1}}{2} \right) H_0(z) \tag{9.155}$$

then $\phi'(t)$ generated by $H_0'(z)$ will have regularity $(r + 1)$ (Rioul, 1992; Mallat, 1998).

The regularity of a wavelet is the same as the regularity of the corresponding scaling function (Rioul, 1992). It can be shown that the regularity of a wavelet imposes the following a priori constraints on the filter banks (Vetterli & Kovačević, 1995)

$$H_0(1) = G_0(1) = \sqrt{2} \tag{9.156}$$
$$H_0(-1) = G_0(-1) = 0 \tag{9.157}$$

Equation (9.156) implies that the filters $H_0(z)$, $H_1(z)$, $G_0(z)$, and $G_1(z)$ have to be normalized in order to generate a wavelet transform.

It is interesting to note that when deriving the wavelet transform from the octave-band filter bank in Subsection 9.10.2, it was supposed that the lowpass filters had enough zeros at $z = -1$. In fact, what was meant there was that the wavelets should be regular.

In Figure 9.56, we see examples of wavelets with different regularities. For example, Figure 9.56a corresponds to the analysis wavelet generated by the filter bank described by equations (9.45)–(9.48), and Figure 9.56b corresponds to the analysis wavelet generated by the filter bank described by equations (9.49)–(9.52).

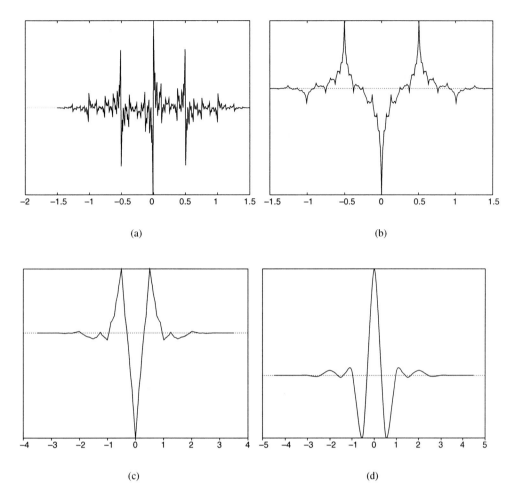

Figure 9.56 Examples of wavelets with different regularities. (a) regularity $= -1$; (b) regularity $=$ 0; (c) regularity $= 1$; (d) regularity $= 2$.

9.10.7 Examples

Every 2-band perfect reconstruction filter bank with $H_0(z)$ having enough zeros at $z = -1$, has corresponding analysis and synthesis wavelets and scaling functions. For example, the filter bank described by equations (9.17)–(9.20), normalized such that equation (9.156) is satisfied, generates the so-called Haar wavelet. It is the only orthogonal wavelet that has linear phase (Vetterli & Kovačević, 1995; Mallat, 1998). The scaling function and wavelets are shown in Figure 9.57.

The wavelets and scaling functions corresponding to the symmetric short-kernel filter bank, described by equations (9.45)–(9.48) are depicted in Figure 9.58.

A good example of an orthogonal wavelet is the Daubechies wavelet, whose filters

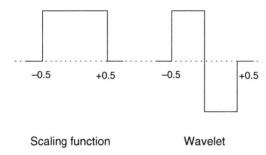

Scaling function   Wavelet

Figure 9.57 Haar wavelet and scaling function.

have length 4. They are also an example of CQF banks, seen in Section 9.6. The filters are (Daubechies, 1988)

$$H_0(z) = +0.482\,9629 + 0.836\,5163z^{-1} + 0.224\,1439z^{-2} - 0.129\,4095z^{-3} \quad (9.158)$$
$$H_1(z) = -0.129\,4095 - 0.224\,1439z^{-1} + 0.836\,5163z^{-2} - 0.482\,9629z^{-3} \quad (9.159)$$
$$G_0(z) = -0.129\,4095 + 0.224\,1439z^{-1} + 0.836\,5163z^{-2} + 0.482\,9629z^{-3} \quad (9.160)$$
$$G_1(z) = -0.482\,9629 + 0.836\,5163z^{-1} - 0.224\,1439z^{-2} - 0.129\,4095z^{-3} \quad (9.161)$$

Since the wavelet transform is orthogonal, the analysis and synthesis scaling functions and wavelets are the same. These are depicted in Figure 9.59.

It is important to notice that, unlike the biorthogonal wavelets in Figure 9.58, these orthogonal wavelets are nonsymmetric, and, therefore, do not have linear phase.

When implementing a wavelet transform using the scheme in Figure 9.48, it is essential that the delay introduced by each analysis/synthesis stage is compensated. Failure to do so may result in the loss of the perfect reconstruction property.

9.11 Filter banks and wavelets with MATLAB

The functions described below are from the MATLAB Wavelet toolbox. This toolbox includes many pre-defined wavelets, divided into families. For example, among others we have the Daubechies, the biorthogonal, the Coiflet and the Symmlet families (Mallat, 1998). Most functions require the desired wavelet in a family to be specified. The functions which involve wavelet transform computation also permit the specification of the filter bank coefficients.

- dyaddown: Dyadic downsampling.
  Input parameters:

  - The input vector or matrix x;
  - The integer evenodd; if evenodd is even, the output corresponds to the even samples of the input; if evenodd is odd, the output corresponds to the odd samples of the input;

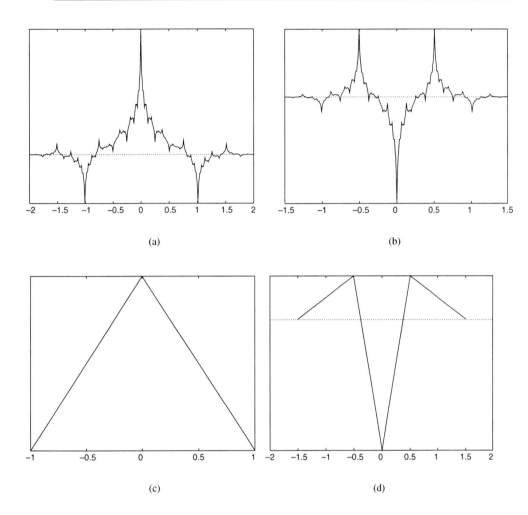

Figure 9.58 The 'symmetric short-kernel' wavelet transform (equations (9.45)–(9.48)): (a) analysis scaling function; (b) analysis wavelet; (c) synthesis scaling function; (d) synthesis wavelet.

– The parameter `'type'`, when x is a matrix; if `'type'='c'`, only the rows are downsampled; if `'type'='r'`, only the columns are downsampled; if `'type'='m'`, both rows and columns are downsampled;

Output parameter: The vector or matrix y containing the downsampled signal.
Example:
```
x=[1.0 2.7 2.4 5.0]; evenodd=1;
y=dyaddown(x,evenodd);
```

• `dyadup`: Dyadic upsampling.
For a complete list of input and output parameters, see command `dyadown`.
Example:

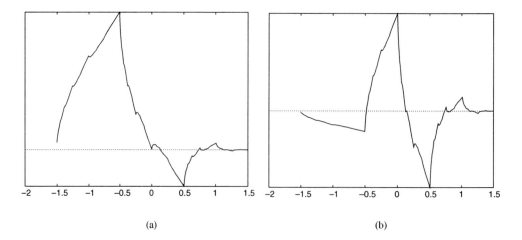

(a)   (b)

Figure 9.59 Daubechies wavelet transform (equations (9.158)–(9.161)): (a) scaling function;
(b) wavelet.

```
x=[0.97 -2.1 ; 3.7 -0.03]; evenodd=0;
y=dyadup(x,'m',evenodd);
```

- qmf: Generates a quadrature mirror filter of the input.
  Input parameters:

  - The input vector or matrix x;
  - The integer p. If p is even, the output corresponds to x in reversed order with
    the signs changed for the even index entries. If p is odd, the output corresponds
    to x in reversed order, with the signs of the odd index entries changed.

  Output parameter: The vector y containing the output filter coefficients.
  Example:
  ```
  y=qmf(x,0);
  ```

- waveinfo: Gives information on wavelet families.
  Input parameter: The name wfname of the wavelet family. Use the command
  waveinfo with no parameters to see a list of names and a brief description of
  the wavelet families.
  Example:
  ```
  wfname='coif'; waveinfo('wfname');
  ```

- wfilters: Computes the coefficients of the analysis and synthesis filters given a
  wavelet transform.
  Input parameter: The name wname of the wavelet transform. Available wavelet
  names are:

– Daubechies:	`'db1'` or `'haar'`, `'db2'`, ... ,`'db50'`
– Coiflet:	`'coif1'`, ... ,`'coif5'`
– Symmlet:	`'sym2'`, ... ,`'sym8'`
– Biorthogonal:	`'bior1.1'`, `'bior1.3'`, `'bior1.5'`,
	`'bior2.2'`, `'bior2.4'`, `'bior2.6'`, `'bior2.8'`,
	`'bior3.1'`, `'bior3.3'`, `'bior3.5'`, `'bior3.7'`,
	`'bior3.9'`, `'bior4.4'`, `'bior5.5'`, `'bior6.8'`.

Output parameters:

- A vector `Lo_D` containing the decomposition lowpass filter coefficients;
- A vector `Hi_D` containing the decomposition highpass filter coefficients;
- A vector `Lo_R` containing the reconstruction lowpass filter coefficients;
- A vector `Hi_R` containing the reconstruction highpass filter coefficients.

Example:

```
wname='db5';
[Lo_D,Hi_D,Lo_R,Hi_R]=wfilters(wname);
```

- `dbwavf`: Computes the coefficients of the two-scale difference equation (Exercise 9.23) given a wavelet transform from the Daubechies family.
  Input parameter: The name `wname` of the wavelet transform from the Daubechies family. See description of the function `wfilters` for a list of available names.
  Output parameter: A vector `F` containing the coefficients of the two-scale difference equation.
  Example:

```
wname='db5';
F=dbwavf(wname);
```

- The functions `coifwavf` and `symwavf` are the equivalent to the function `dbwavf` for the Coiflet and Symmlet families. Please refer to the MATLAB Wavelet toolbox documentation for details.

- `orthfilt`: Computes the coefficients of the analysis and synthesis filters given the coefficients of the two-scale difference equation of an orthogonal wavelet (see Exercise 9.23).
  Input parameter: A vector `W` containing the coefficients of the two-scale difference equation.
  Output parameters: See command `wfilters`.
  Example:

```
wname='coif4'; W=coifwavf(wname);
[Lo_D,Hi_D,Lo_R,Hi_R]=orthfilt(F);
```

- `biorwavf`: Computes the coefficients of the analysis and synthesis two-scale difference equations (see Exercise 9.23) given a wavelet transform from the

Biorthogonal family.

Input parameter: The name **wname** of the wavelet transform from the biorthogonal family. See description of the function **wfilters** for a list of available names.

Output parameters:

– A vector **RF** containing the coefficients of the two-scale synthesis difference equation;

– A vector **DF** containing the coefficients of the two-scale analysis difference equation.

Example:
```
wname='bior2.2';
[RF,DF]=biorwavf(wname);
```

• **biorfilt**: Computes the coefficients of the analysis and synthesis filters given the coefficients of the analysis and synthesis two-scale difference equations of a biorthogonal wavelet (see Exercise 9.23).

Input parameters:

– A vector **DF** containing the coefficients of the two-scale analysis difference equation.

– A vector **RF** containing the coefficients of the two-scale synthesis difference equation.

Output parameters: See command **wfilters**.

Example:
```
wname='bior3.5'; [RF,DF]=biorwavf(wname);
[Lo_D,Hi_D,Lo_R,Hi_R]=biorfilt(DF,RF);
```

• **dwt**: Single-stage discrete one-dimensional wavelet decomposition.

Input parameters:

– A vector **x** containing the input signal;

– A vector **Lo_D** containing the analysis lowpass filter;

– A vector **Hi_D** containing the analysis highpass filter;

– There is the possibility of providing, instead of **Lo_D** and **Hi_D**, **wname**, the name of the wavelet transform. See description of the function **wfilters** for a list of available names.

Output parameters:

– A vector **cA**, containing the approximation coefficients (lowpass band);

– A vector **cD**, containing the detail coefficients (highpass band).

Example:
```
Lo_D=[-0.0625 0.0625 0.5 0.5 0.0625 -0.0625];
Hi_D=[-0.5 0.5];
load leleccum; x=leleccum;
[cA,cD]=dwt(x,Lo_D,Hi_D);
```

- `idwt`: Single-stage discrete one-dimensional wavelet reconstruction.
  Input parameters:

  - A vector `cA` containing the approximation coefficients (lowpass band);
  - A vector `cD` containing the detail coefficients (highpass band);
  - A vector `Lo_D` containing the analysis lowpass filter;
  - A vector `Hi_D` containing the analysis highpass filter;
  - There is the possibility of providing, instead of `Lo_D` and `Hi_D`, wname, the name of the wavelet transform. See description of the function `wfilters` for a list of available names.

  Output parameter: A vector `x` containing the output signal.
  Example:
  ```
  x=idwt(cA,cD,'bior2.2');
  ```

- `dwtmode`: Sets the type of signal extension at the signal boundaries for wavelet transform calculations.
  Input parameters:

  - `'zpd'` sets the extension mode to zero-padding (default mode);
  - `'sym'` sets the extension mode to symmetric (boundary value replication);
  - `'spd'` sets the extension mode to smooth padding (first derivative interpolation at the edges).

  Example:
  ```
  dwtmode('sym'); load leleccum; x=leleccum;
  [cA,cD]=dwt(x,'bior1.3');
  ```

- `dwtper`, `idwtper`: Similar to functions `dwt` and `idwt`, but assuming periodic signal extension at the boundaries.

- `wavedec`: Performs multiple-stage one-dimensional wavelet decomposition.
  Input parameters:

  - A vector `x` containing the input signal;
  - The number of stages `n`;
  - A vector `Lo_D` containing the analysis lowpass filter;
  - A vector `Hi_D` containing the analysis highpass filter;

- There is the possibility of providing, instead of Lo_D and Hi_D, wname, the name of the wavelet transform. See the description of the function **wfilters** for a list of available names.

Output parameters:

- A vector c containing the full wavelet decomposition;
- A vector l containing the number of elements of each band in the vector c.

Example:
```
load sumsin; x=sumsin; n=3;
[c,l] = wavedec(x,n,'dB1');
indx1=1+l(1)+l(2); indx2=indx1+l(3)-1;
plot(c(indx1:indx2)); %Plots 2nd stage coefficients
plot(c(1:l(1))); %Plots lowpass coefficients
```

- **waverec**: Performs multiple-stage one-dimensional wavelet reconstruction.
Input parameters:

- A vector c containing the full wavelet decomposition;
- A vector l containing the number of elements of each band in c;
- A vector Lo_D containing the analysis lowpass filter;
- A vector Hi_D containing the analysis highpass filter;
- There is the possibility of providing, instead of Lo_D and Hi_D, wname, the name of the wavelet transform. See the description of the function **wfilters** for a list of available names.

Output parameter: A vector x containing the output signal.
Example:
```
x = waverec(c,l,'dB1');
```

- **upwlev**: Performs single-stage reconstruction of a multiple-stage one-dimensional wavelet decomposition.
Input parameters: See function **waverec**.
Output parameters:

- The vector nc containing the full wavelet decomposition of the output coefficients obtained after 1 stage of wavelet reconstruction;
- The vector nl containing the number of elements of each band in the vector nc.

Example:
```
[nc,nl] = upwler(c,l,'dB1');
```

- The interested reader may also refer to the MATLAB documentation commands `upcoef` and `wrcoef`, as well as to the commands `appcoef2`, `detcoef2`, `dwt2`, `dwtper2`, `idwt2`, `idwtper2`, `upcoef2`, `upwlev2`, `wavedec2`, `waverec2`, and `wrcoef2`, which apply to two-dimensional signals.

- `wavefun`: Generates approximations of wavelets and scaling functions given a wavelet transform.

  Input parameters:

  – The name `wname` of the wavelet transform. See description of the function `wfilters` for a list of available names;
  – The number of iterations `iter` used in the approximation.

  Output parameters:

  – A vector `phi1` containing the analysis scaling function;
  – A vector `psi1` containing the analysis wavelet;
  – A vector `phi2` containing the synthesis scaling function;
  – A vector `psi2` containing the synthesis wavelet
     (in the case of orthogonal wavelets, only the analysis scaling functions and wavelets are output);
  – A grid `xval` containing 2^{iter} points.

  Example:
  ```
  iter=7; wname='dB5';
  [phi,psi,xval]=wavefun(wname,iter);
  plot(xval,psi);
  ```

- The MATLAB Wavelet toolbox provides the command `wavemenu` that integrates most of the commands above into a graphics user interface.

- The command `wavedemo` gives a tour through most of the commands. For further information, type '`help` *command*', or refer to the Wavelet toolbox reference manual.

9.12 Summary

In this chapter, we have discussed the concept of filter banks. They are used in several digital signal processing applications, for instance, in signal analysis, coding, and transmission. Several examples of filter banks were considered including the QMF, CQF, cosine-modulated, and biorthogonal perfect reconstruction filter banks.

Several tools for signal analysis were discussed in the framework of filter banks with special emphasis given to block transforms, lapped orthogonal transforms and wavelet transforms. Also, functions from the MATLAB Wavelet toolbox were described.

This chapter was not intended to give an extensive coverage of such a broad topic, but it is expected that its study will equip the reader to follow the more advanced literature in the subject without difficulty.

9.13 Exercises

9.1 Deduce the two noble identities (equations (9.1) and (9.2)) using an argument in the time domain.

9.2 Prove equations (9.41)–(9.43).

9.3 Prove that for a linear-phase filter whose impulse response has length ML, its M polyphase components (equation (9.6)) should satisfy

$$E_j(z) = \pm z^{-(L-1)} E_{M-1-j}(z^{-1})$$

9.4 Show that a perfect reconstruction linear-phase filter bank with causal filters must be such that $H_0(z)H_1(-z)$ has an odd number of coefficients and that all but one of its odd powers of z must be zero.

9.5 Prove, using an argument in the frequency domain, that the system depicted in Figure 9.14 has a transfer function equal to z^{-M+1}.

9.6 Find the matrices $\mathbf{E}(z)$ and $\mathbf{R}(z)$ for the filter bank described by equations (9.45)–(9.48), and verify that their product is equal to a pure delay.

9.7 Modify the causal analysis and synthesis filters of the filter bank in equations (9.45)–(9.48) so that they constitute a zero-delay perfect reconstruction filter bank. Do the same for the filter banks in equations (9.49)–(9.52). Suggest a general method to transform any perfect reconstruction causal filter bank into a zero-delay filter bank.

9.8 Consider a 2-band linear-phase filter bank whose product filter $P(z) = H_0(z)H_1(-z)$ of order $(4M - 2)$ can be expressed as

$$P(z) = z^{-2M+1} + \sum_{k=0}^{M-1} a_{2k}\left(z^{-2k} + z^{-4M+2+2k}\right)$$

Show that such a $P(z)$:

(a) Can have at most $2M$ zeros at $z = -1$.

(b) Has exactly $2M$ zeros at $z = -1$ provided that its coefficients satisfy the following set of M equations:

$$\sum_{k=0}^{M-1} a_{2k}(2M - 1 - 2k)^{2n} = \frac{1}{2}\delta(n), \quad n = 0, \ldots, M - 1$$

Use the fact that a polynomial root has multiplicity p if it is also the root of the first $p - 1$ derivatives of the polynomial.

9.9   Using the relation deduced in Exercise 9.8, design a 2-band linear-phase perfect reconstruction filter bank such that the lowpass analysis filter has z transform equal to $(1 + z^{-1})^5$ and the highpass analysis filter has order 5 with one zero at $z = 1$. Plot the frequency responses of the filters and comment on how 'good' this filter bank is.

9.10   Design a Johnston 2-band filter bank with 64 coefficients.

9.11   Design a Johnston 2-band filter bank with at least 60 dB of stopband attenuation.

9.12   Design a CQF bank with at least 55 dB of stopband attenuation.

9.13   Design a filter bank using the cosine-modulated filter bank with $n = 5$ sub-bands and at least 40 dB of stopband attenuation.

9.14   Design a filter bank using the cosine-modulated filter bank with $n = 15$ sub-bands with at least 20 dB of stopband attenuation.

9.15   Express the DFT as an M-band filter bank. Plot the magnitude responses of the analysis filters for $M=8$.

9.16   Design a fast LOT-based filter bank with at least 8 sub-bands.

9.17   Prove the relationship in equation (9.94).

9.18   Show that the relations in equations (9.129)–(9.131) are valid.

9.19   Show that if a filter bank has linear-phase analysis and synthesis filters with the same lengths $N = LM$, then the following relations for the polyphase matrices are valid

$$\mathbf{E}(z) = z^{-L+1}\mathbf{D}\mathbf{E}(z^{-1})\mathbf{J}$$
$$\mathbf{R}(z) = z^{-L+1}\mathbf{J}\mathbf{R}(z^{-1})\mathbf{D}$$

where $\mathbf{D}$ is a diagonal matrix whose entries are 1, if the corresponding filter is symmetric, and -1 otherwise.

9.20   Consider an M-band linear-phase filter bank with perfect reconstruction with all analysis and synthesis filters having the same length $N = LM$. Show that for a polyphase $\mathbf{E}_1(z)$ corresponding to a linear-phase filter bank, then $\mathbf{E}(z) = \mathbf{E}_2(z)\mathbf{E}_1(z)$ will also lead to a linear-phase perfect reconstruction filter bank, if

$$\mathbf{E}_2(z) = z^{-L}\mathbf{D}\mathbf{E}_2(z^{-1})\mathbf{D}$$

where $\mathbf{D}$ is a diagonal matrix with entries 1 or -1, as described in Exercise 9.19, and L is the order of $\mathbf{E}_2(z)$. Also show that

$$\mathbf{E}_{2,i} = \mathbf{D}\mathbf{E}_{2,L-i}\mathbf{D}$$

9.21   Deduce equations (9.137)–(9.140).

9.22   If one wants to compute the wavelet transform of a continuous signal, one has to assume that its digital representation is derived as in equation (9.145) and Figure 9.55. However, the digital representation of a signal is usually obtained by filtering it with a band-limiting filter and sampling it at or above the Nyquist frequency. Supposing that the continuous signal, in order to be digitized, is filtered with a filter arbitrarily close to the ideal one, find an expression for the mean-square error made in the computation of the wavelet transform. Discuss the implications of this result in the accuracy of signal processing operations carried out in the wavelet transform domain.

9.23　Wavelets and scaling functions are known to satisfy the so-called two-scale difference equations:

$$\phi(t) = \sum_{k=-\infty}^{\infty} c_k \sqrt{2}\phi(2t - k)$$

$$\psi(t) = \sum_{k=-\infty}^{\infty} d_k \sqrt{2}\phi(2t - k)$$

$$\overline{\phi}(t) = \sum_{k=-\infty}^{\infty} \overline{c}_k \sqrt{2}\overline{\phi}(2t - k)$$

$$\overline{\psi}(t) = \sum_{k=-\infty}^{\infty} \overline{d}_k \sqrt{2}\overline{\phi}(2t - k)$$

These equations mean that the wavelet or scaling functions at a larger scale can be represented as a linear combination of shifts of the scaling function at a smaller scale. Show that equations (9.150)–(9.153) can be derived from the above equations, and determine the coefficients of the analysis and synthesis filter banks in terms of c_k, $\overline{c}_k$, d_k, and $\overline{d}_k$.

9.24　Given the expressions in Exercise 9.23 and the fact that the set

$$\left\{ 2^{-\frac{m}{2}}\phi(2^{-m}t - n), 2^{-\frac{m}{2}}\psi(2^{-m}t - n) \right\}$$

is orthogonal to the set

$$\left\{ 2^{-\frac{m}{2}}\overline{\phi}(2^{-m}t - n), 2^{-\frac{m}{2}}\overline{\psi}(2^{-m}t - n) \right\}$$

derive equations (9.146)–(9.149).

9.25　Show that the biorthogonality between the analysis and synthesis wavelets and scaling functions is equivalent to the perfect reconstruction conditions (equations (9.41)–(9.44)) of the 2-band filter bank described by equations (9.146)–(9.149). Use the two-scale difference equations introduced in Exercise 9.23.

9.26　Show that, in order for the analysis and synthesis wavelets to be regular, it is necessary that

$$H_0(1) = G_0(1) = \sqrt{2}$$
$$H_0(-1) = G_0(-1) = 0$$

using the fact that regularity demands convergence of the infinite products in equations (9.150)–(9.153).

10 Efficient FIR structures

10.1 Introduction

In this chapter, alternative realizations to those introduced in Chapter 5 for FIR filters are discussed.

We first present the lattice realization, highlighting its application to the design of linear-phase perfect reconstruction filter banks. Then, the polyphase structure is revisited, discussing its application in parallel processing. We also present an FFT-based realization for implementing the FIR filtering operation in the frequency domain. Such a form can be very efficient in terms of computational complexity, and is particularly suitable for offline processing, although widely used in real-time implementations. Next, the so-called recursive running sum is described as a special recursive structure for a very particular FIR filter, which has applications in the design of FIR filters with low arithmetic complexity.

In the case of FIR filters, the main concern is to examine methods which aim to reduce the number of arithmetic operations. These methods lead to more economical realizations with reduced quantization effects. In this chapter, we also present the prefilter, the interpolation, and the frequency masking approaches for designing lowpass and highpass FIR filters with reduced arithmetic complexity. The frequency masking method can be seen as a generalization of the other two schemes, allowing the design of passbands with general widths. For bandpass and bandstop filters, the quadrature approach is also introduced.

10.2 Lattice form

Figure 10.1 depicts the block diagram of a nonrecursive lattice filter of order M, which is formed by concatenating basic blocks of the form shown in Figure 10.2.

To obtain a useful relation between the lattice parameters and the filter impulse response, we must analyze the recurrent relationships that appear in Figure 10.2. These equations are

$$e_i(n) = e_{i-1}(n) + k_i \tilde{e}_{i-1}(n-1) \tag{10.1}$$

$$\tilde{e}_i(n) = \tilde{e}_{i-1}(n-1) + k_i e_{i-1}(n) \tag{10.2}$$

for $i = 1, \ldots, M$, with $e_0(n) = \tilde{e}_0(n) = k_0 x(n)$ and $e_M(n) = y(n)$. In the frequency

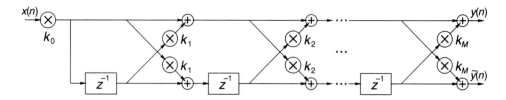

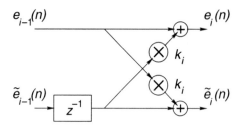

Figure 10.1 Lattice form realization of nonrecursive digital filters.

Figure 10.2 Basic block of the nonrecursive lattice form.

domain, equations (10.1) and (10.2) become

$$
\begin{bmatrix} E_i(z) \\ \tilde{E}_i(z) \end{bmatrix} = \begin{bmatrix} 1 & k_i z^{-1} \\ k_i & z^{-1} \end{bmatrix} \begin{bmatrix} E_{i-1}(z) \\ \tilde{E}_{i-1}(z) \end{bmatrix}
\tag{10.3}
$$

with $E_0(z) = \tilde{E}_0(z) = k_0 X(z)$ and $E_M(n) = Y(z)$.

Defining the auxiliary polynomials

$$
N_i(z) = k_0 \frac{E_i(z)}{E_0(z)}
\tag{10.4}
$$

$$
\tilde{N}_i(z) = k_0 \frac{\tilde{E}_i(z)}{\tilde{E}_0(z)}
\tag{10.5}
$$

one can demonstrate by induction, using equation (10.3), that these polynomials obey the following recurrence formulas:

$$
N_i(z) = N_{i-1}(z) + k_i z^{-i} N_{i-1}(z^{-1})
\tag{10.6}
$$

$$
\tilde{N}_i(z) = z^{-i} N_i(z^{-1})
\tag{10.7}
$$

for $i = 1, \ldots, M$, with $N_0(z) = \tilde{N}_0(z) = k_0$ and $N_M(z) = H(z)$. Therefore, from equations (10.3) and (10.7), we have that

$$
N_{i-1}(z) = \frac{1}{1 - k_i^2} \left(N_i(z) - k_i z^{-i} N_i(z^{-1}) \right)
\tag{10.8}
$$

for $i = 1, \ldots, M$.

Note that the recursion in equation (10.3) implies that $N_i(z)$ has degree i. Therefore, $N_i(z)$ can be expressed as

$$N_i(z) = \sum_{m=0}^{i} h_{m,i} z^{-m} \tag{10.9}$$

Since $N_{i-1}(z)$ has degree $(i-1)$ and $N_i(z)$ has degree i, in equation (10.8) the highest degree coefficient of $N_i(z)$ must be canceled by the highest degree coefficient of $k_i z^{-i} N_i(z^{-1})$. This implies that

$$h_{i,i} - k_i h_{i,0} = 0 \tag{10.10}$$

Since, from equation (10.9), $h_{i,0} = N_i(\infty)$, and, from equation (10.3), $N_i(\infty) = N_{i-1}(\infty) = \cdots = N_0(\infty) = 1$, we have that equation (10.10) is equivalent to $h_{i,i} = k_i$. Therefore, given the filter impulse response, the k_i coefficients are determined by successively computing the polynomials $N_{i-1}(z)$ from $N_i(z)$ using equation (10.8) and making

$$k_i = h_{i,i}, \text{ for } i = M, \ldots, 0 \tag{10.11}$$

To determine the filter impulse response $N_M(z) = H(z) = \sum_{i=0}^{M} h_i z^{-i}$ from the set of lattice coefficients k_i, we must first use equation (10.6) to compute the auxiliary polynomials $N_i(z)$, starting with $N_0(z) = k_0$. The desired coefficients are then given by

$$h_i = h_{i,M}, \text{ for } i = 0, \ldots, M \tag{10.12}$$

10.2.1 Filter banks using the lattice form

Structures similar to the lattice form are useful for the realization of critically decimated filter banks. They are referred to as lattice realizations of filter banks. We now show one example of such structures, which implement linear-phase 2-band perfect reconstruction filter banks.

In order for a 2-band filter bank to have linear phase, all its analysis and synthesis filters, $H_0(z)$, $H_1(z)$, $G_0(z)$, and $G_1(z)$, have to have linear phase. From equation (4.15), Subsection 4.2.3, if we suppose that all the filters in the filter bank have the same order M, then they have linear phase if

$$\left. \begin{array}{l} H_i(z) = \pm z^{-M} H_i(z^{-1}) \\ G_i(z) = \pm z^{-M} G_i(z^{-1}) \end{array} \right\} \tag{10.13}$$

for $i = 0, 1$.

The perfect reconstruction property holds if the analysis and synthesis polyphase matrices are related by equation (9.36) in Section 9.4,

$$\mathbf{R}(z) = z^{-\Delta} \mathbf{E}^{-1}(z) \tag{10.14}$$

where the analysis and synthesis polyphase matrices are defined by equations (9.11) and (9.12) as

$$
\begin{bmatrix} H_0(z) \\ H_1(z) \end{bmatrix} = \mathbf{E}(z^2) \begin{bmatrix} 1 \\ z^{-1} \end{bmatrix}
\tag{10.15}
$$

$$
\begin{bmatrix} G_0(z) \\ G_1(z) \end{bmatrix} = \mathbf{R}^{\mathrm{T}}(z^2) \begin{bmatrix} z^{-1} \\ 1 \end{bmatrix}
\tag{10.16}
$$

Suppose, now, that we define the polyphase matrix of the analysis filter, $\mathbf{E}(z)$, as having the following general form:

$$
\mathbf{E}(z) = \mathcal{K} \begin{bmatrix} 1 & 1 \\ -1 & 1 \end{bmatrix} \left(\prod_{i=1}^{M} \begin{bmatrix} 1 & 0 \\ 0 & z^{-1} \end{bmatrix} \begin{bmatrix} 1 & k_i \\ k_i & 1 \end{bmatrix} \right)
\tag{10.17}
$$

with

$$
\mathcal{K} = -\frac{1}{2} \prod_{i=1}^{M} \left(\frac{1}{1 - k_i^2} \right)
\tag{10.18}
$$

If we define the polyphase matrix of the synthesis filters, $\mathbf{R}(z)$, as

$$
\mathbf{R}(z) = \left(\prod_{i=M}^{1} \begin{bmatrix} 1 & -k_i \\ -k_i & 1 \end{bmatrix} \begin{bmatrix} z^{-1} & 0 \\ 0 & 1 \end{bmatrix} \right) \begin{bmatrix} 1 & -1 \\ 1 & 1 \end{bmatrix}
\tag{10.19}
$$

then we have that

$$
\mathbf{R}(z) = z^{-M} \mathbf{E}^{-1}(z)
\tag{10.20}
$$

and perfect reconstruction is guaranteed irrespective of the values of k_i.

Also, from equations (10.15) and (10.16), as well as equations (10.17) and (10.19), we have that

$$
\left. \begin{aligned}
H_0(z) &= z^{-M} H_0(z^{-1}) \\
H_1(z) &= -z^{-M} H_1(z^{-1}) \\
G_0(z) &= z^{-M} G_0(z^{-1}) \\
G_1(z) &= -z^{-M} G_1(z^{-1})
\end{aligned} \right\}
\tag{10.21}
$$

and linear phase is also guaranteed irrespective of the values of k_i.

The realizations of the analysis and synthesis filters are shown in Figures 10.3a and 10.3b, respectively. One should note that such realizations are restricted to when analysis and synthesis filters have the same order.

One important property of such realizations is that both perfect reconstruction and

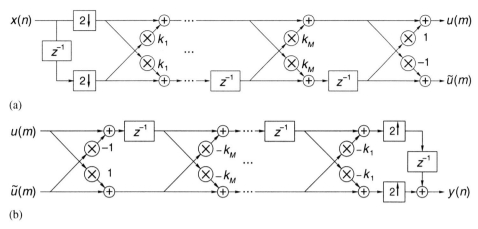

(a)

(b)

Figure 10.3 Lattice form realization of a linear-phase filter bank: (a) analysis filters; (b) synthesis filters.

linear phase are guaranteed irrespective of the values of k_i.[1] They are often referred to as having structurally induced perfect reconstruction and linear phase. Therefore, one possible design strategy for such filter banks is to carry out a multi-variable optimization on k_i, for $i = 1, \ldots, M$, using a chosen objective function.

For example, one could minimize the L_2 norm of the deviation of both the lowpass filter $H_0(z)$ and the highpass filter $H_1(z)$ in both their passbands and stopbands using the following objective function

$$\xi(k_1, k_2, \ldots, k_M) = \int_0^{\omega_p} \left(1 - \left|H_0(e^{j\omega})\right|\right)^2 d\omega + \int_{\omega_r}^{\pi} \left|H_0(e^{j\omega})\right|^2 d\omega$$
$$+ \int_{\omega_r}^{\pi} \left(1 - \left|H_1(e^{j\omega})\right|\right)^2 d\omega + \int_0^{\omega_p} \left|H_1(e^{j\omega})\right|^2 d\omega \qquad (10.22)$$

Note that this result essentially corresponds to a least-squares error function.

10.3 Polyphase form

A nonrecursive transfer function of the form

$$H(z) = \sum_{n=0}^{M} h(n) z^{-n} \qquad (10.23)$$

---

[1] There are rare cases of perfect reconstruction FIR filter banks where the corresponding lattices cannot be synthesized as some coefficient may be $k_i = 1$.

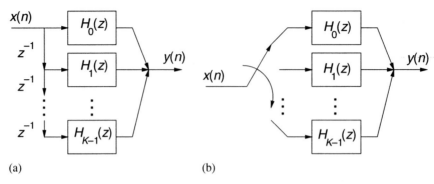

(a)   (b)

Figure 10.4 Block diagrams of alternative realizations of equation (10.24): (a) parallel form (using tapped delay line); (b) polyphase form (using rotating switch).

when M is a multiple of $\bar{M}$ (Subsection 9.3.2), can also be expressed as

$$
\begin{aligned}
H(z) &= \sum_{n=0}^{\bar{M}} h(Kn)z^{-Kn} + \sum_{n=0}^{\bar{M}} h(Kn+1)z^{-Kn-1} + \cdots \\
&\quad + \sum_{n=0}^{\bar{M}} h(Kn+K-1)z^{-Kn-K+1} \\
&= \sum_{n=0}^{\bar{M}} h(Kn)z^{-Kn} + z^{-1}\sum_{n=0}^{\bar{M}} h(Kn+1)z^{-Kn} + \cdots \\
&\quad + z^{-K+1}\sum_{n=0}^{\bar{M}} h(Kn+K-1)z^{-Kn}
\end{aligned}
\tag{10.24}
$$

where K is any natural number such that $\bar{M} = M/K$. This last form of writing $H(z)$ can be directly mapped onto the realization shown in Figure 10.4a, where each $H_i(z)$ is given by

$$
H_i(z) = z^i \sum_{n=0}^{\bar{M}} h(Kn+i)z^{-Kn}
\tag{10.25}
$$

for $i = 0, \ldots, K-1$. An alternative realization to the one shown in Figure 10.4a is given in Figure 10.4b, where a rotating switch substitutes the tapped delay line by connecting the input to consecutive branches at different instants of time. Such a realization is referred to as the polyphase realization and it finds a large range of applications in the study of multirate systems and filter banks (Chapter 9).

10.4   Frequency-domain form

The output of an FIR filter corresponds to the linear convolution of the input signal with the finite-length impulse response of the filter $h(n)$. Therefore, from Section 3.4 we have that, if the input signal $x(n)$ of a nonrecursive digital filter is known for all n, and null for $n < 0$ and $n > L$, an alternative approach to computing the output $y(n)$ can be derived using the fast Fourier transform (FFT). By completing these sequences with the necessary number of zeros (zero-padding procedure) and determining the resulting $(N + L)$-element FFTs of $h(n)$, $x(n)$, and $y(n)$, we have that

$$\text{FFT}\{y(n)\} = \text{FFT}\{h(n)\}\text{FFT}\{x(n)\} \tag{10.26}$$

and then

$$y(n) = \text{FFT}^{-1}\{\text{FFT}\{h(n)\}\text{FFT}\{x(n)\}\} \tag{10.27}$$

Using this approach, we are able to compute the entire sequence $y(n)$ with a number of arithmetic operations per output sample proportional to $\log_2(L + N)$. In the case of direct evaluation, the number of operations is of the order of NL. Clearly, for large values of N and L, the FFT method is the most efficient one.

In the above approach, the entire input sequence must be available to allow one to compute the output signal. In this case, if the input is extremely long, the complete computation of $y(n)$ can result in a long computational delay, which is undesirable in several applications. For such cases, the input signal can be sectioned and each data block processed separately using the so-called overlap-and-save and overlap-and-add methods, as described in Chapter 3.

10.5   Recursive running sum form

The direct realization of the transfer function

$$H(z) = \sum_{i=0}^{M} z^{-i} \tag{10.28}$$

where all the multiplier coefficients are equal to one, requires a large number of additions. An alternative way to implement such a transfer function results from interpreting it as the sum of the terms of a geometric series. This yields

$$H(z) = \frac{1 - z^{-M-1}}{1 - z^{-1}} \tag{10.29}$$

This equation leads to the realization in Figure 10.5, widely known as the recursive running sum (RRS) (Adams & Willson, 1983).

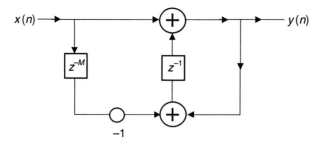

Figure 10.5 Block diagram of the recursive running sum (RRS) realization.

The RRS corresponds to a very simple lowpass filter, comprising $(M + 1)$ delays and only 2 additions per output sample. The RRS passband can be controlled by appropriately choosing the value of M (see Exercise 10.9). In d.c., the RRS gain is equal to $(M+1)$. To compensate for that, a scaling factor of $\frac{1}{(M+1)}$ should be employed at the filter input, which generates an output roundoff noise with variance of about $[(M+1)q]^2/12$. As an alternative for the scaling factor, we could perform the internal computations of the RRS with more precision, by increasing the number of bits in the most significant part of the binary word. In this case, the output-noise variance is reduced to $q^2/12$, because the signal is quantized only at the output. To guarantee this, the number of extra bits must be the smallest integer larger than $\log_2(M + 1)$.

10.6 Realizations with reduced number of arithmetic operations

The main drawback of FIR filters is the large number of arithmetic operations required to satisfy practical specifications. However, especially in the case of filters with narrow passbands or transition bands, there is a correlation among the values of filter multiplier coefficients. This fact can be exploited in order to reduce the number of arithmetic operations required to satisfy such specifications (van Gerwen et al., 1975; Adams & Willson, 1983, 1984; Benvenuto et al., 1984; Neüvo et al., 1984; Saramäki et al., 1988; Lim, 1986). This section presents some of the most widely used methods for designing FIR filters with reduced computational complexity.

10.6.1 Prefilter approach

The main idea of the prefilter method consists of generating a simple FIR filter, with reduced multiplication and addition counts, whose frequency response approximates the desired response as far as possible. Then this simple filter is cascaded with an amplitude equalizer, designed so that the overall filter satisfies the prescribed specifications (Adams & Willson, 1983). The reduction in the computational complexity results from the fact that the prefilter greatly relieves the specifications for the equalizer.

This happens because the equalizer has wider transition bands to approximate, thus requiring a smaller order.

Several prefilter structures are given in the literature (Adams & Willson, 1984), and choosing the best one for a given specification is not an easy task. A very simple lowpass prefilter is the RRS filter, seen in the previous section. From equation (10.29), the frequency response of the Mth-order RRS filter is given by

$$H(e^{j\omega}) = \frac{\sin\left[\frac{\omega(M+1)}{2}\right]}{\sin\left(\frac{\omega}{2}\right)} e^{-\frac{j\omega M}{2}} \tag{10.30}$$

This indicates that the RRS frequency response has several ripples at the stopband with decreasing magnitudes as ω approaches π. The first zero in the RRS frequency response occurs at

$$\omega_{z1} = \frac{2\pi}{M+1} \tag{10.31}$$

Using the RRS as a prefilter, this first zero must be placed above and as close as possible to the stopband edge ω_r. In order to achieve this, the RRS order M must be such that

$$M = \left\lfloor \frac{2\pi}{\omega_r} - 1 \right\rfloor \tag{10.32}$$

where $\lfloor x \rfloor$ represents the largest integer less than or equal to x.

More efficient prefilters can be generated by cascading several RRS sections. For example, if we cascade two prefilters, so that the first one satisfies equation (10.32), and the second is designed to cancel the secondary ripples of the first, we could expect a higher stopband attenuation for the resulting prefilter. This would relieve the specifications for the equalizer even further. In particular, the first stopband ripple belonging to the first RRS must be attenuated by the second RRS, without introducing zeros in the passband. Although the design of prefilters by cascading several RRS sections is always possible, practice has shown that there is little to gain in cascading more than three sections.

We show now that the Chebyshev (minimax) approach for designing optimal FIR filters presented in Chapter 5 can be adapted to design the equalizer, by modifying the error function definition, in the following way. The response obtained by cascading the prefilter with the equalizer is given by

$$H(z) = H_p(z)H_e(z) \tag{10.33}$$

where $H_p(z)$ is the prefilter transfer function, and only the coefficients of $H_e(z)$ are to be optimized.

The error function from equation (5.126) can then be rewritten as

$$
\begin{aligned}
|E(\omega)| &= \left| W(\omega) \left(D(\omega) - H_p(e^{j\omega}) H_e(e^{j\omega}) \right) \right| \\
&= \left| W(\omega) H_p(e^{j\omega}) \left(\frac{D(\omega)}{H_p(e^{j\omega})} - H_e(e^{j\omega}) \right) \right| \\
&= \left| W'(\omega) \left(D'(\omega) - H_e(e^{j\omega}) \right) \right|
\end{aligned}
\tag{10.34}
$$

where

$$
D'(\omega) = \begin{cases} \dfrac{1}{H_p(e^{j\omega})}, & \omega \in \text{passbands} \\ 0, & \omega \in \text{stopbands} \end{cases}
\tag{10.35}
$$

$$
W'(\omega) = \begin{cases} \left| H_p(e^{j\omega}) \right|, & \omega \in \text{passbands} \\ \dfrac{\delta_p}{\delta_r} \left| H_p(e^{j\omega}) \right|, & \omega \in \text{stopbands} \end{cases}
\tag{10.36}
$$

It is worth mentioning that $H_p(e^{j\omega})$ often has zeros at some frequencies, which causes problems to the optimization algorithm. One way to circumvent this is to replace $|H_p(e^{j\omega})|$ in the neighborhoods of its zeros by a small number such as 10^{-6}.

EXAMPLE 10.1

Design a highpass filter using the standard minimax and the prefilter methods satisfying the following specifications:

$$
\left.
\begin{aligned}
A_r &= 40\,\text{dB} \\
\Omega_r &= 6600\,\text{Hz} \\
\Omega_p &= 7200\,\text{Hz} \\
\Omega_s &= 16\,000\,\text{Hz}
\end{aligned}
\right\}
\tag{10.37}
$$

SOLUTION

The prefilter approach described above applies only to narrowband lowpass filters. However, it requires only a slight modification in order to be applied to narrowband highpass filter design. The modification consists of designing the lowpass filter approximating $D(\pi - \omega)$ and then replacing z^{-1} by $-z^{-1}$ in the realization. Therefore, the lowpass specifications are

$$
\Omega'_p = \frac{\Omega_s}{2} - \Omega_p = 8000 - 7200 = 800\ \text{Hz}
\tag{10.38}
$$

$$
\Omega'_r = \frac{\Omega_s}{2} - \Omega_r = 8000 - 6600 = 1400\ \text{Hz}
\tag{10.39}
$$

Using the minimax approach, the resulting direct-form filter has order 42, thus requiring 22 multiplications per output sample. Using the prefilter approach, with an

Table 10.1. *Equalizer coefficients.*

	$h(0)$ to $h(17)$	
$-5.752\,477\,25\mathrm{E}{-03}$	$3.903\,874\,03\mathrm{E}{-03}$	$-2.955\,236\,78\mathrm{E}{-03}$
$1.479\,040\,09\mathrm{E}{-04}$	$-5.368\,539\,44\mathrm{E}{-03}$	$7.102\,374\,26\mathrm{E}{-03}$
$-1.405\,843\,25\mathrm{E}{-03}$	$6.192\,752\,35\mathrm{E}{-03}$	$-1.146\,341\,47\mathrm{E}{-02}$
$6.081\,906\,32\mathrm{E}{-04}$	$-5.984\,233\,61\mathrm{E}{-03}$	$1.527\,116\,92\mathrm{E}{-02}$
$6.369\,215\,37\mathrm{E}{-04}$	$4.424\,267\,79\mathrm{E}{-03}$	$-1.785\,284\,03\mathrm{E}{-02}$
$-2.209\,895\,02\mathrm{E}{-03}$	$9.163\,418\,39\mathrm{E}{-04}$	$1.875\,563\,93\mathrm{E}{-02}$

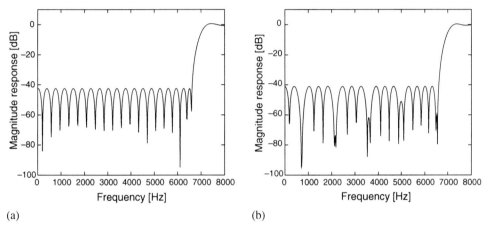

(a) (b)

Figure 10.6 Magnitude responses: (a) direct-form minimax approach; (b) prefilter approach.

RRS of order 10, the resulting equalizer has order 34, requiring only 18 multiplications per output sample. Only half of the equalizer coefficients are shown in Table 10.1, as the other coefficients can be obtained as $h(34 - n) = h(n)$.

The magnitude responses of the direct-form and the prefilter-equalizer filters are depicted in Figure 10.6.

$\triangle$

With the prefilter approach we can also design bandpass filters centered at $\omega_0 = \frac{\pi}{2}$ and with band edges at $(\frac{\pi}{2} - \frac{\omega_p}{2})$, $(\frac{\pi}{2} + \frac{\omega_p}{2})$, $(\frac{\pi}{2} - \frac{\omega_r}{2})$, and $(\frac{\pi}{2} + \frac{\omega_r}{2})$. This can be done by noting that such bandpass filters can be obtained from a lowpass filter with band edges at ω_p and ω_r, by applying the transformation $z^{-1} \to -z^{-2}$.

There are also generalizations of the prefilter approach that allow the design of narrow bandpass filters with central frequency away from $\frac{\pi}{2}$, as well as narrow stopband filters (Neüvo et al., 1987; Cabezas & Diniz, 1990).

Due to the reduced number of multipliers, filters designed using the prefilter method tend to generate less roundoff noise at the output than minimax filters implemented in direct form. Their sensitivity to coefficient quantization is also reduced when compared to direct-form minimax designs, as shown in Adams & Willson (1983).

10.6.2 Interpolation approach

FIR filters with narrow passband and transition bands tend to have adjacent multiplier coefficients, representing their impulse responses, with very close magnitude. This means there is a correlation among them that could be exploited to reduce computational complexity. Indeed, we could think of removing some samples of the impulse response, by replacing them with zeros, and approximating their values by interpolating the remaining nonzero samples. That is, using the terminology of Chapter 8, we would decimate and then interpolate the filter coefficients. This is the main idea behind the interpolation approach.

Consider an initial filter with frequency and impulse responses given by $H_i(e^{j\omega})$ and $h_i(n)$, respectively. If $h_i(n)$ is interpolated by L (see equation (8.17)), then $(L-1)$ null samples are inserted after each sample of $h_i(n)$, and the resulting sequence $h'_i(n)$ is given by

$$h'_i(n) = \begin{cases} h'_i(n), & \text{for } n = iL, \text{ with } i = 0, 1, \ldots \\ 0, & \text{for } n \neq iL \end{cases} \tag{10.40}$$

The corresponding frequency response, $H'_i(e^{j\omega})$, is periodic with period $2\pi/L$. For example, Figure 10.7 illustrates the form of $H'_i(e^{j\omega})$ generated from a lowpass filter with frequency response $H_i(e^{j\omega})$, using $L = 3$.

The filter with frequency response $H'_i(e^{j\omega})$, commonly referred to as the interpolated filter, is then connected in cascade with an interpolator $G(e^{j\omega})$, resulting in a transfer function of the form

$$H(z) = H'_i(z)G(z) \tag{10.41}$$

The function of the interpolator is to eliminate undesirable bands of $H'_i(z)$ (see Figure 10.7), while leaving its lowest frequency band unaffected.

As can be observed, the initial filter has passband and stopband which are L (in Figure 10.7, $L = 3$) times larger than the ones of the interpolated filter. As a consequence, the number of multiplications in the initial filter tends to be approximately L times smaller than the number of multiplications of a filter directly designed with the minimax approach to satisfy the narrowband specifications. An intuitive explanation for this is that a filter with larger passbands, stopbands, and transition bands is easier to implement than one with narrow bands. Indeed, this can be verified by analyzing the formula in Exercise 5.19, which predicts the order of the minimax FIR filter required to satisfy a given specification.

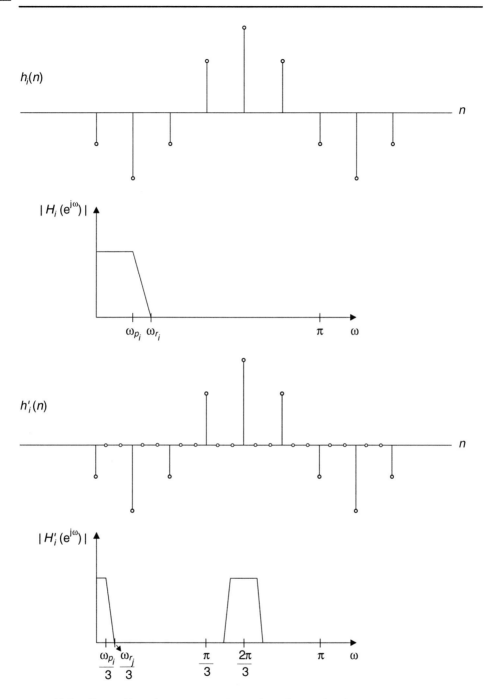

Figure 10.7 Effects of inserting $(L - 1) = 2$ zeros in a discrete-time impulse response.

For lowpass filters, the maximum value of L such that the initial filter satisfies the specifications in the passband and in the stopband is given by

$$L_{\max} = \left\lfloor \frac{\pi}{\omega_r} \right\rfloor \qquad (10.42)$$

where ω_r is the lowest rejection band frequency of the desired filter. This value for L assures that $\omega_{r_i} < \pi$.

For highpass filters, $L_{\max}$ is given by

$$L_{\max} = \left\lfloor \frac{\pi}{\pi - \omega_r} \right\rfloor \qquad (10.43)$$

whereas for bandpass filters, $L_{\max}$ is the largest L such that $\frac{\pi k}{L} \leq \omega_{r_1} < \omega_{r_2} \leq \frac{\pi(k+1)}{L}$, for some k. In practice, L is chosen smaller than $L_{\max}$ in order to relieve the interpolator specifications.

Naturally, to achieve reduction in the computational requirements of the final filter, the interpolator must be as simple as possible, not requiring too many multiplications. For instance, the interpolator can be designed as a cascade of subsections, in which each subsection places zeros on an undesirable passband. For the example in Figure 10.7, if we wish to design a lowpass filter, $G(e^{j\omega})$ should have zeros at $e^{\pm j\frac{2\pi}{3}}$. Alternatively, we can use the minimax method to design the interpolator, with the passband of $G(e^{j\omega})$ coinciding with the specified passband, and with the stopbands located in the frequency ranges of the undesired passband replicas of the interpolated filter (Saramäki et al., 1988).

Once the value of L and the interpolator are chosen, the interpolated filter can be designed such that

$$\left.\begin{aligned} (1 - \delta_p) \leq \left| H_i(e^{j\omega}) G(e^{\frac{j\omega}{L}}) \right| \leq (1 + \delta_p), & \quad \text{for } \omega \in [0, L\omega_p] \\ \left| H_i(e^{j\omega}) G(e^{\frac{j\omega}{L}}) \right| \leq \delta_r, & \quad \text{for } \omega \in [L\omega_r, \pi] \end{aligned}\right\} \qquad (10.44)$$

where the minimax method to design optimal FIR filters can be directly used.

EXAMPLE 10.2

Design the filter specified in Example 10.1 using the interpolation method.

SOLUTION

Using $L = 2$, we obtain the initial filter of order 20, thus requiring 11 multiplications per output sample, whose coefficients are listed in Table 10.2. The required interpolator is given by $G(z) = (1 - z^{-1})^4$, and the resulting magnitude response of the cascade of the initial filter and the interpolator is seen in Figure 10.8.

$\triangle$

Table 10.2. *Initial filter coefficients.*

	$h(n)$	
1.070 319 25E−03	−2.813 089 59E−03	9.580 902 83E−03
7.355 247 75E−04	−3.348 267 60E−03	1.476 842 27E−02
3.982 798 22E−04	−1.268 959 28E−03	1.686 309 64E−02
−1.277 078 79E−03	3.388 196 94E−03	

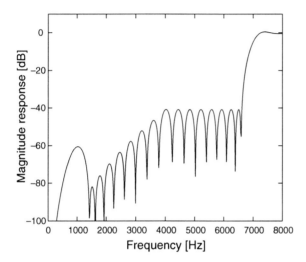

Figure 10.8 Response of the interpolated model filter in cascade with the interpolator.

It is worth observing that the prefilter and the interpolation methods were initially described as effective methods to design narrowband filters of the types lowpass, highpass, and bandpass. However, we can also design both wideband and narrow stopband filters with the interpolation method by noting they can be obtained from a narrowband filter $H(z)$ which is complementary to the desired filter, using

$$H_{FL}(z) = z^{-\frac{M}{2}} - H(z) \qquad (10.45)$$

where M is the order of $H(z)$.

10.6.3 Frequency response masking approach

Another interesting application of interpolation appears in the design of wideband sharp cutoff filters using the so-called frequency masking approach (Lim, 1986). A brief introduction to it was given in Subsection 8.7.2. Such an approach makes use of the concept of complementary filters, which constitute a pair of filters, $H_a(z)$ and

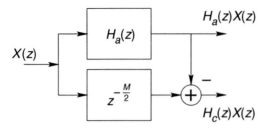

Figure 10.9 Realization of the complementary filter $H_c(z)$.

$H_c(z)$, whose frequency responses add to a constant delay, that is

$$\left| H_a(e^{j\omega}) + H_c(e^{j\omega}) \right| = 1 \tag{10.46}$$

If $H_a(z)$ is a linear-phase FIR filter of even order M, its frequency response can be written as

$$H_a(e^{j\omega}) = e^{-j\frac{M}{2}\omega} A(\omega) \tag{10.47}$$

where $A(\omega)$ is a trigonometric function of ω, as given in Section 5.6, equation (5.125). Therefore, the frequency response of the complementary filter must be of the form

$$H_c(e^{j\omega}) = e^{-j\frac{M}{2}\omega} (1 - A(\omega)) \tag{10.48}$$

and the corresponding transfer functions are such that

$$H_a(z) + H_c(z) = z^{-\frac{M}{2}} \tag{10.49}$$

Hence, given the realization of $H_a(z)$, its complementary filter $H_c(z)$ can easily be implemented by subtracting the output of $H_a(z)$ from the $(M/2)$th delayed version of its input, as seen in Figure 10.9. For an efficient implementation of both filters, the tapped delay line of $H_a(z)$ can be used to form $H_c(z)$, as indicated in Figure 10.10, in which either the symmetry or antisymmetry of $H_a(z)$ is exploited, as we are assuming that this filter has linear phase.

The overall structure of the filter designed with the frequency response masking approach is seen in Figure 10.11. The basic idea is to design a wideband lowpass filter and compress its frequency response by using an interpolation operation. A complementary filter is obtained, following the development seen above. We then use masking filters, $H_{Ma}(z)$ and $H_{Mc}(z)$, to eliminate the undesired bands in the interpolated and complementary filters, respectively. The corresponding outputs are added together to form the desired lowpass filter.

To understand the overall procedure in the frequency domain, consider Figure 10.12. Suppose that $H_a(z)$ corresponds to a lowpass filter of even order M, designed with the standard minimax approach, with passband edge θ and stopband edge ϕ, as seen in

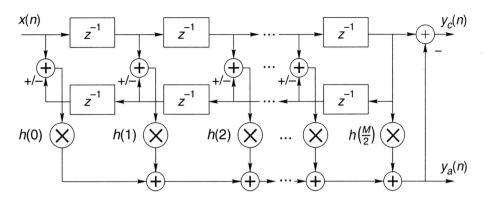

Figure 10.10 Efficient realization of the complementary filter $H_c(z)$.

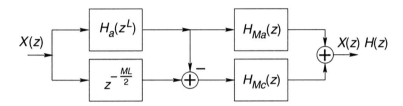

Figure 10.11 Block diagram of the frequency masking approach.

Figure 10.12a. We can then form $H_c(z)$, corresponding to a highpass filter, with θ and ϕ being the respective stopband and passband edges. By interpolating both filters by L, two complementary multiband filters are generated, as represented in Figures 10.12b and 10.12c, respectively. We can then use two masking filters, $H_{Ma}(z)$ and $H_{Mc}(z)$, characterized as in Figures 10.12d and 10.12e, to generate the magnitude responses shown in Figures 10.12f and 10.12g. Adding these two components, the resulting desired filter seen in Figure 10.12h can have a passband of arbitrary width with a very narrow transition band. In Figure 10.12, the positions of the transition band edges are dictated by the $H_{Ma}(z)$ masking filter. An example of the frequency response masking approach where the transition band edges are determined by the $H_{Mc}(z)$ masking filter, is shown in Figure 10.13.

From Figure 10.11, it is easy to see that the product ML must be even to avoid a half-sample delay. This is commonly satisfied by forcing M to be even, as above, thus freeing the parameter L from any constraint. In addition, $H_{Ma}(z)$ and $H_{Mc}(z)$ must have the same group delay, so that they complement each other appropriately in the resulting passband when added together to form the desired filter $H(z)$. This means that they must be both of even order or both of odd order, and that a few delays may be appended before and after either $H_{Ma}(z)$ or $H_{Mc}(z)$, if necessary, to equalize their group delays.

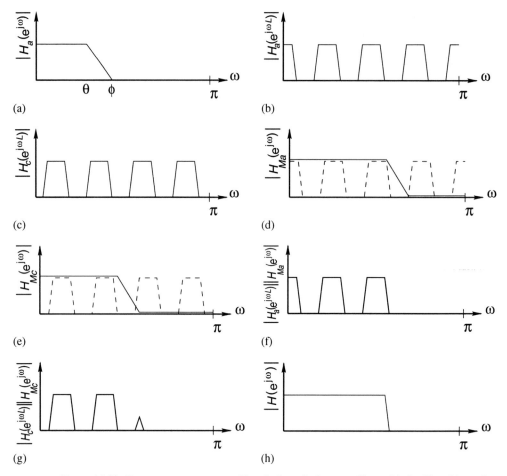

Figure 10.12 Frequency response masking design of a lowpass filter with the $H_{Ma}(z)$ mask determining the passband. (a) Base filter; (b) interpolated filter; (c) complementary to the interpolated filter; (d) masking filter $H_{Ma}(z)$; (e) masking filter $H_{Mc}(z)$; (f) cascade of $H_a(z^L)$ with masking filter $H_{Ma}(z)$; (g) cascade of $H_c(z^L)$ with masking filter $H_{Mc}(z)$; (h) frequency response masking filter.

For a complete description of the frequency response masking approach, we must characterize the filters $H_a(z)$, $H_{Ma}(z)$, and $H_{Mc}(z)$. When the resulting magnitude response is determined mainly by the masking filter $H_{Ma}(z)$, as exemplified in Figure 10.12, then we can conclude that the desired band edges are such that

$$\omega_p = \frac{2m\pi + \theta}{L} \tag{10.50}$$

$$\omega_r = \frac{2m\pi + \phi}{L} \tag{10.51}$$

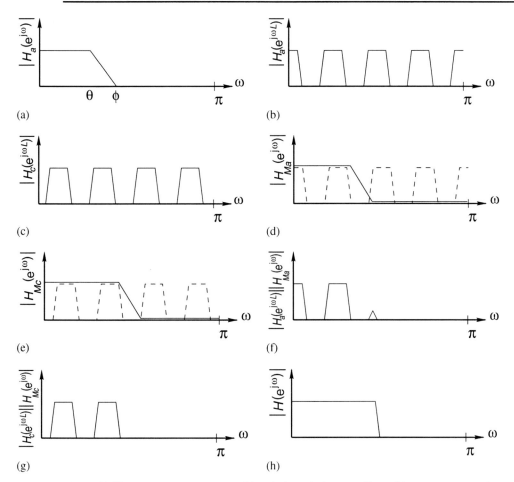

Figure 10.13 Frequency response masking design of a lowpass filter with the $H_{Mc}(z)$ mask determining the passband. (a) Base filter; (b) interpolated filter; (c) complementary to the interpolated filter; (d) masking filter $H_{Ma}(z)$; (e) masking filter $H_{Mc}(z)$; (f) cascade of $H_a(z^L)$ with masking filter $H_{Ma}(z)$; (g) cascade of $H_c(z^L)$ with masking filter $H_{Mc}(z)$; (h) frequency response masking filter.

where m is an integer less than L. Therefore, a solution for the values of m, θ, and ϕ, such that $0 < \theta < \phi < \pi$, is given by

$$m = \left\lfloor \frac{\omega_p L}{2\pi} \right\rfloor \tag{10.52}$$

$$\theta = \omega_p L - 2m\pi \tag{10.53}$$

$$\phi = \omega_r L - 2m\pi \tag{10.54}$$

where $\lfloor x \rfloor$ indicates the largest integer less than or equal to x. With these values, from

Figure 10.12, we can determine the band edges for the masking filters as given by

$$\omega_{p,Ma} = \frac{2m\pi + \theta}{L} \tag{10.55}$$

$$\omega_{r,Ma} = \frac{2(m+1)\pi - \phi}{L} \tag{10.56}$$

$$\omega_{p,Mc} = \frac{2m\pi - \theta}{L} \tag{10.57}$$

$$\omega_{r,Mc} = \frac{2m\pi + \phi}{L} \tag{10.58}$$

where $\omega_{p,Ma}$ and $\omega_{r,Ma}$ are the passband and stopband edges for the $H_{Ma}(z)$ masking filter, respectively, and $\omega_{p,Mc}$ and $\omega_{r,Mc}$ are the passband and stopband edges for the $H_{Mc}(z)$ masking filter, respectively.

When $H_{Mc}(z)$ is the dominating masking filter, as seen in Figure 10.13, we have that

$$\omega_p = \frac{2m\pi - \theta}{M} \tag{10.59}$$

$$\omega_r = \frac{2m\pi - \phi}{M} \tag{10.60}$$

and a solution for m, θ, and ϕ, such that $0 < \theta < \phi < \pi$, is given by

$$m = \left\lceil \frac{\omega_r L}{2\pi} \right\rceil \tag{10.61}$$

$$\theta = 2m\pi - \omega_r L \tag{10.62}$$

$$\phi = 2m\pi - \omega_p L \tag{10.63}$$

where $\lceil x \rceil$ indicates the smallest integer greater than or equal to x. In this case, from Figure 10.13, the band edges for the masking filters are given by

$$\omega_{p,Ma} = \frac{2(m-1)\pi + \phi}{L} \tag{10.64}$$

$$\omega_{r,Ma} = \frac{2m\pi - \theta}{L} \tag{10.65}$$

$$\omega_{p,Mc} = \frac{2m\pi - \phi}{L} \tag{10.66}$$

$$\omega_{r,Mc} = \frac{2m\pi + \theta}{L} \tag{10.67}$$

Given the desired ω_p and ω_r, each value of L will allow only one solution such that $\theta < \phi$, either in the form of equations (10.52)–(10.54) or of equations (10.61)–(10.63). In practice, the determination of the best L, that is the one which minimizes the total number of multiplications per output sample, can be done empirically with the aid of the order estimation given in Exercise 5.19.

The passband ripples and attenuation levels used when designing $H_a(z)$, $H_{Ma}(z)$, and $H_{Mc}(z)$ are determined based on the specifications of the desired filter. As these filters are cascaded, their frequency responses will be added in dB, thus requiring a certain margin to be used in their designs. For the base filter $H_a(z)$, one must keep in mind that its passband ripple δ_p corresponds to the stopband attenuation δ_r of its complementary filter $H_c(z)$, and vice versa. Therefore, when designing $H_a(z)$ one should use the smallest value between δ_p and δ_r, incorporating an adequate margin. In general, a margin of 50% in the values of the passband ripples and stopband attenuations should be used, as in the example that follows.

EXAMPLE 10.3

Design a lowpass filter using the frequency response masking method satisfying the following specifications:

$$\left.\begin{array}{l} A_p = 0.2 \text{ dB} \\ A_r = 60 \text{ dB} \\ \Omega_p = 0.6\pi \text{ rad/s} \\ \Omega_r = 0.61\pi \text{ rad/s} \\ \Omega_s = 2\pi \text{ rad/s} \end{array}\right\} \tag{10.68}$$

Compare your results with the filter obtained using the standard minimax scheme.

SOLUTION

Table 10.3 shows the estimated orders for the base filter and the masking filters for several values of the interpolation factor L. Although these values are probably slight underestimates, they allow a quick decision as to the value of L that minimizes the total number of multiplications required. In this table M_{H_a} is the order of the base filter $H_a(z)$, $M_{H_{Ma}}$ is the order of the masking filter $H_{Ma}(z)$, and $M_{H_{Mc}}$ is the order of the masking filter $H_{Mc}(z)$. Also,

$$\Pi = f(M_{H_a}) + f(M_{H_{Ma}}) + f(M_{H_{Mc}}) \tag{10.69}$$

indicates the total number of multiplications required to implement the overall filter, where

$$f(x) = \begin{cases} \dfrac{x+1}{2}, & \text{if } x \text{ is odd} \\ \dfrac{x}{2} + 1, & \text{if } x \text{ is even} \end{cases} \tag{10.70}$$

and

$$M = LM_{H_a} + \max\{M_{H_{Ma}}, M_{H_{Mc}}\} \tag{10.71}$$

is the effective order of the overall filter designed with the frequency response masking approach. From this table, we predict that $L = 9$ should yield the most efficient filter with respect to the total number of multiplications per output sample.

Table 10.3. *Filter characteristics for several values of the interpolation factor L.*

L	M_{H_a}	$M_{H_{Ma}}$	$M_{H_{Mc}}$	Π	M
2	368	29	0	200	765
3	246	11	49	155	787
4	186	21	29	120	773
5	150	582	16	377	1332
6	124	29	49	103	793
7	108	28	88	115	844
8	94	147	29	137	899
9	**84**	**61**	**49**	**99**	**817**
10	76	32	582	348	1342
11	68	95	51	109	843

Using $L = 9$ in equations (10.52)–(10.67), the corresponding band edges for all filters are given by

$$\left.\begin{aligned} \theta &= 0.5100\pi \\ \phi &= 0.6000\pi \\ \omega_{p,Ma} &= 0.5111\pi \\ \omega_{r,Ma} &= 0.6100\pi \\ \omega_{p,Mc} &= 0.6000\pi \\ \omega_{r,Mc} &= 0.7233\pi \end{aligned}\right\} \tag{10.72}$$

From the filter specifications, we have that $\delta_p = 0.0115$ and $\delta_r = 0.001$. Therefore, the value of δ_r with a margin of 50% was used as the passband ripple and the stopband attenuation for the base filter to generate Table 10.3, that is

$$\delta_{p,a} = \delta_{r,a} = \min\{\delta_p, \delta_r\} \times 50\% = 0.0005 \tag{10.73}$$

corresponding to a passband ripple of 0.0087 dB, and a stopband attenuation of 66.0206 dB. As $\delta_{p,a} = \delta_{r,a}$ the relative weights for the passband and stopband in the minimax design of the base filter are both equal to 1.0000.

For the masking filters, we use

$$\delta_{p,Ma} = \delta_{p,Mc} = \delta_p \times 50\% = 0.00575 \tag{10.74}$$

$$\delta_{r,Ma} = \delta_{r,Mc} = \delta_r \times 50\% = 0.0005 \tag{10.75}$$

corresponding to a passband ripple of 0.0996 dB, and a stopband attenuation of 66.0206 dB. In this case, the relative weights for the minimax design of both masking filters were made equal to 1.0000 and 11.5124 in the passband and stopband, respectively.

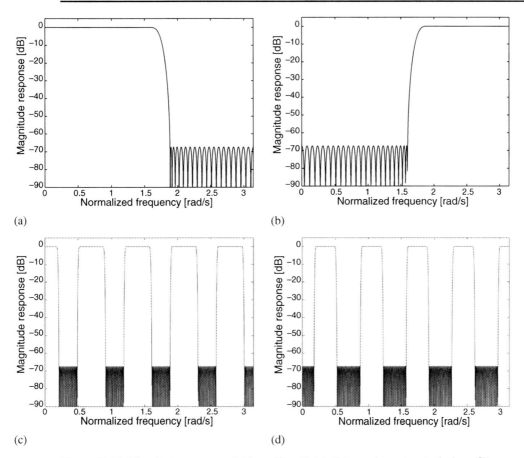

(a)

(b)

(c)

(d)

Figure 10.14 Magnitude responses: (a) base filter $H_a(z)$; (b) complementary to the base filter $H_c(z)$; (c) interpolated base filter $H_a(z^L)$; (d) complementary to the interpolated base filter $H_c(z^L)$.

The magnitude responses of the resulting filters for $L = 9$ are depicted in Figures 10.14 and 10.15. In Figures 10.14a and 10.14b, the base filter and its complementary filter are depicted, and Figures 10.14c and 10.14d show the corresponding interpolated filters. Similarly, Figures 10.15a and 10.15b depict the two masking filters, $H_{Ma}(z)$ and $H_{Mc}(z)$, respectively, and Figures 10.15c and 10.15d show the results at the outputs of these filters, which are added together to form the desired filter. The overall frequency response masking filter is characterized in Figure 10.16 and presents a passband ripple equal to $A_p = 0.0873$ dB and a stopband attenuation of $A_r = 61.4591$ dB.

The resulting minimax filter is of order 504, thus requiring 253 multiplications per output sample. Therefore, in this case, the frequency response masking design represents a saving of about 60% of the number of multiplications required by the standard minimax filter.

$\triangle$

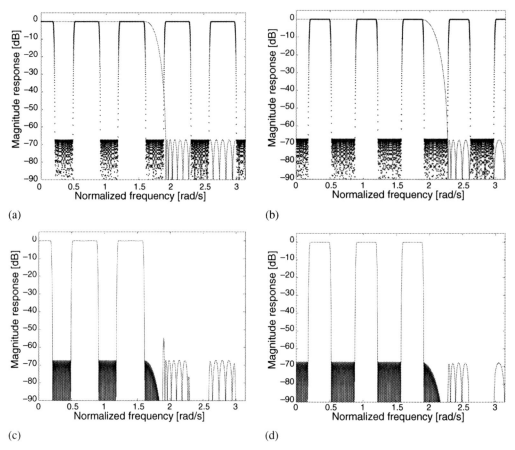

Figure 10.15 Magnitude responses: (a) masking filter $H_{Ma}(z)$; (b) masking filter $H_{Mc}(z)$; (c) combination of $H_a(z^L)H_{Ma}(z)$; (d) combination of $H_c(z^L)H_{Mc}(z)$.

Example 10.3 above shows that a constant ripple margin throughout the whole frequency range $\omega \in [0, \pi]$ is not required. In fact, with the ripple margin of 50% the passband ripple was considerably smaller than necessary, as seen in Figure 10.16a, and the attenuation was higher than required through most of the stopband, as seen in Figure 10.16b. A detailed analysis of the required margins in each band was performed in Lim (1986), where it was concluded that:

- The ripple margin must be of the order of 50% at the beginning of the stopbands of each masking filter.
- For the remaining frequency values, the ripple margin can be set around 15–20%.

It can be verified that such a distribution of the ripple margins results in a more efficient design, yielding an overall filter with a smaller group delay and fewer multiplications per output sample, as illustrated in the following example.

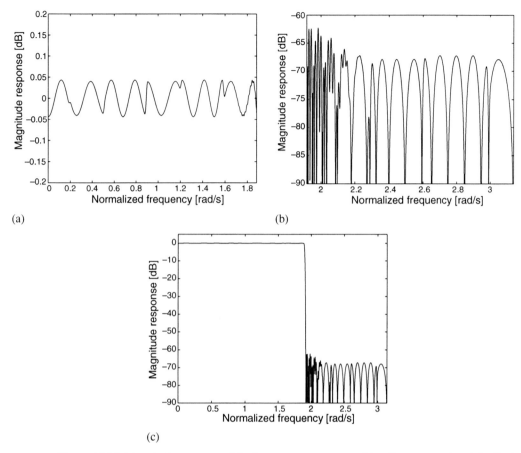

Figure 10.16 Magnitude response of the frequency response masking filter: (a) passband detail; (b) stopband detail; (c) overall response.

EXAMPLE 10.4

Design the lowpass filter specified in Example 10.3 using the frequency response masking method with an efficient assignment of the ripple margin. Compare the results with the filter obtained in Example 10.3.

SOLUTION

The design follows the same procedure as before, except that the relative weights at the beginning of the stopbands of the masking filters is set to 2.5, whereas for the remaining frequency the weight is set to 1.0. That corresponds to ripple margins proportional to 50% and 20% respectively, in these two distinct frequency ranges.

Table 10.4 shows the filter characteristics for several values of the interpolation factor L. As in Example 10.3, the minimum number of multiplications per output sample is obtained when $L = 9$, and the band edges for the base and frequency masking filters are as given in equation (10.72). In this case, however, only 91

Table 10.4. *Filter characteristics for several values of the interpolation factor L.*

L	M_{H_a}	$M_{H_{Ma}}$	$M_{H_{Mc}}$	Π	M
2	342	26	0	186	710
3	228	10	44	144	728
4	172	20	26	112	714
5	138	528	14	343	1218
6	116	26	44	96	740
7	100	26	80	106	780
8	88	134	26	127	838
9	**78**	**55**	**45**	**91**	**757**
10	70	30	528	317	1228
11	64	86	46	101	790

multiplications are required, as opposed to 99 multiplications in Example 10.3.

Figures 10.17a and 10.17b depict the magnitude responses of the two masking filters, $H_{Ma}(z)$ and $H_{Mc}(z)$, respectively, and Figures 10.17c and 10.17d show the magnitude responses which are added together to form the desired filter. From these figures, one can clearly see the effects of the more efficient ripple margin distribution. The overall frequency response masking filter is characterized in Figure 10.18, and presents a passband ripple equal to $A_p = 0.1502$ dB and a stopband attenuation of $A_r = 60.5578$ dB. Notice how these values are closer to the specifications than the values of the filter obtained in Example 10.3.

Table 10.5 presents the first half of the base filter coefficients before interpolation, and the coefficients of the masking filters $H_{Ma}(z)$ and $H_{Mc}(z)$ are given in Tables 10.6 and 10.7, respectively. It must be noted that, as stated before, for a smooth composition of the outputs of the masking filters when forming the filter $H(z)$, both these filters must have the same group delay. To achieve that in this design example, we must add 5 delays before and 5 delays after the masking filter $H_{Mc}(z)$, which has a smaller number of coefficients.

$\triangle$

So far, we have discussed the use of the frequency response masking filters to design wideband lowpass filters. The design of narrowband lowpass filters can also be performed by considering that only one masking filter is necessary. Usually, we consider the branch formed by the base filter $H_a(z)$ and its corresponding masking filter $H_{Ma}(z)$, greatly reducing the overall complexity of the designed filter. In such cases, the frequency response masking approach becomes similar to the prefilter and interpolation approaches seen in previous subsections. The design of highpass filters can be inferred from the design for lowpass filters, or can be performed using the concept of complementary filters seen in the beginning of this subsection. The design of bandpass and bandstop filters with reduced arithmetic complexity is addressed in the next subsection.

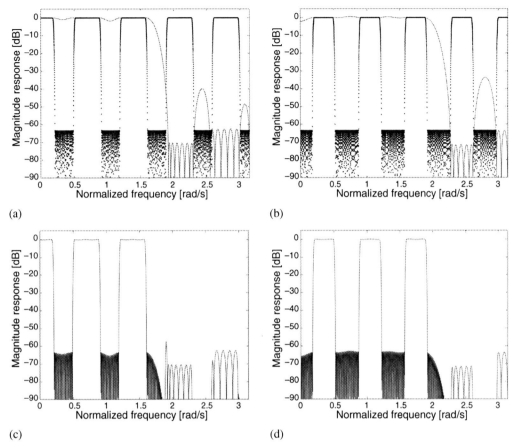

Figure 10.17 Magnitude responses: (a) masking filter $H_{Ma}(z)$; (b) masking filter $H_{Mc}(z)$;
(c) combination of $H_a(z^L)H_{Ma}(z)$; (d) combination of $H_c(z^L)H_{Mc}(z)$.

10.6.4 Quadrature approach

In this subsection, a method for designing symmetric bandpass and bandstop filters
is introduced. For narrowband filters, the so-called quadrature approach uses an FIR
prototype of the form (Neüvo et al., 1987)

$$H_p(z) = H_a(z^L)H_M(z) \tag{10.76}$$

where $H_a(z)$ is the base filter and $H_M(z)$ is the masking filter or interpolator, which
attenuates the undesired spectral images of the passband of $H_a(z^L)$, commonly
referred to as the shaping filter. Such a prototype can be designed using prefilter,
interpolation, or simplified one-branch frequency response masking approaches seen

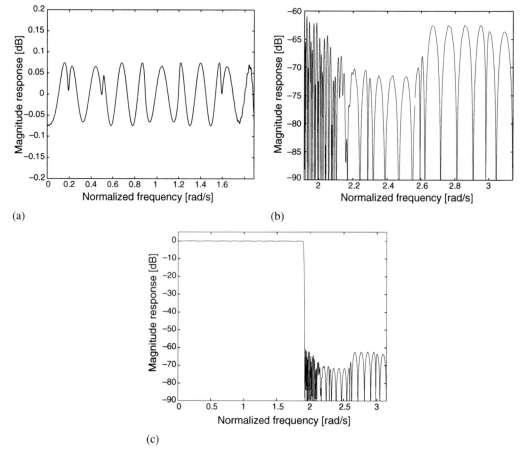

(a)

(b)

(c)

Figure 10.18 Magnitude response of the frequency response masking filter: (a) passband detail; (b) stopband detail; (c) overall response.

above. The main idea of the quadrature approach is to shift the frequency response of the base filter to the desired central frequency ω_o, and then apply the masking filter (interpolator) to eliminate any other undesired passbands.

Consider a linear-phase lowpass filter $H_a(z)$ with impulse response $h_a(n)$, such that

$$H_a(z) = \sum_{n=0}^{M} h_a(n)z^{-n} \tag{10.77}$$

Let the passband ripple and stopband attenuation be equal to δ'_p and δ'_r, and the passband and stopband edges be ω'_p and ω'_r, respectively. If $h_a(n)$ is interpolated by a factor L, and the resulting sequence is multiplied by $e^{j\omega_o n}$, we generate an auxiliary $H_1(z^L)$ as

$$H_1(z^L) = \sum_{n=0}^{M} h_a(n)e^{j\omega_o n}z^{-nL} \tag{10.78}$$

Table 10.5. *Base filter $H_a(z)$ coefficients.*

$h_a(0)$ to $h_a(39)$		
−3.772 751 62E−04	−1.727 512 31E−03	7.761 599 22E−03
−2.725 321 75E−04	−4.317 371 17E−03	−2.695 354 96E−02
6.702 693 56E−04	3.919 198 03E−03	5.056 563 71E−04
−1.122 248 41E−04	4.023 902 43E−03	3.542 882 23E−02
−8.289 487 54E−04	−6.569 755 32E−03	−1.492 737 54E−02
4.126 302 05E−04	−2.575 208 30E−03	−4.321 276 47E−02
1.113 679 43E−03	9.318 231 57E−03	3.981 059 79E−02
−1.091 139 79E−03	−3.438 490 90E−04	4.949 067 57E−02
−1.105 796 38E−03	−1.160 797 21E−02	−9.091 889 87E−02
1.948 002 78E−03	4.907 366 71E−03	−5.356 947 00E−02
7.465 842 08E−04	1.271 185 57E−02	3.130 965 69E−01
−2.942 681 84E−03	−1.108 431 73E−02	5.549 836 14E−01
1.706 283 62E−04	−1.176 133 46E−02	
3.831 458 89E−03	1.860 358 68E−02	

Table 10.6. *Masking filter $H_{Ma}(z)$ coefficients.*

$h_{Ma}(0)$ to $h_{Ma}(27)$		
3.989 370 63E−03	−1.599 326 64E−02	1.806 551 22E−02
5.799 146 25E−03	−4.408 841 07E−03	−4.834 253 44E−02
9.277 055 00E−05	1.612 310 26E−02	−1.221 449 86E−02
−6.143 048 28E−03	4.566 428 30E−03	6.739 082 13E−02
−2.505 907 97E−03	−1.529 152 89E−02	−1.327 670 36E−02
3.121 297 26E−03	1.759 893 96E−03	−1.124 709 98E−01
−8.669 974 01E−04	1.538 857 16E−02	1.053 740 84E−01
−3.800 843 53E−03	−1.132 415 98E−02	4.718 372 83E−01
2.194 996 38E−03	−7.277 442 23E−03	
−3.890 699 34E−03	3.782 551 92E−02	

This implies that the passband of $H_a(z)$ is squeezed by a factor of L, and becomes centered at ω_o. Analogously, using the interpolation operation followed by the modulating sequence $e^{-j\omega_o n}$, we have another auxiliary function such that

$$H_2(z^L) = \sum_{n=0}^{M} h_a(n) e^{-j\omega_o n} z^{-nL} \tag{10.79}$$

with the corresponding squeezed passband centered at $-\omega_o$. We can then use two masking filters, $H_{M1}(z)$ and $H_{M2}(z)$, appropriately centered at ω_o and $-\omega_o$ to eliminate the undesired bands in $H_1(z^L)$ and $H_2(z^L)$, respectively. The addition of

Table 10.7. *Masking filter $H_{Mc}(z)$ coefficients.*

$h_{Mc}(0)$ to $h_{Mc}(22)$		
1.973 509 03E−04	−1.303 122 05E−02	2.309 237 05E−02
−7.004 413 96E−03	3.592 131 20E−03	−5.885 046 73E−02
−7.377 387 41E−03	−4.728 015 88E−03	7.320 822 72E−03
1.931 045 81E−03	−2.572 965 68E−02	5.531 326 77E−02
−3.093 788 17E−04	6.552 836 81E−03	−1.232 612 61E−01
−7.104 725 27E−03	1.174 475 69E−02	9.769 821 21E−03
3.003 895 17E−03	−3.214 725 19E−02	5.401 665 00E−01
5.800 402 66E−04	7.838 465 67E−03	

the two resulting sequences yields a symmetric bandpass filter centered at ω_o.

Clearly, although the two-branch overall bandpass filter will have real coefficients, each branch in this case will present complex coefficients. To overcome this problem, first note that $H_1(z^L)$ and $H_2(z^L)$ have complex conjugate coefficients. If we design $H_{M1}(z)$ and $H_{M2}(z)$ so that their coefficients are complex conjugates of each other, it is easy to verify that

$$
\begin{aligned}
H_1(z^L)H_{M1}(z) &= \left(H_{1,R}(z^L) + jH_{1,I}(z^L) \right) \left(H_{M1,R}(z) + jH_{M1,I}(z) \right) \\
&= \left(H_{1,R}(z^L)H_{M1,R}(z) - H_{1,I}(z^L)H_{M1,I}(z) \right) \\
&\quad + j \left(H_{1,R}(z^L)H_{M1,I}(z) + H_{1,I}(z^L)H_{M1,R}(z) \right)
\end{aligned} \tag{10.80}
$$

$$
\begin{aligned}
H_2(z^L)H_{M2}(z) &= \left(H_{2,R}(z^L) + jH_{2,I}(z^L) \right) \left(H_{M2,R}(z) + jH_{M2,I}(z) \right) \\
&= \left(H_{1,R}(z^L) - jH_{1,I}(z^L) \right) \left(H_{M1,R}(z) - jH_{M1,I}(z) \right) \\
&= \left(H_{1,R}(z^L)H_{M1,R}(z) - H_{1,I}(z^L)H_{M1,I}(z) \right) \\
&\quad - j \left(H_{1,R}(z^L)H_{M1,I}(z) + H_{1,I}(z^L)H_{M1,R}(z) \right)
\end{aligned} \tag{10.81}
$$

where the subscripts R and I indicate the parts of the corresponding transfer function with real and imaginary coefficients, respectively. Therefore,

$$
H_1(z^L)H_{M1}(z) + H_2(z^L)H_{M2}(z) = 2 \left(H_{1,R}(z^L)H_{M1,R}(z) - H_{1,I}(z^L)H_{M1,I}(z) \right) \tag{10.82}
$$

and the structure seen in Figure 10.19 can be used for the real implementation of the quadrature approach for narrowband filters. Disregarding the effects of the masking filters, the resulting quadrature filter is characterized by

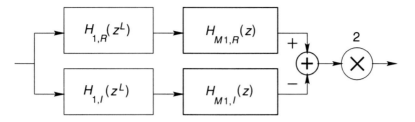

Figure 10.19 Block diagram of the quadrature approach for narrowband filters.

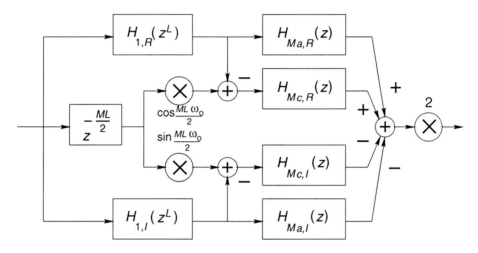

Figure 10.20 Block diagram of the quadrature approach for wideband filters.

$$
\left.
\begin{aligned}
\delta_p &= \delta_p' + \delta_r' \\
\delta_r &= 2\delta_r' \\
\omega_{r_1} &= \omega_o - \omega_r' \\
\omega_{p_1} &= \omega_o - \omega_p' \\
\omega_{p_2} &= \omega_o + \omega_p' \\
\omega_{r_2} &= \omega_o + \omega_r'
\end{aligned}
\right\}
\tag{10.83}
$$

For wideband filters, the prototype filter should be designed with the frequency masking approach. In that case, we have two complete masking filters, and the quadrature implementation involving solely real filters is seen in Figure 10.20, with $H_1(z^L)$ as defined in equation (10.78), and $H_{Ma}(z)$ and $H_{Mc}(z)$ corresponding to the two masking filters, appropriately centered at ω_o and $-\omega_o$, respectively.

For bandstop filters, we may start with a highpass prototype and apply the quadrature design, or design a bandpass filter and then determine its complementary filter (Rajan et al., 1988).

Table 10.8. *Filter characteristics for the interpolation factor L = 8.*

L	M_{H_a}	$M_{H_{Ma}}$	$M_{H_{Mc}}$	Π	M
8	58	34	42	70	506

EXAMPLE 10.5

Design a bandpass filter using the quadrature method satisfying the following specifications:

$$
\left.
\begin{aligned}
A_p &= 0.2 \text{ dB} \\
A_r &= 40 \text{ dB} \\
\Omega_{r_1} &= 0.09\pi \text{ rad/s} \\
\Omega_{p_1} &= 0.1\pi \text{ rad/s} \\
\Omega_{p_2} &= 0.7\pi \text{ rad/s} \\
\Omega_{r_2} &= 0.71\pi \text{ rad/s} \\
\Omega_s &= 2\pi \text{ rad/s}
\end{aligned}
\right\}
\tag{10.84}
$$

SOLUTION

Given the bandpass specifications, the lowpass prototype filter must have a passband half the size of the desired passband width, and a transition bandwidth equal to the minimum transition bandwidth for the bandpass filter. For the passband ripple and stopband attenuation, the values specified for the bandpass filter can be used with a margin of about 40%. Therefore, in this example, the lowpass prototype is characterized by

$$
\left.
\begin{aligned}
\delta'_p &= 0.0115 \times 40\% = 0.0046 \\
\delta'_r &= 0.01 \times 40\% = 0.004 \\
\omega'_p &= \frac{\omega_{p2} - \omega_{p1}}{2} = 0.3\pi \\
\omega'_r &= \omega'_p + \min\left\{(\omega_{p1} - \omega_{r_1}), (\omega_{r_2} - \omega_{p2})\right\} = 0.31\pi
\end{aligned}
\right\}
\tag{10.85}
$$

This filter can be designed using the frequency response masking approach with an efficient ripple margin assignment, seen in the previous subsection. In this case, the interpolation factor that minimizes the total number of multiplications is $L = 8$ and the corresponding filter characteristics are given in Table 10.8, with the resulting magnitude response in Figure 10.21.

The resulting bandpass filter using the quadrature method is shown in Figure 10.22.

For the complete quadrature realization, the total number of multiplications is 140, twice the number of multiplications necessary for the prototype lowpass filter. For this example, the minimax filter would be of order 384, thus requiring 193 multiplications

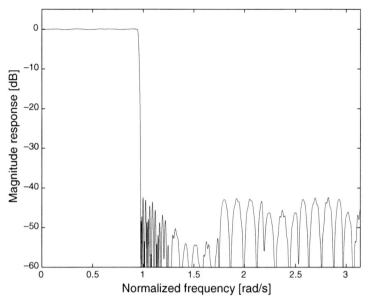

Figure 10.21 Lowpass prototype designed with the frequency response masking approach for the quadrature design of a bandpass filter.

per output sample. Therefore, in this case, the quadrature design represents a saving of about 30% of the number of multiplications required by the standard minimax filter.

△

10.7   Efficient FIR structures with MATLAB

Most of the realizations described in this chapter are somewhat advanced and do not have any related commands in the standard MATLAB versions, with the exception of the lattice and the frequency-domain forms, as seen below:

- `tf2latc`: Converts the direct form into the lattice form, inverting the operation of `latc2tf`. Notice that this command applies to FIR and IIR lattices, according to the transfer function format provided by the user. IIR lattices, however, as will be seen in Chapter 11, may also include a set of ladder parameters.
  Input parameters: Two vectors with the numerator and denominator coefficients of the direct-form transfer function, respectively.
  Output parameters:

  - A vector k with the reflection coefficients;
  - A vector with the ladder coefficients, for IIR lattice filters.

  Example:
  ```
  num=[1 0.6 -0.16]; den=[1 0.7 0.12];
  ```

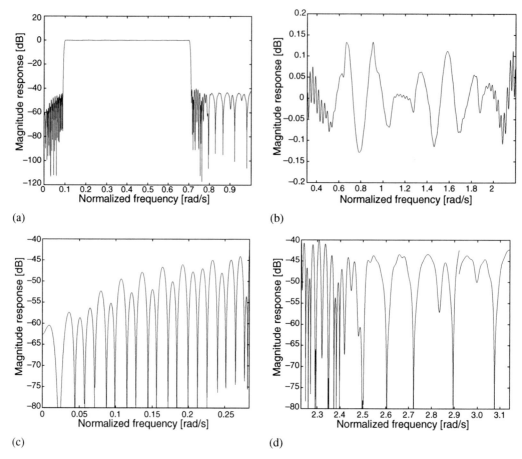

(a)

(b)

(c)

(d)

Figure 10.22 Magnitude responses of the bandpass filter designed with the quadrature approach: (a) overall filter; (b) passband detail; (c) lower stopband detail; (d) upper stopband detail.

```
[k,v]=tf2latc(num,den);
```

- **latc2tf**: Converts the lattice form into the direct form, inverting the operation of **tf2latc**.

  Input parameters:

  - A vector **k** with the reflection coefficients;
  - A vector with the ladder coefficients, for IIR lattice filters;
  - A string **'fir'** (default) or **'iir'** that determines whether the desired lattice is correspondingly FIR or IIR, if only one input vector is provided.

  Output parameters: Two vectors with the numerator and denominator coefficients of the direct-form transfer function, respectively.

  Examples:

  ```
  k=[0.625 0.12];
  ```

```
num=latc2tf(k,'fir'); % FIR case
[num,den]=latc2tf(k,'iir'); % IIR case
```

- `latcfilt`: Performs signal filtering with lattice form.
  Input parameters:

  - A vector k of reflection coefficients;
  - For IIR lattice filters, a vector of ladder coefficients. For FIR lattice filters, use a scalar 1;
  - The input-signal vector x.

  Output parameter: The filtered signal y.
  Example:
  ```
  k=[0.9 0.8 0.7]; x=0:0.1:10;
  y=latcfilt(k,x);
  ```

- `fftfilt`: Performs signal filtering with the FIR frequency-domain form. See Section 3.8.

For all other realizations described in this chapter, MATLAB can be used as a friendly environment for implementing efficient design procedures, and the interested reader is encouraged to do so.

10.8   Summary

In this chapter, several efficient FIR structures were presented as alternatives to the direct form seen in Chapter 5.

The lattice structure was introduced and its applications to the implementation of 2-band biorthogonal filter banks with linear phase was discussed. Other structures included the polyphase form, which is suitable for parallel processing, the frequency domain form, which is suitable for offline filtering operations, and the recursive form, which constitutes a basic block for several algorithms in digital signal processing.

The design of linear-phase FIR filters with reduced arithmetic complexity was presented, including the prefilter, the interpolation, and the frequency response masking approaches. The frequency response masking approach can be seen as a generalization of the first two approaches, allowing the design of wideband filters with very narrow transition bands. The complete frequency response masking design was presented, including the use of an efficient ripple margin assignment. Extensions of the frequency response masking approach include the use of multiple levels of masking, where the frequency response masking concept is applied to the design of one or both of the masking filters (Lim, 1986). The computational complexity can also be reduced by exploiting the similarities of the two masking filters, that is, the same passband ripple and stopband attenuation levels. If these two filters also have similar cutoff frequencies, a single masking filter with intermediate characteristics can be used. Such a filter is then followed by simple equalizers that transform the frequency response of this unique masking filter into the required responses of the original frequency masks (Lim & Lian, 1994). The chapter ended with a description of the quadrature method for bandpass and bandstop filters, which reduces the computational complexity.

10.9   Exercises

10.1   Prove equations (10.6) and (10.7).

10.2   Using equations (10.3)–(10.7), derive the order-recursive relationship given in equation (10.8).

10.3   Write down a MATLAB command that determines the FIR lattice coefficients from the FIR direct-form coefficients.

10.4   For $\mathbf{E}(z)$ and $\mathbf{R}(z)$ defined as in equations (10.17) and (10.19), show that:

(a) $\mathbf{E}(z)\mathbf{R}(z) = z^{-M}$.

(b) The filter bank has linear phase.

10.5   Design a highpass filter using the minimax method satisfying the following specifications:

$$A_p = 0.8 \text{ dB}$$
$$A_r = 40 \text{ dB}$$
$$\Omega_r = 5000 \text{ Hz}$$
$$\Omega_p = 5200 \text{ Hz}$$
$$\Omega_s = 12\,000 \text{ Hz}$$

Determine the corresponding lattice coefficients for the resulting filter using the command you created in Exercise 10.3. Compare the results obtained using the standard command `tf2latt`.

10.6   Use the MATLAB commands `filter` and `filtlatt` with the coefficients obtained in Exercise 10.5 to filter a given input signal. Verify that the output signals are identical. Compare the processing time and the total number of floating-point operations used by each command, using `tic`, `toc`, and `flops`.

10.7   Show that the linear-phase relations in equation (10.21) hold for the analysis filters when their polyphase components are given by equation (10.17).

10.8   Use the MATLAB commands `filter`, `conv`, and `fft` to filter a given input signal, with the impulse response obtained in Exercise 10.5. Verify what must be done to the input signal in each case to force all output signals to be identical. Compare the processing time and the total number of floating point operations used by each command, using `tic`, `toc`, and `flops` (see the MATLAB `help` for the correct usage of these commands).

10.9 Plot the magnitude response of the RRS filter with $M = 2, 4, 6, 8$. Compare the resulting passband widths and d.c. gains in each case.

10.10 Design a lowpass filter using the prefilter and interpolation methods satisfying the following specifications:

$A_p = 0.8$ dB
$A_r = 40$ dB
$\Omega_p = 4500$ Hz
$\Omega_r = 5200$ Hz
$\Omega_s = 12\,000$ Hz

Compare the required number of multiplications per output sample in each design.

10.11 Demonstrate quantitatively that FIR filters based on the prefilter and the interpolation methods have lower sensitivity and reduced output roundoff noise than the minimax filters implemented using the direct form, when they satisfy the same specifications.

10.12 Design a lowpass filter using the frequency masking method satisfying the following specifications:

$A_p = 2.0$ dB
$A_r = 40$ dB
$\omega_p = 0.33\pi$ rad/sample
$\omega_r = 0.35\pi$ rad/sample

Compare the results obtained with and without an efficient ripple margin distribution, with respect to the total number of multiplications per output sample, to the maximum passband ripple, and to the resulting minimum stopband attenuation.

10.13 Design a highpass filter using the frequency masking method satisfying the specifications in Exercise 10.5. Compare the results obtained with and without an efficient ripple margin distribution, with respect to the total number of multiplications per output sample, with the results obtained in Exercise 10.5.

10.14 Design a bandpass filter using the quadrature method satisfying the following specifications:

$A_p = 0.02$ dB
$A_r = 40$ dB
$\Omega_{r_1} = 0.068\pi$ rad/s
$\Omega_{p_1} = 0.07\pi$ rad/s
$\Omega_{p_2} = 0.95\pi$ rad/s
$\Omega_{r_2} = 0.952\pi$ rad/s
$\Omega_s = 2\pi$ rad/s

Compare your results with the standard minimax filter.

10.15 Design a bandstop filter using the quadrature method satisfying the following specifications:

$A_p = 0.2$ dB
$A_r = 60$ dB
$\omega_{p_1} = 0.33\pi$ rad/sample
$\omega_{r_1} = 0.34\pi$ rad/sample
$\omega_{r_2} = 0.46\pi$ rad/sample
$\omega_{p_2} = 0.47\pi$ rad/sample

Compare your results with the standard minimax filter.

11 Efficient IIR structures

11.1 Introduction

The most widely used realizations for IIR filters are the cascade and parallel forms of second-order, and, sometimes, first-order, sections. The main advantages of these realizations come from their inherent modularity, which leads to efficient VLSI implementations, to simplified noise and sensitivity analyses, and to simple limit-cycle control. This chapter presents high-performance second-order structures, which are used as building blocks in high-order realizations. The concept of section ordering for the cascade form, which can reduce roundoff noise in the filter output, is introduced. Then we present a technique to reduce the output roundoff-noise effect known as error spectrum shaping. This is followed by consideration of some closed-form equations for the scaling coefficients of second-order sections for the design of parallel-form filters.

There are also other interesting realizations such as the IIR lattice structures, whose synthesis method is presented. A related class of realizations are the wave digital filters, which have very low sensitivity and also allow the elimination of zero-input and overflow limit cycles. The wave digital filters are derived from analog filter prototypes, employing the concepts of incident and reflected waves. The detailed design of these structures is presented in this chapter.

11.2 IIR parallel and cascade filters

The Nth-order direct forms seen in Chapter 4, Figures 4.11–4.13, have roundoff-noise transfer functions $G_i(z)$ (see Figure 7.3) and scaling transfer functions $F_i(z)$ (see Figure 7.7) whose L_2 or L_∞ norms assume significantly high values. This occurs because these transfer functions have the same denominator polynomial as the filter transfer function, but without the zeros to attenuate the gain introduced by the poles close to the unit circle (see Exercise 7.3).

Also, we have that an Nth-order direct-form filter is in general implemented as the ratio of two high-order polynomials. Since a variation in a single coefficient can cause significant variation on all the polynomial roots, such a filter tends to present high sensitivity to coefficient quantization.

In order to deal with the above problems, it is wise to implement high-order transfer functions through the cascade or parallel connection of second-order building blocks, instead of using the direct-form realization. The second-order building blocks can be selected from a large collection of possibilities (Jackson, 1969; Szczupak & Mitra, 1975; Diniz, 1984). In this section, we deal with cascade and parallel realizations whose sub-filters are direct-form second-order sections. These realizations are canonic in the sense that they require the minimum number of multipliers, adders, and delays to implement a given transfer function.

Another advantage inherent to the cascade and parallel forms is their inherent modularity, which makes them suitable for VLSI implementation.

11.2.1 Parallel form

A canonic parallel realization is shown in Figure 11.1, where the corresponding transfer function is given by

$$H(z) = h_0 + \sum_{i=1}^{m} H_i^p(z) = h_0 + \sum_{i=1}^{m} \frac{\gamma_{0i} z^2 + \gamma_{1i} z}{z^2 + m_{1i} z + m_{2i}} \tag{11.1}$$

Many alternative parallel realizations can be generated by choosing different second-order sections, as seen, for instance, in Jackson (1969); Szczupak & Mitra (1975); Diniz (1984). Another example is considered in Figure 11.52, Exercise 11.1. In general, the performances of the structures presented in Figures 11.1 and 11.52 are among the best for parallel forms employing canonic second-order sections.

It is easy to show that the scaling coefficients to avoid internal signal overflow in the form seen in Figure 11.1 are given by (see Exercise 7.3)

$$\lambda_i = \frac{1}{\|F_i(z)\|_p} \tag{11.2}$$

where

$$F_i(z) = \frac{1}{D_i(z)} = \frac{1}{z^2 + m_{1i} z + m_{2i}} \tag{11.3}$$

Naturally, the numerator coefficients of each section must be divided by λ_i, so that the overall filter transfer function remains unchanged.

Using equation (7.20), one can show that the PSD of the output roundoff noise for the structure in Figure 11.1 is given by

$$P_y(z) = \sigma_e^2 \left(2m + 1 + 3 \sum_{i=1}^{m} \frac{1}{\lambda_i^2} H_i^p(z) H_i^p(z^{-1}) \right) \tag{11.4}$$

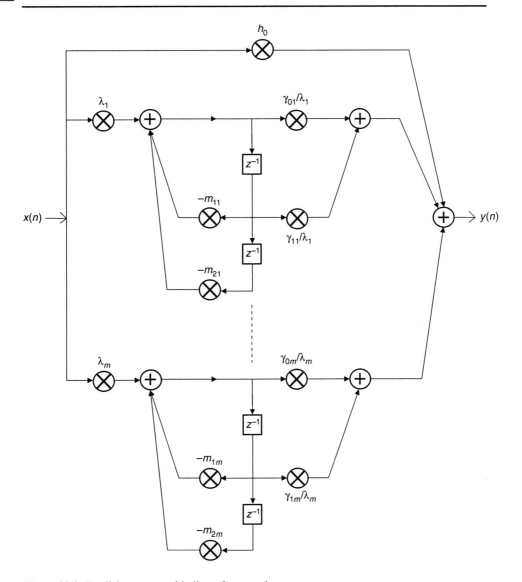

Figure 11.1   Parallel structure with direct-form sections.

when quantizations are performed before the additions. In this case, the output-noise variance, or the average power of the output noise, is then equal to

$$\sigma_o^2 = \sigma_e^2 \left(2m + 1 + 3 \sum_{i=1}^{m} \frac{1}{\lambda_i^2} \| H_i^p (e^{j\omega}) \|_2^2 \right) \tag{11.5}$$

and the relative noise variance becomes

$$\sigma^2 = \frac{\sigma_o^2}{\sigma_e^2} = \left(2m + 1 + 3 \sum_{i=1}^{m} \frac{1}{\lambda_i^2} \| H_i^p (e^{j\omega}) \|_2^2 \right) \tag{11.6}$$

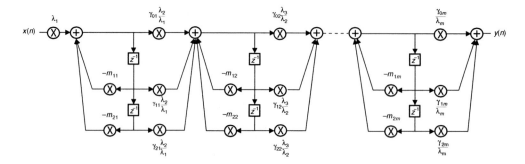

Figure 11.2 Cascade of direct-form sections.

For the cases where quantization is performed after the additions, the PSD becomes

$$P_y(z) = \sigma_e^2 \left(1 + \sum_{i=1}^{m} \frac{1}{\lambda_i^2} H_i^P(z) H_i^P(z^{-1}) \right) \tag{11.7}$$

and then

$$\sigma^2 = \frac{\sigma_o^2}{\sigma_e^2} = \left(1 + \sum_{i=1}^{m} \frac{1}{\lambda_i^2} \| H_i^P(e^{j\omega}) \|_2^2 \right) \tag{11.8}$$

Although only even-order structures have been discussed so far, expressions for odd-order structures (which contain one first-order section) can be obtained in a very similar way.

In the parallel forms, as the positions of the zeros depend on the summation of several polynomials, which involves all coefficients in the realization, the precise positioning of the filter zeros becomes a difficult task. Such high sensitivity of the zeros to coefficient quantization constitutes the main drawback of the parallel forms for most practical implementations.

11.2.2 Cascade form

The cascade connection of direct-form second-order sections, depicted in Figure 11.2, has a transfer function given by

$$H(z) = \prod_{i=1}^{m} H_i(z) = \prod_{i=1}^{m} \frac{\gamma_{0i} z^2 + \gamma_{1i} z + \gamma_{2i}}{z^2 + m_{1i} z + m_{2i}} \tag{11.9}$$

In this structure, the scaling coefficients are calculated as

$$\lambda_i = \frac{1}{\| \prod_{j=1}^{i-1} H_j(z) F_i(z) \|_p} \tag{11.10}$$

with

$$F_i(z) = \frac{1}{D_i(z)} = \frac{1}{z^2 + m_{1i} z + m_{2i}} \tag{11.11}$$

as before. As illustrated in Figure 11.2, the scaling coefficient of each section can be incorporated with the output coefficients of the previous section. This strategy leads not only to a reduction in the multiplier count, but also to a possible decrease in the quantization noise at the filter output, since the number of nodes to be scaled is reduced.

Assuming that the quantizations are performed before the additions, the output PSD for the cascade structure of Figure 11.2 is given by

$$
P_y(z) = \sigma_e^2 \left(3 + \frac{3}{\lambda_1^2} \prod_{i=1}^{m} H_i(z) H_i(z^{-1}) + 5 \sum_{j=2}^{m} \frac{1}{\lambda_j^2} \prod_{i=j}^{m} H_i(z) H_i(z^{-1}) \right) \tag{11.12}
$$

The relative noise variance is then

$$
\sigma^2 = \frac{\sigma_o^2}{\sigma_e^2} = \left(3 + \frac{3}{\lambda_1^2} \left\| \prod_{i=1}^{m} H_i(e^{j\omega}) \right\|_2^2 + 5 \sum_{j=2}^{m} \frac{1}{\lambda_j^2} \left\| \prod_{i=j}^{m} H_i(e^{j\omega}) \right\|_2^2 \right) \tag{11.13}
$$

For the case where quantizations are performed after additions, $P_y(z)$ becomes

$$
P_y(z) = \sigma_e^2 \left(1 + \sum_{j=1}^{m} \frac{1}{\lambda_j^2} \prod_{i=j}^{m} H_i(z) H_i(z^{-1}) \right) \tag{11.14}
$$

and then

$$
\frac{\sigma_o^2}{\sigma_e^2} = \left(1 + \sum_{j=1}^{m} \frac{1}{\lambda_j^2} \left\| \prod_{i=j}^{m} H_i(e^{j\omega}) \right\|_2^2 \right) \tag{11.15}
$$

Two practical problems that need consideration in the design of cascade structures are:

- Which pairs of poles and zeros will form each second-order section (the pairing problem).
- The ordering of the sections.

Both issues have a large effect on the output quantization noise. In fact, the roundoff noise and the sensitivity of cascade form structures can be very high if an inadequate choice for the pairing and ordering is made (Jackson, 1969).

A rule of thumb for the pole–zero pairing in cascade form using second-order sections is to minimize the L_p norm of the transfer function of each section, for either $p = 2$ or $p = \infty$. The pairs of complex conjugate poles close to the unit circle, if not accompanied by zeros which are close to them, tend to generate sections whose norms of $H_i(z)$ are high. As a result, a natural rule is to pair the poles closest to the unit circle with the zeros that are closest to them. Then one should pick the poles second closest to the unit circle and pair them with the zeros, amongst the remaining, that are closest to them, and so on, until all sections are formed. Needless to say, when dealing with filters with real coefficients, most poles and zeros come in complex conjugate pairs, and in those cases the complex conjugate poles (and zeros) are jointly considered in the pairing process.

For section ordering, first we must notice that, for a given section of the cascade structure, the previous sections affect its scaling factor, whereas the following sections affect the noise gain. We then define a peaking factor that indicates how sharp the section frequency response is

$$P_i = \frac{\|H_i(z)\|_\infty}{\|H_i(z)\|_2} \tag{11.16}$$

We now consider two separate cases (Jackson, 1970b):

- If we scale the filter using the L_2 norm, then the scaling coefficients tend to be large, and thus the signal-to-noise ratio at the output of the filter is in general not problematic. In these cases, it is interesting to choose the section ordering so that the maximum value of the output-noise PSD, $\|\text{PSD}\|_\infty$, is minimized. Section i amplifies the $\|\text{PSD}\|_\infty$ originally at its input by $\left(\lambda_i \|H_i(e^{j\omega})\|_\infty\right)^2$. Since in the L_2 scaling, $\lambda_i = \frac{1}{\|H_i(e^{j\omega})\|_2}$, then each section amplifies the $\|\text{PSD}\|_\infty$ by $\left(\frac{\|H_i(e^{j\omega})\|_\infty}{\|H_i(e^{j\omega})\|_2}\right)^2 = P_i^2$, the square of the peaking factor. Since the first sections affect the least number of noise sources, one should order the sections in decreasing order of peaking factors so as to minimize the maximum value of output-noise PSD.

- If we scale the filter using the L_∞ norm, then the scaling coefficients tend to be small, and thus the maximum peak value of the output-noise PSD is in general not problematic. In these cases, it is interesting to choose the section ordering so that the output signal-to-noise ratio is maximized, that is, the output-noise variance σ_o^2 is minimized. Section i amplifies the output-noise variance at its input by $\left(\lambda_i \|H_i(e^{j\omega})\|_2\right)^2$. Since in the L_∞ scaling, $\lambda_i = \frac{1}{\|H_i(e^{j\omega})\|_\infty}$, then each section amplifies the σ_o^2 by $\left(\frac{\|H_i(e^{j\omega})\|_2}{\|H_i(e^{j\omega})\|_\infty}\right)^2 = \frac{1}{P_i^2}$, the inverse of the square of the peaking factor. Since the first sections affect the least number of noise sources, one should order the sections in increasing order of peaking factors so as to minimize σ_o^2.

For other types of scaling, both ordering strategies are considered equally efficient.

Table 11.1. *Parallel structure using direct-form second-order sections.*
Feedforward coefficient: $h_0 = -1.389\,091\,10\text{E}{-}04.$

Coefficient	Section 1	Section 2	Section 3
γ_1	$4.926\,442\,74\text{E}{-}03$	$-1.282\,936\,15\text{E}{-}02$	$7.775\,256\,56\text{E}{-}03$
γ_0	$-7.666\,341\,48\text{E}{-}03$	$1.585\,671\,08\text{E}{-}02$	$-7.916\,850\,32\text{E}{-}03$
m_2	$9.923\,758\,27\text{E}{-}01$	$9.842\,689\,99\text{E}{-}01$	$9.921\,021\,78\text{E}{-}01$
m_1	$-1.626\,784\,34\text{E}{+}00$	$1.605\,434\,10\text{E}{+}00$	$-1.596\,449\,73\text{E}{+}00$

Table 11.2. *Parallel structure using direct-form second-order sections quantized*
with 9 bits. Feedforward coefficient: $h_0 = 0.000\,000\,00.$

Coefficient	Section 1	Section 2	Section 3
γ_1	$5.126\,953\,13\text{E}{-}03$	$-1.269\,531\,25\text{E}{-}02$	$7.781\,982\,42\text{E}{-}03$
γ_0	$-7.812\,500\,00\text{E}{-}03$	$1.586\,914\,06\text{E}{-}02$	$-8.041\,381\,84\text{E}{-}03$
m_2	$9.921\,875\,00\text{E}{-}01$	$9.843\,750\,00\text{E}{-}01$	$9.921\,875\,00\text{E}{-}01$
m_1	$-1.625\,000\,00\text{E}{+}00$	$1.601\,562\,50\text{E}{+}00$	$-1.593\,750\,00\text{E}{+}00$

EXAMPLE 11.1

Design an elliptic bandpass filter satisfying the following specifications:

$$\left.\begin{array}{l} A_p = 0.5 \text{ dB} \\ A_r = 65 \text{ dB} \\ \Omega_{r_1} = 850 \text{ rad/s} \\ \Omega_{p_1} = 980 \text{ rad/s} \\ \Omega_{p_2} = 1020 \text{ rad/s} \\ \Omega_{r_2} = 1150 \text{ rad/s} \\ \Omega_s = 10\,000 \text{ rad/s} \end{array}\right\} \tag{11.17}$$

Realize the filter using the parallel and cascade forms of second-order direct-form sections. Then quantize the coefficients to 9 bits, including the sign bit, and verify the results.

SOLUTION

Tables 11.1 and 11.3 list the multiplier coefficients of the designed filters, and Tables 11.2 and 11.4 list the quantized coefficients. Figure 11.3 depicts the magnitude responses obtained with the quantized parallel and cascade forms using direct-form sections.

For the cascade form, we first performed the pole–zero pairing, then we performed the section ordering according to suitable rules, leading to practical realizations with low roundoff noise at the filter output.

Table 11.3. *Cascade structure using direct-form second-order sections. Constant gain:* $h_0 = 1.346\,099\,35\text{E}-04$.

Coefficient	Section 1	Section 2	Section 3
γ_2	1.000 000 00	−1.000 000 00	1.000 000 00
γ_1	−1.722 654 64	0.000 000 00	−1.479 801 37
γ_0	1.000 000 00	1.000 000 00	1.000 000 00
m_2	0.992 375 83	0.984 269 00	0.992 102 18
m_1	−1.626 784 34	−1.605 434 10	−1.596 449 73

Table 11.4. *Cascade structure using direct-form second-order sections quantized with 9 bits. Constant gain:* $h_0 = 1.364\,052\,30\text{E}-04$.

Coefficient	Section 1	Section 2	Section 3
γ_2	1.000 000 00	−1.000 000 00	1.000 000 00
γ_1	−1.717 391 30	0.000 000 00	−1.477 386 93
γ_0	1.000 000 00	1.000 000 00	1.000 000 00
m_2	0.992 187 50	0.984 375 00	0.992 187 50
m_1	−1.625 000 00	−1.601 562 50	−1.593 750 00

Note that despite the reasonably large number of bits used to represent the coefficients, the magnitude responses moved notably away from the ideal ones. This occurred, in part, because the designed elliptic filter has poles with high selectivity, that is, poles very close to the unit circle. In particular, the performance of the quantized parallel realization became quite poor over the stopbands, as discussed in Subsection 11.2.1.

$\triangle$

11.2.3 Error spectrum shaping

This subsection introduces a technique to reduce the quantization noise effects on digital filters by feeding back the quantization error. This technique is known as error spectrum shaping (ESS) or error feedback.

Consider every adder whose inputs include at least one nontrivial product which is followed by a quantizer. The ESS consists of replacing all these adders by a recursive structure, as illustrated in Figure 11.4, whose purpose is to introduce zeros in the output-noise PSD. Although Figure 11.4 depicts a second-order feedback network for the error signal, in practice the order of this network can assume any value. The ESS coefficients are chosen to minimize the output-noise PSD (Higgins & Munson, 1984; Laakso & Hartimo, 1992). In some cases, these coefficients can be made trivial, such

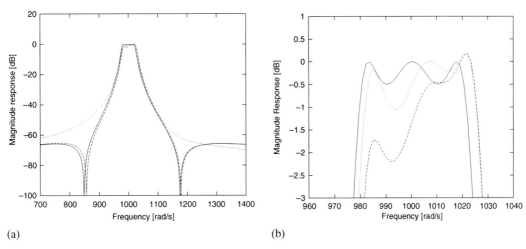

(a)   (b)

Figure 11.3 Coefficient-quantization effects in the cascade and parallel forms, using direct-form second-order sections: (a) overall magnitude response; (b) passband detail.
(Solid line – initial design; dashed line – cascade of direct-form sections (9 bits); dotted line – parallel of direct-form sections (9 bits).)

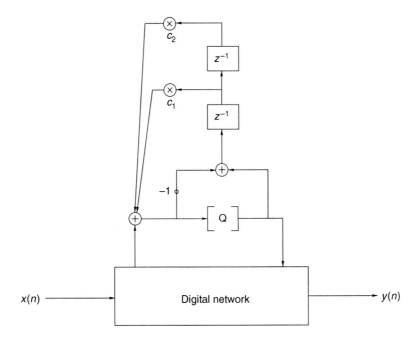

Figure 11.4 Error spectrum shaping structure (Q denotes quantizer).

as $0, \pm 1, \pm 2$, and still achieve sufficient noise reduction. Overall, the ESS approach can be interpreted as a form of recycling the quantization error signal, thus reducing the effects of signal quantization after a particular adder.

ESS can be applied to any digital filter structure and to any internal quantization node. However, since its implementation implies a cost overhead, ESS should be applied only at selected internal nodes, whose noise gains to the filter output are high. Structures having reduced number of quantization nodes (Diniz & Antoniou, 1985), for instance, are particularly suitable for ESS implementation. For example, the direct-form structure of Figure 4.11 requires a single ESS substitution for the whole filter.

For the cascade structure of direct-form second-order sections, as in Figure 11.2, each section j requires an ESS substitution. Let each feedback network be of second order. The values of $c_{1,j}$ and $c_{2,j}$ that minimize the output noise are calculated by solving the following optimization problem (Higgins & Munson, 1984):

$$\min_{c_{1,j}, c_{2,j}} \left\{ \left\| \left(1 + c_{1,j} z^{-1} + c_{2,j} z^{-2}\right) \prod_{i=j}^{m} H_i(e^{j\omega}) \right\|_2^2 \right\} \tag{11.18}$$

The optimal values of $c_{1,j}$ and $c_{2,j}$ are given by

$$c_{1,j} = \frac{t_1 t_2 - t_1 t_3}{t_3^2 - t_1^2} \tag{11.19}$$

$$c_{2,j} = \frac{t_1^2 - t_2 t_3}{t_3^2 - t_1^2} \tag{11.20}$$

where

$$t_1 = \int_{-\pi}^{\pi} \left| \prod_{i=j}^{m} H_i(e^{j\omega}) \right|^2 \cos \omega d\omega \tag{11.21}$$

$$t_2 = \int_{-\pi}^{\pi} \left| \prod_{i=j}^{m} H_i(e^{j\omega}) \right|^2 \cos(2\omega) d\omega \tag{11.22}$$

$$t_3 = \int_{-\pi}^{\pi} \left| \prod_{i=j}^{m} H_i(e^{j\omega}) \right|^2 d\omega \tag{11.23}$$

The complete algebraic development behind equations (11.19) and (11.20) is left as an exercise to the reader.

Using ESS, with the quantization performed after summation, the output relative power spectrum density (RPSD), which is independent of σ_e^2 for the cascade design, is given by

$$\text{RPSD} = 1 + \sum_{j=1}^{m} \left| \frac{1}{\lambda_j} \left(1 + c_{1,j} z^{-1} + c_{2,j} z^{-2}\right) \prod_{i=j}^{m} H_i(e^{j\omega}) \right|^2 \tag{11.24}$$

where λ_j is the scaling factor of section j. This expression explicitly shows how the ESS technique introduces zeros in the RPSD, thus allowing its subsequent reduction.

For a first-order ESS, the optimal value of $c_{1,j}$ would be

$$c_{1,j} = \frac{-t_1}{t_3} \tag{11.25}$$

with t_1 and t_3 as above.

For the cascade realization with ESS, the most appropriate strategy for section ordering is to place the sections with poles closer to the unit circle at the beginning. This is because the poles close to the unit circle contribute to peaks in the noise spectrum, so this ordering strategy forces the noise spectrum to have a narrower band, making the action of the ESS zeros more effective.

As seen in Chang (1981); Singh (1985); Laakso et al. (1992); Laakso (1993), the ESS technique can also be used to eliminate limit cycles. In such cases, absence of limit cycles can only be guaranteed at the quantizer output; hidden limit cycles may still exist in the internal loops (Butterweck et al., 1984).

11.2.4 Closed-form scaling

In most types of digital filter implemented with fixed-point arithmetic, scaling is based on the L_2 and L_∞ norms of the transfer functions from the filter inputs to the inputs of the multipliers. Usually, the L_2 norm is computed through summation of a large number of sample points (of the order of 200 or more) of the squared magnitude of the scaling transfer function. For the L_∞ norm, a search for the maximum magnitude of the scaling transfer function is performed over about the same number of sample points. It is possible, however, to derive simple closed-form expressions for the L_2 and L_∞ norms of second-order transfer functions. Such expressions are useful for scaling the sections independently, and greatly facilitate the design of parallel and cascade realizations of second-order sections (Bomar & Joseph, 1987; Laakso, 1992), and also of noncanonic structures with respect to the number of multipliers (Bomar, 1989).

Consider, for example,

$$H(z) = \frac{\gamma_1 z + \gamma_2}{z^2 + m_1 z + m_2} \tag{11.26}$$

Using, for instance, the pole-residue approach for solving circular integrals, the corresponding L_2 norm is given by

$$\|H(e^{j\omega})\|_2^2 = \frac{\gamma_1^2 + \gamma_2^2 - 2\gamma_1\gamma_2 \dfrac{m_1}{m_2 + 1}}{(1 - m_2^2)\left[1 - \left(\dfrac{m_1}{m_2 + 1}\right)^2\right]} \tag{11.27}$$

For the L_∞ norm, we have to find the maximum of $|H(z)|^2$. By noticing that $|H(e^{j\omega})|^2$ is a function of $\cos\omega$, and $\cos\omega$ is limited to the interval $[-1, 1]$, we have that the maximum is either at the extrema $\omega = 0$ $(z = 1)$, $\omega = \pi$ $(z = -1)$, or at ω_0 such that $-1 \leq \cos \omega_0 \leq 1$. Therefore, the L_∞ norm is given by (Bomar & Joseph, 1987; Laakso, 1992)

$$\|H(e^{j\omega})\|_\infty^2 = \max \left\{ \left(\frac{\gamma_1 + \gamma_2}{1 + m_1 + m_2} \right)^2 , \left(\frac{-\gamma_1 + \gamma_2}{1 - m_1 + m_2} \right)^2 , \frac{\gamma_1^2 + \gamma_2^2 + 2\gamma_1\gamma_2\zeta}{4m_2[(\zeta - \eta)^2 + \upsilon]} \right\}$$

(11.28)

where

$$\eta = \frac{-m_1(1 + m_2)}{4m_2}$$

(11.29)

$$\upsilon = \left(1 - \frac{m_1^2}{4m_2} \right) \frac{(1 - m_2)^2}{4m_2}$$

(11.30)

$$\zeta = \begin{cases} \text{sat}(\eta), & \text{for } \gamma_1\gamma_2 = 0 \\ \text{sat}\left\{ \upsilon \left[\sqrt{\left(1 + \frac{\eta}{\upsilon} \right)^2 + \frac{\upsilon}{\upsilon^2}} - 1 \right] \right\}, & \text{for } \gamma_1\gamma_2 \neq 0 \end{cases}$$

(11.31)

with

$$v = \frac{\gamma_1^2 + \gamma_2^2}{2\gamma_1\gamma_2}$$

(11.32)

and the function $\text{sat}(\cdot)$ being defined as

$$\text{sat}(x) = \begin{cases} 1, & \text{for } x > 1 \\ -1, & \text{for } x < -1 \\ x, & \text{for } -1 \leq x \leq 1 \end{cases}$$

(11.33)

The computations of the L_2 and L_∞ norms of $H(z)$ are left as an exercise for the interested reader.

11.3 State-space sections

The state-space approach allows the formulation of a design method for IIR digital filters with minimum roundoff noise. The theory behind this elegant design method was originally proposed in Mullis & Roberts (1976a,b); Hwang (1977). For a filter of order N, the minimum noise method leads to a realization entailing $(N + 1)^2$ multiplications. This multiplier count is very high for most practical implementations, which induced investigators to search for realizations which could approach the minimum noise performance while employing a reasonable number of multiplications. A good tradeoff is achieved if we realize high-order filters using parallel or cascade forms, where the second-order sections are minimum-noise state-space structures. In this section, we study two commonly used second-order state-space sections suitable for such approaches.

11.3.1 Optimal state-space sections

The second-order state-space structure shown in Figure 11.5 can be described by

$$
\left.\begin{aligned}
\mathbf{x}(n+1) &= \mathbf{A}\mathbf{x}(n) + \mathbf{b}u(n) \\
y(n) &= \mathbf{c}\mathbf{x}(n) + \mathbf{d}u(n)
\end{aligned}\right\}
\tag{11.34}
$$

where $\mathbf{x}(n)$ is a column vector representing the outputs of the delays, $y(n)$ is a scalar, and

$$
\mathbf{A} = \begin{bmatrix} a_{11} & a_{12} \\ a_{21} & a_{22} \end{bmatrix}
\tag{11.35}
$$

$$
\mathbf{b} = \begin{bmatrix} b_1 \\ b_2 \end{bmatrix}
\tag{11.36}
$$

$$
\mathbf{c} = \begin{bmatrix} c_1 & c_2 \end{bmatrix}
\tag{11.37}
$$

$$
\mathbf{d} = \begin{bmatrix} d \end{bmatrix}
\tag{11.38}
$$

The overall transfer function, described as a function of the matrix elements related to the state-space formulation, is given by

$$
H(z) = \mathbf{c}\,[\mathbf{I}z - \mathbf{A}]^{-1}\,\mathbf{b} + \mathbf{d}
\tag{11.39}
$$

The second-order state-space structure can realize transfer functions described by

$$
H(z) = d + \frac{\gamma_1 z + \gamma_2}{z^2 + m_1 z + m_2}
\tag{11.40}
$$

Given $H(z)$ in the form of equation (11.40), an optimal design in the sense of minimizing the output roundoff noise can be derived, since the state-space structure has more coefficients than the minimum required. To explore this feature, we examine, without proof, a theorem first proposed in Mullis & Roberts (1976a,b). The design procedure resulting from the theorem generates realizations with a minimum-variance output noise, provided that the L_2 norm is employed to determine the scaling factor. It is interesting to notice that, despite being developed for filters using L_2 scaling, the minimum-noise design also leads to low-noise filters scaled using the L_∞ norm.

Note that, in the remainder of this subsection, primed variables will indicate filter parameters after scaling.

THEOREM 11.1

The necessary and sufficient conditions to obtain an output noise with minimum variance in a state-space realization are given by

$$
\mathbf{W}' = \mathbf{R}\mathbf{K}'\mathbf{R}
\tag{11.41}
$$

$$
K'_{ii}W'_{ii} = K'_{jj}W'_{jj}
\tag{11.42}
$$

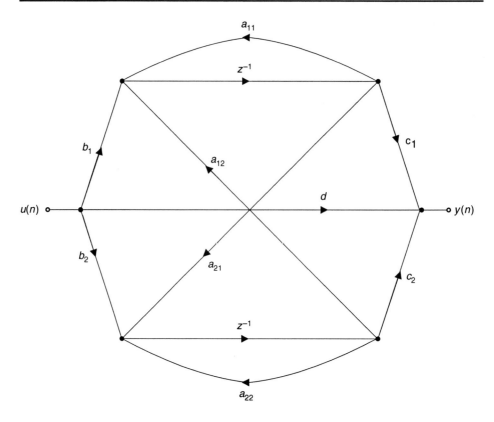

Figure 11.5 Second-order state-space structure.

for $i, j = 1, 2, \ldots, N$, where N is the filter order, $\mathbf{R}$ is an $N \times N$ diagonal matrix, and

$$\mathbf{K}' = \sum_{k=0}^{\infty} \mathbf{A}'^{k} \mathbf{b}' \mathbf{b}'^{H} (\mathbf{A}'^{k})^{H} \tag{11.43}$$

$$\mathbf{W}' = \sum_{k=0}^{\infty} (\mathbf{A}'^{k})^{H} \mathbf{c}'^{H} \mathbf{c}' \mathbf{A}'^{k} \tag{11.44}$$

where the H indicates the conjugate and transpose operations.

$\Diamond$

It can be shown (see Exercise 11.6) that

$$K'_{ii} = \left\| F'_i(e^{j\omega}) \right\|^2 \tag{11.45}$$

$$W'_{ii} = \left\| G'_i(e^{j\omega}) \right\|^2 \tag{11.46}$$

for $i = 1, 2, \ldots, N$, where $F'_i(z)$ is the transfer function from the scaled filter input to the state variable $x_i(k + 1)$, and $G'_i(z)$ is the transfer function from the state variable $x_i(k)$ to the scaled filter output (see Figures 7.7 and 7.3).

Then, from equations (11.45) and (11.46), we have that, in the frequency domain, equation (11.42) is equivalent to

$$\|F_i'\|_2^2 \|G_i'\|_2^2 = \|F_j'\|_2^2 \|G_j'\|_2^2 \tag{11.47}$$

In the case of second-order filters, if the L_2 scaling is performed, then

$$K_{11}' = K_{22}' = \|F_1'\|_2 = \|F_2'\|_2 = 1 \tag{11.48}$$

and then, from Theorem 11.1, the following equality must hold

$$W_{11}' = W_{22}' \tag{11.49}$$

Similarly, we can conclude that we must have

$$\|G_1'\|_2^2 = \|G_2'\|_2^2 \tag{11.50}$$

indicating that the contributions of the internal noise sources to the output-noise variance are identical.

The conditions $K_{ii}' = \|F_i'\|_2^2 = 1$ and $W_{ii}' = W_{jj}'$, for all i and j, show that equation (11.41) can only be satisfied if

$$\mathbf{R} = \alpha\mathbf{I} \tag{11.51}$$

and, as a consequence, the optimality condition of Theorem 11.1 is equivalent to

$$\mathbf{W}' = \alpha^2\mathbf{K}' \tag{11.52}$$

For a second-order filter, since $\mathbf{W}'$ and $\mathbf{K}'$ are symmetric and their respective diagonal elements are identical, equation (11.52) remains valid if we rewrite it as

$$\mathbf{W}' = \alpha^2\mathbf{J}\mathbf{K}'\mathbf{J} \tag{11.53}$$

where $\mathbf{J}$ is the reverse identity matrix defined as

$$\mathbf{J} = \begin{bmatrix} 0 & 1 \\ 1 & 0 \end{bmatrix} \tag{11.54}$$

By employing the definitions of $\mathbf{W}'$ and $\mathbf{K}'$ in equations (11.43) and (11.44), equation (11.53) is satisfied when

$$\mathbf{A}'^{\mathrm{T}} = \mathbf{J}\mathbf{A}'\mathbf{J} \tag{11.55}$$

$$\mathbf{c}'^{\mathrm{T}} = \alpha\mathbf{J}\mathbf{b}' \tag{11.56}$$

or, equivalently

$$a_{11}' = a_{22}' \tag{11.57}$$

$$\frac{b_1'}{b_2'} = \frac{c_2'}{c_1'} \tag{11.58}$$

Then, the following procedure can be derived for designing optimal second-order state-space sections.

(i) For filters with complex conjugate poles, choose an antisymmetric matrix **A** such that

$$\left.\begin{array}{l} a_{11} = a_{22} = \text{real part of the poles} \\ -a_{12} = a_{21} = \text{imaginary part of the poles} \end{array}\right\} \tag{11.59}$$

Note that the first optimality condition (equation (11.57)) is satisfied by this choice of **A**. The coefficients of matrix **A** can be calculated as a function of the coefficients of the transfer function $H(z)$ using

$$\left.\begin{array}{l} a_{11} = -\dfrac{m_1}{2} \\[2mm] a_{12} = -\sqrt{m_2 - \dfrac{m_1^2}{4}} \\[2mm] a_{21} = -a_{12} \\ a_{22} = a_{11} \end{array}\right\} \tag{11.60}$$

Also, compute the parameters b_1, b_2, c_1, and c_2, using

$$\left.\begin{array}{l} b_1 = \sqrt{\dfrac{\sigma + \gamma_2 + a_{11}\gamma_1}{2a_{21}}} \\[2mm] b_2 = \dfrac{\gamma_1}{2b_1} \\[2mm] c_1 = b_2 \\ c_2 = b_1 \end{array}\right\} \tag{11.61}$$

where

$$\sigma = \sqrt{\gamma_2^2 - \gamma_1\gamma_2 m_1 + \gamma_1^2 m_2} \tag{11.62}$$

For real poles, the matrix **A** must be of the form

$$\mathbf{A} = \left[\begin{array}{cc} a_1 & a_2 \\ a_2 & a_1 \end{array}\right] \tag{11.63}$$

where

$$\left.\begin{array}{l} a_1 = \frac{1}{2}(p_1 + p_2) \\[2mm] a_2 = \pm\frac{1}{2}(p_1 - p_2) \end{array}\right\} \tag{11.64}$$

with p_1 and p_2 denoting the real poles. The elements of vectors **b** and **c** are given by

$$b_1 = \pm\sqrt{\frac{\pm\sigma + \gamma_2 + a_1\gamma_1}{2a_2}}$$

$$b_2 = \frac{\beta_2}{2b_1}$$

$$c_1 = b_2$$

$$c_2 = b_1$$ \right\} \qquad (11.65)

with σ as before.

This synthesis approach is only valid if $(\gamma_2^2 - \gamma_1\gamma_2 m_1 + \gamma_1^2 m_2) > 0$, which may not occur for real poles in some designs. Solutions to this problem have been proposed in Kim (1980). However, it is worth observing that real poles can be implemented separately in first-order sections. In practice, we very seldom find more than one real pole in filter approximations.

(ii) Scale the filter using L_2 norm, through the following similarity transformation

$$(\mathbf{A'}, \mathbf{b'}, \mathbf{c'}, d) = (\mathbf{T}^{-1}\mathbf{A}\mathbf{T}, \mathbf{T}^{-1}\mathbf{b}, \mathbf{c}\mathbf{T}, d) \qquad (11.66)$$

where

$$\mathbf{T} = \begin{bmatrix} \|F_1\|_2 & 0 \\ 0 & \|F_2\|_2 \end{bmatrix} \qquad (11.67)$$

For the L_∞-norm scaling, use the following scaling matrix

$$\mathbf{T} = \begin{bmatrix} \|F_1\|_\infty & 0 \\ 0 & \|F_2\|_\infty \end{bmatrix} \qquad (11.68)$$

In this case the resulting second-order section is not optimal in the L_∞ sense. Nevertheless, practical results indicate that the solution is close to the optimal one.

Defining the vectors $\mathbf{f}(z) = [F_1(z), F_2(z)]^\mathsf{T}$ and $\mathbf{g}(z) = [G_1(z), G_2(z)]^\mathsf{T}$, the effects of the scaling matrix on these vectors are

$$\mathbf{f'}(z) = \left[z\mathbf{I} - \mathbf{A'} \right]^{-1}\mathbf{b'} = \mathbf{T}^{-1}\mathbf{f}(z) \qquad (11.69)$$

$$\mathbf{g'}(z) = \left[z\mathbf{I} - \mathbf{A'}^\mathsf{T} \right]^{-1}\mathbf{c'}^\mathsf{T} = (\mathbf{T}^{-1})^\mathsf{T}\mathbf{g}(z) \qquad (11.70)$$

The transfer functions $F_i(z)$ from the input node, where $u(n)$ is inserted, to the state variables $x_i(n)$ of the system $(\mathbf{A}, \mathbf{b}, \mathbf{c}, d)$ are given by

$$F_1(z) = \frac{b_1 z + (b_2 a_{12} - b_1 a_{22})}{z^2 - (a_{11} + a_{22})z + (a_{11}a_{22} - a_{12}a_{21})} \qquad (11.71)$$

$$F_2(z) = \frac{b_2 z + (b_1 a_{21} - b_2 a_{11})}{z^2 - (a_{11} + a_{22})z + (a_{11}a_{22} - a_{12}a_{21})} \qquad (11.72)$$

The expressions for the transfer functions from the internal nodes, that is, from the signals $x_i(n+1)$ to the section output node are

$$G_1(z) = \frac{c_1 z + (c_2 a_{21} - c_1 a_{22})}{z^2 - (a_{11} + a_{22})z + (a_{11}a_{22} - a_{12}a_{21})} \tag{11.73}$$

$$G_2(z) = \frac{c_2 z + (c_1 a_{12} - c_2 a_{11})}{z^2 - (a_{11} + a_{22})z + (a_{11}a_{22} - a_{12}a_{21})} \tag{11.74}$$

The output roundoff-noise PSD, considering quantization before the adders, for the section-optimal state-space structure in cascade form can be expressed as

$$P_y(z) = 3\sigma_e^2 \sum_{j=1}^{m} \prod_{l=j+1}^{m} H_l(z)H_l(z^{-1}) \left(1 + \sum_{i=1}^{2} G'_{ij}(z)G'_{ij}(z^{-1})\right) \tag{11.75}$$

where G'_{ij}, for $i = 1, 2$, are the noise transfer functions of the jth scaled section, and we consider $\prod_{l=m+1}^{m} H_l(z)H_l(z^{-1}) = 1$.

The scaling in the state-space sections of the cascade form is performed internally, using the transformation matrix $\mathbf{T}$. In order to calculate the elements of matrix $\mathbf{T}$, we can use the same procedure as in the cascade direct form.

In the case of the parallel form, the expression for the output roundoff-noise PSD, assuming quantization before additions, is

$$P_y(z) = \sigma_e^2 \left(2m + 1 + 3 \sum_{j=1}^{m} \sum_{i=1}^{2} G'_{ij}(z)G'_{ij}(z^{-1})\right) \tag{11.76}$$

The expressions for the output roundoff-noise PSD assuming quantization after additions can be easily derived.

11.3.2 State-space sections without limit cycles

This section presents a design procedure for a second-order state-space section that is free from constant-input limit cycles.

The transition matrix related to the section-optimal structure, described in the previous subsection (see equation (11.57)), has the following general form

$$\mathbf{A} = \begin{bmatrix} a & -\frac{\zeta}{\sigma} \\ \zeta\sigma & a \end{bmatrix} \tag{11.77}$$

where a, ζ, and σ are constants. This form is the most general for $\mathbf{A}$ that allows the realization of complex conjugate poles and the elimination of zero-input limit cycles.

As studied in Subsection 7.6.3, one can eliminate zero-input limit cycles on a recursive structure if there is a positive-definite diagonal matrix $\mathbf{G}$, such that

$(\mathbf{G} - \mathbf{A}^T\mathbf{G}\mathbf{A})$ is positive semidefinite. For second-order sections, this condition is satisfied if

$$a_{12}a_{21} \geq 0 \tag{11.78}$$

or

$$a_{12}a_{21} < 0 \quad \text{and} \quad |a_{11} - a_{22}| + \det\{\mathbf{A}\} \leq 1 \tag{11.79}$$

In the section-optimal structures, the elements of matrix $\mathbf{A}$ automatically satisfy equation (11.78), since $a_{11} = a_{22}$ and $\det(\mathbf{A}) \leq 1$, for stable filters.

Naturally, the quantization performed at the state variables still must be such that

$$\left|[x_i(k)]_Q\right| \leq |x_i(k)|, \ \forall k \tag{11.80}$$

where $[x]_Q$ denotes the quantized value of x. This condition can be easily guaranteed by using, for example, magnitude truncation and saturation arithmetic to deal with overflow.

If we also want to eliminate constant-input limit cycles, according to Theorem 7.3, the values of the elements of $\mathbf{p}u_0$, where $\mathbf{p} = (\mathbf{I} - \mathbf{A})^{-1}\mathbf{b}$, must be machine representable. In order to guarantee this condition independently of u_0, the column vector $\mathbf{p}$ must assume one of the forms

$$\mathbf{p} = \begin{cases} [\pm 1 \ 0]^T & \text{(Case I)} \\ [0 \ \pm 1]^T & \text{(Case II)} \\ [\pm 1 \ \pm 1]^T & \text{(Case III)} \end{cases} \tag{11.81}$$

For each case above, vector $\mathbf{b}$ of the state-space structure must be appropriately chosen to ensure the elimination of constant-input limit cycles, as given by

- Case I:

$$\left. \begin{aligned} b_1 &= \pm(1 - a_{11}) \\ b_2 &= \mp a_{21} \end{aligned} \right\} \tag{11.82}$$

- Case II:

$$\left. \begin{aligned} b_1 &= \mp a_{12} \\ b_2 &= \pm(1 - a_{22}) \end{aligned} \right\} \tag{11.83}$$

- Case III:

$$\left. \begin{aligned} b_1 &= \mp a_{12} \pm (1 - a_{11}) \\ b_2 &= \mp a_{21} \pm (1 - a_{22}) \end{aligned} \right\} \tag{11.84}$$

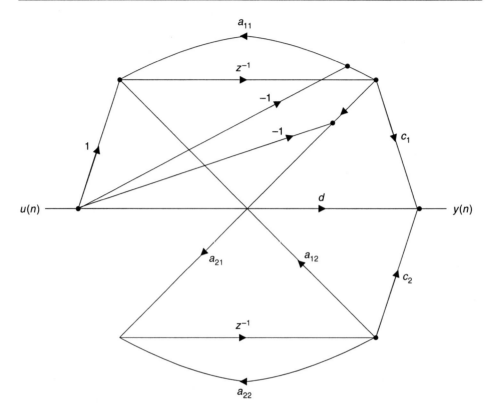

Figure 11.6 State-space structure free from limit cycles.

Based on the values of b_1 and b_2, for each case, it is possible to generate three structures (Diniz & Antoniou, 1986), henceforth referred to as Structures I, II, and III. Figure 11.6 depicts Structure I, where it can be seen that b_1 and b_2 are formed without actual multiplications. As a consequence, the resulting structure is more economical than the optimal second-order state-space structure. Similar results apply to all three structures. In fact, Structures I and II have the same complexity, whereas Structure III requires 5 extra additions, if we consider the adders needed for the elimination of constant-input limit cycles. For that reason, in what follows, we present the design for Structure I. If desired, the complete design for Structure III can be found in Sarcinelli & Camponêz (1997, 1998).

For Structure I, we have that

$$
\left.
\begin{aligned}
a_{11} &= a \\
a_{12} &= -\frac{\zeta}{\sigma} \\
a_{21} &= \sigma\zeta \\
a_{22} &= a_{11}
\end{aligned}
\right\}
\tag{11.85}
$$

and

$$b_1 = 1 - a_{11}$$
$$b_2 = -a_{21}$$
$$c_1 = \frac{\gamma_1 + \gamma_2}{1 + m_1 + m_2}$$
$$c_2 = -\frac{(m_1 + 2m_2)\gamma_1 + (2 + m_1)\gamma_2}{2\sigma\zeta(1 + m_1 + m_2)}$$

$$\qquad (11.86)$$

where

$$a = -\frac{m_1}{2} \qquad (11.87)$$

$$\zeta = \sqrt{\left(m_2 - \frac{m_1^2}{4}\right)} \qquad (11.88)$$

and σ is a free parameter whose choice is explained below. From the above equations,

$$\frac{b_1}{b_2} = -\frac{2 + m_1}{2\sigma\zeta} \qquad (11.89)$$

$$\frac{c_2}{c_1} = \frac{-(m_1 + 2m_2)\gamma_1 + (2 + m_1)\gamma_2}{2\sigma\zeta(\gamma_1 + \gamma_2)} \qquad (11.90)$$

Therefore, in this case, the optimality condition derived from Theorem 11.1 (equations (11.57) and (11.58)) is only met if

$$\frac{\gamma_1}{\gamma_2} = \frac{m_1 + 2}{m_2 - 1} \qquad (11.91)$$

Usually, this condition is violated, showing that the state-space structure which is free from constant-input limit cycles does not lead to minimum output noise. In practice, however, it is found that the performance of Structure I is very close to the optimal. More specifically, in the case where the zeros of $H(z)$ are placed at $z = 1$, the values of γ_1 and γ_2 are

$$\left.\begin{array}{l} \gamma_1 = -\gamma_0(2 + m_1) \\ \gamma_2 = \gamma_0(1 - m_2) \end{array}\right\} \qquad (11.92)$$

which satisfies equation (11.91). Hence, in the special case of filters with zeros at $z = 1$, Structure I also leads to minimum output noise.

The signal-to-noise ratio in a digital filter implemented using fixed-point arithmetic increases by increasing the internal signal dynamic range, which in turn increases by equalizing the maximum signal levels at the input of the quantizers. For the three structures, the maximum signal levels must be equalized at the state variables. Parameter σ is usually used to optimize the dynamic range of the state variables.

For Structure I, the transfer functions from the input node $u(k)$ to the state variables $x_i(k)$ are given by

$$F_1(z) = \frac{(1-a)z + (\zeta^2 - a + a^2)}{z^2 - 2az + (a^2 + \zeta^2)} \tag{11.93}$$

$$F_2(z) = \sigma F_2''(z) \tag{11.94}$$

where

$$F_2''(z) = \frac{-\zeta z + \zeta}{z^2 - 2az + (a^2 + \zeta^2)} \tag{11.95}$$

Equalization of the maximum signal level at the state variables is then achieved by forcing

$$\|F_1(z)\|_p - \|\sigma F_2''(z)\|_p \tag{11.96}$$

where $p = \infty$ or $p = 2$. Consequently, we must have

$$\sigma = \frac{\|F_1(z)\|_p}{\|F_2''(z)\|_p} \tag{11.97}$$

The transfer functions from the state variables $x_i(k+1)$ to the output in the structure of Figure 11.6 can be expressed as

$$G_1(z) = \frac{c_1}{2} \frac{2z + (m_1 + 2\xi)}{z^2 + m_1 z + m_2} \tag{11.98}$$

$$G_2(z) = \frac{c_2}{2} \frac{2z + (\alpha_1 + \frac{2\zeta^2}{\xi})}{z^2 + \alpha_1 z + \alpha_2} \tag{11.99}$$

where

$$\xi = \frac{-(\alpha_1 + 2\alpha_2)\beta_1 + (2 + \alpha_1)\beta_2}{2(\beta_1 + \beta_2)} \tag{11.100}$$

The RPSD expression for Structure I is then

$$\begin{aligned} \text{RPSD} &= 2\left|G_1'(e^{j\omega})\right|^2 + 2\left|G_2'(e^{j\omega})\right|^2 + 3 \\ &= 2\frac{1}{\lambda^2}\left|G_1''(e^{j\omega})\right|^2 + 2\frac{1}{\lambda^2\sigma^2}\left|G_2''(e^{j\omega})\right|^2 + 3 \end{aligned} \tag{11.101}$$

where $G_1'(e^{j\omega})$ and $G_2'(e^{j\omega})$ are the noise transfer functions for the scaled filter, λ is the scaling factor, and $G_1''(e^{j\omega})$ and $G_2''(e^{j\omega})$ are functions generated from $G_1'(e^{j\omega})$ and $G_2'(e^{j\omega})$, when we remove the parameters σ and λ from them.

We now show that choosing σ according to equation (11.97) leads to the minimization of the output noise. From equation (11.101), we can infer that the output noise can be minimized when σ and λ are maximized, with the scaling coefficient given by

$$\lambda = \frac{1}{\max\left\{\|F_1(z)\|_p, \|F_2(z)\|_p\right\}} \tag{11.102}$$

However, $F_1(z)$ is not a function of σ, and, as a consequence, the choice of $\|F_2(z)\|_p = \|F_1(z)\|_p$ leads to a maximum value for λ. On the other hand, the maximum value that σ can assume, without reducing the value of λ, is given by

$$\sigma = \frac{\|F_1(e^{j\omega})\|_p}{\|F_2''(e^{j\omega})\|_p} \tag{11.103}$$

from which we can conclude that this choice for σ minimizes the roundoff noise at the filter output.

In order to design a cascade structure without limit cycles which realizes

$$H(z) = \prod_{i=1}^{m} H_i(z) = H_0 \prod_{i=1}^{m} \frac{z^2 + \gamma_{1i}' z + \gamma_{2i}'}{z^2 + \alpha_{1i} z + \alpha_{2i}} = \prod_{i=1}^{m} (d_i + H_i'(z)) \tag{11.104}$$

with $H_i'(z)$ described in the form of the first term of the right side of equation (11.39), one must adopt the following procedure for Structure I:

(i)   Calculate σ_i and λ_i for each section using

$$\sigma_i = \frac{\left\| F_{1i}(z) \prod_{j=1}^{i-1} H_j(z) \right\|_p}{\left\| F_{2i}''(z) \prod_{j=1}^{i-1} H_j(z) \right\|_p} \tag{11.105}$$

$$\lambda_i = \frac{1}{\left\| F_{2i}(z) \prod_{j=1}^{i-1} H_j(z) \right\|_p} \tag{11.106}$$

(ii) Determine a and ζ from equations (11.87) and (11.88).
(iii) Compute the coefficients of **A**, **b**, and **c** using equations (11.85) and (11.86).
(iv) Calculate the multiplier coefficients d_i by

$$d_i = \begin{cases} \dfrac{1}{\left\| \prod_{j=1}^{i} H_j(z) \right\|_p}, & \text{for } i = 1, 2, \ldots, (m-1) \\[2em] \dfrac{H_0}{\prod_{j=1}^{m-1} d_j}, & \text{for } i = m \end{cases} \tag{11.107}$$

in order to satisfy overflow constraints at the output of each section.
(v)   Incorporate the scaling multipliers of sections $2, 3, \ldots, m$ into the output multipliers of sections $1, 2, \ldots, (m-1)$, generating

$$\left. \begin{aligned} c_{1i}' &= c_{1i} \frac{\lambda_{i+1}}{\lambda_i} \\ c_{2i}' &= c_{2i} \frac{\lambda_{i+1}}{\lambda_i} \\ d_i' &= d_i \frac{\lambda_{i+1}}{\lambda_i} \end{aligned} \right\} \tag{11.108}$$

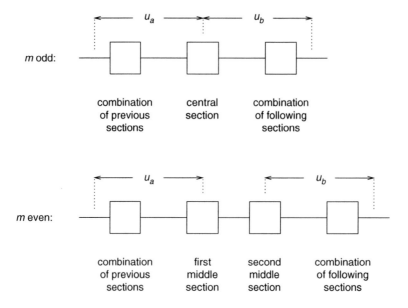

Figure 11.7 Ordering of state-space sections.

The cascade form design procedures employing the section-optimal and the limit-cycle-free state-space structures use the same strategy for pairing the poles and zeros as those employing direct-form sections. The section ordering depends on the definition of a parameter u_j, given by

$$u_j = \sum_{i=1}^{2} \frac{\max\left\{\left|F_{ij}(e^{j\omega})\right|\right\}}{\min\left\{\left|F_{ij}(e^{j\omega})\right|\right\}} \tag{11.109}$$

where the maximum is computed for all ω, while the minimum is calculated solely within the passband. According to Figure 11.7, for an odd value of sections, m, the ordering consists of placing the section with highest value for u_j to be the central section. For an even number of sections, the two with largest u_j are placed in the central positions, these are called first and second middle sections. For odd m, the previous and following sections to the central block are chosen from the remaining sections so as to minimize the summation of u_a and u_b (see Figure 11.7), one referred to the combination of the central section and all previous ones, and the other referred to the combination of the central section and all the following ones. For even m, the sections before and after the central sections are chosen, among the remaining ones, in order to minimize the summation of u_a and u_b (see Figure 11.7), one referred to the combination of the first middle section and all previous sections, and the other referred to the combination of the second middle section and the sections following it (Kim, 1980). This approach is continuously employed until all second-order blocks have been ordered.

Table 11.5. *Cascade structure using optimal state-space second-order sections.*

Coefficient	Section 1	Section 2	Section 3
a_{11}	8.027 170 50E−01	8.133 921 70E−01	7.982 248 63E−01
a_{12}	−5.909 402 84E−01	−5.780 858 57E−01	−6.066 663 28E−01
a_{21}	5.752 092 83E−01	5.721 797 21E−01	5.850 650 18E−01
a_{22}	8.027 170 50E−01	8.133 921 70E−01	7.982 248 63E−01
b_1	7.871 071 80E−02	5.785 102 31E−03	2.973 376 61E−02
b_2	1.544 577 30E−01	−1.667 467 43E−02	9.074 625 11E−03
c_1	8.873 568 59E−01	−9.035 059 50E−01	2.783 031 72E−02
c_2	4.521 916 47E−01	3.134 618 57E−01	9.118 835 57E−02
d	8.701 010 58E−02	1.090 405 31E−01	1.418 794 38E−02

The output roundoff-noise PSD of the state-space structure without limit cycles in cascade form is expressed by

$$P_y(z) = \sigma_e^2 \sum_{i=1}^{m} \frac{1}{\lambda_{i+1}^2} \left(2G'_{1i}(z)G'_{1i}(z^{-1}) + 2G'_{2i}(z)G'_{2i}(z^{-1}) + 3 \right) \qquad (11.110)$$

where $G'_{1i}(z)$ and $G'_{2i}(z)$ are the noise transfer functions of the scaled sections and $\lambda_{m+1} = 1$.

The design procedure for the state-space structure without limit cycles in the parallel form, which is rather simple, is provided in Diniz & Antoniou (1986). In this case, the expression for the roundoff-noise PSD at the output is

$$P_y(z) = \sigma_e^2 \left[1 + \sum_{i=1}^{m} \left(2G'_{1i}(z)G'_{1i}(z^{-1}) + 2G'_{2i}(z)G'_{2i}(z^{-1}) + 2 \right) \right] \qquad (11.111)$$

EXAMPLE 11.2
Repeat Example 11.1 using the cascade of optimal and limit-cycle-free state-space sections. Quantize the coefficients to 9 bits, including the sign bit, and verify the results.

SOLUTION
Tables 11.5–11.8 list the coefficients of all the designed filters. Figure 11.8 depicts the magnitude responses obtained by the cascade of optimal state-space sections and of state-space sections without limit cycles. In all cases the coefficients were quantized to 9 bits including the sign bit.

Note that with the same given number of bits used for quantization, the magnitude responses did not move away as much as in Example 11.1. That indicates that the present state-space sections have better sensitivity properties than the second-order direct-form sections used in Example 11.1.

Table 11.6. *Cascade structure using optimal state-space second-order sections quantized with 9 bits.*

Coefficient	Section 1	Section 2	Section 3
a_{11}	8.007 812 50E−01	8.125 000 00E−01	7.968 750 00E−01
a_{12}	−5.898 437 50E−01	−5.781 250 00E−01	−6.054 687 50E−01
a_{21}	5.742 187 50E−01	5.703 125 00E−01	5.859 375 00E−01
a_{22}	8.007 812 50E−01	8.125 000 00E−01	7.968 750 00E−01
b_1	7.812 500 00E−02	3.906 250 00E−03	3.125 000 00E−02
b_2	1.562 500 00E−01	−1.562 500 00E−02	7.812 500 00E−03
c_1	8.867 187 50E−01	−9.023 437 50E−01	2.734 375 00E−02
c_2	4.531 250 00E−01	3.125 000 00E−01	8.984 375 00E−02
d	8.593 750 00E−02	1.093 750 00E−01	1.562 500 00E−02

Table 11.7. *Cascade structure without limit cycles using optimal state-space second-order sections.* $\lambda = 2.720\,208\,87\text{E}{-}01$.

Coefficient	Section 1	Section 2	Section 3
a_{11}	8.027 170 50E−01	8.133 921 70E−01	7.982 248 63E−01
a_{12}	−5.828 867 98E−01	−5.782 313 97E−01	−5.848 591 58E−01
a_{21}	5.831 566 92E−01	5.720 357 05E−01	6.068 798 63E−01
a_{22}	8.027 170 50E−01	8.133 921 70E−01	7.982 248 63E−01
b_1	1.972 829 50E−01	1.866 078 30E−01	2.017 751 37E−01
b_2	−5.831 566 92E−01	−5.720 357 05E−01	−6.068 798 63E−01
c_1	−9.251 643 18E−03	−4.422 848 65E−02	9.189 037 75E−02
c_2	−2.859 968 02E−02	1.628 075 39E−02	−2.555 692 36E−02
d	9.251 643 18E−03	1.832 324 92E−01	2.919 130 42E−01

Table 11.8. *Cascade structure without limit cycles using optimal state-space second-order sections quantized with 9 bits.* $\lambda = 2.734\,375\,00\text{E}{-}01$.

Coefficient	Section 1	Section 2	Section 3
a_{11}	8.007 812 50E−01	8.125 000 00E−01	7.968 750 00E−01
a_{12}	−5.820 312 50E−01	−5.781 250 00E−01	−5.859 375 00E−01
a_{21}	5.820 312 50E−01	5.703 125 00E−01	6.054 687 50E−01
a_{22}	8.007 812 50E−01	8.125 000 00E−01	7.968 750 00E−01
b_1	1.992 187 50E−01	1.875 000 00E−01	2.031 250 00E−01
b_2	−5.820 312 50E−01	−5.703 125 00E−01	−6.054 687 50E−01
c_1	−7.812 500 00E−03	−4.296 875 00E−02	9.375 000 00E−02
c_2	−2.734 375 00E−02	1.562 500 00E−02	−2.734 375 00E−02
d	7.812 500 00E−03	1.835 937 50E−01	2.929 687 50E−01

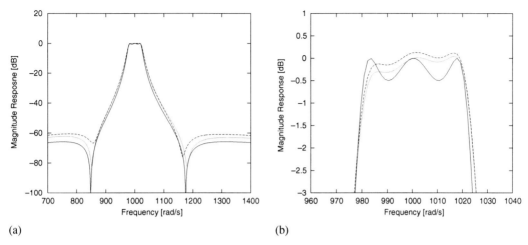

Figure 11.8 Coefficient-quantization effects in the cascade forms, using state-space second-order sections: (a) overall magnitude response; (b) passband detail.
(Solid line – initial design; dashed line – cascade of optimal state-space sections (9 bits); dotted line – cascade of limit-cycle-free state-space sections (9 bits).

Once again, in all designs, we first performed the pole–zero pairing, and then we performed the section ordering according to the adequate rule for the state-space sections.

$\triangle$

11.4   Lattice filters

Consider a general IIR transfer function written in the form

$$H(z) = \frac{N_M(z)}{D_N(z)} = \frac{\displaystyle\sum_{i=0}^{M} b_{i,M} z^{-i}}{1 + \displaystyle\sum_{i=1}^{N} a_{i,N} z^{-i}} \tag{11.112}$$

In the lattice construction, we concentrate first on the realization of the denominator polynomial through an order-reduction strategy. For that, we define the auxiliary Nth-order polynomial, obtained by reversing the order of the coefficients of the denominator $D_N(z)$, as given by

$$z B_N(z) = D_N(z^{-1}) z^{-N} = z^{-N} + \sum_{i=1}^{N} a_{i,N} z^{i-N} \tag{11.113}$$

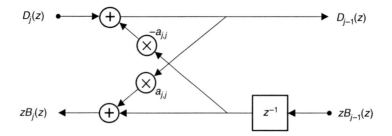

Figure 11.9 Two-multiplier network TP$_j$ implementing equation (11.117).

We can then calculate a reduced order polynomial as

$$(1 - a_{N,N}^2)D_{N-1}(z) = D_N(z) - a_{N,N}zB_N(z)$$
$$= (1 - a_{N,N}^2) + \cdots + (a_{N-1,N} - a_{N,N}a_{1,N})z^{-N+1} \quad (11.114)$$

where we can also express $D_{N-1}(z)$ as $1 + \sum_{i=1}^{N-1} a_{i,N-1}z^{-i}$. Note that the first and last coefficients of $D_N(z)$ are 1 and $a_{N,N}$, whereas for the polynomial $zB_N(z)$ they are $a_{N,N}$ and 1, respectively. This strategy to achieve the order reduction guarantees a monic $D_{N-1}(z)$, that is, $D_{N-1}(z)$ having the coefficient of z^0 equal to 1. By induction, this order-reduction procedure can be performed repeatedly, thus yielding

$$zB_j(z) = D_j(z^{-1})z^{-j} \quad (11.115)$$

$$D_{j-1}(z) = \frac{1}{1 - a_{j,j}^2}\left(D_j(z) - a_{j,j}zB_j(z)\right) \quad (11.116)$$

for $j = N, N-1, \ldots, 1$, with $zB_0(z) = D_0(z) = 1$. It can be shown that the above equations are equivalent to the following expression:

$$\begin{bmatrix} D_{j-1}(z) \\ B_j(z) \end{bmatrix} = \begin{bmatrix} 1 & -a_{j,j} \\ a_{j,j}z^{-1} & (1 - a_{j,j}^2)z^{-1} \end{bmatrix}\begin{bmatrix} D_j(z) \\ B_{j-1}(z) \end{bmatrix} \quad (11.117)$$

The above equation can be implemented, for example, by the two-port network TP$_j$ shown in Figure 11.9.

The advantage of this representation arises from the fact that, by cascading two-port networks TP$_j$ as in Figure 11.10, for $j = N, \ldots, 1$, we can implement $\frac{1}{D_N(z)}$, where $D_N(z)$ is the denominator of the transfer function. This can be easily understood by looking at the input $X(z)$ as $\frac{X(z)D_N(z)}{D_N(z)}$. If we do so, then at the right of TP$_1$ we will end up having $\frac{X(z)D_0(z)}{D_N(z)} = \frac{X(z)}{D_N(z)}$.

Since at the lower branches of the two-port networks in Figure 11.10 we have the signals $\frac{zB_j(z)}{D_N(z)}$ available, then a convenient way to form the desired numerator is to apply weights to the polynomials $zB_j(z)$, such that

$$N_M(z) = \sum_{j=0}^{M} v_j zB_j(z) \quad (11.118)$$

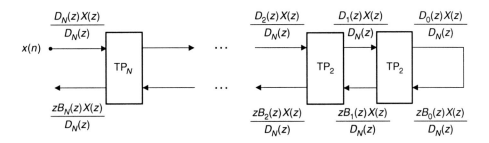

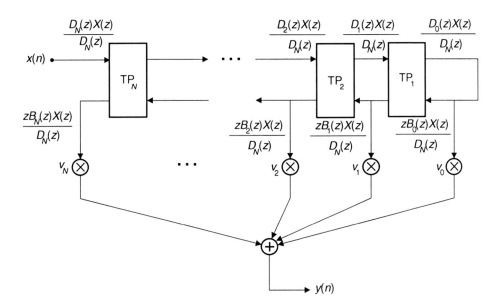

Figure 11.10 Generation of the denominator of the IIR lattice digital filter structure.

Figure 11.11 General IIR lattice digital filter structure.

where the tap coefficients v_j are calculated through the following order-reduction recursion

$$N_{j-1}(z) = N_j(z) - z v_j B_j(z) \tag{11.119}$$

for $j = M, M - 1, \ldots, 1$, with $v_M = b_{M,M}$ and $v_0 = b_{0,0}$.

Then, a way of implementing the overall IIR transfer function $H(z) = \frac{N(z)}{D(z)} = \frac{\sum_{j=0}^{M} v_j z B_j(z)}{D_N(z)}$ is to use the structure in Figure 11.11, which is called the IIR lattice realization.

From the above, we have a simple procedure for obtaining the lattice network given the direct-form transfer function, $H(z) = \frac{N_M(z)}{D_N(z)}$:

(i) Obtain, recursively, the polynomials $B_j(z)$ and $D_j(z)$, as well as the lattice coefficient $a_{j,j}$, for $j = N, N - 1, \ldots, 1$, using equations (11.115) and (11.116).

(ii) Compute the coefficients v_j, for $j = N, N - 1, \ldots, 1$, using the recursion in equation (11.119).

Conversely, if given the lattice realization we want to compute the direct-form transfer function, we can use the following procedure:

(i)   Start with $zB_0(z) = D_0(z)=1$.

(ii)   Compute recursively $B_j(z)$ and $D_j(z)$ for $j = 1, \ldots, N$, using the following relation, that can be derived from equations (11.115) and (11.116):

$$\left[\begin{array}{c} D_j(z) \\ B_j(z) \end{array} \right] = \left[\begin{array}{cc} 1 & a_{j,j} \\ a_{j,j}z^{-1} & z^{-1} \end{array} \right] \left[\begin{array}{c} D_{j-1}(z) \\ B_{j-1}(z) \end{array} \right] \tag{11.120}$$

(iii) Compute $N_M(z)$ using equation (11.118).

(iv)   The direct-form transfer function is then $H(z) = \frac{N_M(z)}{D_N(z)}$.

There are some important properties related to the lattice realization which should be mentioned. If $D_N(z)$ has all the roots inside the unit circle the lattice structure will have all coefficients $a_{j,j}$ with magnitude less than one. Otherwise, $H(z) = \frac{N(z)}{D(z)}$ represents an unstable system. This straightforward stability condition makes the lattice realizations useful for implementing time-varying filters. In addition, the polynomials $zB_j(z)$, for $j = 0, \ldots, M$, form an orthogonal set. This property justifies the choice of these polynomials to form the desired numerator polynomial $N_M(z)$, as described in equation (11.119).

Since, in Figure 11.10, the two-port system consisting of section TP$_j$ and all the sections to its right, which relates the output signal $\frac{zB_j(z)X(z)}{D_N(z)}$ to the input signal $\frac{D_j(z)X(z)}{D_N(z)}$ is linear, its transfer function remains unchanged if we multiply its input signal by λ_j and divide its output by the same amount. Therefore, $\frac{zB_N(z)}{D_N(z)}$ will not change if we multiply the signal entering the upper-left branch of section TP$_j$ by λ_j and divide the signal leaving the lower-left branch by λ_j. This is equivalent to scaling the section TP$_j$ by λ_j. If we do this for every branch j, the signals entering and leaving at the left of section N remain unchanged, the signals entering and leaving at the left of section $(N - 1)$ will be scaled by λ_N, the signals entering and leaving at the left of section $(N - 2)$ will be multiplied by $\lambda_N\lambda_{N-1}$, and so on, leading to the scaled signals $\frac{\overline{D}_j(z)X(z)}{D_N(z)}$ and $\frac{z\overline{B}_j(z)X(z)}{D_N(z)}$ at the left of section TP$_j$, such that

$$\overline{D}_j(z) = \left(\prod_{i=N}^{j+1} \lambda_i \right) D_j(z) \tag{11.121}$$

$$\overline{B}_j(z) = \left(\prod_{i=N}^{j+1} \lambda_i \right) B_j(z) \tag{11.122}$$

for $j = (N - 1), \ldots, 1$, with $\overline{D}_N(z) = D_N(z)$ and $\overline{B}_N(z) = B_N(z)$. Therefore, in order to maintain the transfer function of the scaled lattice realization unchanged, we

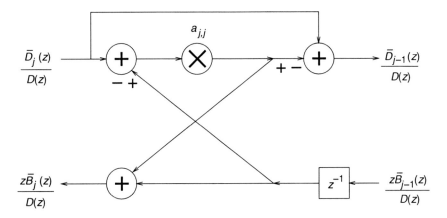

Figure 11.12   The one-multiplier network for equation (11.117).

must make

$$\overline{v}_j = \frac{v_j}{\left(\displaystyle\prod_{i=N}^{j+1} \lambda_i\right)} \tag{11.123}$$

for $j = (N-1), \ldots, 1$, with $\overline{v}_N = v_N$.

Based on the above property, we can derive a more economical two-port network using a single multiplier, as shown in Figure 11.12, where the plus-or-minus signs indicate that two different realizations are possible. The choice of these signs can vary from section to section, aiming at the reduction of the quantization noise at the filter output. Note that this network is equivalent to the one in Figure 11.9 scaled using $\lambda_j = 1 \pm a_{j,j}$, and therefore the coefficients $\overline{v}_j$ should be computed using equation (11.123). The plus sign in the computation of λ_j corresponds to summation being performed on the adder on the right and subtraction performed on the adder on the left, and vice versa for the minus sign.

Another important realization for the two-port network results when the scaling parameters λ_i are chosen such that all the internal nodes of the lattice network have a transfer function from the input of unit L_2 norm. The appropriate scaling can be derived by first noting that at the left of section TP_j, the norms of the corresponding transfer functions are given by

$$\left\| \frac{\overline{D}_j(z)}{\overline{D}_N(z)} \right\|_2 = \left\| \frac{z\overline{B}_j(z)}{\overline{D}_N(z)} \right\|_2 \tag{11.124}$$

since, from equation (11.115), $z B_j(z) = D_j(z^{-1})z^{-j}$.

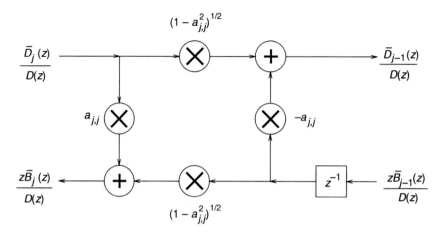

Figure 11.13 The normalized network for equation (11.117).

From the above equations, if we want unit L_2 norm at the internal nodes of the lattice network, we must have

$$\left\| \frac{z\overline{B}_0(z)}{\overline{D}_N(z)} \right\|_2 = \cdots = \left\| \frac{z\overline{B}_{N-1}(z)}{\overline{D}_N(z)} \right\|_2 = \left\| \frac{z\overline{B}_N(z)}{\overline{D}_N(z)} \right\|_2 = \left\| \frac{\overline{D}_N(z)}{\overline{D}_N(z)} \right\|_2 = 1 \qquad (11.125)$$

Then, using λ_j from equation (11.122), it can be derived that (Gray & Markel, 1973, 1975)

$$\lambda_j = \frac{\left\| \dfrac{z\overline{B}_j(z)}{\overline{D}_N(z)} \right\|_2}{\left\| \dfrac{z\overline{B}_{j-1}(z)}{\overline{D}_N(z)} \right\|_2} = \sqrt{1 - a_{j,j}^2} \qquad (11.126)$$

It is easy to show that section TP_j of the normalized lattice can be implemented as depicted in Figure 11.13. The most important feature of the normalized lattice realization is that, since all its internal nodes have transfer function with unit L_2 norm, it presents an automatic scaling in the L_2-norm sense. This explains the low roundoff noise generated by the normalized lattice realization as compared with the other forms of the lattice realization. Note that the coefficients $\overline{v}_j$ have to be computed using equation (11.123).

EXAMPLE 11.3

Repeat Example 11.1 using the one-multiplier, two-multiplier, and normalized lattice forms. Quantize the coefficients of the normalized lattice using 9 bits, including the sign bit, and verify the results.

Table 11.9. *Two-multiplier lattice.*

Section j	$a_{j,j}$	v_j
0		$-2.152\,1407\text{E}-06$
1	$8.093\,7536\text{E}-01$	$-1.187\,8594\text{E}-06$
2	$-9.998\,1810\text{E}-01$	$9.382\,1221\text{E}-06$
3	$8.090\,3062\text{E}-01$	$3.401\,0244\text{E}-06$
4	$-9.996\,9784\text{E}-01$	$8.872\,0978\text{E}-05$
5	$8.088\,3630\text{E}-01$	$-2.332\,5690\text{E}-04$
6	$-9.690\,5393\text{E}-01$	$-1.436\,2176\text{E}-04$

Table 11.10. *One-multiplier lattice.*

Section j	$a_{j,j}$	$\overline{v}_j$
0		$-2.137\,0673\text{E}+02$
1	$8.093\,7536\text{E}-01$	$-2.134\,2301\text{E}+02$
2	$-9.998\,1810\text{E}-01$	$3.066\,3279\text{E}-01$
3	$8.090\,3062\text{E}-01$	$2.010\,8199\text{E}-01$
4	$-9.996\,9784\text{E}-01$	$1.584\,9720\text{E}-03$
5	$8.088\,3630\text{E}-01$	$-7.537\,5299\text{E}-03$
6	$-9.690\,5393\text{E}-01$	$-1.436\,2176\text{E}-04$

Table 11.11. *Normalized lattice.*

Section j	$\sqrt{1 - a_{j,j}^2}$	$a_{j,j}$	$\overline{v}_j$
0			$-9.161\,4472\text{E}-02$
1	$5.872\,9169\text{E}-01$	$8.093\,7536\text{E}-01$	$-2.969\,6982\text{E}-02$
2	$1.907\,2863\text{E}-02$	$-9.998\,1810\text{E}-01$	$4.473\,6732\text{E}-03$
3	$5.877\,6650\text{E}-01$	$8.090\,3062\text{E}-01$	$9.531\,8622\text{E}-04$
4	$2.458\,0919\text{E}-02$	$-9.996\,9784\text{E}-01$	$6.112\,1280\text{E}-04$
5	$5.880\,3388\text{E}-01$	$8.088\,3630\text{E}-01$	$-9.449\,3725\text{E}-04$
6	$2.468\,4909\text{E}-01$	$-9.690\,5393\text{E}-01$	$-1.436\,2176\text{E}-04$

SOLUTION

Tables 11.9–11.12 list the multiplier coefficients of the designed filters. Figure 11.14 depicts the magnitude responses obtained by the original and quantized normalized lattices. Note that the magnitude responses of the normalized lattice structure are significantly different to the ideal one, especially when compared to the results shown in Examples 11.1 and 11.2.

It must be added that the normalized lattice performs much better than the two- and

Table 11.12. *Normalized lattice quantized with 9 bits.*

Section j	$\sqrt{1-a_{j,j}^2}$	$a_{j,j}$	$\overline{v}_j$
0			$-8.984\,3750E-02$
1	$5.859\,3750E-01$	$8.085\,9375E-01$	$-2.734\,3750E-02$
2	$1.562\,5000E-02$	$-9.960\,9375E-01$	$3.906\,2500E-03$
3	$5.859\,3750E-01$	$8.085\,9375E-01$	$0.000\,0000E+00$
4	$2.343\,7500E-02$	$-9.960\,9375E-01$	$0.000\,0000E+00$
5	$5.859\,3750E-01$	$8.085\,9375E-01$	$0.000\,0000E+00$
6	$2.460\,9375E-01$	$-9.687\,5000E-01$	$0.000\,0000E+00$

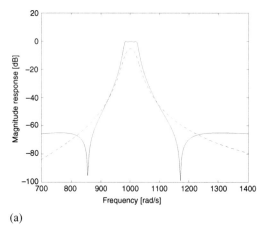
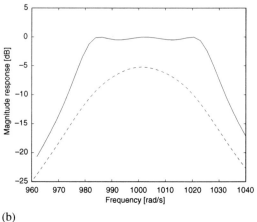

(a)   (b)

Figure 11.14 Coefficient-quantization effects in the normalized lattice form: (a) overall magnitude response; (b) passband detail. Solid line – initial design; dashed line – normalized lattice (9 bits).

one-multiplier lattices with respect to quantization effects. It is also worth mentioning that in the two-multiplier structure, the feedforward coefficients assume very small values, forcing the use of more than 9 bits for their representation, this normally happens when designing a filter having poles very close to the unit circle, as is the case in this example.

△

11.5 Wave filters

In classical analog filter design, it is widely known that doubly terminated LC lossless filters have zero sensitivity of the transfer function, with respect to the lossless L's and C's components, at frequencies where the maximal power is transferred to the load. Filter transfer functions that are equiripple in the passband, such as Chebyshev and elliptic filters, have several frequencies of maximal power transfer. Since the ripple values are usually kept small within the passband, the sensitivities of the transfer function to variations in the filter components remain small over the frequency range consisting of the entire passband. This is the reason why several methods have been proposed to generate realizations that attempt to emulate the internal operations of the doubly terminated lossless filters.

In digital-filter design, the first attempt to derive a realization starting from an analog prototype consisted of applying the bilinear transformation to the continuous-time transfer function, establishing a direct correspondence between the elements of the analog prototype and of the resulting digital filter. However, the direct simulation of the internal quantities, such as voltages and currents, of the analog prototype in the digital domain leads to delay-free loops, as will be seen below. These loops cannot be computed sequentially, since not all node values in the loop are initially known (Antoniou, 1993).

An alternative approach results from the fact that any analog n-port network can be characterized using the concepts of incident and reflected waves' quantities known from distributed parameter theory (Belevitch, 1968). Through the application of the wave characterization, digital filter realizations without delay-free loops can be obtained from passive and active analog filters, using the bilinear transformation, as originally proposed in Fettweis (1971a, 1986). The realizations obtained using this procedure are known as wave digital filters (Fettweis, 1971a, 1986; Sedlmeyer & Fettweis, 1973; Fettweis et al., 1974; Fettweis & Meerkötter, 1975a,b; Antoniou & Rezk, 1977, 1980; Diniz & Antoniou, 1988). The name wave digital filter derives from the fact that wave quantities are used to represent the internal analog signals in the digital-domain simulation. The possible wave quantities are voltage, current, and power quantities. The choice between voltage or current waves is irrelevant, whereas power waves lead to more complicated digital realizations. Traditionally, voltage wave quantities are the most widely used, and, therefore, we base our presentation on that approach.

Another great advantage of the wave digital filters, when imitating doubly terminated lossless filters, is their inherent stability under linear conditions (infinite-precision arithmetic), as well as in the nonlinear case, where the signals are subjected to quantization. Also, if the states of a wave digital filter structure, imitating a passive analog network, are quantized using magnitude truncation and saturation arithmetic, no zero-input or overflow limit cycles can be sustained.

The wave digital filters are also adequate to simulate certain analog systems, such as power systems, due to the topological equivalence with their analog counterparts (Roitman & Diniz, 1995, 1996).

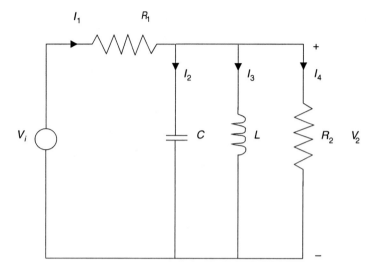

Figure 11.15 Doubly terminated LC network.

11.5.1 Motivation

The transformation of a transfer function $T(s)$ representing a continuous-time system into a discrete-time transfer function $H(z)$ may be performed using the bilinear transformation in the following form

$$H(z) = T(s)|_{s=\frac{2}{T}\frac{z-1}{z+1}} \tag{11.127}$$

Given the doubly terminated LC network depicted in Figure 11.15, if we use voltage and current variables to simulate the analog components, we have that

$$
\left.
\begin{aligned}
I_1 &= \frac{V_i - V_2}{R_1} \\
I_2 &= \frac{V_2}{Z_C} \\
I_3 &= \frac{V_2}{Z_L} \\
I_4 &= \frac{V_2}{R_2} \\
I_1 &= I_2 + I_3 + I_4
\end{aligned}
\right\} \tag{11.128}
$$

The possible representations for an inductor in the z plane will be in one of the forms shown in Figure 11.16, that is

$$
\left.
\begin{aligned}
V &= sLI = \frac{2L}{T}\frac{z-1}{z+1}I \\
I &= \frac{V}{sL} = \frac{T}{2L}\frac{z+1}{z-1}V
\end{aligned}
\right\} \tag{11.129}
$$

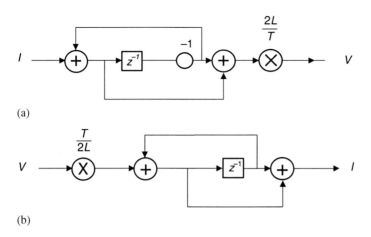

(a)

(b)

Figure 11.16 Two possible inductor realizations.

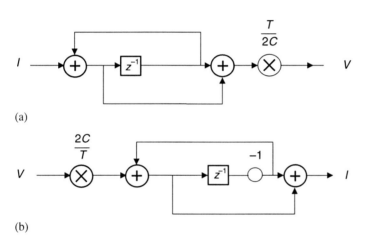

(a)

(b)

Figure 11.17 Two possible capacitor realizations.

For a capacitor, the resulting possible representations are depicted in Figure 11.17, such that

$$\left. \begin{array}{l} V = \dfrac{I}{sC} = \dfrac{T}{2C}\dfrac{z+1}{z-1}I \\[3mm] I = sCV = \dfrac{2C}{T}\dfrac{z-1}{z+1}V \end{array} \right\} \tag{11.130}$$

The sources and the loads are represented in Figure 11.18. Therefore, using Figures 11.16–11.18, the digital simulation of the doubly terminated LC network of Figure 11.15 leads to the digital network shown in Figure 11.19, where we notice the existence of delayless loops. In the next subsection, we show how these loops can be avoided using the concept of wave digital filters.

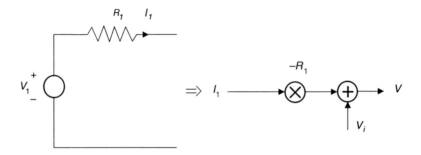

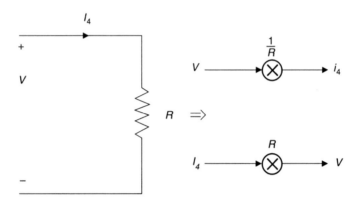

Figure 11.18 Termination realizations.

11.5.2 Wave elements

As discussed in the previous subsection, the direct simulation of the branch elements of the analog network introduces delayless loops, generating a non-computable digital network. This problem can be circumvented by simulating the analog network using the wave equations that represent multiport networks instead of representing the voltages and currents in a straightforward manner.

As shown in Figure 11.20, an analog one-port network can be described in terms of a wave characterization as a function of the variables

$$
\left.\begin{aligned}
a &= v + Ri \\
b &= v - Ri
\end{aligned}\right\} \tag{11.131}
$$

where a and b are the incident and reflected voltage wave quantities, respectively, and R is the port resistance assigned to the one-port network. In the frequency domain the

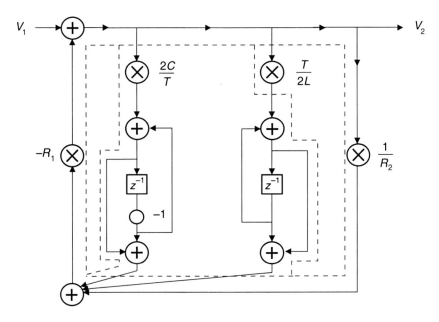

Figure 11.19 Digital network with delay-free loops.

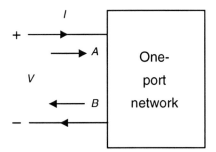

Figure 11.20 Convention for the incident and reflected waves.

wave quantities are A and B, such that

$$\left. \begin{array}{l} A = V + RI \\ B = V - RI \end{array} \right\}$$
(11.132)

Notice how the voltage waves consist of linear combinations of the voltage and current of the one-port network.

The value of R is a positive parameter called port resistance. A proper choice of R leads to simple multiport network realizations. In the following, we examine how to represent several analog elements using the incident and reflected waves.

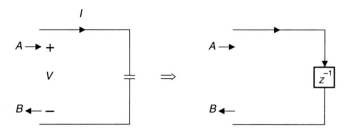

Figure 11.21   Wave realization for a capacitor.

One-port elements

For a capacitor, the following equations apply

$$
\left.
\begin{aligned}
V &= \frac{1}{sC} I \\[2mm]
B &= A \frac{V - RI}{V + RI}
\end{aligned}
\right\}
\tag{11.133}
$$

thus

$$
B = A \frac{\frac{1}{sC} - R}{\frac{1}{sC} + R}
\tag{11.134}
$$

By applying the bilinear transformation, we find

$$
B = \frac{\frac{T}{2C}(z + 1) - R(z - 1)}{\frac{T}{2C}(z + 1) + R(z - 1)} A
\tag{11.135}
$$

The value of R that leads to a significant simplification in the implementation of B as a function of A, is

$$
R = \frac{T}{2C}
\tag{11.136}
$$

and then

$$
B = z^{-1} A
\tag{11.137}
$$

The realization of B as a function of A is done as shown in Figure 11.21.

Following a similar reasoning, the digital representation of several other one-port elements can be derived, as shown in Figure 11.22, along with the respective wave equations.

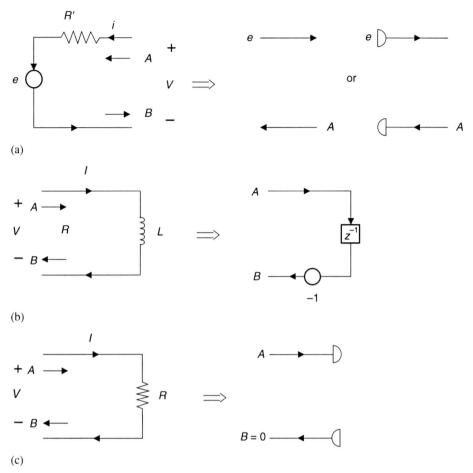

Figure 11.22 Wave realization for several one-port networks: (a) series connection of a voltage source with resistor: $e = V - R'I = V - RI = B$, for $R' = R$; (b) inductor: $R = 2L/T$; (c) resistor: $B = 0$, $A = 2RI$. (cont. overleaf)

Voltage generalized immittance converter

The voltage generalized immittance converter (VGIC) (Diniz & Antoniou, 1988), depicted in Figure 11.23, is a two-port network characterized by

$$\left.\begin{array}{l} V_1(s) = r(s)V_2(s) \\ I_1(s) = -I_2(s) \end{array}\right\} \tag{11.138}$$

where $r(s)$ is the so-called conversion function, and the pairs (V_1, I_1) and (V_2, I_2) are the VGIC voltages and currents at ports 1 and 2, respectively.

The VGICs are not employed in the design of analog circuits due to difficulties in implementation when using conventional active devices such as transistors and operational amplifiers. However, there is no difficulty in utilizing VGICs in the design of digital filters.

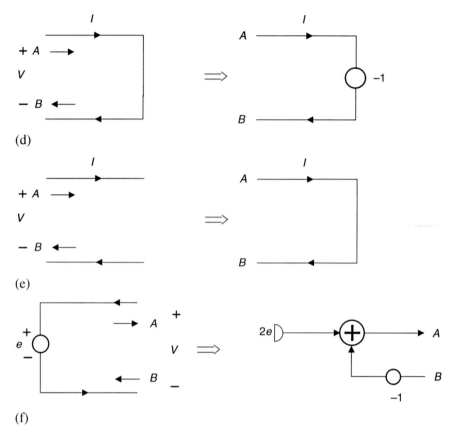

(d)

(e)

(f)

Figure 11.22 (cont.) Wave realization for several one-port networks: (d) short circuit: $A = RI$, $B = -RI$; (e) open circuit: $A = V$, $B = V$; (f) voltage source: $A = 2e - B$.

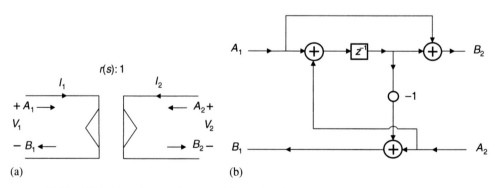

(a)   (b)

Figure 11.23 VGIC: (a) analog symbol; (b) digital realization: $r(s) = Ts/2$.

The VGIC of Figure 11.23 may be described in terms of wave equations by

$$
\left.
\begin{aligned}
A_1 &= V_1 + \frac{I_1}{G_1} \\
A_2 &= V_2 + \frac{I_2}{G_2} \\
B_1 &= V_1 - \frac{I_1}{G_1} \\
B_2 &= V_2 - \frac{I_2}{G_2} \\
V_1(s) &= r(s)V_2(s) \\
I_1(s) &= -I_2(s)
\end{aligned}
\right\}
\tag{11.139}
$$

where A_i and B_i are the incident and reflected waves of each port, respectively, and G_i represents the conductance of port i, for $i = 1, 2$.

After some algebraic manipulation, we can calculate the values of B_1 and B_2, as functions of A_1, A_2, G_1, G_2, and $r(s)$, as

$$
\left.
\begin{aligned}
B_1 &= \frac{r(s)G_1 - G_2}{r(s)G_1 + G_2}A_1 + \frac{2r(s)G_2}{r(s)G_1 + G_2}A_2 \\
B_2 &= \frac{2G_1}{r(s)G_1 + G_2}A_1 + \frac{G_2 - r(s)G_1}{r(s)G_1 + G_2}A_2
\end{aligned}
\right\}
\tag{11.140}
$$

By applying the bilinear transformation, and by choosing $G_2 = G_1$ and $r(s) = \frac{T}{2}s$, which lead to a simple digital realization, as seen in Figure 11.23b, the following relations result:

$$
\left.
\begin{aligned}
B_1 &= -z^{-1}A_1 + (1 - z^{-1})A_2 \\
B_2 &= (1 + z^{-1})A_1 + z^{-1}A_2
\end{aligned}
\right\}
\tag{11.141}
$$

Current generalized immittance converter

The current generalized immittance converter CGIC (Antoniou & Rezk, 1977) is described by

$$
\left.
\begin{aligned}
V_1 &= V_2 \\
I_1 &= -h(s)I_2
\end{aligned}
\right\}
\tag{11.142}
$$

Choosing $G_1 = \frac{2G_2}{T}$ and $h(s) = s$, a simple realization for the CGIC results, as illustrated in Figure 11.24.

Transformer

A transformer with a turn ratio of $n : 1$, and with port resistances R_1 and R_2, with $\frac{R_2}{R_1} = \frac{1}{n^2}$, has a digital representation as shown in Figure 11.25.

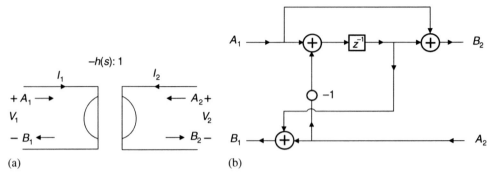

(a) (b)

Figure 11.24 CGIC: (a) analog symbol; (b) digital realization: $h(s) = s$.

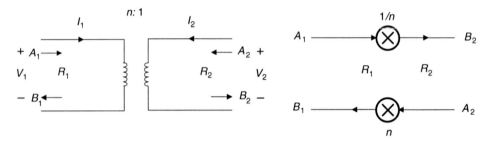

Figure 11.25 Transformer digital representation: (a) analog symbol; (b) digital realization: $R_2/R_1 = 1/n^2$.

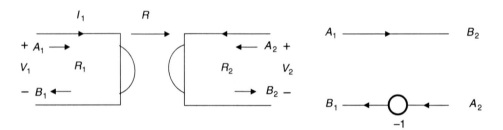

Figure 11.26 Gyrator digital representation.

Gyrator

A gyrator is a lossless two-port element described by

$$\left. \begin{array}{l} V_1 = -RI_2 \\ V_2 = RI_1 \end{array} \right\} \tag{11.143}$$

It can be easily shown in this case that $B_2 = A_1$ and $B_1 = -A_2$, with $R_1 = R_2 = R$. The digital realization of a gyrator is depicted in Figure 11.26.

We are now equipped with the digital representations of the main analog elements which serve as the basic building blocks for the realization of wave digital filters. However, to achieve our goal, we still have to learn how to interconnect these building

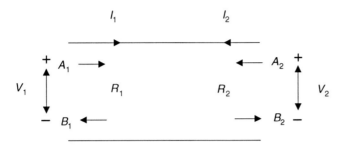

Figure 11.27 Two-port adaptor.

blocks. To avoid delay-free loops, this must be done in the same way as these blocks are interconnected in the reference analog filter. Since the port resistances of the various elements are different, there is also a need to derive the so-called adaptors to allow the interconnection. Such adaptors guarantee that the current and voltage Kirchoff laws are satisfied at all series and parallel interconnections of ports with different port resistances.

Two-port adaptors

Consider the parallel interconnection of two elements with port resistances given by R_1 and R_2, respectively, as shown in Figure 11.27. The wave equations in this case are given by

$$\left.\begin{aligned} A_1 &= V_1 + R_1 I_1 \\ A_2 &= V_2 + R_2 I_2 \\ B_1 &= V_1 - R_1 I_1 \\ B_2 &= V_2 - R_2 I_2 \end{aligned}\right\} \tag{11.144}$$

Since $V_1 = V_2$ and $I_1 = -I_2$, we have that

$$\left.\begin{aligned} A_1 &= V_1 + R_1 I_1 \\ A_2 &= V_1 - R_2 I_1 \\ B_1 &= V_1 - R_1 I_1 \\ B_2 &= V_1 + R_2 I_1 \end{aligned}\right\} \tag{11.145}$$

Eliminating V_1 and I_1 from the above equations, we obtain

$$\left.\begin{aligned} B_1 &= A_2 + \alpha\,(A_2 - A_1) \\ B_2 &= A_1 + \alpha\,(A_2 - A_1) \end{aligned}\right\} \tag{11.146}$$

where $\alpha = (R_1 - R_2)/(R_1 + R_2)$. A realization for this general two-port adaptor is depicted in Figure 11.28a.

Expressing B_1 as a function of B_2 in equation (11.146), we get

$$\left.\begin{aligned} B_1 &= B_2 - A_1 + A_2 \\ B_2 &= A_1 + \alpha\,(A_2 - A_1) \end{aligned}\right\} \tag{11.147}$$

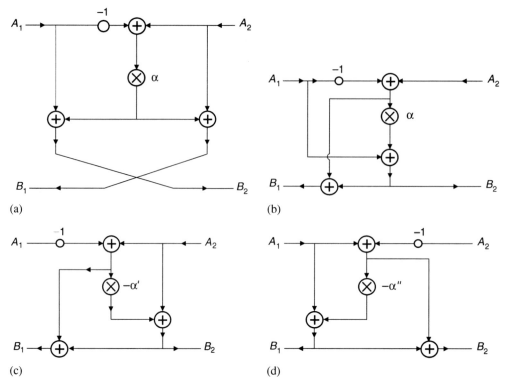

Figure 11.28 Possible digital realizations of the general two-port adaptors based on:
(a) equation (11.146); (b) equation (11.147); (c) equation (11.148); (d) equation (11.149).

leading to a modified version of the two-port adaptor, as shown in Figure 11.28b.

Other alternative forms of two-port adaptors are generated by expressing the equations for B_1 and B_2 in different ways, such as

$$\left.\begin{array}{l} B_1 = B_2 - A_1 + A_2 \\ B_2 = A_2 - \alpha'\,(A_2 - A_1) \end{array}\right\} \tag{11.148}$$

where $\alpha' = 2R_2/(R_1 + R_2)$, generating the structure seen in Figure 11.28c, or

$$\left.\begin{array}{l} B_1 = A_1 - \alpha''\,(A_1 - A_2) \\ B_2 = B_1 + A_1 - A_2 \end{array}\right\} \tag{11.149}$$

where $\alpha'' = 2R_1/(R_1 + R_2)$, leading to the structure of Figure 11.28d.

It is worth observing that the incident and reflected waves in any port could be expressed in the time domain, that is, through the instantaneous signal values ($a_i(k)$ and $b_i(k)$), or in the frequency domain ($A_i(z)$ and $B_i(z)$), corresponding to the steady-state description of the wave signals.

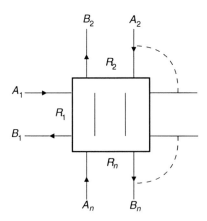

Figure 11.29 Symbol of the *n*-port parallel adaptor.

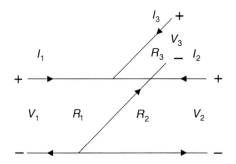

Figure 11.30 The three-port parallel adaptor.

n-port parallel adaptor

In cases where we need to interconnect *n* elements in parallel, with port resistances given by $R_1, R_2, \ldots, R_n$, it is necessary to use an *n*-port parallel adaptor. The symbol to represent the *n*-port parallel adaptor is shown in Figure 11.29. Figure 11.30 illustrates a three-port parallel adaptor.

The wave equation on each port of a parallel adaptor is given by

$$\left. \begin{aligned} A_k &= V_k + R_k I_k \\ B_k &= V_k - R_k I_k \end{aligned} \right\} \tag{11.150}$$

for $k = 1, 2, \ldots, n$. As all ports are connected in parallel, we then have that

$$\left. \begin{aligned} V_1 &= V_2 = \cdots = V_n \\ I_1 + I_2 &+ \cdots + I_n = 0 \end{aligned} \right\} \tag{11.151}$$

After some algebraic manipulation to eliminate V_k and I_k, we have that

$$B_k = (A_0 - A_k) \tag{11.152}$$

where

$$A_0 = \sum_{k=1}^{n} \alpha_k A_k \qquad (11.153)$$

with

$$\alpha_k = \frac{2G_k}{G_1 + G_2 + \cdots + G_n} \qquad (11.154)$$

and

$$G_k = \frac{1}{R_k} \qquad (11.155)$$

From equation 11.154, we note that

$$\alpha_1 + \alpha_2 + \cdots + \alpha_n = 2 \qquad (11.156)$$

and thus, we can eliminate an internal multiplication of the adaptor, since the computation related to one of the α_i is not required. If we calculate α_n as a function of the remaining α_i, we can express A_0 as

$$
\begin{aligned}
A_0 &= \sum_{k=1}^{n-1} \alpha_k A_k + \alpha_n A_n \\
&= \sum_{k=1}^{n-1} \alpha_k A_k + \left[2 - (\alpha_1 + \alpha_2 + \cdots + \alpha_{n-1}) \right] A_n \\
&= 2A_n + \sum_{k=1}^{n-1} \alpha_k (A_k - A_n) \qquad (11.157)
\end{aligned}
$$

where only $(n-1)$ multipliers are required to calculate A_0. In this case, port n is called the dependent port. It is also worth observing that if we have several port resistances R_k with the same value, the number of multiplications can be further reduced. If, however, $\sum_{k=1}^{n-1} \alpha_k \approx 2$, the error in computing α_n may be very high due to quantization effects. In this case, it is better to choose another port k, with α_k as large as possible, to be the dependent one.

In practice, the three-port adaptors are the most widely used in wave digital filters. A possible implementation for a three-port parallel adaptor is shown in Figure 11.31a, which corresponds to the direct realization of equation (11.152), with A_0 calculated using equation (11.157).

Substituting equation (11.157) into equation (11.152), with $k = n$, then

$$
\left.
\begin{aligned}
B_n &= (A_0 - A_n) = A_n + \sum_{k=1}^{n-1} \alpha_k (A_k - A_n) \\
B_k &= (A_0 - A_k) = B_n + A_n - A_k
\end{aligned}
\right\} \qquad (11.158)
$$

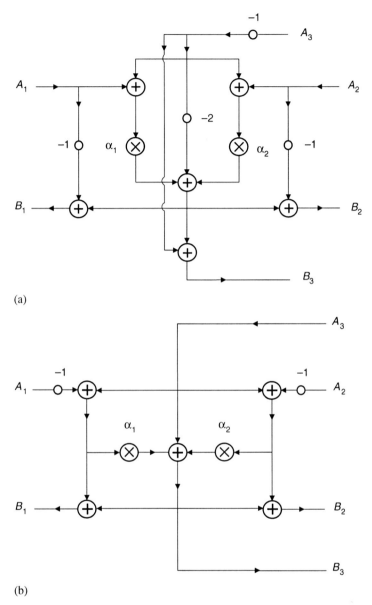

(a)

(b)

Figure 11.31 Possible digital realizations of the three-port parallel adaptor based on:
(a) equation (11.157); (b) equation (11.158).

for $k = 1, 2, \ldots, (n-1)$, and we end up with an alternative realization for the three-port parallel adaptor, as seen in Figure 11.31b.

Analyzing equation (11.158), we observe that the reflection wave B_i is directly dependent on the incident wave A_i at the same port. Hence, if two arbitrary adaptors are directly interconnected, a delay-free loop will appear between the two adaptors, as shown in Figure 11.32. A solution to this problem is to choose one of the α

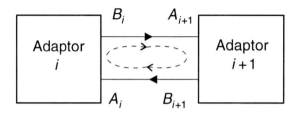

Figure 11.32 Adaptor interconnection.

in the adaptor to be equal to 1. For example, $\alpha_n = 1$. In this case, according to equations (11.152), (11.154), and (11.157), the equations describing the parallel adaptor become

$$
\left.
\begin{aligned}
G_n &= G_1 + G_2 + \cdots + G_{n-1} \\
B_n &= \sum_{k=1}^{n-1} \alpha_k A_k
\end{aligned}
\right\}
\tag{11.159}
$$

and the expression for B_n becomes independent of A_n, thus eliminating the delay-free loops at port n. In this case, equation (11.156) becomes

$$
\alpha_1 + \alpha_2 + \cdots + \alpha_{n-1} = 1
\tag{11.160}
$$

which still allows one of the α_i, for $i = 1, 2, \ldots, (n-1)$, to be expressed as a function of the others, thus eliminating one multiplication.

It is worth observing that choosing one of the α equal to 1 does not imply any loss of generality in the wave digital filter design. In fact, at the ports corresponding to these coefficients, the port resistances can be chosen arbitrarily, since they are used only for interconnection, and therefore they do not depend on any component value of the analog prototype. For example, the resistance of the ports common to two interconnected adaptors must be the same but can have arbitrary values. For instance, in the case of a three-port parallel adaptor, the describing equations would be

$$
\left.
\begin{aligned}
\alpha_2 &= 1 - \alpha_1 \\
B_3 &= \alpha_1 A_1 + (1 - \alpha_1) A_2 \\
B_2 &= (A_0 - A_2) = \alpha_1 A_1 + (1 - \alpha_1) A_2 + A_3 - A_2 = \alpha_1 (A_1 - A_2) + A_3 \\
B_1 &= \alpha_1 A_1 + (1 - \alpha_1) A_2 + A_3 - A_1 = (1 - \alpha_1)(A_2 - A_1) + A_3
\end{aligned}
\right\}
\tag{11.161}
$$

the realization of which is depicted in Figure 11.33. Note that the port with no possibility of delay-free loops is marked with (⊢), and is known as the reflection-free port.

A parallel adaptor, as illustrated in Figure 11.34, can be interpreted as a parallel connection of n ports with $(n - 2)$ auxiliary ports, which are introduced to provide separation among several external ports. The same figure also shows the symbolic representation of the n-port parallel adaptor consisting of several three-port adaptors.

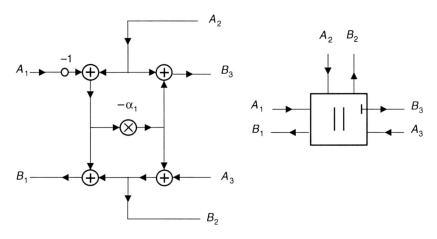

Figure 11.33 Reflection-free parallel adaptor at port 3.

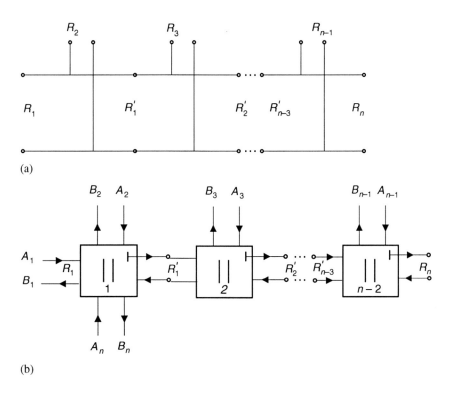

(a)

(b)

Figure 11.34 The *n*-port parallel adaptor: (a) equivalent connection; (b) interpretation as several three-port parallel adaptors.

n-port series adaptor

In the situation where we need to interconnect n elements in series with distinct port resistances $R_1, R_2, \ldots, R_n$, we need to use an n-port series adaptor, whose symbol is

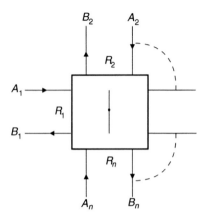

Figure 11.35   Symbol of the *n*-port series adaptor.

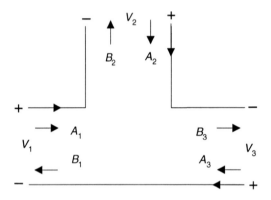

Figure 11.36   The three-port series adaptor.

shown in Figure 11.35. In this case, the wave equations for each port are

$$
\left.
\begin{aligned}
A_k &= V_k + R_k I_k \\
B_k &= V_k - R_k I_k
\end{aligned}
\right\}
\tag{11.162}
$$

for $k = 1, 2, \ldots, n$. We then must have that

$$
\left.
\begin{aligned}
V_1 + V_2 + \cdots + V_n &= 0 \\
I_1 = I_2 = \cdots = I_n &= I
\end{aligned}
\right\}
\tag{11.163}
$$

Figure 11.36 depicts a possible three-port series adaptor.
Since

$$
\sum_{i=1}^{n} A_i = \sum_{i=1}^{n} V_i + \sum_{i=1}^{n} R_i I_i = I \sum_{i=1}^{n} R_i
\tag{11.164}
$$

it follows that

$$B_k = A_k - 2R_k I_k = A_k - \frac{2R_k}{\sum\limits_{i=1}^{n} R_i} \sum_{i=1}^{n} A_i = A_k - \beta_k A_0 \tag{11.165}$$

where

$$\left. \begin{aligned} \beta_k &= \frac{2R_k}{\sum\limits_{i=1}^{n} R_i} \\[2em] A_0 &= \sum_{i=1}^{n} A_i \end{aligned} \right\} \tag{11.166}$$

From equation (11.166), we observe that

$$\sum_{k=1}^{n} \beta_k = 2 \tag{11.167}$$

Therefore, it is possible to eliminate a multiplication, as previously done for the parallel adaptor, by expressing β_n as a function of the remaining β, and then

$$B_n = A_n + (-2 + \beta_1 + \beta_2 + \cdots + \beta_{n-1})A_0 \tag{11.168}$$

Since from equation (11.165)

$$\sum_{k=1}^{n-1} B_k = \sum_{k=1}^{n-1} A_k - A_0 \sum_{k=1}^{n-1} \beta_k \tag{11.169}$$

then

$$B_n = A_n - 2A_0 + \sum_{k=1}^{n-1} A_k - \sum_{k=1}^{n-1} B_k = -A_0 - \sum_{k=1}^{n-1} B_k \tag{11.170}$$

where port n is the so-called dependent port.

The three-port adaptor realized from the equations above is shown in Figure 11.37. The specific equations describing this adaptor are given by

$$\left. \begin{aligned} B_1 &= A_1 - \beta_1 A_0 \\ B_2 &= A_2 - \beta_2 A_0 \\ B_3 &= -(A_0 + B_1 + B_2) \end{aligned} \right\} \tag{11.171}$$

where $A_0 = A_1 + A_2 + A_3$.

For the series adaptors, we avoid the delay-free loops by choosing one of the β equal to 1. For example, if we choose $\beta_n = 1$, we have that

$$\left. \begin{aligned} R_n &= R_1 + R_2 + \cdots + R_{n-1} \\ B_n &= (A_n - \beta_n A_0) = -(A_1 + A_2 + \cdots + A_{n-1}) \end{aligned} \right\} \tag{11.172}$$

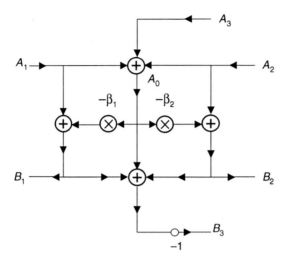

Figure 11.37 Digital realization of the three-port series adaptor.

Equation (11.167) can now be replaced by

$$\beta_1 + \beta_2 + \cdots + \beta_{n-1} = 1 \tag{11.173}$$

which allows one of the β_i to be calculated from the remaining ones.

A three-port series adaptor having $\beta_3 = 1$ and β_2 described as a function of β_1 is described by

$$\left.\begin{array}{l} \beta_2 = 1 - \beta_1 \\ B_1 = A_1 - \beta_1 A_0 \\ B_2 = A_2 - (1 - \beta_1) A_0 = A_2 - (A_1 + A_2 + A_3) + \beta_1 A_0 = \beta_1 A_0 - (A_1 + A_3) \\ B_3 = (A_3 - A_0) = -(A_1 + A_2) \end{array}\right\} \tag{11.174}$$

and its implementation is depicted in Figure 11.38. Note again that the port which avoids delay-free loops is marked with (⊢).

Consider now a series connection with inverted odd-port orientations, starting from port 3 and with $(n-2)$ auxiliary ports used for separation, as depicted in Figure 11.39a. We can easily show that such a series connection can be implemented through several elementary series adaptors, as shown in Figure 11.39b.

11.5.3 Lattice wave digital filters

In some designs, it is preferable to implement a wave digital filter from a lattice analog network rather than from a ladder analog network. This is because, when implemented digitally, the lattice structures have low sensitivity in the filter passband and high sensitivity in the stopband. There are two explanations for the sensitivity properties of

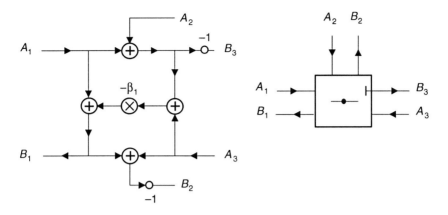

Figure 11.38 Reflection-free series adaptor at port 3.

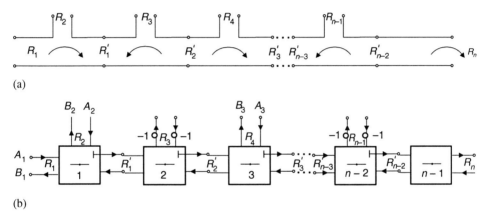

(a)

(b)

Figure 11.39 The n-port series adaptor: (a) equivalent connection; (b) interpretation as several three-port series adaptors.

the lattice realization. First, any changes in the coefficients belonging to the adaptors of the lattice structure do not destroy its symmetry, whereas in the symmetric ladder this symmetry can be lost. The second reason applies to the design of filters with zeros on the unit circle. In the lattice structures, quantization usually moves these zeros away from the unit circle, whereas in the ladder structures the zeros always move around the unit circle.

Fortunately, all symmetric ladder networks can be transformed into lattice networks by applying the so-called Bartlett bisection theorem (van Valkenburg, 1974). This subsection deals with the implementation of lattice wave digital filters.

Given the symmetric analog lattice network of Figure 11.40, where Z_1 and Z_2 are the lattice impedances, it can be shown that the incident and reflected waves are related

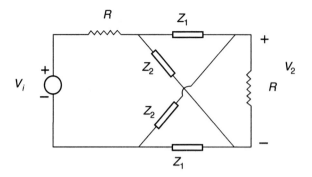

Figure 11.40 Analog lattice network.

by

$$B_1 = \frac{S_1}{2}(A_1 - A_2) + \frac{S_2}{2}(A_1 + A_2)$$

$$B_2 = -\frac{S_1}{2}(A_1 - A_2) + \frac{S_2}{2}(A_1 + A_2)$$

$$\left.\right\} \qquad (11.175)$$

where

$$S_1 = \frac{Z_1 - R}{Z_1 + R}$$

$$S_2 = \frac{Z_2 - R}{Z_2 + R}$$

$$\left.\right\} \qquad (11.176)$$

with S_1 and S_2 being the reflectances of the impedances Z_1 and Z_2, respectively. That is, S_1 and S_2 correspond to the ratio between the reflected and incident waves at ports Z_1 and Z_2 with port resistance R, since

$$\frac{B}{A} = \frac{V - RI}{V + RI} = \frac{Z - R}{Z + R} \qquad (11.177)$$

The lattice realization then consists solely of impedance realizations, as illustrated in Figure 11.41. Note that, in this figure, since the network is terminated by a resistance, then $A_2 = 0$.

EXAMPLE 11.4
Realize the lowpass filter represented in Figure 11.42, using a wave lattice network.

SOLUTION
The circuit in Figure 11.42 is a symmetric ladder network, as is made clear when it is redrawn as in Figure 11.43.

Bartlett's bisection theorem states that, when you have a two-port network composed of two equal half networks connected by any number of wires, as illustrated in

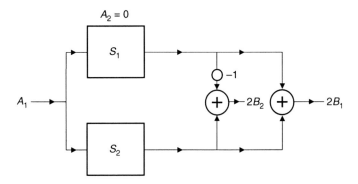

Figure 11.41 Wave digital lattice representation.

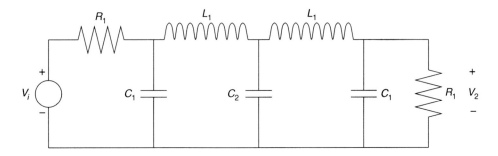

Figure 11.42 Lowpass RLC network.

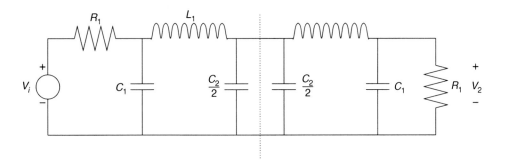

Figure 11.43 Symmetric ladder network.

Figure 11.44, then it is equivalent to a lattice network as in Figure 11.40. Figure 11.43 is a good example of a symmetric ladder network.

The impedance Z_1 is equal to the input impedance of any half-network when the connection wires to the other half are short circuited, and Z_2 is equal to the input impedance of any half-network when the connection wires to the other half are open. This is illustrated in Figure 11.45 below, where the determinations of the impedances

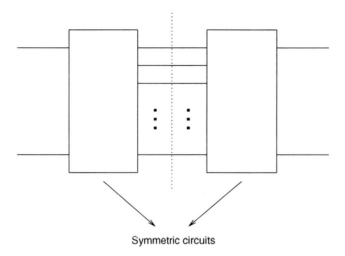

Symmetric circuits

Figure 11.44 Generic symmetric ladder network.

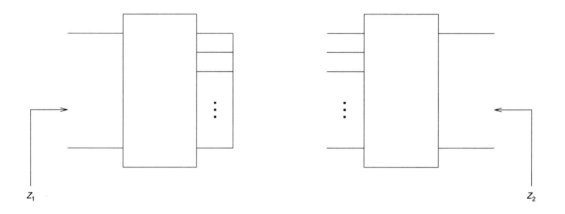

Z_1 Z_2

Figure 11.45 Computation of the lattice's impedances Z_1 and Z_2 for the generic case.

Z_1 and Z_2 of the equivalent lattice are shown. Figure 11.46 shows the computation of Z_1 and Z_2 for this example.

From the resulting lattice network, the final wave filter is then as represented in Figure 11.47, where

$$\alpha_1 = \frac{2G_1}{G_1 + \frac{2C_1}{T} + \frac{T}{2L_1}} \tag{11.178}$$

$$\alpha_2 = \frac{2\frac{2C_1}{T}}{G_1 + \frac{2C_1}{T} + \frac{T}{2L_1}} \tag{11.179}$$

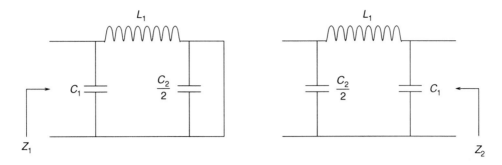

Figure 11.46 Computation of the lattice's impedances Z_1 and Z_2 for Example 11.4.

$$\alpha_3 = \frac{2G_1}{G_1 + \frac{2C_1}{T} + G_3} + \frac{2G_1}{2G_1 + \frac{4C_1}{T}} = \frac{G_1}{G_1 + \frac{2C_1}{T}} \tag{11.180}$$

$$G_3 = G_1 + \frac{2C_1}{T} \tag{11.181}$$

$$\beta_1 = \frac{2R_3}{R_3 + \frac{T}{C_2} + \frac{2L_1}{T}} \tag{11.182}$$

$$\beta_2 = \frac{\frac{2T}{C_2}}{R_3 + \frac{T}{C_2} + \frac{2L_1}{T}} \tag{11.183}$$

One should note that, since we want a network that generates the output voltage at the load, then B_2 is the only variable of interest to us, and therefore B_1 does not need to be computed.

$\triangle$

EXAMPLE 11.5
Realize the ladder filter represented in Figure 11.48 using a wave network.

SOLUTION
The connections among the elements can be interpreted as illustrated in Figure 11.49. The resulting wave filter should be represented as in Figure 11.50, where the choice of the reflection-free ports is arbitrary.

The equations below describe how to calculate the multiplier coefficient of each adaptor, and Figure depicts the resulting wave digital filter realization.

$$G_1' = G_1 + \frac{2C_1}{T} \tag{11.184}$$

$$\alpha_1 = \frac{2G_1}{G_1 + \frac{2C_1}{T} + G_1'} = \frac{2G_1}{2G_1 + \frac{4C_1}{T}} = \frac{G_1}{G_1 + \frac{2C_1}{T}} \tag{11.185}$$

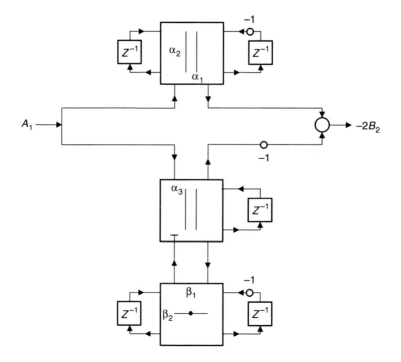

Figure 11.47 Resulting digital lattice network.

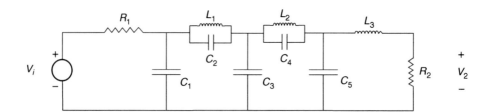

Figure 11.48 Ladder RLC network.

$$G'_6 = \frac{2C_2}{T} + \frac{T}{2L_1} \tag{11.186}$$

$$\alpha_2 = \frac{2\frac{2C_2}{T}}{\frac{2C_2}{T} + \frac{T}{2L_1} + G'_6} = \frac{\frac{2C_2}{T}}{\frac{2C_2}{T} + \frac{T}{2L_1}} \tag{11.187}$$

$$R'_2 = R'_1 + R'_6 \tag{11.188}$$

$$\beta_1 = \frac{R'_1}{R'_1 + R'_6} = \frac{1}{1 + \frac{G_1 + \frac{2C_1}{T}}{\frac{2C_2}{T} + \frac{T}{2L_1}}} \tag{11.189}$$

$$G'_3 = G'_2 + \frac{2C_3}{T} \tag{11.190}$$

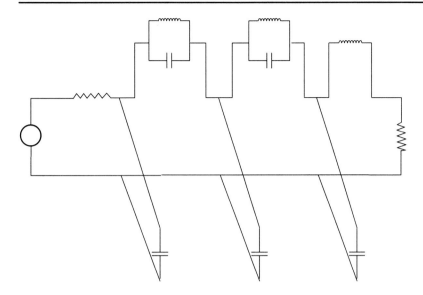

Figure 11.49   Component connections.

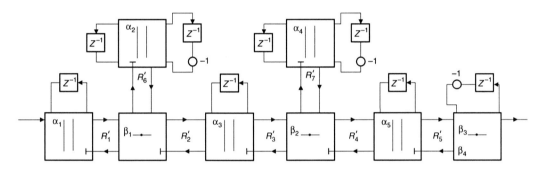

Figure 11.50   Resulting digital wave network.

$$\alpha_3 = \frac{G_2'}{G_2' + \frac{2C_3}{T}} \tag{11.191}$$

$$G_7' = \frac{2C_4}{T} + \frac{T}{2L_2} \tag{11.192}$$

$$\alpha_4 = \frac{\frac{2C_4}{T}}{\frac{2C_4}{T} + \frac{T}{2L_2}} \tag{11.193}$$

$$R_4' = R_3' + R_7' \tag{11.194}$$

$$\beta_2 = \frac{R_3'}{R_3' + R_7'} \tag{11.195}$$

$$G_5' = G_4' + \frac{2C_5}{T} \tag{11.196}$$

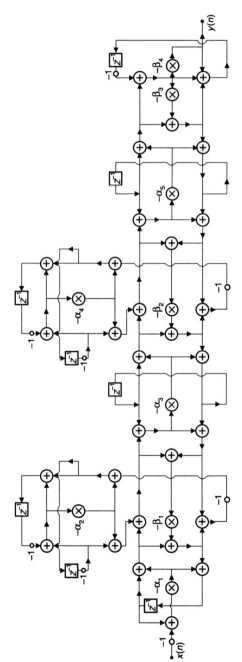

Figure 11.51 Resulting wave filter realization.

$$\alpha_5 = \frac{2G'_4}{G'_4 + \frac{2C_5}{T}} \tag{11.197}$$

$$\beta_3 = \frac{2R'_5}{R'_5 + \frac{2L_3}{T} + R_2} \tag{11.198}$$

$$\beta_4 = \frac{2R_2}{R'_5 + \frac{2L_3}{T} + R_2} \tag{11.199}$$

$\triangle$

11.6 Efficient IIR structures with MATLAB

MATLAB has several functions related to the structures described in this chapter. These functions have already been described in Chapters 4 and 10. In order to make reference easier, we present below a list of these functions.

- `sos2tf`: Converts the cascade form into the direct form. See Chapter 4 for a list of parameters. Notice that there is no direct command that reverses this operation.

- `residuez`: Performs the partial-fraction expansion in the z domain, if there are two input parameters. This command considers complex roots. To obtain a parallel expansion of a given transfer function, one should combine such roots in complex conjugate pairs with the `cplxpair` command (see Chapter 4) to form second-order sections with solely real coefficients. The `residuez` command also converts the partial-fraction expansion back to the original direct form. See Chapter 4 for a list of parameters.

- `tf2ss` and `ss2tf`: Convert the direct form into the state-space form, and vice versa. See Chapter 4 for a list of parameters.

- `sos2ss` and `ss2sos`: Convert the cascade form into the state-space form, and vice versa. See Chapter 4 for a list of parameters.

- `tf2latc` and `latc2tf`: Convert the direct form into the lattice form, and vice versa. See Chapter 10 for a list of parameters.

11.7 Summary

This chapter introduced several efficient structures for the realization of IIR digital filters.

For IIR filters, the designs of cascade and parallel forms were presented in detail. The emphasis was given to the cases where the second-order sections were comprised of direct-form sections, section-optimal state-space sections, and state-space structures which are free from limit cycles. These structures are modular and should always be considered as one of the main candidates for practical implementations.

We then addressed the issue of reducing the roundoff error at the filter output by introducing zeros to the noise transfer function by feeding back the quantization error. Such a technique is widely known as error spectrum shaping. We also introduced closed-form equations to calculate the scaling factor of second-order sections facilitating the design of parallel-form filters of noncanonic structures with respect to the number of multipliers.

The lattice structures were presented in some detail. In particular, the structures discussed here utilize two-port sections with one, two, and four multipliers. These structures can be shown to be free from zero-input limit cycles, and their design methods rely on a polynomial-reduction strategy.

The last class of structures presented in this chapter was the one comprised of the wave digital filters, which are designed from analog prototype filters. This wave strategy yields digital filters with very low sensitivity which are free from zero-input limit cycles.

11.8 Exercises

11.1   Consider the alternative parallel structure seen in Figure 11.52, whose transfer function is

$$H(z) = h'_0 + \sum_{k=1}^{m} \frac{\gamma'_{1k}z + \gamma'_{2k}}{z^2 + m_{1k}z + m_{2k}} \tag{11.200}$$

Discuss the scaling procedure and determine the relative noise variance for this realization.

11.2   Derive the expressions in equations (11.19) and (11.20).

11.3   Show that the L_2 and L_∞ norms of a second-order $H(z)$ given by

$$H(z) = \frac{\gamma_1 z + \gamma_2}{z^2 + m_1 z + m_2}$$

are as given in equations (11.27) and (11.28), respectively.

11.4   Derive the expressions for the output-noise PSD, as well as their variances, for odd-order cascade filters realized with the following second-order sections:

(a) Direct-form.

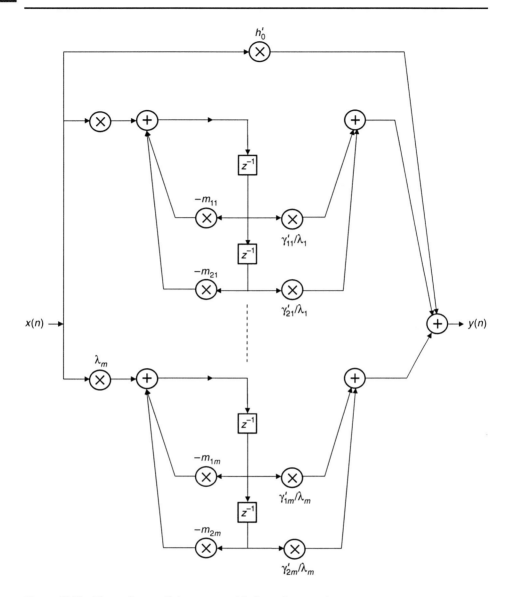

Figure 11.52 Alternative parallel structure with direct-form sections.

(b) Optimal state space.

(c) State-space free of limit cycles.

Assume fixed-point arithmetic.

11.5 Repeat Exercise 11.4 assuming floating-point arithmetic.

11.6 From equations (11.43), (11.44), and the definitions of $F_i'(z)$ and $G_i'(z)$ in Figures 7.7 and 7.3, respectively, derive equations (11.45) and (11.46).

11.7   Verify that equations (11.60)–(11.61) correspond to a state-space structure with minimum output noise.

11.8   Show that when the poles of a second-order section with transfer function

$$H(z) = d + \frac{\gamma_1 z + \gamma_2}{z^2 + m_1 z + m_2}$$

are complex conjugate, then the parameter σ defined in equation (11.62) is necessarily real.

11.9   Show that the state-space section free of limit cycles, Structure I, as determined by equations (11.85) and (11.86), has minimum roundoff noise only if

$$\frac{\gamma_1}{\gamma_2} = \frac{m_1 + 2}{m_2 - 1}$$

as indicated by equation (11.91).

11.10   Design an elliptic filter satisfying the specifications below:

$$A_p = 0.4 \text{ dB}$$
$$A_r = 50 \text{ dB}$$
$$\Omega_{r_1} = 1000 \text{ rad/s}$$
$$\Omega_{p_1} = 1150 \text{ rad/s}$$
$$\Omega_{p_2} = 1250 \text{ rad/s}$$
$$\Omega_{r_2} = 1400 \text{ rad/s}$$
$$\Omega_s = 10\,000 \text{ rad/s}$$

Realize the filter using:

(a) Parallel direct-form sections.
(b) Cascade of direct-form sections.
(c) Cascade of optimal state-space sections.
(d) Cascade of state-space sections without limit cycles.

The specifications must be satisfied when the coefficients are quantized in the range of 18 to 20 bits, including the sign bit.

11.11   Repeat Exercise 11.10 with the specifications below:

$$A_p = 1.0 \text{ dB}$$
$$A_r = 70 \text{ dB}$$
$$\omega_p = 0.025\pi \text{ rad/sample}$$
$$\omega_r = 0.04\pi \text{ rad/sample}$$

Table 11.13. *Coefficients of a Chebyshev filter.*
Scaling factor: $\lambda = 2.616669 \times 10^{-2}$.

Coefficient	Section 1	Section 2	Section 3
γ_0	0.025 15	0.088 33	0.999 99
γ_1	−0.050 29	−0.176 77	−1.999 98
γ_2	0.025 15	0.088 33	0.999 99
m_0	1.575 10	1.648 93	1.546 50
m_1	0.780 87	0.701 19	0.917 22

11.12 Design an elliptic filter satisfying the specifications of Exercise 11.10, using a cascade of direct-form structures with the ESS technique.

11.13 Design an elliptic filter satisfying the specifications of Exercise 11.10, using parallel connection of second-order direct-form structures, with L_2 and L_∞ norms for scaling, employing the closed-form equations of Subsection 11.2.4.

11.14 Derive equation (11.120).

11.15 Derive the scaling factor utilized in the two-multiplier lattice of Figure 11.9, in order to generate the one-multiplier lattice of Figure 11.12.

11.16 Derive the scaling factor to be used in the normalized lattice of Figure 11.13, in order to generate a three-multiplier lattice.

11.17 Design a two-multiplier lattice structure to implement the filter described in Exercise 11.10.

11.18 Derive a realization for the circuit of Figure 11.42, using the standard wave digital filter method, and compare it with the wave lattice structure obtained in Figure 11.47, with respect to overall computational complexity.

11.19 Table 11.13 shows the multiplier coefficients of a scaled Chebyshev filter designed with the following specifications

$A_p = 0.6$ dB

$A_r = 48$ dB

$\Omega_r = 3300$ rad/s

$\Omega_p = 4000$ rad/s

$\Omega_s = 10\,000$ rad/s

The structure used is the cascade of second-order direct-form structures.

(a) Plot the output-noise RPSD in dB.

(b) Determine the filter coefficient wordlength, using the sign-magnitude representation, that obtains at most 1.0 dB as the passband ripple.

(c) Plot the filter magnitude response when the filter coefficients are quantized with the number of bits obtained in item (b). Compare the results with the nonquantized magnitude response.

12 Implementation of DSP systems

12.1 Introduction

This chapter addresses some implementation methods for the several digital filtering algorithms and structures seen in the previous chapters.

The implementation of any building block of digital signal processing can be performed using a software routine on a simple personal computer. In this case, the designer's main concern becomes the description of the desired filter as an efficient algorithm that can be easily converted into a piece of software. In such cases, the hardware concerns tend to be noncritical, except for some details such as memory size, processing speed, and data input/output.

Another implementation strategy is based on specific hardware, especially suitable for the application to hand. In such cases, the system architecture must be designed within the speed constraints at a minimal cost. This form of implementation is mainly justified in applications that require high processing speed or in large-scale production. The four main forms of appropriate hardware for implementing a given system are:

- The development of a specific architecture using basic commercial electronic components and integrated circuits (Jackson et al., 1968; Peled & Liu, 1974, 1985; Freeny, 1975; Rabiner & Gold, 1975; Wanhammar, 1981).

- The use of programmable logic devices (PLD), such as field-programmable gate arrays (FPGA), which represent an intermediate integrated stage between discrete hardware and full-custom integrated circuits or digital signal processors (Skahill, 1996).

- The design of a dedicated integrated circuit for the application to hand using computer-automated tools for a very large scale integration (VLSI) design. The VLSI technology allows several basic processing systems to be integrated in a single silicon chip. The basic goal in designing application-specific integrated circuits (ASIC) is to obtain a final design satisfying very strict specifications with respect to, for example, chip area, power consumption, processing speed, testability, and overall production cost, (Skahill, 1996). These requirements are often achieved with the aid of computer-automated design systems. The first general purpose semi-automated design tools were the silicon compilers developed mainly in the mid-1980s. In these systems, a mid-level description language of the desired signal processing system is converted into the final design using a library describing the physical implementations of standard elements such as adders, memory units and multipliers (Denyer & Renshaw, 1985; Pope, 1985; Ulbrich, 1985). Nowadays, the silicon compilers have evolved to have a more modern approach for ASIC design where a high-level representation form, commonly referred to as a behavioral description, is employed (Kung, 1988; Weste & Eshraghian, 1993; Bayoumi & Swartzlander, 1994; Wanhammar, 1999; Parhi, 1999).

- The use of a commercially available general-purpose digital signal processor (DSP) to implement the desired system. The complete hardware system must also include external memory, input- and output-data interfaces, and sometimes analog-to-digital and digital-to-analog converters. There are several commercial DSPs available today, which include features such as fixed- or floating-point operation, several ranges of clock speed and price, internal memory, and very fast multipliers (Texas Instruments Inc., 1990, 1991; Motorola Inc., 1990).

We start our presentation in this chapter by introducing the basic elements necessary for the implementation of systems for digital signal processing, in particular the digital filters extensively covered by this book. Distributed arithmetic is presented as a design alternative for digital filters that eliminates the necessity of multiplier elements. Following that, the use of PLDs such as FPGA to implement digital signal processing techniques is discussed. The VLSI approach is then addressed by presenting the general approach for achieving an efficient practical design (Kung, 1988; Weste & Eshraghian, 1993; Bayoumi & Swartzlander, 1994; Wanhammar, 1999; Parhi, 1999). We give a few techniques to be used in the implementation of some specific systems, based on CAD tools such as the silicon compiler (Denyer & Renshaw, 1985; Pope, 1985) or the behavioral compilers. We then compare the main commercial families of DSP chips, emphasising their major differences and strengths. We finish the chapter by giving an example of digital filter implementation using a commercial DSP.

Naturally, the presentation of this chapter gives only an overview, since any attempt to give a completely up-to-date and in-depth presentation may quickly become obsolete, considering the speed of changes in the area.

12.2 Basic elements

12.2.1 Properties of the two's-complement representation

Since the two's-complement representation is vital to the understanding of the arithmetic operations described in the following sections, here we supplement the introduction given in Chapter 7, giving some other properties of the two's-complement representation.

Given a positive number x, its two's-complement representation is equal to its sign-magnitude representation. Also, we represent $-x$ by inverting all its bits and adding 1 to the least significant position. In addition, the sign bit of positive numbers is $s_x = 0$ and of negative number is $s_x = 1$. In that manner, if the two's-complement representation of x is given by

$$[x]_{2c} = s_x.x_1x_2\cdots x_n \tag{12.1}$$

then, the value of x is determined by

$$x = -s_x + x_12^{-1} + x_22^{-2} + \cdots + x_n2^{-n} \tag{12.2}$$

One of the advantages of the two's-complement representation is that, if A and B are represented in two's complement, then $C = A - B$, represented in two's complement, can be computed by adding A to the two's complement of B. Also, given x as in equation (12.2), we have that

$$\frac{x}{2} = -s_x2^{-1} + x_12^{-2} + x_22^{-3} + \cdots + x_n2^{-n-1} \tag{12.3}$$

and since $-s_x2^{-1} = -s_x + s_x2^{-1}$, then equation (12.3) can be rewritten as

$$\frac{x}{2} = -s_x + s_x2^{-1} + x_12^{-2} + x_22^{-3} + \cdots + x_n2^{-n-1} \tag{12.4}$$

that is, the two's-complement representation of $\frac{x}{2}$ is given by

$$\left[\frac{x}{2}\right]_{2c} = s_x.s_xx_1x_2\cdots x_n \tag{12.5}$$

This property implies that dividing a number represented in two's complement by 2 is equivalent to shifting all of its bits one position to the right, with the extra care of repeating the sign bit. This property is very important in the development of the multiplication algorithm, which involves multiplications by 2^{-j}.

12.2.2 Serial adder

A very economic implementation of digital filters is achieved through the use of so-called serial arithmetic. Such an approach performs a given operation by processing the bits representing a given binary number one by one serially. The overall result tends to be very efficient in terms of required hardware, power consumption, modularity, and ease of interconnection between serial-bit cells. The main drawback is clearly the associated processing speed, which tends to be very slow, when compared with other approaches.

An important basic element for the implementation of all signal processing systems is the full adder whose symbol and respective logic circuit are shown in Figure 12.1. Such a system presents two output ports, one equal to the sum of the two input bits A and B, and the other, usually referred to as the carry bit, corresponding to the

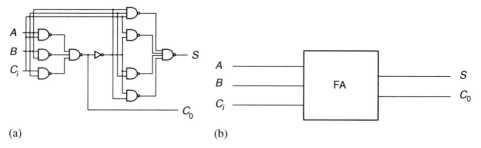

(a)

(b)

Figure 12.1 Full adder: (a) symbol; (b) logic circuit.

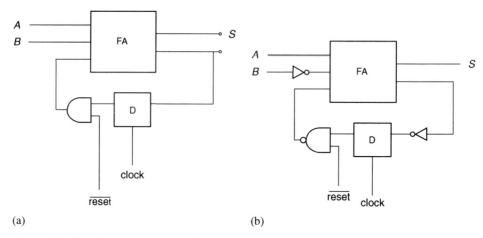

(a)

(b)

Figure 12.2 Serial implementations using the full adder as a basic block: (a) adder; (b) subtractor.

possible extra bit generated by the addition operation. A third input port C_i is used to allow the carry bit from a previous addition to be taken into account in the current addition.

A simple implementation of a serial adder for the two's-complement arithmetic, based on the full adder, is illustrated in Figure 12.2a. In such a system, the two input words A and B must be serially fed to the adder starting with their least-significant bit. A D-type flip-flop (labeled D) is used to store the carry bit from a given 1-bit addition, saving it to be used in the addition corresponding to the next significant bit. A reset signal is used to clear this D-type flip-flop before starting the addition of two other numbers, forcing the carry input to be zero at the beginning of the summation.

Figure 12.2b depicts the serial subtractor for the two's-complement arithmetic. This structure is based on the fact that $A - B$ can be computed by adding A to the two's complement of B, which can be determined by inverting B and summing 1 in the least-significant bit. The representation of B in two's complement is commonly done with an inverter at the input of B and with a NAND gate, substituting the AND gate, at the carry input. Then, at the beginning of the summation, the signal **reset** is asserted, and the carry input becomes 1. An extra inverter must be placed at the carry output because, with the use of the NAND gate, the carry output fed back to the D-type flip-flop must be inverted.

12.2.3 Serial multiplier

The most complex basic element for digital signal processing is the multiplier. In general, the product of two numbers is determined as a sum of partial products, as performed in the usual multiplication algorithm. Naturally, partial products involving a bit equal to zero do not need to be performed or taken into account. For that matter, there are several filter designs in the literature that attempt to represent all filter coefficients as the sum of a minimum number of nonzero bits (Pope, 1985; Ulbrich, 1985).

Let A and B be two numbers of m and n bits, respectively, that can be represented using two's-complement arithmetic as

$$[A]_{2c} = s_A.a_1 a_2 \cdots a_m \tag{12.6}$$
$$[B]_{2c} = s_B.b_1 b_2 \cdots b_n \tag{12.7}$$

such that, from equation (12.2), A and B are given by

$$A = -s_A + a_1 2^{-1} + a_2 2^{-2} + \cdots + a_m 2^{-m} \tag{12.8}$$
$$B = -s_B + b_1 2^{-1} + b_2 2^{-2} + \cdots + b_n 2^{-n} \tag{12.9}$$

Using two's-complement arithmetic, the product $P = AB$ is given by

$$
\begin{aligned}
P &= (-s_A + a_1 2^{-1} + \cdots + a_m 2^{-m})(-s_B + b_1 2^{-1} + \cdots + b_n 2^{-n}) \\
&= (-s_A + a_1 2^{-1} + \cdots + a_m 2^{-m})b_n 2^{-n} \\
&\quad + (-s_A + a_1 2^{-1} + \cdots + a_m 2^{-m})b_{n-1} 2^{-n+1} \\
&\quad \vdots \\
&\quad + (-s_A + a_1 2^{-1} + \cdots + a_m 2^{-m})b_1 2^{-1} \\
&\quad - (-s_A + a_1 2^{-1} + \cdots + a_m 2^{-m})s_B
\end{aligned}
\tag{12.10}
$$

We can perform this multiplication in a step-by-step manner by first summing the terms multiplied by b_n and b_{n-1}, taking the result and adding to the term multiplied by b_{n-2}, and so on. Let us develop this reasoning a bit further. The sum of the first two terms can be written as

$$C = b_n 2^{-n} A + b_{n-1} 2^{-n+1} A$$
$$= (-s_A + a_1 2^{-1} + \cdots + a_m 2^{-m}) b_n 2^{-n}$$
$$+ (-s_A + a_1 2^{-1} + \cdots + a_m 2^{-m}) b_{n-1} 2^{-n+1}$$
$$= 2^{-n+1} [b_n 2^{-1} (-s_A + a_1 2^{-1} + \cdots + a_m 2^{-m})$$
$$+ b_{n-1} (-s_A + a_1 2^{-1} + \cdots + a_m 2^{-m})] \tag{12.11}$$

From equation (12.4), the above equation becomes

$$C = 2^{-n+1} \Big[b_n (-s_A + s_A 2^{-1} + a_1 2^{-2} + \cdots + a_m 2^{-m-1})$$
$$+ b_{n-1} (-s_A + a_1 2^{-1} + \cdots + a_m 2^{-m}) \Big] \tag{12.12}$$

which can be represented in the form of the multiplication algorithm as

	(	s_A	s_A	a_1	...	a_{m-1}	a_m	)	$\times$	b_n
+	(	s_A	a_1	a_2	...	a_m		)	$\times$	b_{n-1}
	s_C	c_1	c_2	c_3	...	c_{m+1}	c_{m+2}			

that is

$$C = 2^{-n+2} (-s_C + c_1 2^{-1} + c_2 2^{-2} + c_3 2^{-3} + \cdots + c_{m+1} 2^{-m-1} + c_{m+2} 2^{-m-2}) \tag{12.13}$$

Note that the sum of two positive numbers is always positive, and the sum of two negative numbers is always negative. Therefore, $s_C = s_A$. If fact, since in two's-complement arithmetic the sign has to be extended, the more compact representation for the summation above would be

	(	...	s_A	s_A	s_A	s_A	a_1	...	a_{m-1}	a_m	)	$\times$	b_n
+	(	...	s_A	s_A	s_A	a_1	a_2	...	a_m		)	$\times$	b_{n-1}
		...	s_A	s_A	c_1	c_2	c_3	...	c_{m+1}	c_{m+2}			

In the next step, C should be added to $b_{n-2} 2^{-n+2} A$. This yields

$$D = (-s_A + c_1 2^{-1} + \cdots + c_{m+2} 2^{-m-2}) 2^{-n+2}$$
$$+ (-s_A + a_1 2^{-1} + \cdots + a_m 2^{-m}) b_{n-2} 2^{-n+2} \tag{12.14}$$

which can be represented in the compact form of the multiplication algorithm as

	(	...	s_A	s_A	c_1	c_2	...	c_m	c_{m+1}	c_{m+2}	)		
+	(	...	s_A	s_A	a_1	a_2	...	a_m			)	$\times$	b_{n-2}
		...	s_A	d_1	d_2	d_3	...	d_{m+1}	d_{m+2}	d_{m+3}			

which is equivalent to

$$D = 2^{-n+3}(-s_A + d_1 2^{-1} + d_2 2^{-2} + \cdots + d_{m+2} 2^{-m-2} + d_{m+3} 2^{-m-3}) \qquad (12.15)$$

This process goes on until we obtain the next-to-last partial sum, Y. If B is positive, s_B is zero, and the final product, Z, is equal to Y, that is

$$Z = Y = -s_A + y_1 2^{-1} + \cdots + y_m 2^{-m} + y_{m+1} 2^{-m-1} + \cdots + y_{m+n} 2^{-m-n} \quad (12.16)$$

such that the two's complement of Z is given by

$$[Z]_{2c} = s_A . z_1 z_2 \cdots z_m z_{m+1} \cdots z_{m+n} = s_A . y_1 y_2 \cdots y_m y_{m+1} \cdots y_{m+n} \qquad (12.17)$$

On the other hand, if B is negative, then $s_B = 1$, and Y still needs to be subtracted from $s_B A$. This can be represented as

	(	s_A	y_1	y_2	$\cdots$	y_m	y_{m+1}	$\cdots$	y_{m+n}	)		
$-$	(	s_A	a_1	a_2	$\cdots$	a_m				)	$\times$	s_B
		s_Z	z_1	z_2	$\cdots$	z_m	z_{m+1}	$\cdots$	z_{m+n}			

The full precision two's-complement multiplication of A, with $(m + 1)$ bits, by B, with $(n + 1)$ bits, should be represented with $(m + n + 1)$ bits. If we want to represent the final result using just the same number of bits as A, that is, $(m + 1)$, we can use either rounding or truncation. For truncation, it suffices to disregard the bits $z_{m+1}, \ldots, z_{m+n}$. For rounding, we must add to the result, prior to truncation, a value equal to $\Delta = 2^{-m-1}$, such that

$$[\Delta]_{2c} = 0.\underbrace{0 \cdots 0}_{m \text{ zeros}} 1 \qquad (12.18)$$

By looking at the algorithmic representation of the last partial sum above, one sees that to add Δ to the result is equivalent to summing bit 1 at position $(m + 1)$, which is the nth position from the rightmost bit, z_{m+n}. Since it does not matter whether this bit is summed in the last or the first partial sum, it suffices to sum this bit during the first partial sum. Then the rounding can be performed by summing the number

$$[Q]_{2c} = 1.\underbrace{0 \cdots 0}_{n-1 \text{ zeros}} \qquad (12.19)$$

to the first partial sum. In addition, since only $(m + 1)$ bits of the product will be kept, we need to keep, from each partial sum, its $(m + 1)$ most significant bits.

The main idea behind a serial multiplier is to perform each partial sum using a serial adder such as the one depicted in Figure 12.1, taking care to introduce proper delays between the serial adder (SA) blocks, to align each partial sum properly with $b_j A$. This scheme is depicted in Figure 12.3, where the rounding signal Q is introduced in the first partial sum, and the least significant bits of A and Q are input first.

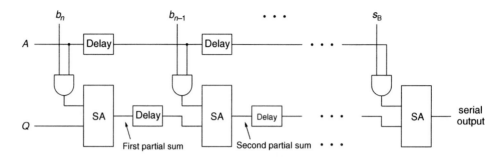

Figure 12.3 Schematic representation of a serial multiplier.

In the scheme of Figure 12.3, depending on the values of the delay elements, one serial adder can begin to perform one partial sum as soon as the first bit of the previous partial sum becomes available. When this happens, we have what is called a pipelined architecture. The main idea behind the concept of pipelining is to use delay elements at strategic points of the system that allow one operation to start before the previous operation has finished.

Figure 12.4 shows a pipelined serial multiplier. In this multiplier, the input A is in two's-complement format, while the input B is assumed to be positive. In this figure, the cells labeled D are D-type flip-flops, all sharing the same clock line, which is omitted for convenience, and the cells labeled SA represent the serial adder depicted in Figure 12.2a. The latch element used is shown in Figure 12.5. In this case, if the enable signal is high, after the clock, the output y becomes equal to the input x; otherwise, the output y keeps its previous value.

In Figure 12.4, since B is assumed to be positive, then $s_B = 0$. Also, the rounding signal Q is such that

$$[Q]_{2c} = \begin{cases} 00 \cdots 00, & \text{for truncation} \\ \cdots 01 \underbrace{0 \cdots 0}_{n+1 \text{ zeros}}, & \text{for rounding} \end{cases} \tag{12.20}$$

In the case of rounding, it is important to note that $[Q]_{2c}$ has two more zeros to the right than the one in equation (12.19). This is so because this signal must be synchronized with input $b_n A$, which is delayed by two bits before entering the serial adder of the first cell. Finally, the control signal CT is equal to

$$CT = 00 \underbrace{1 \cdots 1}_{m \text{ ones}} 0 \tag{12.21}$$

It should be noticed that all signals A, B, Q, CT should be input from right to left, that is, starting from their least-significant bits. Naturally, the output signal is generated serially, also starting from its least-significant bit, the first bit coming out after $(2n+3)$ clock cycles and the entire word taking a total of $(2n + m + 3)$ cycles to be calculated.

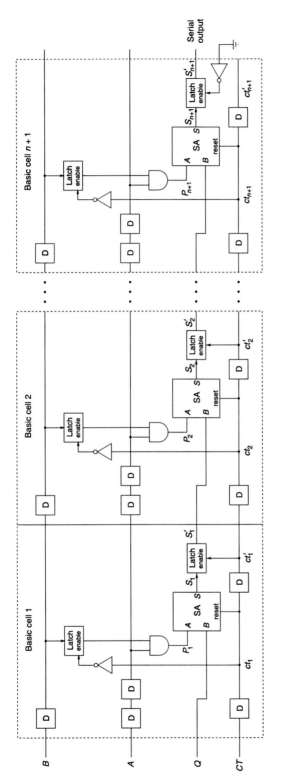

Figure 12.4 Basic architecture of the pipelined serial multiplier. Clock lines have been suppressed for convenience.

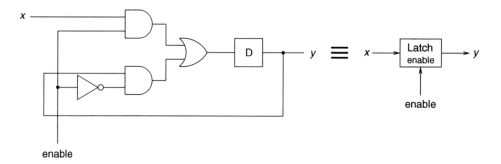

Figure 12.5 Latch element.

In the literature, it has been shown that the multiplication of two words of lengths $(m+1)$ and $(n+1)$ using a serial or serial–parallel multiplier takes at least $(m+n+2)$ clock cycles (Hwang, 1979), even in the cases where the final result is quantized with $(m+1)$ bits. Using a pipelined architecture, it is possible to determine the $(m+1)$-bit quantized product at every $(m+1)$ clock cycles.

In the serial multiplier of Figure 12.4, each basic cell performs one partial sum of the multiplication algorithm. As the result of cell j is being generated, it is directly fed to cell $(j+1)$, indicating the pipelined nature of the multiplier. A detailed explanation of the behavior of the circuit is given below:

(i) After the $(2j+1)$th clock cycle, b_{m-j} will be at the input of the upper latch of cell $j+1$. The first 0 of the control signal CT will be at input ct_{j+1} at this time, which will reset the serial adder and enable the upper latch of cell $j+1$. Therefore, after the $(2j+2)$th clock cycle, b_{m-j} will be at the output of the upper latch, and will remain there for m clock cycles (the number of 1s of the signal CT).

(ii) After the $(2j+2)$th clock cycle, a_m will be at the input of the AND gate of cell $j+1$, and therefore $a_m b_{m-j}$ will be the P_{j+1} input of the serial adder. Since at this time the first 0 of the control signal CT will be at input ct'_{j+1}, the lower latch becomes on hold. Therefore, although the first bit of the $(j+1)$th partial sum is at S_{j+1} after the $(2j+2)$th clock cycle, it will not be at S'_{j+1} after the $(2j+3)$th clock cycle, and it will be discarded. Since in the next $m+1$ clock cycles the bit 1 will be at ct'_{j+1}, the remaining $m+1$ bits of the $(j+1)$th partial sum will pass from S_{j+1} to S'_{j+1}, which is input to the next basic cell.

(iii) After the $(2j+2+k)$th clock cycle, $a_{m-k} b_{m-j}$ will be the P_{j+1} input of the serial adder of cell $j+1$. Therefore, after the $(2j+2+m)$th clock cycle, $s_A b_{m-j}$ will be at the P_{j+1} input of the serial adder of cell $j+1$. This is equivalent to saying that the last bit of the $(j+1)$th partial sum will be at the input of the lower latch of cell $j+1$ at this time. Since then ct'_{j+1} is still 1, then, after the $(2j+3+m)$th clock cycle the last bit of the partial sum of cell $j+1$ will be at the output of the lower latch of cell $j+1$. But from this time on, ct'_{j+1} will be zero, and therefore the output of the latch of cell $j + 1$ will hold the last bit of the $(j + 1)$th partial sum. Since this last bit represents the sign of the $(j+1)$th partial sum, it performs the sign extension necessary in two's-complement arithmetic.

(iv) Since there is no need to perform sign extension of the last partial sum, the lower latch of the last cell is always enabled, as indicated in Figure 12.4.

(v) Since each basic cell but the last one only outputs m bits apart from the sign extensions, then the serial output at the last cell will contain the product either truncated or rounded to $m + 1$ bits, depending on the signal Q.

EXAMPLE 12.1

Verify how the pipelined serial multiplier depicted in Figure 12.4 processes the product of the binary numbers $A = 1.1001$ and $B = 0.011$.

SOLUTION

We have that $m = 4$ and $n = 3$. Therefore, the serial multiplier has four basic cells. We expect the least-significant bit of the truncated product to be output after $(2n + 3) = 9$ clock cycles and its last bit after $(2n + m + 3) = 13$ clock cycles. Supposing that the quantization signal Q is zero, which corresponds to a truncated result, we have that (the variable t indicates the clock cycle after which the values of the signals are given; $t = 0$ means the time just before the first clock pulse):

t	13	12	11	10	9	8	7	6	5	4	3	2	1	0
Q	0	0	0	0	0	0	0	0	0	0	0	0	0	0
P_1	0	0	0	0	0	0	0	1	1	0	0	1	0	0
S_1	0	0	0	0	0	0	0	1	1	0	0	1	0	0
ct_1'	0	0	0	0	0	0	0	1	1	1	1	0	0	0
S_1'	1	1	1	1	1	1	1	1	0	0	0	0	0	0
P_2	0	0	0	0	0	1	1	0	0	1	0	0	0	0
S_2	0	0	0	0	0	1	0	1	0	1	0	0	0	0
ct_2'	0	0	0	0	0	1	1	1	1	0	0	0	0	0
S_2'	1	1	1	1	1	0	1	0	0	0	0	0	0	0
P_3	0	0	0	0	0	0	0	0	0	0	0	0	0	0
S_3	1	1	1	1	1	0	1	0	0	0	0	0	0	0
ct_3'	0	0	0	1	1	1	1	0	0	0	0	0	0	0
S_3'	1	1	1	1	0	1	0	0	0	0	0	0	0	0
P_4	0	0	0	0	0	0	0	0	0	0	0	0	0	0
S_4	1	1	1	1	0	1	0	0	0	0	0	0	0	0
ct_4'	1	1	1	1	1	1	1	1	1	1	1	1	1	1
S_4'	**1**	**1**	**1**	**0**	**1**	0	0	0	0	0	0	0	0	0

And the computed product is 1.1101.

For rounding, since $n = 3$, we have, from equation (12.20), that $[Q]_{2c} = \cdots 0010000$. Thus, the operation of the serial multiplier is as follows:

t	13	12	11	10	9	8	7	6	5	4	3	2	1	0
Q	0	0	0	0	0	0	0	0	0	1	0	0	0	0
P_1	0	0	0	0	0	0	0	1	1	0	0	1	0	0
S_1	0	0	0	0	0	0	0	1	1	1	0	1	0	0
ct_1'	0	0	0	0	0	0	0	1	1	1	1	0	0	0
S_1'	1	1	1	1	1	1	1	1	1	0	0	0	0	0
P_2	0	0	0	0	0	1	1	0	0	1	0	0	0	0
S_2	0	0	0	0	0	1	0	1	1	1	0	0	0	0
ct_2'	0	0	0	0	0	1	1	1	1	0	0	0	0	0
S_2'	1	1	1	1	1	0	1	1	0	0	0	0	0	0
P_3	0	0	0	0	0	0	0	0	0	0	0	0	0	0
S_3	1	1	1	1	1	0	1	1	0	0	0	0	0	0
ct_3'	0	0	0	1	1	1	1	0	0	0	0	0	0	0
S_3'	1	1	1	1	0	1	0	0	0	0	0	0	0	0
P_4	0	0	0	0	0	0	0	0	0	0	0	0	0	0
S_4	1	1	1	1	0	1	0	0	0	0	0	0	0	0
ct_4'	1	1	1	1	1	1	1	1	1	1	1	1	1	1
S_4'	**1**	**1**	**1**	**0**	**1**	0	0	0	0	0	0	0	0	0

That is, the rounded product is also 1.1101.

$\triangle$

The general multiplier for two's-complement arithmetic (without the positive restriction in one of the factors) can be obtained from the multiplier seen in Figure 12.4 by a slight modification of the connections between the last two basic cells. In fact, from the representation given in equation (12.2), we note that the general multiplication of any two numbers represented in two's complement is equal to the multiplication of a positive number and a two's-complement number, except that in the last step we must perform the subtraction of the product between the data and the coefficient sign bit from the partial result obtained at that point. Then, we must perform a subtraction to obtain $S_{n+1} = S_n' - s_A A$. It can be shown that, using two's-complement arithmetic, $X - Y = \overline{Y + \overline{X}}$, where $\overline{X}$ represents the inversion of all bits of X (see Exercise 12.2). Therefore, in the last cell, we should invert S_n' before it is input to the $(n+1)$th serial adder, and then invert its output. Thus, the connections to the last basic cell should be modified as shown in Figure 12.6. An alternative implementation for the rounding quantization, instead of using the Q signal, consists of forcing the carry signal in serial adder of the nth basic cell to 1 while it is performing the addition of $a_m b_1$ to the $(n-1)$th partial sum. Other versions of the pipelined serial multiplier can be found in Lyon (1976).

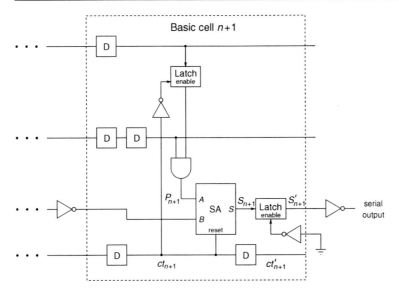

Figure 12.6 Tail-end of the modified general multiplier for two's-complement arithmetic.

So far, we have focused on single-precision multipliers, that is, the final product is quantized, either by rounding or truncation. As seen in previous chapters, in many cases it is desirable that the final product presents double precision to avoid, for instance, nonlinear oscillations. Such operations can be easily performed with the basic multipliers seen so far, artificially doubling the precision of the two multiplying factors by juxtaposing the adequate number of bits 0 to the right of each input number. For example, the exact multiplication of two inputs with $(n + 1)$ bits each is obtained by doubling the number of bits in each input by adding $(n + 1)$ bits 0 to the right of each number. Naturally, in this case, we need $(n + 1)$ more basic cells, and the complete operation takes twice as long to be implemented as basic single-precision multiplication.

For binary representations other than the two's complement, the implementation of the corresponding serial multipliers can be found, for instance, in Jackson et al. (1968); Freeny (1975); Wanhammar (1981, 1999).

12.2.4 Parallel adder

Parallel adders can be easily constructed by interconnecting several full adders as shown in Figure 12.7, where we can observe that the carry signal in each cell propagates to the next cell through the overall adder structure. The main time-consuming operation in this realization is the propagation of the carry signal that must go through all the full adders to form the desired sum. This problem can be reduced, as proposed in Lyon (1976), at the cost of an increase in the hardware complexity.

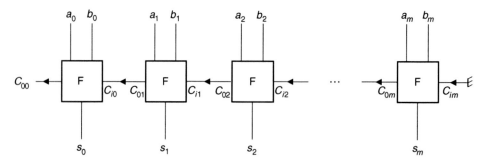

Figure 12.7 Block diagram of the parallel adder.

12.2.5 Parallel multiplier

A parallel multiplier is usually implemented as a matrix of basic cells (Rabiner & Gold, 1975; Freeny, 1975), in which the internal data propagates horizontally, vertically, and diagonally in an efficient and ordered way. In general, such multipliers occupy a very large area of silicon and consume significant power. For these reasons they are only used in cases where time is a major factor in the overall performance of the given digital processing system, as in major DSP chips of today.

Since a digital filter is basically composed of delays, adders, and multipliers (using serial or parallel arithmetic), we now have all the tools necessary to implement a digital filter. The implementation is carried out by properly interconnecting such elements according to a certain realization, which should be chosen from among the ones seen in the previous chapters. If in the resulting design, the sampling rate multiplied by the number of bits used to represent the signals is below the reachable processing speed, multiplexing techniques can be incorporated to optimize the use of hardware resources, as described in Jackson et al. (1968). There are two main multiplexing approaches. In the first one, a single filter processes several input signals that are appropriately multiplexed. In the second one, the filter coefficients are multiplexed, resulting in several distinct transfer functions. It is always possible to combine both approaches, if desired, as suggested in Figure 12.8, where a memory element stores several sets of coefficients corresponding to different digital filters to be multiplexed.

12.3   Distributed arithmetic implementation

An alternative approach for implementing digital filters, the so-called distributed arithmetic (Peled & Liu, 1974, 1985; Wanhammar, 1981), avoids the explicit implementation of the multiplier element. Such a concept first appeared in the open literature in 1974 (Peled & Liu, 1974), although a patent describing it had been issued in December 1973 (Croisier et al., 1973). The basic idea behind the concept of distributed

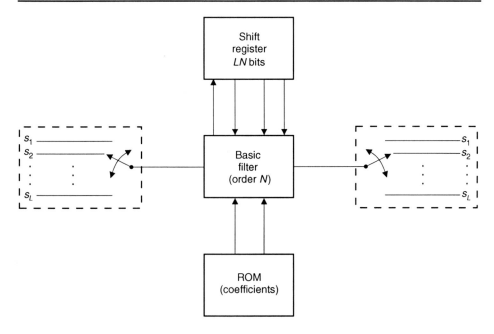

Figure 12.8 Fully multiplexed realization of digital filters.

arithmetic is to perform the summation of the products between filter coefficients and internal signals without using multipliers.

For example, suppose one wishes to calculate the following expression:

$$y = \sum_{i=1}^{N} c_i X_i \tag{12.22}$$

where c_i are the filter coefficients and X_i are a set of signals represented in two's complement with $(b + 1)$ bits. Assuming that the X_i are properly scaled such that $|X_i| < 1$, we can rewrite equation (12.22) as

$$y = \sum_{i=1}^{N} c_i \left(\sum_{j=1}^{b} x_{i_j} 2^{-j} - x_{i_0} \right) \tag{12.23}$$

where x_{i_j} corresponds to the jth bit of x_i, and x_{i_0} to its sign. By reversing the summation order in this equation, the following relationship applies

$$y = \sum_{j=1}^{b} \left(\sum_{i=1}^{N} c_i x_{i_j} \right) 2^{-j} - \sum_{i=1}^{N} c_i x_{i_0} \tag{12.24}$$

If we define the function $s(\cdot)$ of N binary variables $z_1, z_2, \ldots, z_N$, as

$$s(z_1, z_2, \ldots, z_N) = \sum_{i=1}^{N} c_i z_i \tag{12.25}$$

then equation (12.24) can be written as

$$y = \sum_{j=1}^{b} s(x_{1_j}, x_{2_j}, \ldots, x_{N_j}) 2^{-j} - s(x_{1_0}, x_{2_0}, \ldots, x_{N_0}) \tag{12.26}$$

Using the notation $s_j = s(x_{1_j}, x_{2_j}, \ldots, x_{N_j})$, then the above equation can be written as

$$y = \{\ldots [(s_b 2^{-1} + s_{b-1}) 2^{-1} + s_{b-2}] 2^{-1} + \cdots + s_1\} 2^{-1} - s_0 \tag{12.27}$$

The value of s_j depends on the jth bit of all signals that determine y. Equation (12.27) gives a methodology to compute y: we first compute s_b, then divide the result by 2 with a right-shift operation, add the result to s_{b-1}, then divide the result by 2, add it to s_{b-2}, and so on. In the last step, s_0 must be subtracted from the last partial result.

Overall, the function $s(\cdot)$ in equation (12.25) can assume at most 2^N possible distinct values, since all of its N variables are binary. Thus, an efficient way to compute $s(\cdot)$ is to pre-determine all its possible values, which depend on the values of the coefficients c_i, and store them in a memory unit whose addressing logic must be based on the synchronized data content.

The distributed arithmetic implementation of equation (12.22) is as shown in Figure 12.9. In this implementation, the words $X_1, X_2, \ldots, X_N$ are stored in shift-registers $SR_1, SR_2, \ldots, SR_N$, respectively, each one with $(b+1)$ bits. The N outputs of the shift-registers are used to address a ROM unit. Then, once the words are loaded in the shift-registers, after the jth right-shift, the ROM will be addressed with $x_{1_{b-j}} x_{2_{b-j}} \cdots x_{N_{b-j}}$, and the ROM will output s_{b-j}. This value is then loaded into register A to be added to the partial result from another register, B, which stores the previous accumulated value. The result is divided by 2 (see equation (12.27)). For each calculation, register A is initialized with s_b and register B is reset to contain only zeros. Register C is used to store the final sum, obtained from the subtraction of s_0 from the next-to-last sum. Naturally, all the above calculations could have been implemented without register A, whose importance lies in the possibility of accessing the memory unit simultaneously to the adder/subtractor operation, thus forming a pipelined architecture.

Using the implementation illustrated in Figure 12.9, the complete time interval for computing one sample of the output y entails $(b+1)$ clock cycles, and therefore depends solely on the number of bits of each word, and is not a function of the number, N, of partial products involved to form y. The duration of the clock cycle, in this case, is determined by the maximum value between the times of a memory access and a computation of an addition or subtraction operation. The number of bits, b, used to represent each word greatly affects the memory size, along with the number, N, of partial sums, which should not be made too large in order to limit the memory access time. The value of b depends basically on the dynamic range necessary to represent

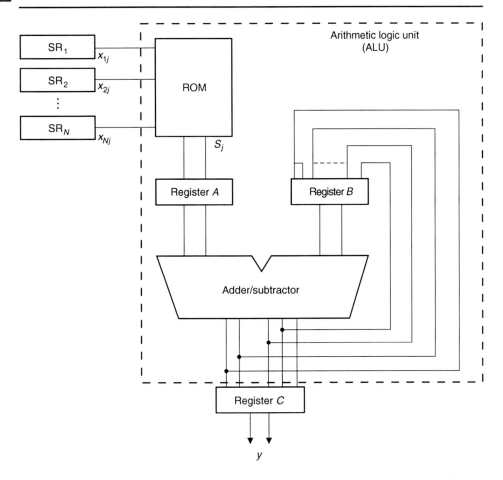

Figure 12.9 Basic architecture for implementing equation (12.22) using distributed arithmetic.

$|s_j|$ and the precision required to represent the coefficients c_i to avoid large coefficient quantization effects.

We can greatly increase the overall processing speed for computing y as in equation (12.22), by reading the desired values of s_j in parallel, and using several adders in parallel to determine all necessary partial sums, as described in Peled & Liu (1974).

This distributed arithmetic can be used to implement several of the digital filter structures presented in this book. For instance, the second-order direct-form realization implementing the following transfer function

$$H(z) = \frac{b_0 z^2 + b_1 z + b_2}{z^2 + a_1 z + a_2} \tag{12.28}$$

which corresponds to the following difference equation

$$y(n) = b_0 x(n) + b_1 x(n-1) + b_2 x(n-2) - a_1 y(n-1) - a_2 y(n-2) \tag{12.29}$$

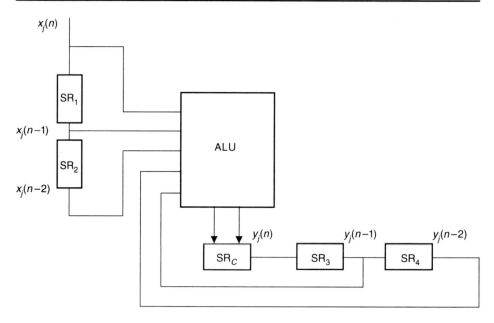

Figure 12.10 Second-order direct-form filter implemented with distributed arithmetic.

can be mapped onto the implementation seen in Figure 12.10, where the present and past values of the input signal are used in conjunction with the past values of the output signal to address the memory unit. In other words, we make $X_i = x(n - i)$, for $i = 0, 1, 2$, $X_3 = y(n - 1)$, and $X_4 = y(n - 2)$. The function $s(\cdot)$, which determines the content of the memory unit, is then given by

$$s(z_1, z_2, z_3, z_4, z_5) = b_0 z_1 + b_1 z_2 + b_2 z_3 - a_1 z_4 - a_2 z_5 \qquad (12.30)$$

Therefore, at a given instant, the quantity s_j is

$$s_j = b_0 x_j(n) + b_1 x_j(n - 1) + b_2 x_j(n - 2) - a_1 y_j(n - 1) - a_2 y_j(n - 2) \qquad (12.31)$$

As the number of partial sums required to calculate y is $N = 5$, the memory unit in this example should have 32 positions of L bits, where L is the number of bits established to represent s_j.

In general, all coefficients b_i and a_i in equation (12.30) should be already quantized. This is because it is easier to predict the quantization effects at the stage of filter design, following the theory presented in Chapter 7, than to analyze such effects after quantizing s_j as given in equation (12.30). In fact, performing the quantization on s_j can even introduce nonlinear oscillations at the output of a given structure which was initially shown to be immune to limit cycles.

For the second-order state-variable realization as given in Figure 11.5, the distributed arithmetic implementation is shown in Figure 12.11. In this case, the contents

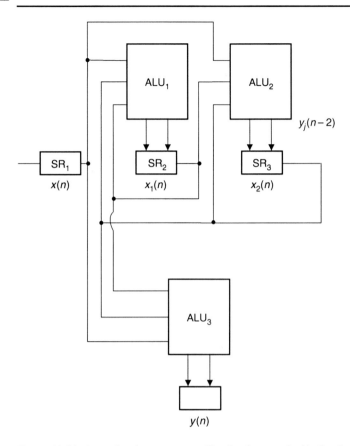

Figure 12.11 Second-order state-space filter implemented with distributed arithmetic.

of the memory units are generated by

$$s_{1j} = a_{11}x_{1j}(n) + a_{12}x_{2j}(n) + b_1x_j(n); \text{ for the ROM of the ALU}_1 \tag{12.32}$$

$$s_{2j} = a_{22}x_{2j}(n) + a_{21}x_{1j}(n) + b_2x_j(n); \text{ for the ROM of the ALU}_2 \tag{12.33}$$

$$s_{3j} = c_1x_{1j}(n) + c_2x_{2j}(n) + dx_j(n); \text{ for the ROM of the ALU}_3 \tag{12.34}$$

Each memory unit has $8 \times L$ words, where L is the established wordlength for the given s_{ij}.

In a regular implementation, to eliminate zero-input limit cycles in the state-space realization, the state variables $x_1(n)$ and $x_2(n)$ must be calculated by ALU_1 and ALU_2 in double precision, and then properly quantized, before being loaded into the shift-registers SR_2 and SR_3. However, it can be shown that no double-precision computation is necessary to avoid zero-input limit cycles when implementing the state-space realization with the distributed arithmetic approach (de la Vega et al., 1995).

To eliminate constant-input limit cycles, the state-space realization shown in Figure 12.12 requires that the state variable $x_1(n)$ must be computed by ALU_1 in

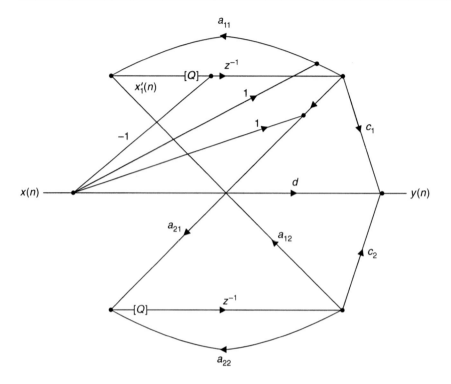

Figure 12.12 State-space realization immune to constant-input limit cycles.

double precision, and then properly quantized to be subtracted from the input signal. To perform this subtraction, register A in Figure 12.9 must be multiplexed with another register that contains the input signal $x(n)$, in order to guarantee that the signal arriving at the adder, at the appropriate instant of time, is the complemented version of $x(n)$, instead of a signal coming from the memory. In such a case, the content of the ROM of ALU_1 must be generated as

$$s'_{1j} = a_{11}x_{1j}(n) + a_{12}x_{2j}(n) + a_{11}x_j(n); \text{ for the ROM of the ALU}_1 \tag{12.35}$$

while ALU_3 is filled in the same fashion as given in equation and (12.34) and for ALU_2 the content is the same as in equaiton (12.33) with b_2 replaced by a_{21}.

The implementation of high-order filters using distributed arithmetic is simple when the parallel and cascade forms seen in Chapter 4 are used. Both forms can be implemented using a single block realization, whose coefficients are multiplexed in time to perform the computation of a specific second-order block. The cascade version of this type of implementation is represented in Figure 12.13. Such an approach is very efficient in terms of chip area and power consumption. Faster versions, however, are possible using both the parallel and cascade forms. For instance, the fast parallel approach is depicted in Figure 12.14, where all numerical calculations can

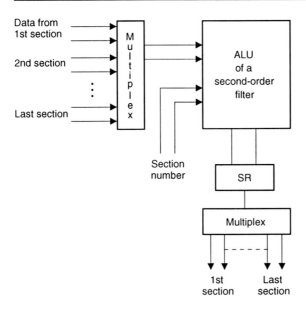

Figure 12.13 Implementation of the cascade form using distributed arithmetic.

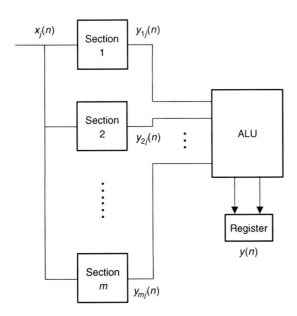

Figure 12.14 Fast implementation of the parallel form using distributed arithmetic.

be performed in parallel. The fast cascade form can be made more efficient if delay elements are added in between the blocks to allow pipelined processing.

The distributed arithmetic technique presented in this section can also be used to implement other digital filter structures (Wanhammar, 1981), as well as some specific processors for the computation of the FFT (Liu & Peled, 1975).

12.4   Programmable logic devices

In the mid-1970s the TTL series of logic circuits by Texas Instruments was the backbone for designing digital logic circuits such as multiplexers, encoders, decoders, adders, and state machines. Such families of integrated circuits included Boolean logic (NOR, NAND, AND, OR, inverter gates), flip-flops, full adders, and counters. Using these chips to implement a digital process, the design procedure involved Boolean algebra or Karnaugh maps to convert a truth table into a digital logic circuit.

More complex applications, however, have made the implementation process using TTL integrated circuits rather intricate. Also, using this technology, the resulting system tended to become quite large, containing possibly hundreds of integrated circuits, besides requiring a long debugging time and large power consumption. Using multifunctional devices such as a programmable logic device (PLD), which integrate several logic functions within a single chip, provided more flexibility to the system designer to implement today's very demanding digital signal processing algorithms.

Possibly the earliest PLDs are the programmable array logic (PAL) devices, still commercially available today. These integrated circuits consist of two levels of digital logic, the first one involving an array of AND gates connected to several input signals, and the second one including a logic OR stage. Each input signal of the PAL is internally inverted so that both versions of the signal, the direct and the inverted, are fed to the AND array in the first stage. The programming portion of the PAL is the selection of which inputs are connected or not to each AND gate. With this structure, a PAL is able to implement relatively complex digital logic functions.

An advance on the PAL device is the so-called complex programmable logic device (CPLD), which extends the concept of PLD to a higher level of integration. A CPLD contains several PAL elements, each with an additional logic block at the output of the OR stage, referred to as a macrocell, that is capable of performing more intricate logic operations. The contents of a macrocell vary from one CPLD manufacturer to another. Most, however, include a flip-flop and a polarity control unit that enables the implementation of either direct or inverse digital logic, giving the system designer the power to choose whichever logic uses the fewest circuit resources. A programmable interconnect unit within the CPLD allows the communications between several of these enhanced-PAL units. Figure 12.15 depicts the block diagram of a CPLD, detailing the internal structure of the AND and OR levels of digital logic elements.

Another evolution stage of PLDs is the field-programmable gate array (FPGA) architecture. An FPGA device consists of an array of logic macrocells that are interconnected through several horizontal and vertical communication channels, as shown in Figure 12.16. The macrocells in an FPGA are, in general, simpler than those in a CPLD. This is compensated for in the FPGA device by the integration of a larger number of cells and the possibility of grouping them in series, thus enabling the system designer to implement very complex digital logic. The programming

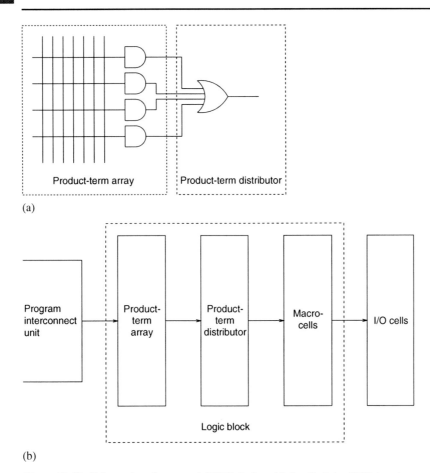

Figure 12.15 Schematics of a general CPLD device: (a) detail of the AND (product-term array) and OR (product-term distribution) stages; (b) general block diagram.

portion of the FPGA includes the communication channels, dictating how each cell is connected to each other one, and the function of each macrocell. The characteristics of the FPGA technology that make it suitable for modern digital signal processing algorithms include low cost, low power consumption, small size, ease of use (allowing fast marketing of the product), and in-system reprogrammability (Skahill, 1996).

Most vendors of CPLDs and FPGAs provide high level software tools for the design of digital systems. In addition, they usually provide toolboxes especially suited to digital signal processing.

12.5   ASIC implementation

In this section, we discuss the implementation of application-specific integrated circuits (ASIC) for digital signal processing using modern VLSI technology. We begin by

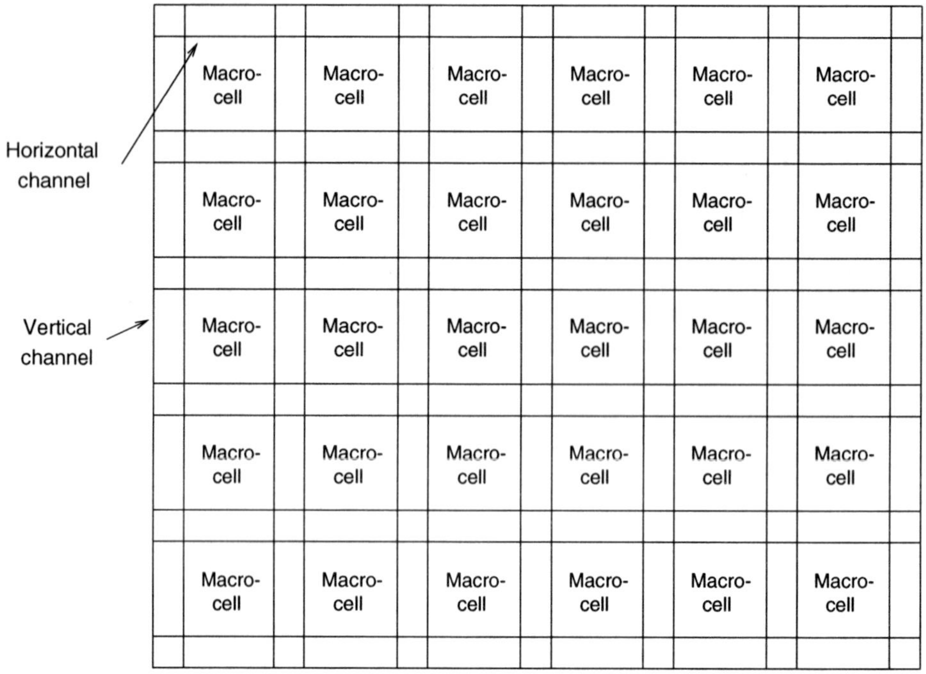

Figure 12.16 Schematic of a general FPGA device.

introducing the main issues that must be considered to achieve efficient design of VLSI systems (Weste & Eshraghian, 1993; Bayoumi & Swartzlander, 1994; Wanhammar, 1999; Parhi, 1999), namely hierarchy, regularity, modularity, and locality.

Hierarchy is related to the idea of dividing a design process into several subprocesses, so that all individual subprocesses have an appropriate level of difficulty. Naturally, all steps in the VLSI design procedure should be clearly identified, forming a regular hierarchy of stages that should be solved in an ordered fashion.

While hierarchy is related to dividing a system into a series of subsystems, regularity is related to the concept of dividing the hierarchy into a set of similar blocks. In fact, we can make the overall design procedure more efficient if we can form partial stages that are identical or at least similar in nature. This property allows solutions devised for a given stage to be transposed to other tasks, thus resulting in an improvement in productivity, as the number of different designs is greatly reduced.

The concept of modularity is related to the fact that not only is it very important that all design stages should be clearly defined, but also that all their interrelationships should be well identified. This allows each stage to be clearly devised and solved in an orderly fashion.

With the modularity concept well perceived, each stage can be individually optimized and solved, if only local variables or problems are addressed. In fact, with the idea of locality, the internal issues related to a stage are unimportant to any other stage.

In this way, one is able to reduce the apparent complexity in each stage of the overall design procedure.

Naturally, when dealing with the implementation of a complete system, based on one or more chips, we must employ a powerful computer-automated design (CAD) system for integrated circuits. Such a tool will design systems which are significantly less prone to errors, in a very time-efficient manner. The CAD system starts with some description of the desired system and generates the circuit mask to be integrated, allowing the system designer, who may not be a VLSI specialist, to follow the whole process, starting at the system specification and ending with the final chip implementation.

It is common practice that, before using the CAD tool, the designer tests the project with the help of a system simulator. When the simulator indicates that the designed system satisfies the desired characteristics, the designer then uses the CAD system to perform the actual design. Sometimes, the designer must redo the project to obtain more efficient implementation in terms of overall silicon area, power consumption, and so on. It should be noted that generating the circuit mask does not constitute the final stage of the design process, although it is basically the last one in which the designer intervenes. Once the masks are designed, these are sent to a silicon foundry and then converted into integrated circuits, which are finally mounted and thoroughly tested (Denyer & Renshaw, 1985). Following the concepts of hierarchy, regularity, modularity, and locality, Figure 12.17 illustrates the complete procedure for designing an integrated circuit using VLSI technology.

The CAD tools for integrated circuit designs are commonly referred to as silicon compilers. In such systems, description of the desired signal processing system is done with a mid-level functional language. One of the first silicon compilers for the implementation of digital signal processing systems was the FIRST, developed at the University of Edinburgh (Denyer & Renshaw, 1985; Murray & Denyer, 1985). The FIRST compiler was based on a serial architecture and was restricted to the implementation of signal processing algorithms. The FIRST designs presented, at the time of its development, an integration density similar to the ones achieved by full custom designs, with the natural advantages of faster design and being mostly free from design errors.

The basic structure of a general silicon compiler is depicted in Figure 12.18. Initially, the desired system must be described using a mid-level language whose syntax is related to the basic elements for digital signal processing. We can then define new elements that can be the basic elements to other more complex systems, and so on. In the end, the compiler describes the desired system with a list of primitive basic elements and their respective interconnections.

The system simulator, using the models for each primitive basic element, verifies the system description provided and activates the testing pattern generator (TPG). Meanwhile, the primitive generator processes information about the geometry of

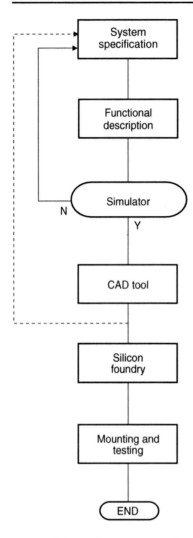

Figure 12.17 Design procedure using VLSI technology.

all basic cells and their corresponding connections, creating two sets of primitive elements, the standard ones and the special (pads) ones. The main idea in the basic silicon compilers was that the primitive basic cells had their geometry fixed, while other basic elements could be rearranged to achieve a more efficient design. The two sets of primitive elements were then processed in the last step of the design, generating the desired circuit mask.

Several design examples using the FIRST compiler can be found in Denyer & Renshaw (1985); Murray & Denyer (1985). Other similar silicon compilers include, for instance, the LAGER macrocell compiler (Pope, 1985; Rabaey et al., 1985), the Inpact (Yassa et al., 1987), and the very successful Cathedral series (Jain et al., 1986; Note et al., 1988).

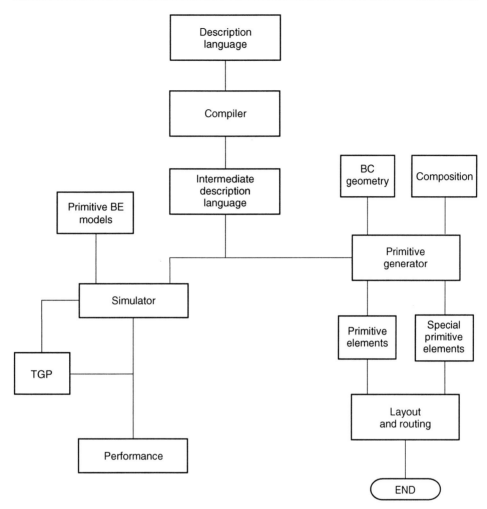

Figure 12.18 Block diagram of the FIRST silicon compiler (Denyer & Renshaw, 1985). (BE – basic element; BC – basic cell.)

Most of the concepts inherent to all these silicon compilers are still used in modern CAD systems. In fact, the main change in today's systems seems to be the incorporation of a high-level description language, evolving to a design technique referred to as behavioral synthesis. Today, at the beginning of the twenty-first century we identify three main levels of circuit description (Weste & Eshraghian, 1993; Wanhammar, 1999): behavioral, structural, and physical.

In the behavioral description, a high-level description language is used to describe the signal processing algorithm to be implemented in a VLSI circuit. Emulation and debugging are performed in a user-friendly manner. Good examples of systems that use behavioral description are the LAGER (Pope, 1985; Rabaey et al., 1985) and the Cathedral (Jain et al., 1986; Note et al., 1988) compilers. Such systems currently provide excellent implementations for very important applications, and most certainly will continue to expand their applicability in designing signal processing systems.

In the structural description, the signal processing algorithm is converted into a set of basic elements referred to as cells. All cells are seen together, and the complete cell placement and routing procedures are performed, usually in an automatic fashion, to meet timing and area constraints. There are four types of basic cells in digital signal processing systems:

- Data path elements: encoders/decoders, adders/subtractors, and multipliers.
- Memory elements: RAM, ROM, and registers.
- Control elements: shifters, latches, and clock circuitry.
- I/O elements: data buses, buffers, and ports.

An efficient approach for generating the structural description is to use the register-transfer logic (RTL) programs. These programs convert the initial specification, commonly given with a hardware description language such as Verilog® or VHDL, into a set of interconnected registers and combinatorial logic (Weste & Eshraghian, 1993; Skahill, 1996).

In the physical or data-flow description, each cell is replaced, with the help of a specific library, by its actual integration description consisting of gates, switches, transistors, etc. In this manner, a geometric description results, allowing the generation of the system's physical layout. Sometimes, further routing and placement can be performed at this level to generate even better VLSI designs.

12.6   Digital signal processors

As discussed in the previous section, ASIC design can yield very fast devices using very little chip area and power consumption. The main disadvantage of ASIC is the total amount of time to design the complete system, which tends to increase its overall production cost. Furthermore, long manufacturing stages can lead to the loss of commercial opportunities in today's ever-changing global economy. When this is the case, the use of DSP chips, as presented in this section, seems to be the most efficient alternative, if it is possible for the given application.

A digital signal processor (DSP) is a programmable general-purpose processor that represents a very low-cost and time-efficient way of implementing the most complex and interesting digital signal processing algorithms (Texas Instruments Inc., 1990, 1991; Motorola Inc., 1990). Most DSPs are based on the so-called Harvard architecture, where the data path (including the bus and memory units) is made distinct from the program path, allowing an instruction search to be performed simultaneously with another instruction execution and other tasks.

Today, most DSPs have a very fast parallel multiplier unit that performs the multiplication operation in a single clock cycle. The instruction set of most DSPs is similar to the ones for standard microprocessors, and the implementation of an algorithm is basically a matter of translating the desired mathematical operations into an ordered list of commands belonging to the complete set of instructions of the DSP.

The main advantages of using DSPs include their reasonable cost and the reduced amount of time required to implement a desired algorithm. In fact, all major DSP families have specific tools that allow programming and debugging using high-level languages such as ANSI C. The disadvantages of DSP include the necessity of additional hardware as interface I/O ports, analog-to-digital and digital-to-analog data converters, internal memory units. Also, due to their generality, DSP circuits are often suboptimal with respect to computational speed when compared to dedicated hardware for a given application.

In a nutshell, the interesting hardware characteristics of today's commercially available DSPs include:

- Either fixed-point arithmetic for simpler applications, or floating-point arithmetic for more complex and demanding applications.
- Relatively large wordlengths. Most modern DSPs use at least 12 or 16 bits to represent the internal data and also to allow double-precision arithmetic.
- An expanded ALU with a complete set of instructions for double-precision or bitwise operations.
- Several system buses for faster memory access and less probability of addressing conflicts.
- A fast parallel multiplier that can perform multiply-and-accumulate operations in a single processing cycle, ideal for calculating vector multiplications, commonly found in digital filtering algorithms.
- A relatively large internal RAM unit for data, instructions, and look-up tables.
- Instruction cache memory.
- Power-saving operating mode, when only peripheral and memory units are kept active.

Amongst the desired characteristics for programming the DSP, we note:

- Mnemonic assembly language.
- Definition of macroinstructions by the chip programmer, and use of subroutines.
- Possibility of programming using a high-level language.
- Availability of a fully operational debugger system.
- Special addressing modes such as circular addressing, for recursive algorithms, and bit-reversal addressing, for FFT computation.

- Parallelism of instructions: ability to perform more than one operation (multiply-and-accumulate, memory access, instructions fetch, or address increment/decrement) in a single processor cycle.

In the following, we describe some of the main families of commercial DSP chips by Analog Devices (www.analogdevices.com), Motorola (www.motorola.com), and Texas Instruments (www.ti.com). Other DSP manufacturers include AT&T (www.att.com), Fujitsu (www.fujitsu.com), and NEC (www.nec.com). The interested reader may find information about the DSPs manufactured by AT&T, Fujitsu, and NEC on their respective homepages.

12.6.1   Analog Devices DSPs

Analog Devices Inc. has several families of commercially available DSPs, such as:

ADSP-21XX^{TM}

General-purpose family of DSPs. The third-generation ADSP-219X^{TM} subfamily includes 24-bit data and addressing buses and reaches performances beyond 300 million operations per second (MOPS).

SHARC®

Powerful family of more than fifty 32-bit DSPs, with top performance, in the range of 5 billion operations per second.

TigerSHARC^{TM}

Top-of-the-line series, at the time of writing, of Analog Devices' DSPs capable of up to 2 billion multiply-and-accumulate operations in the 250 MHz version.

12.6.2   Motorola DSPs

Motorola Inc. has several commercial families of widely used DSPs such as:

DSP5600X^{TM}

Family of 24-bit fixed-point processors with clock rates 40–88 MHz, yielding up to 44 million instructions per second (MIPS); extremely suitable for audio applications.

DSP5630X^{TM}

Code-compatible and upgraded versions of the DSP5600X^{TM} family, this family of 24-bit DSPs reach up to 66–120 MIPS of computing performance with clock rates of 66–150 MHz.

DSP566XX™

Family of low-voltage DSPs, specifically designed for wireless applications.

MSC81XX™

Last generation family of DSPs, at the time of writing, having only one member, the MSC8101™ a 32- or 64-bit processor with a clock rate of 300 MHz and performance of 1200 MIPS.

12.6.3 Texas Instruments DSPs

The TMS320 is a very successful family of DSPs manufactured by Texas Instruments Inc. which currently includes the following members:

TMS320C10™

A low-cost 16-bit fixed-point processor, first introduced in 1983, based on a single-bus architecture. It operates with a 20 MHz clock and has an instruction cycle of 200 ns, yielding a throughput of 5 MIPS. A faster version, the TMS320C10-25™ is available with an instruction cycle of 160 ns.

TMS320C20X™

A complete series of processors similar to the TMS320C10™, but with larger data and program memory units and the ability to perform multiply-and-accumulate operations in a single cycle. The clock ranges 40–80 MHz, yielding instruction cycles of 25–50 ns, corresponding to a processing speed of 20–40 MIPS. The TMS320C25™ is a compatible processor whose 40 MHz version has an instruction cycle of 100 ns.

TMS320C50™

This is an improved version of the TMS320C25™ with a larger internal memory unit, a clock frequency of 66 MHz, and a cycle time of 30 ns, yielding a maximum throughput of 33 MIPS.

TMS320C3X™

A 32-bit floating-point series of processors with multiple buses, serial and parallel ports, and a memory unit of 16 Mwords. Parallel processing yields up to 825 million operations per second (MOPS) with the TMS320C33-150™ member of this family.

TMS320C40™

A 32-bit floating-point processor that provides extensive support for parallel operation.

TMS320C8X™

Powerful family of second-generation processors suitable for parallel processing.

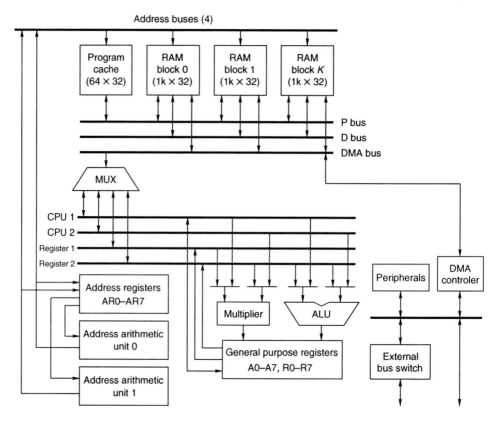

Figure 12.19 Internal architecture of the TMS320C30™.

TMS320C54X™ and TMS320C55X™

Power-efficient, 16-bit DSPs especially suitable for fixed-point mobile applications, achieving performances of up to 800 MIPS.

TMS320C62X™ and TMS320C64X™

High-performance, 16-bit fixed-point processors with the most recently developed achieving processing rates of up to 2400 MIPS.

TMS320C67X™

Last-generation 32-bit floating-point processor capable of performing up to 1 GFLOPS (giga floating-point operations per second).

EXAMPLE 12.2

The architecture of the TMS320C30™ is depicted in Figure 12.19. Write down an assembly program for this DSP implementing a basic FIR filter of order N.

Table 12.1. *Assembly commands for the TMS320C30^{TM}.*

Command	Parameters	Function
LDI	x, Ra	Ra=x (integer)
LDF	f, Ra	Ra=f (floating point)
STF	f, *ARa	Memory position pointed at by ARa receives f
RPTS	Ra	Repeat single instruction Ra times
MPYF	Ra, Rb, Rc	Rc = Ra * Rb
ADDF	Ra, Rb	Rb = Ra + Rb

SOLUTION

The important characteristics of the TMS320C30^{TM} architecture in Figure 12.19, for the purposes of this example, are the set of addressing registers AR0–AR7 that point to memory positions and the set of auxiliary registers R0–R7 that store intermediate results. Notice also the multiplier and ALU elements are in parallel, allowing multiply-and-accumulate operations to be performed in a single cycle.

A few assembly commands which are relevant for this case are included in Table 12.1, and an excerpt of the proposed program is listed in Table 12.2. Usually execution of the FIR filtering routine is controlled with interrupt calls, forcing a synchronous mode of operation between the DSP and its client. In that manner, a main program should include a dummy infinite loop that is interrupted every time new input data is available.

In the filtering routine, the single-cycled multiply-and-accumulate operations are invoked by the double vertical bar (pipe) just underneath the MPYF command. For this routine to work properly, AR2, AR4, and AR5 must be initialized, such that AR2 points to the delayed-by-order value of the input signal, AR4 points to the highest-order filter coefficient, and AR5 to the memory position where the current value of the input signal is. Also, R3 should be loaded with the filter order. At the end of the routine, register R2 should contain the current output signal.

$\triangle$

12.7 Summary

In this chapter, some implementation procedures for digital signal processing systems were discussed.

Initially, the basic elements for such systems were described and their implementations were presented with emphasis given to the two's-complement arithmetic. In addition, the so-called distributed arithmetic was seen as a possible alternative to eliminate the use of multiplier elements in the implementation of practical digital filters.

Table 12.2. *Extract of FIR filtering program for the TMS320C30^{TM}.*

```
 :
LDF      *AR5,R6                      ;load input signal into R6
STF      R6,*AR2- -(1)%               ;store R6 in address pointed to by AR2
LDF      0.0,R0                       ;initialize R0 with zero
LDF      0.0,R2                       ;initialize R2 with zero
RPTS     R3                           ;repeat following instruction R3 times
MPYF     *AR2- -(1)%,*AR4- -(1)%,R0   ;parallel multiply-and-...
|| ADDF  R0,R2                        ;...-accumulate operations
ADDF     R0,R2                        ;final addition operation
 :
```

A programmable logic device (PLD) was seen as a step further in integration from the discrete hardware implementation. In this technique for implementing digital signal processing algorithms, multi-functional devices are programmed to perform certain tasks necessary in the overall processing. Main commercial PLDs include complex-PLD and field-programmable gate array (FPGA) devices.

ASIC design was seen as an implementation alternative for very demanding applications with respect to computational speed or power consumption. The major concepts related to an efficient VLSI design, namely hierarchy, regularity, modularity, and locality, were introduced. We then considered the use of computer-automated design for VLSI, starting with the concept of silicon compiler systems, which have evolved to modern behavioral compilers.

The use of DSP chips was considered as a possible cost- and time-efficient alternative to ASIC implementations. The main commercial families available today were discussed and comparisons were made describing the major advantages of the most widely used DSP chips with respect to cost, computational speed, internal architecture, and so on.

12.8   Exercises

12.1   Design a circuit that determines the two's complement of a binary number in a serial fashion.

12.2   Show, using equation (12.2), that, if X and Y are represented in two's-complement arithmetic, then

(a) $X - Y = X + c[Y]$, where $c[Y]$ is the two's complement of Y

(b) $X - Y = \overline{Y + \overline{X}}$, where $\overline{X}$ represents the number obtained by inverting all bits of X.

Table 12.3. *Coefficients of cascade realization of three second-order state-space blocks.*

Coefficient	Section 1	Section 2	Section 3
a_{11}	0.802 734 3750	0.798 828 1250	0.812 500 0000
a_{12}	−0.582 031 2500	−0.591 796 8750	−0.585 937 5000
a_{21}	0.583 984 3750	0.599 609 3750	0.585 937 5000
a_{22}	0.802 734 3750	0.798 828 1250	0.812 500 0000
c_1	−0.009 765 6250	0.046 875 0000	−0.085 937 5000
c_2	−0.027 343 7500	−0.013 671 8750	0.033 203 1250
d	0.009 765 6250	0.105 039 0625	0.359 375 0000

12.3 Describe an architecture for the parallel multiplier where the coefficients are repre-sented in the two's-complement format.

12.4 Describe the internal circuitry of the multiplier in Figure 12.4 for 2-bit input data. Then perform the multiplication of several combinations of data, determining the internal data obtained at each step of the multiplication operation.

12.5 Describe an implementation of the FIR digital filter seen in Figure 4.3 using a single multiplier and a single adder multiplexed in time.

12.6 Describe the implementation of the FIR digital filter in Figure 4.3 using the distributed arithmetic technique. Design the architecture in detail, specifying the dimensions of the memory unit as a function of the filter order.

12.7 Determine the content of the memory unit in a distributed arithmetic implementation of the direct-form realization of a digital filter, whose coefficients are given by

$$b_0 = b_2 = 0.078\,64$$
$$b_1 = -0.148\,58$$
$$a_1 = -1.936\,83$$
$$a_2 = 0.951\,89$$

Use 8 bits to represent all internal signals.

12.8 Determine the content of the memory unit in a distributed arithmetic implementation of the cascade realization of three second-order state-space blocks, whose coefficients are given in Table 12.3: (scaling factor: $\lambda = 0.283\,203\,1250$).

12.9 Browse through several of the official sites of DSP manufacturers, as given in Section 12.6, and compare the prices of similar devices using fixed- and floating-point arithmetic. Discuss the practical aspects involved when choosing one of these two types of binary representations.

References

Adams, J. W. (1991a). FIR digital filters with least-squares stopbands subject to peak-gain constraints. *IEEE Transactions on Circuits and Systems*, **39**, 376–388.

Adams, J. W. (1991b). A new optimal window. *IEEE Transactions on Signal Processing*, **39**, 1753–1769.

Adams, J. W. & Willson, Jr., A. N. (1983). A new approach to FIR digital filters with fewer multipliers and reduced sensitivity. *IEEE Transactions on Circuits and Systems*, **CAS-30**, 277–283.

Adams, J. W. & Willson, Jr., A. N. (1984). Some efficient digital prefilter structures. *IEEE Transactions on Circuits and Systems*, **CAS-31**, 260–265.

Ahmed, N., Natarajan, T., & Rao, K. R. (1974). Discrete cosine transform. *IEEE Transactions on Computers*, **C-23**, 90–93.

Akansu, A. N. & Medley, M. J. (Eds.). (1999). *Wavelets, Subband and Block Transforms in Communications and Multimedia*. Boston, MA: Kluwer Academic Publishers.

Ansari, R. & Liu, B. (1983). Efficient sampling rate alteration using recursive (IIR) digital filters. *IEEE Transactions on Acoustics, Speech, and Signal Processing*, **ASSP-31**, 1366–1373.

Antoniou, A. (1982). Accelerated procedure for the design of equiripple nonrecursive digital filters. *IEE Proceedings – Part G*, **129**, 1–10.

Antoniou, A. (1983). New improved method for design of weighted-Chebyschev, nonrecursive, digital filters. *IEEE Transactions on Circuits and Systems*, **CAS-30**, 740–750.

Antoniou, A. (1993). *Digital Filters: Analysis, Design, and Applications* (2nd ed.). New York, NY: McGraw-Hill.

Antoniou, A. & Rezk, M. G. (1977). Digital filters synthesis using concept of generalized-immitance converter. *IEE Journal of Electronics Circuits*, **1**, 207–216.

Antoniou, A. & Rezk, M. G. (1980). A comparison of cascade and wave fixed-point digital-filter structures. *IEEE Transactions on Circuits and Systems*, **CAS-27**, 1184–1194.

Avenhaus, E. (1972). On the design of digital filters with coefficients of limited wordlength. *IEEE Transactions on Audio and Electroacoustics*, **AU-20**, 206–212.

Bauer, P. H. & Wang, J. (1993). Limit cycle bounds for floating-point implementations of second-order recursive digital filters. *IEEE Transactions on Circuits and Systems II: Analog and Digital Signal Processing*, **39**, 493–501. For corrections see ibid **41**,176, February 1994.

Bayoumi, M. A. & Swartzlander, E. E. J. (1994). *VLSI Signal Processing Technology*. Boston, MA: Kluwer Academic Publishers.

Belevitch, V. (1968). *Classical Network Theory*. San Francisco, CA: Holden-Day.

Benvenuto, N., Franks, L. E., & Hill, Jr., F. S. (1984). On the design of FIR filters with power-of-two coefficients. *IEEE Transactions on Communications*, **COM-32**, 1299–1307.

Bhaskaran, V. & Konstantinides, K. (1997). *Image and Video Compression Standards: Algorithms and Architectures*. Boston, MA: Kluwer Academic Publishers.

Bomar, B. W. (1989). On the design of second-order state-space digital filter sections. *IEEE Transactions on Circuits and Systems*, **36**, 542–552.

Bomar, B. W. & Joseph, R. D. (1987). Calculation of L_∞ norms in second-order state-space digital filter sections. *IEEE Transactions on Circuits and Systems*, **CAS-34**, 983–984.

Bomar, B. W., Smith, L. M., & Joseph, R. D. (1997). Roundoff noise analysis of state-space digital filters implemented on floating-point digital signal processors. *IEEE Transactions on Circuits and Systems II: Analog and Digital Signal Processing*, **44**, 952–955.

Bracewell, R. N. (1984). The fast Hartley transform. *Proceedings of the IEEE*, **72**, 1010–1018.

Bracewell, R. N. (1994). Aspects of the Hartley transform. *Proceedings of the IEEE*, **82**, 381–386.

Burrus, C. S. & Parks, T. W. (1970). Time domain design of recursive digital filters. *IEEE Transactions on Audio and Electroacoustics*, **AU-18**, 137–141.

Butterweck, H. J. (1975). Suppression of parasitic oscillations in second-order digital filters by means of a controlled-rounding arithmetic. *Archiv Elektrotechnik und Übertragungstechnik*, **29**, 371–374.

Butterweck, H. J., van Meer, A. C. P., & Verkroost, G. (1984). New second-order digital filter sections without limit cycles. *IEEE Transactions on Circuits and Systems*, **CAS-31**, 141–146.

Cabezas, J. C. E. & Diniz, P. S. R. (1990). FIR filters using interpolated prefilters and equalizers. *IEEE Transactions on Circuits and Systems*, **37**, 17–23.

Chang, T.-L. (1981). Supression of limit cycles in digital filters designed with one magnitude-truncation quantizer. *IEEE Transactions on Circuits and Systems*, **CAS-28**, 107–111.

Charalambous, C. & Antoniou, A. (1980). Equalisation of recursive digital filters. *IEE Proceedings – Part G*, **127**, 219–225.

Chen, W.-H., Smith, C. H., & Fralick, S. C. (1977). A fast computational algorithm for the discrete cosine transform. *IEEE Transactions on Communications*, **COM-25**, 1004–1009.

Cheney, E. W. (1966). *Introduction to Approximation Theory*. New York, NY: McGraw-Hill.

Churchill, R. V. (1975). *Complex Variables and Applications*. New York, NY: McGraw-Hill.

Claasen, T. A. C. M., Mecklenbräuker, W. F. G., & Peek, J. B. H. (1975). On the stability of the forced response of digital filters with overflow nonlinearities. *IEEE Transactions on Circuits and Systems*, **CAS-22**, 692–696.

Cochran, W. T., Cooley, J. W., Favin, D. L., Helms, H. H., Kaenel, R. A., Lang, W. W. et al. (1967). What is the fast Fourier transform? *IEEE Transactions on Audio and Electroacoustics*, **AU-15**, 45–55.

Cohen, A., Daubechies, I., & Feauveau, J. C. (1992). Biorthogonal bases of compactly supported wavelets. *Communications on Pure and Applied Mathematics*, **XLV**, 485–560.

Constantinides, A. G. (1970). Spectral transformations for digital filters. *IEE Proceedings*, **117**, 1585–1590.

Cooley, J. W. & Tukey, J. W. (1965). An algorithm for the machine computation of complex Fourier series. *Mathematics of Computation*, **19**, 297–301.

Cortelazzo, G. & Lightner, M. R. (1984). Simultaneous design of both magnitude and group delay of IIR and FIR filters based on multiple criterion optimization. *IEEE Transactions on Acoustics, Speech, and Signal Processing*, **ASSP-32**, 9949–9967.

Crochiere, R. E. & Oppenheim, A. V. (1975). Analysis of linear digital networks. *Proceedings of the IEEE*, **63**, 581–593.

Crochiere, R. E. & Rabiner, L. R. (1983). *Multirate Digital Signal Processing*. Englewood Cliffs, NJ: Prentice-Hall.

Croisier, A., Esteban, D., & Galand, C. (1976). Perfect channel splitting by use of interpolation/decimation/tree decomposition techniques. In *Proceedings of International Symposium on Information, Circuits and Systems*, Patras, Greece.

Croisier, A., Esteban, D. J., Levilion, M. E., & Riso, V. (1973). *Digital filter for PCM encoded signals.* U.S. Patent no. 3777130.

Daniels, R. W. (1974). *Approximation Methods for Electronic Filter Design.* New York, NY: McGraw-Hill.

Daubechies, I. (1988). Orthonormal bases of compactly supported wavelets. *Communications on Pure and Applied Mathematics,* **XLI,** 909–996.

Daubechies, I. (1991). *Ten Lectures on Wavelets.* Philadelphia, PA: Society for Industrial and Applied Mathematics.

de la Vega, A. S., Diniz, P. S. R., Mesquita, Fo., A. C., & Antoniou, A. (1995). A modular distributed arithmetic implementation of the inner product with application to digital filters. *Journal of VLSI Signal Processing,* **10,** 93–106.

de Queiroz, R. L., Nguyen, T. Q., & Rao, K. R. (1996). The GenLOT: generalized linear-phase lapped orthogonal transform. *IEEE Transactions on Signal Processing,* **44,** 497–507.

Deczky, A. G. (1972). Synthesis of recursive digital filters using the minimum p error criterion. *IEEE Transactions on Audio and Electroacoustics,* **AU-20,** 257–263.

Denyer, P. & Renshaw, D. (1985). *VLSI Signal Processing: A Bit-Serial Approach.* Boston, MA: Addison-Wesley.

Diniz, P. S. R. (1984). *New Improved Structures for Recursive Digital Filters.* PhD thesis, Concordia University, Montreal, Canada.

Diniz, P. S. R. (1988). Elimination of constant-input limit cycles in passive lattice digital filters. *IEEE Transactions on Circuits and Systems,* **35,** 1188–1190.

Diniz, P. S. R. (1997). *Adaptive Filtering: Algorithms and Practical Implementation.* Boston, MA: Kluwer Academic Publishers.

Diniz, P. S. R. & Antoniou, A. (1985). Low sensitivity digital filter structures which are amenable to error-spectrum shaping. *IEEE Transactions on Circuits and Systems,* **CAS-32,** 1000–1007.

Diniz, P. S. R. & Antoniou, A. (1986). More economical state-space digital-filter structures which are free of constant-input limit cycles. *IEEE Transactions on Acoustics, Speech, and Signal Processing,* **ASSP-34,** 807–815.

Diniz, P. S. R. & Antoniou, A. (1988). Digital filter structures based on the concept of voltage generalized immitance converter. *Canadian Journal of Electrical and Computer Engineering,* **13,** 90–98.

Diniz, P. S. R. & Netto, S. L. (1999). On WLS-Chebyshev FIR digital filters. *Journal of Circuits, Systems, and Computers,* **9,** 155–168.

Elliott, D. F. & Rao, K. R. (1982). *Fast Transforms: Algorithms, Analyses, and Applications.* New York, NY: Academic Press.

Evans, A. G. & Fischel, R. (1973). Optimal least squares time domain synthesis of recursive digital filters. *IEEE Transactions on Audio and Electroacoustics,* **AU-21,** 61–65.

Fettweis, A. (1971a). Digital filters structures related to classical filters networks. *Archiv Elektrotechnik und Übertragungstechnik,* **25,** 79–89.

Fettweis, A. (1971b). A general theorem for signal-flow networks. *Archiv Elektrotechnik und Übertragungstechnik,* **25,** 557–561.

Fettweis, A. (1986). Wave digital filters: Theory and practice. *Proceedings of IEEE,* **74,** 270–327.

Fettweis, A., Levin, H., & Sedlmeyer, A. (1974). Wave digital lattice filters. *International Journal of Circuit Theory and Applications,* **2,** 203–211.

Fettweis, A. & Meerkötter, K. (1975a). On adaptors for wave digital filters. *IEEE Transactions on Acoustics, Speech, and Signal Processing,* **ASSP-23,** 516–525.

Fettweis, A. & Meerkötter, K. (1975b). Suppression of parasitic oscillations in wave digital filters. *IEEE Transactions on Circuits and Systems,* **CAS-22,** 239–246.

Fletcher, R. (1980). *Practical Methods of Optimization*. Chichester, UK: John Wiley.

Fliege, N. J. (1994). *Multirate Digital Signal Processing*. Chichester, UK: John Wiley.

Freeny, S. L. (1975). Special-purpose hardware for digital filtering. *Proceedings of the IEEE*, **63**, 633–648.

Gabel, R. A. & Roberts, R. A. (1980). *Signals and Linear Systems* (2nd ed.). New York, NY: John Wiley and Sons.

Gersho, A. & Gray, R. M. (1992). *Vector Quantization and Signal Compression*. Boston, MA: Kluwer Academic Publishers.

Godsill, S. J. & Rayner, J. W. (1998). *Digital Audio Restoration*. London, England: Springer Verlag.

Gold, B. & Jordan, Jr., K. L. (1969). A direct search procedure for designing finite impulse response filters. *IEEE Transactions on Audio and Electroacoustics*, **AU-17**, 33–36.

Gray, Jr., A. H. & Markel, J. D. (1973). Digital lattice and ladder filter synthesis. *IEEE Transactions on Audio and Electroacoustics*, **AU-21**, 491–500.

Gray, Jr., A. H. & Markel, J. D. (1975). A normalized digital filter structure. *IEEE Transactions on Acoustics, Speech, and Signal Processing*, **ASSP-23**, 268–277.

Ha, Y. H. & Pearce, J. A. (1989). A new window and comparison to standard windows. *IEEE Transactions on Acoustics, Speech, and Signal Processing*, **37**, 298–301.

Harmuth, H. F. (1970). *Transmission of Information by Orthogonal Signals*. New York, NY: Springer Verlag.

Herrmann, O. (1971). On the approximation problem in nonrecursive digital filter design. *IEEE Transactions on Circuit Theory*, **CT-19**, 411–413.

Hettling, K. J., Saulnier, G. J., Akansu, A. N., & Lin, X. (1999). *Transmultiplexers: A Unifying Time-Frequency Tool for TDMA, FDMA and CDMA Communications*. Boston, MA: Kluwer Academic Publishers.

Higgins, W. E. & Munson, Jr., D. C. (1984). Optimal and suboptimal error spectrum shaping for cascade-form digital filters. *IEEE Transactions on Circuits and Systems*, **CAS-31**, 429–437.

Holford, S. & Agathoklis, P. (1996). The use of model reduction techniques for designing IIR filters with linear phase in the passband. *IEEE Transactions on Signal Processing*, **44**, 2396–2403.

Hwang, K. (1979). *Computer Arithmetic: Principles, Architecture and Design*. New York, NY: John Wiley.

Hwang, S. Y. (1977). Minimum uncorrelated unit noise in state space digital filtering. *IEEE Transactions on Acoustics, Speech, and Signal Processing*, **ASSP-25**, 273–281.

Ifeachor, E. C. & Jervis, B. W. (1993). *Digital Signal Processing: A Practical Approach*. Reading, MA: Addison-Wesley.

Jackson, L. B. (1969). *An Analysis of Roundoff Noise in Digital Filters*. DSc. thesis, Stevens Institute of Technology, Hoboken, NJ.

Jackson, L. B. (1970a). On the interaction of roundoff noise and dynamic range in digital filters. *Bell Systems Technical Journal*, **49**, 159–185.

Jackson, L. B. (1970b). Roundoff-noise analysis for fixed-point digital filters realized in cascade or parallel form. *IEEE Transactions on Audio and Electroacoustics*, **AU-18**, 107–122.

Jackson, L. B. (1996). *Digital Filters and Signal Processing* (3rd ed.). Boston, MA: Kluwer Academic Publishers.

Jackson, L. B., Kaiser, J. F., & McDonald, H. S. (1968). An approach to the implementation of digital filters. *IEEE Transactions on Audio and Electroacoustics*, **AU-16**, 413–421.

Jain, A. K. (1989). *Fundamentals of Digital Image Processing*. Englewood Cliffs, NJ: Prentice-Hall.

Jain, R., Catthoor, F., Vanhoof, J., de Loore, B. J. S., Goossens, G., Goncalves, N. F. et al. (1986).

Custom design of a VLSI PCM-FDM transmultiplexer from system specifications to circuit layout using a computer-aided design system. *IEEE Journal of Solid-State Circuits*, **SC-21**, 73–84.

Jayant, N. S. & Noll, P. (1984). *Digital Coding of Waveforms*. Englewood Cliffs, NJ: Prentice-Hall.

Johnston, J. D. (1980). A filter family designed for use in quadrature mirror filter banks. In *Proceedings of the International Conference on Acoustics, Speech and Signal Processing 80*, Denver, CO, pp. 291–294.

Jury, E. I. (1973). *Theory and Application of the z-Transform Method*. Huntington, NY: R. E. Krieger.

Kahrs, M. & Brandenburg, K. (1998). *Applications of Digital Signal Processing to Audio and Acoustics*. Boston, MA: Kluwer Academic Publishers.

Kailath, T., Sayed, A., & Hassibi, B. (2000). *Linear Estimation*. Englewood Cliffs, NJ: Prentice-Hall.

Kaiser, J. F. (1974). Nonrecursive digital filter design using the I_0–sinh window function. *Proceedings of IEEE International Symposium on Circuits and Systems*, San Francisco, CA, pp. 20–23.

Kalliojärvi, K. & Astola., J. (1996). Roundoff errors in block-floating-point systems. *IEEE Transactions on Signal Processing*, **44**, 783–791.

Kay, S. & Smith, D. (1999). An optimal sidelobeless window. *IEEE Transactions on Signal Processing*, **47**, 2542–2546.

Kim, Y. (1980). *State Space Structures for Digital Filters*. PhD thesis, University of Rhode Island, Kingston, RI.

Koilpillai, R. D. & Vaidyanathan, P. P. (1992). Cosine-modulated FIR filter banks satisfying perfect reconstruction. *IEEE Transactions on Signal Processing*, **40**, 770–783.

Kreider, D. L., Kuller, R. G., Ostberg, D. R., & Perkins, F. W. (1966). *An Introduction to Linear Analysis*. Reading, MA: Addison-Wesley.

Kreyszig, E. (1979). *Advanced Engineering Mathematics* (4th ed.). New York, NY: Wiley.

Kung, S. (1988). *VLSI Array Processors*. Englewood Cliffs, NJ: Prentice-Hall.

Laakso, T. I. (1992). Comments on 'calculation of L_∞ norms in second-order state-space digital filter sections'. *IEEE Transactions on Circuits and Systems II: Analog and Digital Signal Processing*, **39**, 256.

Laakso, T. I. (1993). Elimination of limit cycles in direct form digital filters using error feedback. *International Journal of Circuit Theory and Applications*, **21**, 141–163.

Laakso, T. I., Diniz, P. S. R., Hartimo, I., & Macedo, Jr., T. C. (1992). Elimination of zero-input limit cycles in single-quantizer recursive filter structures. *IEEE Transactions on Circuits and Systems II: Analog and Digital Signal Processing*, **39**, 638–645.

Laakso, T. I. & Hartimo, I. (1992). Noise reduction in recursive digital filters using high-order error feedback. *IEEE Transactions on Signal Processing*, **40**, 141–163.

Laakso, T. I., Zeng, B., Hartimo, I., & Neüvo, Y. (1994). Elimination of limit cycles in floating-point implementation of recursive digital filters. *IEEE Transactions on Circuits and Systems II: Analog and Digital Signal Processing*, **41**, 308–312.

Lawson, C. L. (1968). *Contribution to the Theory of Linear Least Maximum Approximations*. PhD thesis, University of California, Los Angeles, CA.

Le Gall, D. (1992). The MPEG video compression algorithm. *Image Communication*, **4**, 129–140.

Le Gall, D. & Tabatabai, A. (1988). Sub-band coding of digital images using symmetric short kernel filters and arithmetic coding techniques. In *Proceedings of IEEE International Conference on Acoustics, Speech, and Signal Processing*, New York, NY, pp. 761–764.

Lim, Y. C. (1986). Frequency-response masking approach for synthesis of sharp linear phase digital filters. *IEEE Transactions on Circuits and Systems*, **CAS-33**, 357–364.

Lim, Y. C., Lee, J. H., Chen, C. K., & Yang, R. H. (1992). A weighted least squares algorithm for quasi-equiripple FIR and IIR digital filter design. *IEEE Transactions on Signal Processing*, **40**, 551–558.

Lim, Y. C. & Lian, Y. (1994). Frequency-response masking approach for digital filter design: Complexity reduction via masking filter factorization. *IEEE Transactions on Circuits and Systems II: Analog and Digital Signal Processing*, **41**, 518–525.

Liu, B. & Peled, A. (1975). A new hardware realization of high-speed fast Fourier transformers. *IEEE Transactions on Acoustics, Speech, and Signal Processing*, **ASSP-23**, 543–547.

Liu, K. J. R., Chiu, C. T., Kolagotla, R. K., & JáJá, J. F. (1994). Optimal unified architectures for the real-time computation of time-recursive discrete sinusoidal transforms. *IEEE Transactions on Circuits and Systems for Video Technology*, **4**, 168–180.

Liu, K. S. & Turner, L. E. (1983). Stability, dynamic-range and roundoff noise in a new second-order recursive digital filter. *IEEE Transactions on Circuits and Systems*, **CAS-30**, 815–821.

Lu, W.-S. (1999). Design of stable digital filters with equiripple passbands and peak-constrained least-squares stopbands. *IEEE Transactions on Circuits and Systems II: Analog and Digital Signal Processing*, **46**, 1421–1426.

Luenberger, D. G. (1984). *Introduction to Linear and Nonlinear Programming* (2nd ed.). Boston, MA: Addison-Wesley.

Lyon, R. F. (1976). Two's complement pipeline multipliers. *IEEE Transactions on Communications*, **COM-24**, 418–425.

Mallat, S. G. (1998). *A Wavelet Tour of Signal Processing*. San Diego, CA: Academic Press.

Malvar, H. S. (1986). Fast computation of discrete cosine transform through fast Hartley transform. *Electronics Letters*, **22**, 352–353.

Malvar, H. S. (1987). Fast computation of discrete cosine transform and the discrete Hartley transform. *IEEE Transactions on Accoustics, Speech and Signal Processing*, **ASSP-35**, 1484–1485.

Malvar, H. S. (1992). *Signal Processing with Lapped Transforms*. Norwood, MA: Artech House.

Manolakis, D. G., Ingle, V., & Kogon, S. (2000). *Statistical and Adaptive Signal Processing: Spectral Estimation, Signal Modeling, Adaptive Filtering and Array Processing*. New York, NY: McGraw-Hill.

Martinez, H. G. & Parks, T. W. (1979). A class of infinite duration impulse response digital filters for sampling rate reduction. *IEEE Transactions on Acoustics, Speech, and Signal Processing*, **ASSP-27**, 154–162.

McClellan, J. H. & Parks, T. W. (1973). A unified approach to the design of optimum FIR linear-phase digital filters. *IEEE Transactions on Circuit Theory*, **CT-20**, 190–196.

McClellan, J. H., Parks, T. W., & Rabiner, L. R. (1973). A computer program for designing optimum FIR linear-phase digital filters. *IEEE Transactions on Audio and Electroacoustics*, **AU-21**, 506–526.

McClellan, J. H. & Rader, C. M. (1979). *Number Theory in Digital Signal Processing*. Englewood Cliffs, NJ: Prentice-Hall.

Meerkötter, K. (1976). Realization of limit cycle-free second-order digital filters. *Proceedings of IEEE International Symposium on Circuits and Systems*, Munich, Germany, pp. 295–298.

Meerkötter, K. & Wegener, W. (1975). A new second-order digital filter without parasitic oscillations. *Archiv Elektrotechnik und Übertragungstechnik*, **29**, 312–314.

Merchant, G. A. & Parks, T. W. (1982). Efficient solution of a Toeplitz-plus-Hankel coefficient matrix system of equations. *IEEE Transactions on Acoustics, Speech, and Signal Processing*, **ASSP-30**, 40–44.

Mills, W. L., Mullis, C. T., & Roberts, R. A. (1978). Digital filters realizations without overflow oscillations. *IEEE Transactions on Acoustics, Speech, and Signal Processing*, **ASSP-26**, 334–338.

Mitra, S. K. (1998). *Digital Signal Processing: A Computer Based Approach*. New York, NY: McGraw-Hill.

Motorola Inc. (1990). *DSP 56000/DSP 56001 — Digital Signal Processor: User's Manual*.

Mullis, C. T. & Roberts, R. A. (1976a). Roundoff noise in digital filters: Frequency transformations and invariants. *IEEE Transactions on Acoustics, Speech, and Signal Processing*, **ASSP-24**, 538–550.

Mullis, C. T. & Roberts, R. A. (1976b). Synthesis of minimum roundoff noise fixed point digital filters. *IEEE Transactions on Circuits and Systems*, **CAS-23**, 551–562.

Munson, D. C., Strickland, J. H., & Walker, T. P. (1984). Maximum amplitude zero-input limit cycles in digital filters. *IEEE Transactions on Circuits and Systems*, **CAS-31**, 266–275.

Murray, A. L. & Denyer, P. B. (1985). A CMOS design strategy for bit serial signal processing. *IEEE Transactions on Solid-State Circuits*, **SC-20**, 747–753.

Neüvo, Y., Cheng-Yu, D., & Mitra, S. K. (1984). Interpolated finite impulse response filters. *IEEE Transactions on Acoustics, Speech, and Signal Processing*, **ASSP-32**, 563–570.

Neüvo, Y., Rajan, G., & Mitra, S. K. (1987). Design of narrow-band FIR filters with reduced arithmetic complexity. *IEEE Transactions on Circuits and Systems*, **34**, 409–419.

Nguyen, T. Q. & Koilpillai, R. D. (1996). The theory and design of arbitrary-length cosine-modulated FIR filter banks and wavelets, satisfying perfect reconstruction. *IEEE Transactions on Signal Processing*, **44**, 473–483.

Nguyen, T. Q., Laakso, T. I., & Koilpillai, R. D. (1994). Eigenfilter approach for the design of allpass filters approximating a given phase response. *IEEE Transactions on Signal Processing*, **42**, 2257–2263.

Note, S., van Meerbergen, J., Catthoor, F., & De Man, H. J. (1988). Automated synthesis of a high-speed CORDIC algorithm with the Cathedral-III compilation system. *Proceedings of IEEE International Symposium on Circuits and Systems*, Espoo, Finland, pp. 581–584.

Nussbaumer, H. J. (1982). *Fast Fourier Transform and Convolution Algorithms*. Berlin, Germany: Springer Verlag.

Nuttall, A. H. (1981). Some windows with very good sidelobe behavior. *IEEE Transactions on Acoustics, Speech, and Signal Processing*, **ASSP-29**, 84–91.

Olejniczak, K. J. & Heydt, G. T. (1994). Scanning the special section on the Hartley transform. *Proceedings of the IEEE*, **82**, 372–380.

Oppenheim, A. V. & Schafer, R. W. (1975). *Digital Signal Processing*. Englewood Cliffs, NJ: Prentice-Hall.

Oppenheim, A. V. & Schafer, R. W. (1989). *Discrete-Time Signal Processing*. Englewood Cliffs, NJ: Prentice-Hall.

Oppenheim, A. V., Willsky, A. S., & Young, I. T. (1983). *Signals and Systems*. Englewood Cliffs, NJ: Prentice-Hall.

Papoulis, A. (1965). *Probability, Random Variables, and Stochastic Processes*. New York, NY: McGraw-Hill.

Parhi, K. K. (1999). *VLSI Digital Signal Processing Systems: Design and Implementation*. New York, NY: John Wiley and Sons.

Peled, A. & Liu, B. (1974). A new hardware realization of digital filters. *IEEE Transactions on Acoustics, Speech, and Signal Processing*, **ASSP-22**, 456–462.

Peled, A. & Liu, B. (1985). *Digital Signal Processing*. Malabar, FL: R. E. Krieger.

Pope, S. P. (1985). *Automatic Generation of Signal Processing Integrated Circuits*. PhD thesis, University of California, Berkeley, CA.

Proakis, J. G. & Manolakis, D. G. (1996). *Digital Signal Processing: Principles, Algorithms and Applications* (3rd ed.). Englewood Cliffs, NJ: Prentice-Hall.

Rabaey, J. M., Pope, S. P., & Brodersen, R. W. (1985). An integrated automated layout generation system for DSP circuits. *IEEE Transactions on Computer-Aided Design*, **CAD-4**, 285–296.

Rabiner, L. R. (1979). *Programs for Digital Signal Processing*. New York, NY: IEEE Press.

Rabiner, L. R. & Gold, B. (1975). *Theory and Application of Digital Signal Processing*. Englewood Cliffs, NJ: Prentice-Hall.

Rabiner, L. R., Gold, B., & McGonegal, G. A. (1970). An approach to the approximation problem for nonrecursive digital filters. *IEEE Transactions on Audio and Electroacoustics*, **AU-18**, 83–106.

Rabiner, L. R., McClellan, J. H., & Parks, T. W. (1975). FIR digital filter design techniques using weighted Chebyshev approximation. *Proceedings of the IEEE*, **63**, 595–610.

Rajan, G., Neüvo, Y., & Mitra, S. K. (1988). On the design of sharp cutoff wide-band FIR filters with reduced arithmetic complexity. *IEEE Transactions on Circuits and Systems*, **35**, 1447–1454.

Ralev, K. R. & Bauer, P. H. (1999). Realization of block floating-point digital filters and application to block implementations. *IEEE Transactions on Signal Processing*, **47**, 1076–1087.

Rice, J. R. & Usow, K. H. (1968). The Lawson algorithm and extensions. *Mathematics of Computation*, **22**, 118–127.

Rioul, O. (1992). Simple regularity criteria for subdivision schemes. *SIAM Journal on Mathematical Analysis*, **23**, 1544–1576.

Roberts, R. A. & Mullis, C. T. (1987). *Digital Signal Processing*. Reading, MA: Addison-Wesley.

Roitman, M. & Diniz, P. S. R. (1995). Power system simulation based on wave digital filters. *IEEE Transactions on Power Delivery*, **11**, 1098–1104.

Roitman, M. & Diniz, P. S. R. (1996). Simulation of non-linear and switching elements for transient analysis based on wave digital filters. *IEEE Transactions on Power Delivery*, **12**, 2042–2048.

Saramäki, T. (1993). In *Finite-Impulse Response Filter Design*. Mitra, S. K. & Kaiser, J. F. (Eds.), chapter 4. New York, NY: Wiley.

Saramäki, T. & Neüvo, Y. (1984). Digital filters with equiripple magnitude and group delay. *IEEE Transactions on Acoustics, Speech, and Signal Processing*, **ASSP-32**, 1194–1200.

Saramäki, T., Neüvo, Y., & Mitra, S. K. (1988). Design of computationally efficient interpolated FIR filters. *IEEE Transactions on Circuits and Systems*, **CAS-35**, 70–88.

Sarcinelli, Fo., M. & Camponêz, M. d. O. (1997). A new low roundoff noise second order digital filter section which is free of constant-input limit cycles. In *Proceedings of IEEE Midwest Symposium on Circuits and Systems*, volume 1, Sacramento, CA, pp. 468–471.

Sarcinelli, Fo., M. & Camponêz, M. d. O. (1998). Design strategies for constant-input limit cycle-free second-order digital filters. In *Proceedings of IEEE Midwest Symposium on Circuits and Systems*, Notre Dame, IN, pp. 464–467.

Sarcinelli, Fo., M. & Diniz, P. S. R. (1990). Tridiagonal state-space digital filter structures. *IEEE Transactions on Circuits and Systems*, **CAS-36**, 818–824.

Sedlmeyer, A. & Fettweis, A. (1973). Digital filters with true ladder configuration. *International Journal of Circuit Theory and Applications*, **1**, 5–10.

Sedra, A. S. & Brackett, P. O. (1978). *Filter Theory and Design: Active and Passive*. Champaign,

IL: Matrix Publishers.

Selesnick, I. W., Lang, M., & Burrus, C. S. (1996). Constrained least square design of FIR filters without specified transition bands. *IEEE Transactions on Signal Processing*, **44**, 1879–1892.

Selesnick, I. W., Lang, M., & Burrus, C. S. (1998). A modified algorithm for constrained least square design of multiband FIR filters without specified transition bands. *IEEE Transactions on Signal Processing*, **46**, 497–501.

Singh, V. (1985). Formulation of a criterion for the absence of limit cycles in digital filters designed with one quantizer. *IEEE Transactions on Circuits and Systems*, **CAS-32**, 1062–1064.

Singleton, R. C. (1969). An algorithm for computing the mixed radix fast Fourier transform. *IEEE Transactions on Audio and Electroacoustics*, **AU-17**, 93–103.

Skahill, K. (1996). *VHDL for Programmable Logic*. Reading, MA: Addison-Wesley.

Smith, L. M., Bomar, B. W., Joseph, R. D., & Yang, G. C.-J. (1992). Floating-point roundoff noise analysis of second-order state-space digital filter structures. *IEEE Transactions on Circuits and Systems II: Analog and Digital Signal Processing*, **39**, 90–98.

Smith, M. J. T. & Barnwell, T. P. (1986). Exact reconstruction techniques for tree-structured subband coders. *IEEE Transactions on Acoustics, Speech, and Signal Processing*, **34**, 434–441.

Sorensen, H. V. & Burrus, C. S. (1993). In *Fast DFT and Convolution Algorithms*. Mitra, S. K. & Kaiser, J. F. (Eds.), chapter 7. New York, NY: Wiley.

Strang, G. (1980). *Linear Algebra and Its Applications* (2nd ed.). New York: Academic Press.

Strang, G. & Nguyen, T. Q. (1996). *Wavelets and Filter Banks*. Wellesley, MA: Wellesley Cambridge Press.

Stüber, G. L. (1996). *Principles of Mobile Communications*. Norwell, MA: Kluwer Academic Publishers.

Sullivan, J. L. & Adams, J. W. (1998). PCLS IIR digital filters with simultaneous frequency response magnitude and group delay specifications. *IEEE Transactions on Signal Processing*, **46**, 2853–2862.

Szczupak, J. & Mitra, S. K. (1975). Digital filter realization using successive multiplier-extraction approach. *IEEE Transactions on Acoustics, Speech, and Signal Processing*, **ASSP-23**, 235–239.

Texas Instruments Inc. (1990). *TMS 320C5x: User's Guide*.

Texas Instruments Inc. (1991). *TMS 320C3x: User's Guide*.

Tran, T. D., de Queiroz, R. L., & Nguyen, T. Q. (2000) Linear phase perfect reconstruction filter bank: Lattice structure, design, and application in image coding. *IEEE Transactions on Signal Processing*, **48**, 133–147.

Ulbrich, W. (1985). In *MOS Digital Filter Design*. Tsividis, Y. and Antognetti, P. (Eds.), chapter 8. Englewood Cliffs, NJ: Prentice-Hall.

Vaidyanathan, P. P. (1984). On maximally flat linear phase FIR filters. *IEEE Transactions on Circuits and Systems*, **CAS-31**, 830–832.

Vaidyanathan, P. P. (1985). Efficient and multiplierless design of FIR filters with very sharp cutoff via maximally flat building blocks. *IEEE Transactions on Circuits and Systems*, **CAS-32**, 236–244.

Vaidyanathan, P. P. (1987). Eigenfilters: A new approach to least-squares FIR filter design and applications including Nyquist filters. *IEEE Transactions on Circuits and Systems*, **CAS-34**, 11–23.

Vaidyanathan, P. P. (1993). *Multirate Systems and Filter Banks*. Englewood Cliffs, NJ: Prentice-Hall.

van Gerwen, P. J., Mecklenbräuker, W. F., Verhoeckx, N. A. M., Snijders, F. A. M., & van Essen, H. A. (1975). A new type of digital filter for data transmission. *IEEE Transactions on*

Communications, **COM-23**, 222–234.

van Valkenburg, M. E. (1974). *Network Analysis* (3rd ed.). Englewood Cliffs, NJ: Prentice-Hall.

Verdu, S. (1998). *Multiuser Detection*. Cambridge, England: Cambridge University Press.

Verkroost, G. (1977). A general second-order digital filter with controlled rounding to exclude limit cycles for constant input signals. *IEEE Transactions on Circuits and Systems*, **CAS-24**, 428–431.

Verkroost, G. & Butterweck, H. J. (1976). Suppression of parasitic oscillations in wave digital filters and related structures by means of controlled rounding. *Archiv Elektrotechnik und Übertragungstechnik*, **30**, 181–186.

Vetterli, M. & Herley, C. (1992). Wavelets and filters banks: Theory and design. *IEEE Transactions on Signal Processing*, **40**, 2207–2232.

Vetterli, M. & Kovačević, J. (1995). *Wavelets and Subband Coding*. Englewood Cliffs, NJ: Prentice-Hall.

Wanhammar, L. (1981). *An Approach to LSI Implementation of Wave Digital Filters*. PhD thesis, Linköping University, Linköping, Sweden.

Wanhammar, L. (1999). *DSP Integrated Circuits*. New York, NY: Academic Press.

Webster, R. J. (1985). C^∞-windows. *IEEE Transactions on Acoustics, Speech, and Signal Processing*, **ASSP-33**, 753–760.

Weste, N. H. E. & Eshraghian, K. (1993). *Principles of CMOS VLSI Design: A Systems Perspective* (2nd ed.). Reading, MA: Addison-Wesley.

Whitaker, J. (1999). *DTV: The Digital Video Revolution* (2nd ed.). New York, NY: McGraw-Hill.

Willems, J. L. (1970). *Stability Theory of Dynamical Systems*. London, England: Thomas Nelson and Sons.

Winston, W. L. (1991). *Operations Research – Applications and Algorithms* (2nd ed.). Boston, MA: PWS-Kent.

Yang, R. H. & Lim, Y. C. (1991). Grid density for design of one- and two-dimensional FIR filters. *IEE Electronics Letters*, **27**, 2053–2055.

Yang, R. H. & Lim, Y. C. (1993). Efficient computational procedure for the design of FIR digital filters using WLS techniques. *IEE Proceedings – Part G*, **140**, 355–359.

Yang, R. H. & Lim, Y. C. (1996). A dynamic frequency grid allocation scheme for the efficient design of equiripple FIR filters. *IEEE Transactions on Signal Processing*, **44**, 2335–2339.

Yang, S. & Ke, Y. (1992). On the three coefficient window family. *IEEE Transactions on Signal Processing*, **40**, 3085–3088.

Yassa, F. F., Jasica, J. R., Hartley, R. I., & Noujaim, S. E. (1987). A silicon compiler for digital signal processing: Methodology, implementation, and applications. *Proceedings of the IEEE*, **75**, 1272–1282.

Yun, I. D. & Lee, S. U. (1995). On the fixed-point analysis of several fast IDCT algorithms. *IEEE Transactions on Circuits and Systems II: Analog and Digital Signal Processing*, **42**, 685–693.

Index

Printed in the United Kingdom
by Lightning Source UK Ltd.
124581UK00001B/63/A